D1200314

Methods in Enzymology

Volume 211
DNA STRUCTURES
Part A
Synthesis and Physical Analysis of DNA

METHODS IN ENZYMOLOGY

EDITORS-IN-CHIEF

John N. Abelson   Melvin I. Simon

DIVISION OF BIOLOGY
CALIFORNIA INSTITUTE OF TECHNOLOGY
PASADENA, CALIFORNIA

FOUNDING EDITORS

Sidney P. Colowick and Nathan O. Kaplan

Methods in Enzymology

Volume 211

DNA Structures

Part A

Synthesis and Physical Analysis of DNA

EDITED BY

David M. J. Lilley

DEPARTMENT OF BIOCHEMISTRY
THE UNIVERSITY
DUNDEE, SCOTLAND

James E. Dahlberg

DEPARTMENT OF BIOMOLECULAR CHEMISTRY
UNIVERSITY OF WISCONSIN-MADISON
MADISON, WISCONSIN

ACADEMIC PRESS, INC.

Harcourt Brace Jovanovich Publishers

San Diego   New York   Boston
London   Sydney   Tokyo   Toronto

Academic Press, Inc.
1250 Sixth Avenue, San Diego, California 92101-4311

United Kingdom Edition published by
Academic Press Limited
24–28 Oval Road, London NW1 7DX

Library of Congress Catalog Number: 54-9110

International Standard Book Number: 0-12-182112-9

PRINTED IN THE UNITED STATES OF AMERICA
92 93 94 95 96 97   EB   9 8 7 6 5 4 3 2 1

Table of Contents

Section I. Chemical Synthesis of DNA

Section II. Nonstandard DNA Structures and Their Analysis

Section III. Spectroscopic Methods for Analysis of DNA

Section IV. Structural Methods of DNA Analysis

Contributors to Volume 211

Article numbers are in parentheses following the names of contributors.
Affiliations listed are current.

MARC ADRIAN (26), *Laboratoire d'Analyse Ultrastructurale, Université de Lausanne, CH-1015 Lausanne, Switzerland*

CORNELIS ALTONA (15), *Gorlaeus Laboratories, Leiden University, 2300 RA Leiden, The Netherlands*

PATRICIA G. ARSCOTT (25), *Department of Biochemistry, University of Minnesota, St. Paul, Minnesota 55108*

G. BEATON (1), *Amgen, Inc., Boulder, Colorado 80301*

VICTOR A. BLOOMFIELD (25), *Department of Biochemistry, University of Minnesota, St. Paul, Minnesota 55108*

KENNETH J. BRESLAUER (28), *Department of Chemistry, Rutgers, The State University of New Jersey, New Brunswick, New Jersey 08901*

DORCAS J. S. BROWN (2), *Department of Chemistry, Meriot-Watt University, Riccarton, Edinburgh EM14 4AS, Scotland*

TOM BROWN (2), *Department of Chemistry, Meriot-Watt University, Riccarton, Edinburgh EM14 4AS, Scotland*

M. H. CARUTHERS (1), *Department of Chemistry and Biochemistry, University of Colorado, Boulder, Colorado 80309*

ROBERT M. CLEGG (18), *Department of Molecular Biology, Max-Planck-Institute for Biophysical Chemistry, D-W-3400 Göttingen, Germany*

B. A. CONNOLLY (3), *Department of Biochemistry and Genetics, The University, Newcastle-upon-Tyne NE2 4HH, England*

RICHARD E. DICKERSON (5), *Molecular Biology Institute, University of California, Los Angeles, Los Angeles, California 90024*

JACQUES DUBOCHET (26), *Laboratoire d'Analyse Ultrastructurale, Université de Lausanne, CH-1015 Lausanne, Switzerland*

ISABELLE DUSTIN (26), *Laboratoire d'Analyse Ultrastructurale, Université de Lausanne, CH-1015 Lausanne, Switzerland*

ELISABETH EVERTSZ (17), *Department of Chemistry and Institute of Molecular Biology, University of Oregon, Eugene, Oregon 97403*

JULI FEIGON (13), *Department of Chemistry and Biochemistry, University of California, Los Angeles, Los Angeles, California 90024*

MAXIM D. FRANK-KAMENETSKII (9, 23), *Institute of Molecular Genetics, Russian Academy of Sciences, Moscow 123182, Russia*

ERNESTO FREIRE (28), *Biocalorimetry Center, Department of Biology, The Johns Hopkins University, Baltimore, Maryland 21218*

PATRICK FURRER (26), *Laboratoire d'Analyse Ultrastructurale, Université de Lausanne, CH-1015 Lausanne, Switzerland*

DARA E. GILBERT (13), *Department of Chemistry and Biochemistry, University of California, Los Angeles, Los Angeles, California 90024*

WALTER GILBERT (10), *Department of Cellular and Developmental Biology, Harvard University, Cambridge, Massachusetts 02138*

DAVID G. GORENSTEIN (14), *Department of Chemistry, Purdue University, West Lafayette, Indiana 47907*

DONALD M. GRAY (19), *Program in Molecular and Cell Biology, The University of Texas at Dallas, Richardson, Texas 75083*

JACK GRIFFITH (24), *Lineberger Comprehensive Cancer Center, University of North Carolina at Chapel Hill, Chapel Hill, North Carolina 27599*

BERNARD D. HUCKRIEDE (15), *Gorlaeus Laboratories, Leiden University, 2300 RA Leiden, The Netherlands*

WILLIAM N. HUNTER (12), *Department of Chemistry, University of Manchester, Manchester M13 9PL, England*

JOHANNES H. IPPEL (15), *Gorlaeus Laboratories, Leiden University, 2300 RA Leiden, The Netherlands*

VALERY I. IVANOV (6), *Institute of Molecular Biology, Russian Academy of Sciences, Moscow 117984, Russia*

BRIAN H. JOHNSTON (7), *Cell and Molecular Biology Laboratory, SRI International, Menlo Park, California 94025*

THOMAS M. JOVIN (11), *Department of Molecular Biology, Max-Planck-Institute for Biophysical Chemistry, D-W-3400 Göttingen, Germany*

ULRIKE KAPP (21), *European Molecular Biology Laboratory, Grenoble Outstation, F-38042 Grenoble Cedex, France*

WERNER KREMER (21), *European Molecular Biology Laboratory, Grenoble Outstation, F-38042 Grenoble Cedex, France*

DMITRY YU. KRYLOV (6), *Institute of Molecular Biology, Russian Academy of Sciences, Moscow 117984, Russia*

JÖRG LANGOWSKI (21), *European Molecular Biology Laboratory, Grenoble Outstation, F-38042 Grenoble Cedex, France*

DAVID M. J. LILLEY (8), *Department of Biochemistry, The University, Dundee DD1 4HN, Scotland*

J. LIQUIER (16), *Laboratoire de Spectroscopie Biomoléculaire, U.F.R. Santé Médecine et Biologie Humaine, Université Paris Nord, F-93012 Bobigny Cedex, France*

ROMÁN F. MACAYA (13), *Department of Chemistry and Biochemistry, University of California, Los Angeles, Los Angeles, California 90024*

PAUL S. MILLER (4), *Department of Biochemistry, School of Hygiene and Public Health, The Johns Hopkins University, Baltimore, Maryland 21205*

DINA MORAS (20), *Laboratoire de Cristallographie Biologique, Institute de Biologie Moléculaire et Cellulaire du CNRS, 67084 Strasbourg Cedex, France*

ALASTAIR I. H. MURCHIE (8), *Department of Biochemistry, The University, Dundee DD1 4HN, Scotland*

WARNER L. PETICOLAS (17), *Department of Chemistry and Institute of Molecular Biology, University of Oregon, Eugene, Oregon 97403*

ROBERT L. RATLIFF (19), *Life Sciences Division, Los Alamos National Laboratory, Los Alamos, New Mexico 87545*

KARSTEN RIPPE (11), *Department of Molecular Biology, Max-Planck-Institute for Biophysical Chemistry, D-W-3400 Göttingen, Germany*

PETER SCHULTZE (13), *Department of Chemistry and Biochemistry, University of California, Los Angeles, Los Angeles, California 90024*

DIPANKAR SEN (10), *Department of Chemistry, Simon Fraser University, Burnaby, British Columbia, Canada V5A 1S6*

VLADIMÍR SKLENÁŘ (13), *Department of Radiofrequency Spectroscopy, Institute of Scientific Instruments, Czechoslovak Academy of Sciences, Brno, Czechoslovakia*

ANDRZEJ STASIAK (26), *Laboratoire d'Analyse Ultrastructurale, Université de Lausanne, CH-1015 Lausanne, Switzerland*

MARTIN STRAUME (28), *Biocalorimetry Center, Department of Biology, The Johns Hopkins University, Baltimore, Maryland 21218*

E. TAILLANDIER (16), *Laboratoire de Spectroscopie Biomoléculaire, U.F.R. Santé Médecine et Biologie Humaine, Université Paris Nord, F-93012 Bobigny Cedex, France*

RANDY THRESHER (24), *Lineberger Comprehensive Cancer Center, University of North Carolina at Chapel Hill, Chapel Hill, North Carolina 27599*

YOURI TIMSIT (20), *Laboratoire de Cristallographie Biologique, Institute de Biologie Moléculaire et Cellulaire de CNRS, 67084 Strasbourg Cedex, France*

JIM TORBET (27), *Department of Biochemistry, University of Edinburgh, Edinburgh EH8 9XD, Scotland*

MARILYN R. VAUGHAN (19), *Program in Molecular and Cell Biology, The University of Texas at Dallas, Richardson, Texas 75083*

ALEXANDER V. VOLOGODSKII (23), *Institute of Molecular Genetics, Russian Academy of Sciences, Moscow 123182, Russia*

EDMOND WANG (13), *Department of Chemistry and Biochemistry, University of California, Los Angeles, Los Angeles, California 90024*

W. WIESLER (1), *Department of Chemistry and Biochemistry, University of Colorado, Boulder, Colorado 80309*

J. V. WU (1), *Department of Chemistry and Biochemistry, University of Colorado, Boulder, Colorado 80309*

JOHN VAN WIJK (15), *Gorlaeus Laboratories, Leiden University, 2300 RA Leiden, The Netherlands*

EBERHARD VON KITZING (22), *Department of Cell Physiology, Max-Planck-Institute for Medical Research, D-6900 Heidelberg, Germany*

Preface

The structure of DNA has been known in general terms for about forty years now. Yet in the last decade we have become much more attuned to its subtleties and how they depend on the interplay of local base sequence, environmental conditions, covalent modifications, and topology. For this reason the use of the plural form for structure in the titles of Volumes 211 and 212 is significant.

Structural alterations of DNA may be relatively subtle — essentially small (though very important, nevertheless) modifications to the standard B-DNA structure — or they may be very great. Until the late 1970s DNA was regarded as an invariant, straight, right-handed B-form double helix. In the intervening period, DNA structures have been described that depart from virtually all the conventional patterns. DNA can be left-handed (Z-DNA), and base pairing can be rearranged (cruciform) or remodeled (H-triplex). The axis can be bent or kinked, and triple and quadruple helices are possible. This has changed our perspective of DNA structure to one of a highly polymorphic molecule. Not all of the structures that are possible may be biologically relevant, but some undoubtedly are. Some structures may even be conformational traps that must be avoided. But it is clear that if we are to understand the structures and dynamics of DNA and its interactions with ligands, a complete analysis of its conformational flexibility is required.

Progress in a number of areas had led to our current view of DNA as a more structurally adventurous molecule. Initially it was the advent of new methods for the synthesis of oligonucleotides of defined sequence and high purity and the consequent crystallographic analysis of a number of DNA structures at near atomic resolution that revealed the sequence dependence of DNA structure. This also gave us the first major surprise — left-handed Z-DNA. Short, well-defined DNA fragments may also be studied by many other physical techniques, and nuclear magnetic resonance spectroscopy has been particularly useful.

But in parallel with these general methods, a number of powerful techniques that are more specific to DNA have emerged, including the ability to synthesize any DNA sequence of interest, the facility to clone these sequences into plasmids, and the opportunity to manipulate DNA as a closed circle. Many of the perturbed structures that DNA can adopt are stabilized by negative supercoiling, and this can be exploited to analyze DNA structure and energetics. Thus band-shift methods and two-dimensional gel electrophoresis are now important methods for the study of DNA structure. In addition, probing methods have become vital to the analysis of local structure in DNA, and an armory of chemical and enzyme probes are available for this task. Chemical probes are now being extended to the analysis of DNA inside the cell, an important aspect of their use if we are to assess the biological role of DNA structural polymorphism.

In view of the amount of activity in this area in the last decade and the new methods developed, we felt it was time to collect all the methods and approaches in one place, and these volumes are the result. Originally it was our intention to produce a single volume, which, thanks to the hard work of all our authors, eventually became two. We are most grateful to all the authors for their efforts, and we hope that their reward will be the existence of these books that should be a valuable resource for all of us in the field. We also thank the staff of Academic Press for their assistance in the preparation of these volumes.

DAVID M. J. LILLEY
JAMES E. DAHLBERG

METHODS IN ENZYMOLOGY

VOLUME 80. Proteolytic Enzymes (Part C)
Edited by LASZLO LORAND

VOLUME 81. Biomembranes (Part H: Visual Pigments and Purple Membranes, I)
Edited by LESTER PACKER

VOLUME 82. Structural and Contractile Proteins (Part A: Extracellular Matrix)
Edited by LEON W. CUNNINGHAM AND DIXIE W. FREDERIKSEN

VOLUME 83. Complex Carbohydrates (Part D)
Edited by VICTOR GINSBURG

VOLUME 84. Immunochemical Techniques (Part D: Selected Immunoassays)
Edited by JOHN J. LANGONE AND HELEN VAN VUNAKIS

VOLUME 85. Structural and Contractile Proteins (Part B: The Contractile Apparatus and the Cytoskeleton)
Edited by DIXIE W. FREDERIKSEN AND LEON W. CUNNINGHAM

VOLUME 86. Prostaglandins and Arachidonate Metabolites
Edited by WILLIAM E. M. LANDS AND WILLIAM L. SMITH

VOLUME 87. Enzyme Kinetics and Mechanism (Part C: Intermediates, Stereochemistry, and Rate Studies)
Edited by DANIEL L. PURICH

VOLUME 88. Biomembranes (Part I: Visual Pigments and Purple Membranes, II)
Edited by LESTER PACKER

VOLUME 89. Carbohydrate Metabolism (Part D)
Edited by WILLIS A. WOOD

VOLUME 90. Carbohydrate Metabolism (Part E)
Edited by WILLIS A. WOOD

VOLUME 91. Enzyme Structure (Part I)
Edited by C. H. W. HIRS AND SERGE N. TIMASHEFF

Section I

Chemical Synthesis of DNA

[1] Chemical Synthesis of Deoxyoligonucleotides and Deoxyoligonucleotide Analogs

By M. H. Caruthers, G. Beaton, J. V. Wu, and W. Wiesler

Introduction

Over the past several years, improvements in nucleic acid chemistry have led to many applications for oligonucleotides in biology and biochemistry. These include the use of synthetic DNA to clone and synthesize genes,[1] as primers for various polymerase chain reaction (PCR) applications,[2] to sequence DNA,[3] for mutagenesis of genes in a site-specific manner,[4] as probes for examining how proteins interact with polynucleotides,[5] and for studies on nucleic acid structure.[6] The chemical methodologies leading to these very efficient procedures have been published previously in this series[7-9] and will therefore not be repeated. Instead advances since the last article will be emphasized. These include the development of new protecting group strategies and procedures for the attachment of reporter groups to DNA. As a result of these achievements, the time required to synthesize DNA has been reduced, and an even larger repertoire of applications for synthetic deoxyoligonucleotides has been achieved. An attempt will be made to review some of these many excellent contributions, but only one detailed protocol for labeling DNA is included. This is a relatively recent, unpublished procedure for attaching reporter groups to a new DNA analog called dithioate DNA.

Deoxyoligonucleotides having a deoxynucleoside$-OPS_2O-$deoxynucleoside linkage (dithioate DNA) are excellent mimics of natural DNA. At least partially this is because they are isosteric and isoelectronic with normal DNA. They are, however, completely resistant to snake venom, calf spleen, and P_1 exonucleases, all the nucleases found in HeLa nuclear and cytoplasmic extracts, and even the very potent exonucleolytic activity

[1] G. Groger, G. Ramalho-Ortigo, H. Steil, and H. Seliger, *Nucleic Acids Res.* **16**, 7761 (1988).
[2] N. Arnhelm and C. H. Levenson, *Chem. Eng. News* **68**, 36 (1990).
[3] M. Hunkapillar, S. Kent, M. Caruthers, W. Dreyer, J. Firca, C. Giffin, S. Horvath, T. Hunkapiller, P. Tempst, and L. Hood, *Nature (London)* **310**, 105 (1984).
[4] R. J. Leatherbarrow and A. R. Fersht, *Protein Eng.* **1**, 7 (1986).
[5] S. C. Harrison and A. K. Aggarwal, *Annu. Rev. Biochem.* **59**, 933 (1990).
[6] O. Kennard and W. N. Hunter, *Q. Rev. Biophys.* **22**, 327 (1989).
[7] E. L. Brown, R. Belagaje, M. J. Ryan, and H. G. Khorana, this series, Vol. 68, p. 109.
[8] S. A. Narang, H. M. Hsiung, and R. Brousseau, this series, Vol. 68, p. 90.
[9] M. H. Caruthers, A. D. Barone, S. L. Beaucage, D. R. Dodds, E. F. Fisher, L. J. McBride, M. Matteucci, Z. Stabinsky, and J. Y. Tang, this series, Vol. 154, p. 287.

of T4 DNA polymerase.[10-12] These attractive features combined with the ability of dithioate DNA to form stable DNA duplexes have led to several potential applications. For example, we have demonstrated that dithioate DNA stimulates the RNase H activity found in HeLa cell nuclear extracts. These results, when combined with the nuclease stability of this analog, suggest that it could prove quite useful for a large number of applications in basic research and in the use of DNA as a therapeutic drug. Other research conducted in our laboratory has demonstrated that dithioate DNA can be used to probe protein–DNA interactions, to inhibit various polymerases such as human immunodeficiency virus reverse transcriptase, and to chelate heavy metal ions such as silver and mercury for various applications in X-ray crystallography. Because of all these potential uses plus many others currently being investigated, we anticipate that dithioate DNA will become an extremely valuable deoxyoligonucleotide analog in years to come. For these reasons, we have included a detailed protocol for synthesizing this new DNA analog.

Recent Advances in DNA Synthesis

The current, most commonly used approach[9] for preparing synthetic DNA involves adding activated mononucleotides to a growing DNA segment that is covalently linked to an insoluble, silica-based solid support (Fig. 1, structures **1**–**6**). Addition of a mononucleotide requires the following four steps: (1) removal of the dimethoxytrityl protecting group from **1** (Fig. 1) with acid to form **2**; (2) condensation with **6**, a protected deoxynucleoside 3′-phosphoramidite, to yield **3**; (3) acylation or capping of unreactive deoxynucleoside (**5**) or deoxyoligonucleotide; and (4) oxidation of the phosphite triester (**3**) to the phosphate triester (**4**). Thus the synthesis proceeds stepwise in a 3′ to 5′ direction by addition of one deoxynucleotide per cycle until the required DNA has been prepared. Reagents, starting materials, and side products are removed after each synthesis step simply by filtration. After completion of the total synthesis, the DNA segment is chemically freed of blocking groups, hydrolyzed from the support, and purified to homogeneity by polyacrylamide gel electrophoresis (PAGE) or high-performance liquid chromatography (HPLC).

Although this methodology is extremely efficient, with yields exceeding 99% per mononucleotide addition, careful investigations of all the synthe-

---

[10] A. Grandas, W. S. Marshall, J. Nielsen, and M. H. Caruthers, *Tetrahedron Lett.* **30**, 543 (1989).

[11] G. M. Porritt and C. B. Reese, *Tetrahedron Lett.* **31**, 1319 (1990).

[12] M. H. Caruthers, G. Beaton, L. Cummins, D. Graff, Y.-X. Ma, W. S. Marshall, H. Sasmor, P. Norris, and E. K. Yau, *Ciba Found. Symp.* **158**, 158 (1991).

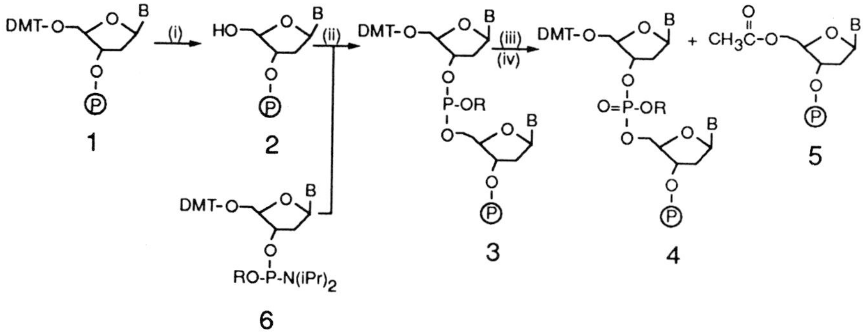

FIG. 1. Steps in the synthesis of DNA on a polymer support. Ⓟ, Silica support; DMT, dimethoxytrityl; R, methyl or β-cyanoethyl; B, thymine, 6-N-benzoyladenosine, 4-N-benzoylcytosine, or 2-N-isobutyrylguanine; iPr, isopropyl. Reagents: *(i)* Trichloroacetic acid; *(ii)* tetrazole; *(iii)* acetic anhydride; *(iv)* iodine–water.

sis steps have led to certain improvements in the quality of synthetic DNA. Perhaps the most significant in recent years has been the substitution of N-methylimidazole for N,N-dimethylaminopyridine in the capping solution.[13] This change eliminates a small but detectable amount of 2,6-diaminopurine, a modification of guanine that is responsible for the CG to TA transition observed with cloned, synthetic DNA.

Another potential problem is the susceptibility of thymine and uracil to methylation. For example, Urdea *et al.* reported that synthetic deoxyoligonucleotides 50–120 bases in length had approximately 30% of the thymine bases converted to 3-N-methylthymine when deoxynucleoside 3′-methylphosphoramidites are used as synthons.[14] However, more recent data from at least two laboratories[15,16] have demonstrated that 0.2% or less of the thymine bases are converted to 3-N-methylthymine under DNA synthesis conditions. These disparaging results can be traced to the deprotection procedures. Urdea *et al.* treated the completely protected DNA directly with concentrated ammonium hydroxide even though strong basic conditions are known[17] to cause methylation of N-3. In contrast when methyl phosphate protected DNA is first treated with thiophenoxide to

[13] J. S. Eadie and D. S. Davidson, *Nucleic Acids Res.* **15**, 8333 (1987).

[14] M. S. Urdea, L. Ku, T. Horn, Y. G. Gee, and B. D. Warner, *Nucleic Acids Res. Symp. Ser.* **16**, 257 (1985).

[15] L. J. McBride, J. S. Eadie, J. W. Efcavitch, and W. A. Andrus, *Nucleosides Nucleotides* **6**, 297 (1987).

[16] A. Andrus and S. L. Beaucage, *Tetrahedron Lett.* **29**, 5479 (1988).

[17] X. Gao, B. L. Gaffney, M. Senior, R. R. Riddle, and R. N. Jones, *Nucleic Acids Res.* **13**, 573 (1985).

remove the methyl group and then fully deprotected with ammonium hydroxide as published originally for this chemistry,[18-20] less than 0.2% N-3 methylated thymine is observed. Such results clearly indicate that caution should be exercised before modifying established protocols. This is the main lesson from these results as the deoxynucleoside 3'-β-cyanoethyl-phosphoramidites[21] have now replaced the corresponding methylphos-phoramidites. The β-cyanoethyl derivatives, despite their reduced reactivity, have one major advantage. Thiophenoxide can be safely eliminated from the deprotection protocol because the β-cyanoethyl group is removed using basic conditions[22,23]—the standard reagent for excising DNA from the support and deprotection of the purines and pyrimidines.

There have been several other significant changes in DNA synthesis procedures over the past few years. Most have focused on reducing the time required to deprotect and purify synthetic DNA. As a consequence, an experiment-ready deoxyoligonucleotide can now be prepared in less than a day (2 days previously[9]). The major change has been to introduce more labile protecting groups for the purine and pyrimidine exocyclic moieties. Schulhof et al.[24] have developed a new set of amide protecting groups that can be removed within 4 hr using mild ammonia treatment (20°/29% ammonium hydroxide, whereas the previous set required 55°/ concentrated ammonium hydroxide for 16 hr). This development is clearly an improvement but it is not enough.

One of the main, current challenges in this area is to design protecting groups for adenine, and, to a lesser extent, for guanine, that do not lead to depurination during the cyclic acid detritylation step (Fig. 1). The new derivatives proposed by Schulhof et al. (6-N-phenoxyacetyldeoxyadeno-sine and 2-N-phenoxyacetyldeoxyguanosine) are only marginally better in this regard than the previously used protecting groups. For example, 6-N-phenoxyacetyldeoxyadenosine is only about 20% more stable toward de-purination than 6-N-benzoyldeoxyadenosine. Clearly this is important as the current estimate of mutagenesis with the classic protecting groups on synthetic DNA is approximately 1.3%/base.[25] Although the chemical basis for this mutagenesis is not entirely clear, sensitivity to piperidine suggests depurination as a major contributor. Hence the development of protecting

[18] M. D. Matteucci and M. H. Caruthers, J. Am. Chem. Soc. 103, 3185 (1981).
[19] S. L. Beaucage and M. H. Caruthers, Tetrahedron Lett. 22, 1859 (1981).
[20] M. H. Caruthers, Science 230, 281 (1985).
[21] N. D. Sinha, J. Biernat, and H. Koster, Tetrahedron Lett. 24, 5843 (1983).
[22] G. M. Tener, J. Am. Chem. Soc. 83, 159 (1961).
[23] R. L. Letsinger and V. Mahadevan, J. Am. Chem. Soc. 88, 5319 (1966).
[24] J. C. Schulhof, D. Molko, and R. Teoule, Nucleic Acids Res. 15, 397 (1987).
[25] I. K. Farrance, J. S. Eadie, and R. Ivarie, Nucleic Acids Res. 17, 1231 (1989).

groups that are severalfold more stable toward depurination than 6-N-benzoyldeoxyadenosine is an important objective that is not met by the phenoxyacetyl group.

The second class of new protecting groups are the amidines, but here there are also problems. The N,N-dimethylacetyl and N,N-dibutylformamidine protected purines are approximately 20- to 25-fold more stable toward acid depurination than the classic amide protected bases (e.g. $t_{1/2}$ for depurination of N-benzoyl-, N,N-dimethylacetyl-, and N,N-dibutylformyldeoxyadenosine is 1.7, 34, and 30 hr, respectively).[26,27] Thus, these protecting groups are superior to amide protection relative to depurination. The problem is that removal of these amidines occurs under conditions comparable to those used for the classic amide protecting groups (10–24 hr at 55°). Recently N,N-dimethylformamidine has been proposed as the protecting group of choice for deoxyguanosine and deoxyadenosine, as it renders the purines resistant to depurination and can be removed very rapidly (1 hr at 55°).[28] These results are encouraging, but the N,N-dimethylformamidine derivative is labile to the acid conditions used for detritylation.[26,29] Thus, during the course of synthesizing even a 20-mer, a significant fraction of deoxyadenosine must lose the exocyclic amine protection. This is important because these exocyclic amino groups either protected or unprotected react with deoxynucleoside 3'-phosphoramidites to form base adducts which, on oxidation, lead to stable P(V) derivatives (phosphoramidates and phosphate triesters). As phosphoramidite or phosphite base adducts, these derivatives in protected form are labile and can be removed by the capping reagent or aqueous lutidine.[13,18,19,30,31] Unfortunately the exocyclic amino base adduct of unprotected deoxyadenosine is quite stable to these conditions and requires 24 hr with aqueous lutidine for complete removal.[32] Thus, in order to use the N,N-dimethylformamidine derivative of deoxyadenosine for routine synthesis, an additional hydrolytic (or perhaps aminolysis) step should be included to retain high quality DNA with the formamidine protecting groups.

[26] L. J. McBride, R. Kierzek, S. L. Beaucage, and M. H. Caruthers, *J. Am. Chem. Soc.* **108,** 2040 (1986).

[27] B. C. Froehler and M. D. Matteucci, *Nucleic Acids Res.* **11,** 8031 (1983).

[28] H. Vu, C. McCollum, K. Jacobson, P. Theisen, R. Vinayak, E. Spiess, and A. Andrus, *Tetrahedron Lett.* **31,** 7269 (1990).

[29] J. Zemlicka, S. Chladek, A. Holy, and J. Smrt, *Collect. Czech. Chem. Commun.* **31,** 3198 (1966).

[30] D. Drahos, G. R. Galluppi, M. H. Caruthers, and W. Szybalski, *Gene* **18,** 343 (1982).

[31] M. H. Caruthers, L. J. McBride, L. P. Bracco, and J. W. Dubendorff, *Nucleosides Nucleotides* **4,** 95 (1985).

[32] S. Beaucage and M. H. Caruthers, unpublished results, 1981.

Despite the limitations, both N-phenoxyacetylamide and N,N-dimethylformamidine represent an improvement over protection with the classic N-benzoylamide on deoxyadenosine as DNA can be synthesized more rapidly. However, until such time as careful mutagenesis studies have been completed using DNA prepared with these new blocking reagents, an overall conclusion on the quality of the resulting synthetic DNA must be deferred.

During the past few years, the H-phosphonate method for DNA synthesis, which was first described by Todd and co-workers,[33] has been reexamined by two laboratories.[34,35] The results are quite positive and merit further examination. With this procedure, 5'-dimethoxytrityldeoxynucleoside 3'-H-phosphonates are activated with sterically hindered carbonyl chlorides (adamantoyl and pivaloyl chlorides) and condensed with the 5'-hydroxyl of a growing DNA segment attached to a polymer support. The resulting phosphite diester is capped, detritylated with acid, and condensed with the next 5'-dimethoxytrityldeoxynucleoside 3'-H-phosphonate. After synthesis of the DNA segment, the phosphite diester internucleotide linkages are oxidized with $tert$-butyl hydroperoxide or iodine to the phosphate diester.

The H-phosphonate approach has the advantage that no phosphate triester protecting group is required, alleviating the problems associated with its removal. One step in the synthesis cycle is also eliminated, as oxidation of all internucleotide linkages occurs postsynthesis and simultaneously. There are two problems. The yields are not as high as with the deoxynucleoside phosphoramidites, and, from the published literature, it is not clear that complete oxidation occurs under the prescribed conditions. Incomplete oxidation leads to degradation of DNA during the ammonium hydroxide deprotection procedures. The main advantage of this approach may be as a method for synthesizing DNA containing all phosphorothioate[36] and phosphoramidate[37] internucleotide linkages. These analogs can be generated, respectively, by oxidation of the phosphite diesters with sulfur or amines under anhydrous conditions.

Synthesis of Modified DNA

For a short time after synthetic DNA became readily available, most laboratories were content to use these deoxyoligonucleotides for various

[33] R. H. Hall, A. Todd, and R. F. Webb, *J. Chem. Soc.,* 3291 (1957).
[34] B. C. Froehler and M. D. Matteucci, *Tetrahedron Lett.* **27,** 469 (1986).
[35] P. J. Garegg, T. Regbert, J. Stawinski, and R. Strömberg, *Chem. Scr.* **26,** 59 (1986).
[36] M. Fujii, K. Ozaki, A. Kume, M. Sekine, and T. Hata, *Tetrahedron Lett.* **26,** 935 (1986).
[37] B. Froehler, P. Ng, and M. Matteucci, *Nucleic Acids Res.* **16,** 4831 (1988).

biochemical and biological studies. However, it soon became apparent that what many really wanted was modified, synthetic DNA having reporter groups or chemically reactive moieties and, in some cases, possessing nuclease resistance. Developing such analogs has been a major research objective of numerous laboratories over the past few years. As a result of this research, procedures are currently available for introducing various molecular conjugates such as reporter groups, intercalators, cross-linking and DNA cutting reagents, steroids, sugars, and peptides.[38,38a] However, most of the procedures are designed for introducing conjugates at the ends of deoxyoligonucleotides or on the bases. There are few procedures for attaching these moieties at predetermined internucleotide linkages. Currently the most general are alkylation of phosphorothioates[39,40] and the use of a combined phosphoramidite/*H*-phosphonate chemistry for introducing a reporter group through a phosphoramidate internucleotide linkage.[41]

Because positioning reporter groups at internucleotide phosphorus has certain attractive features (independence from sequence, solvent accessibility, minimal influence on duplex stability), much recent effort has focused on developing procedures directed toward this goal. Here we outline a new method for accomplishing this objective. The approach involves, as a first step, the synthesis of deoxyoligonucleotides having a phosphorodithioate internucleotide linkage at any predetermined position. This site can then be alkylated postsynthesis with an appropriate reporter group and used for various biochemical studies. The concept is therefore similar to the method for labeling phosphorothioate DNA. However, phosphorodithioate internucleotide linkages are labeled much faster and to higher yields than the phosphorothioates.

As outlined in the introduction, the attractiveness of dithioate DNA for a large number of applications in biology and biochemistry has stimulated considerable effort directed toward developing synthetic methodologies.[10,11,42–45] In our laboratory, deoxynucleoside diamidites, thioamidites,

[38] E. Uhlmann and A. Peyman, *Chem. Rev.* **90,** 543 (1990).

[38a] U. Englisch and D. H. Gauss, *Angew. Chem. Int. Ed. Engl.* **30,** 613 (1991).

[39] J. A. Fidanza and L. W. McLaughlin, *J. Am. Chem. Soc.* **111,** 9117 (1989).

[40] S. Agrawal and P. C. Zamecnik, *Nucleic Acids Res.* **18,** 5419 (1990).

[41] R. L. Letsinger, C. N. Singman, G. Histand, and M. Salunkhe, *J. Am. Chem. Soc.* **110,** 4470 (1988).

[42] J. Nielsen, W. K.-D. Brill, and M. H. Caruthers, *Tetrahedron Lett.* **29,** 2911 (1988).

[43] W. K.-D. Brill, J.-Y. Tang, Y.-X. Ma, and M. H. Caruthers, *J. Am. Chem. Soc.* **111,** 2321 (1989).

[44] J. Stawinski, M. Thelin, and R. Zain, *Tetrahedron Lett.* **30,** 2157 (1989).

[45] B. H. Dahl, K. Bjergårde, V. B. Sommer, and O. Dahl, *Acta Chem. Scand.* **43,** 896 (1989).

H-phosphonodithioates, and dithiophosphate triesters have been studied as synthons for preparing dithioate DNA. Our research to date suggests that deoxynucleoside phosphorothioamidites are the most attractive and versatile. They react rapidly in high yield with a growing DNA segment and are compatible with current methods for synthesizing DNA with deoxynucleoside 3′-phosphoramidites. Because of these features, deoxyoligonucleotides can be prepared with any combination of natural and dithioate internucleotide linkages. Our current methodology is summarized in Fig. 2 (structure 1–8) and Table I.

Briefly, a deoxynucleoside 3′-phosphorothioamidite (4, Fig. 2) is prepared via a one-pot, two-step procedure from a suitably protected deoxynucleoside (1), tris(pyrrolidino)phosphine (2), and tetrazole. The resulting deoxynucleoside diamidite (3) is converted without isolation to the deoxynucleoside phosphorothioamidite by addition of 2,4-dichlorobenzylmercaptan. Following an aqueous workup and drying over sodium sulfate, these synthons are isolated by precipitation and stored as dry powders under an inert gas atmosphere without decomposition.

Synthesis of dithioate DNA begins by treating a 5′-*O*-dimethoxytrityldeoxynucleoside linked to a silica support (5, Fig. 2) with 3% trichloroacetic acid (w/v) to yield 6, a compound having a free 5′-hydroxyl accessible

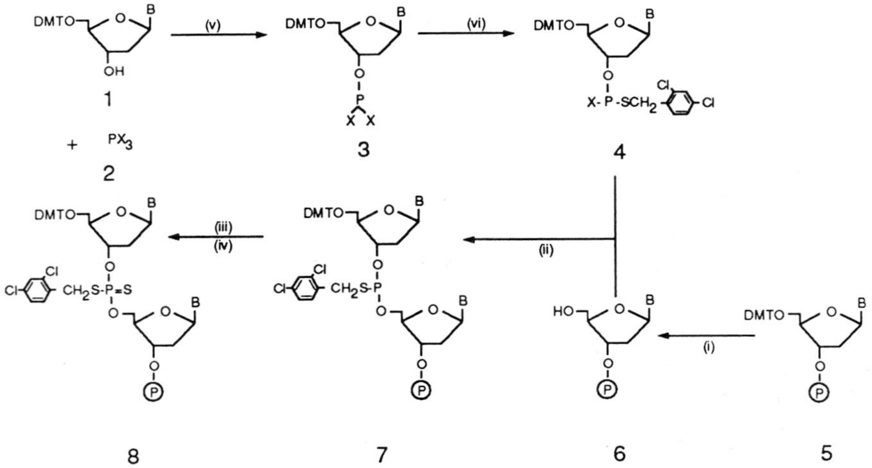

FIG. 2. Steps in the synthesis of DNA on a polymer support. Ⓟ, Silica support; DMT, dimethoxytrityl; X, pyrrolidino; B, thymine, 6-*N*-benzoyladenosine, 4-*N*-benzoylcytosine, or 2-*N*-isobutyrylguanine. Reagents: *(i)* Trichloroacetic acid; *(ii)* tetrazole; *(iii)* sulfur; *(iv)* acetic anhydride; *(v)* tetrazole; *(vi)* 2,4-dichlorobenzylmercaptan.

TABLE 1
CHEMICAL STEPS FOR SYNTHESIS OF DITHIOATE DNA
ON SOLID SUPPORT[a]

Step	Reagent or solvent[b]	Purpose	Time (min)
(i)	Trichloroacetic acid in CH_2Cl_2 (3%, w/v)	Detritylation	0.50
	CH_2Cl_2	Wash	0.50
	Acetonitrile	Wash	0.50
	Dry acetonitrile	Wash	0.50
(ii)	Activated nucleotide in acetonitrile[c]	Add nucleotide	0.75
	Reactivated nucleotide in acetonitrile	Complete nucleotide addition	1.50
(iii)	Sulfur in CS_2–pyridine–TEA (95:95:10, v/v/v)[d]	Oxidation	1.00
	CS_2	Wash	0.50
	CH_3OH	Wash	0.50
	CH_2Cl_2	Wash	0.50
(iv)	NMI–THF (30:70, v/v)[e]; acetic anhydride–lutidine–THF (2:2:15, v/v/v)	Capping reaction	0.50
	CH_2Cl_2	Wash	0.50

[a] See Fig. 2 for an explanation of steps (i)–(iv).

[b] Multiple washes with the same solvent are possible.

[c] For each micromole of deoxynucleoside attached to silica, 0.48 M tetrazole (0.125 ml) and 0.15 M deoxynucleoside phosphorothioamidite (0.125 ml) are premixed in acetonitrile. During this coupling step, activated nucleotide and tetrazole are flushed from the lines leading into the reaction chamber. This procedure reduces contaminating phosphorothioate internucleotide linkages.

[d] 5% sulfur by weight in CS_2–pyridine–TEA (95:95:10, v/v/v); TEA, triethylamine.

[e] NMI, N-Methylimidazole; THF, tetrahydrofuran.

for polynucleotide synthesis. Deoxynucleoside 3′-phosphorothioamidites are then activated with tetrazole and condensed with **6** to yield a thiophosphite triester (**7**). This step is followed by sulfurization using elemental sulfur to yield the protected phosphorodithioate derivative (**8**), capping or acylating the unreactive silica-linked deoxynucleoside with acetic anhydride, and detritylation with trichloroacetic acid. Further repetitions of this cycle using either deoxynucleoside phosphorothioamidites or deoxynucleoside phosphoramidites and tetrazole as an activator yield deoxyoligonucleotides having normal phosphate diester and phosphorodithioate diester linkages in any combination. Condensation yields of 96–99% per cycle (based on dimethoxytrityl cation released during detritylation) and 97–

98% dithioate per linkage (the remaining 2–3% is phosphorothioate) have been obtained. Protecting groups are removed from the deoxyoligonucleotide using standard procedures[9] and the DNA segment purified by HPLC or PAGE.

Attachment of Reporter Groups to Dithioate DNA

Because phosphorodithioate internucleotide linkages are very good nucleophiles relative to the natural phosphate diester, various alkyl halides as reporter groups (Fig. 3) react rapidly and selectively with this DNA analog. For example, a 19-mer containing one dithioate linkage is rapidly alkylated using an excess of the dye monobromobimane at a DNA concentration of 4 μM in aqueous solution (Fig. 4). After 1.5 hr, greater than 95% of the oligomer converts to the labeled product. Interestingly, the phosphorothioate analog of this 19-mer reacted more slowly under identical conditions to yield 90% alkylated DNA after 12 hr. Increasing the phosphorothioate DNA concentration reduced the time for relatively complete reaction (6 hr, 90% as expected from the published procedure[40]). Unmodified DNA did not react under these conditions.

As has been observed with phosphorothioates,[39] reaction of the phosphorodithioate moiety with labels containing the iodoacetamide function

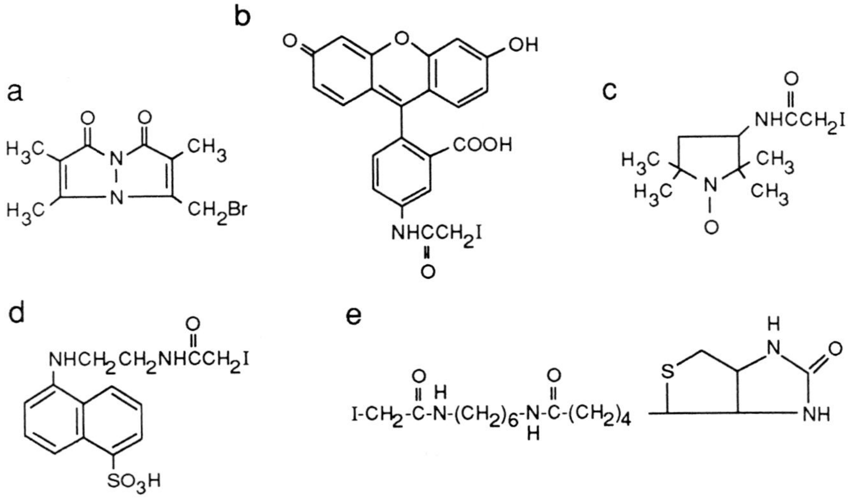

FIG. 3. Reagents used in alkylation studies of phosphorodithioate DNA. (a) Monobromobimane; (b) 5-iodoacetamidofluorescein; (c) 3-(2-iodoacetamido)-PROXYL; (d) 1,5-I-AEDAENS; (e) iodoacetyl-LC-biotin.

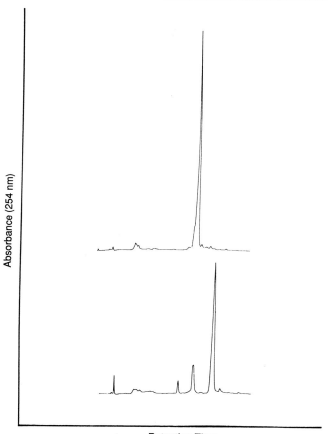

Retention Time

FIG. 4. Reaction of a deoxyoligonucleotide dithioate with monobromobimane. The top trace shows the deoxyoligonucleotide alone, and the bottom trace shows the reaction mixture injected after 30 min. The peak eluting prior to starting material arises from hydrolysis of the dye. In this gradient, the dye elutes much later (not shown). The deoxyoligonucleotide sequence is included in the section on experimental procedures.

proved to be much slower than reaction with monobromobimane. Incubation of the dithioate-containing oligomer with 3-(2-iodoacetamido)-2,2,5,5-tetramethyl-l-pyrrolidinyloxy [3-(2-iodoacetamido)-PROXYL] under conditions similar to those used for monobromobimane showed a smooth conversion to the monoalkylated product over 24 hr as judged by HPLC. Complete labeling could be accelerated to 2.5 hr by increasing both DNA and dye concentrations to 400 μM and 40 mM, respectively. In an analogous reaction, 5-iodoacetamidofluorescein could be joined covalently to the oligomer in 1 hr (95% by HPLC, Fig. 5). In the latter two labeling experiments, the dye-conjugated DNA separates by HPLC as a doublet, which probably reflects a partial separation of the two diastereomers of the *S*-alkylphosphorodithioate triesters. Other labels such as biotin and *N*-(5-sulfo-1-naphthyl)ethylenediamine have also been linked covalently via

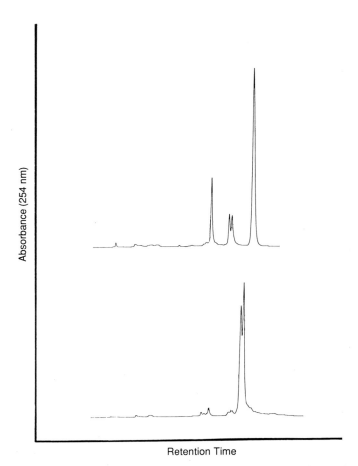

FIG. 5. Labeling of a deoxyoligonucleotide dithioate with iodoacetamido-PROXYL and 5-iodoacetamidofluorescein. The top trace shows the reaction of the deoxyoligonucleotide as described in the experimental section with the spin label after 30 min where the initial peak corresponds to starting material, the second, doublet peak to the reaction product, and the third to the spin label. The bottom trace shows the reaction with fluorescein after 1 hr. Only a peak corresponding to the labeled DNA remains because excess dye is removed by precipitation (see Experimental Procedures).

their iodoacetamido derivatives to dithioate DNA. These results generally suggest the utility of attaching reporter groups or other molecular conjugates to DNA through phosphorodithioate internucleotide linkages. For certain selected analogs, these derivatives have been used in further experiments. For example, fluorescein-tagged *lac* operator DNA has been used to study the *lac* repressor – *lac* operator interaction.[12]

Experimental Procedures

Solvents, Reagents, and Methods

Because polynucleotide synthesis requires highly purified and anhydrous solvents in order to maximize yields, a first step is usually distillation. Dichloromethane and pyridine are distilled from calcium hydride. Tetrahydrofuran is dried over sodium and benzophenone prior to distillation. Acetonitrile must be distilled under argon from phosphorus pentoxide and then calcium hydride. Triethylamine is distilled from *p*-toluenesulfonyl chloride and calcium hydride.

Monobromobimane (Calbiochem, La Jolla, CA), 5-iodoacetamidofluorescein (IAF, Molecular Probes, Eugene, OR), 3-(2-iodoacetamido)-PROXYL (Aldrich, Milwaukee, WI), *N*-iodoacetyl-*N'*-(5-sulfo-1-naphthyl)ethylenediamine (1,5-I-AEDANS, Aldrich), and trimethylsilyl-pyrrolidine (Aldrich) are purchased from commercial sources. For all sulfur oxidations, sulfur (10 g) is dissolved in pyridine – carbon disulfide (1:1; 200 ml) and the solution stored in the dark. Prior to use, triethylamine (2.5 ml) is added to the sulfur solution (37.5 ml). Protected deoxynucleosides, deoxynucleoside 3'-phosphoramidites, and deoxynucleosides attached to silica are either purchased commercially or prepared using published procedures.[9]

^{31}P NMR spectra are recorded on either a Jeol Fx90Q or a Varian VXR-300 with 85% phosphoric acid as the external standard. Quoted chemical shift values are with deuterated chloroform as solvent. HPLC is performed using a Waters (Milford, MA) system and a 5 μm Lichrosorb RP-18 column.

Tris(pyrrolidino)phosphine. Phosphorus trichloride (50.4 g) is placed in a 1-liter flask with 400 ml anhydrous ether. The flask is fitted with a 250-ml addition funnel containing 173.5 g trimethylsilylpyrrolidine (3.3 Eq). The flask is cooled to $-10°$ (ice – brine) under argon with magnetic stirring, and trimethylsilylpyrrolidine is added dropwise over the course of 1.5 hr. Stirring is maintained for an additional 1 hr and then stopped in order to allow precipitates to settle. The reaction mixture is filtered through a medium frit sintered glass funnel, and the salts are washed with an additional 50 ml anhydrous ether. Ether and trimethylsilyl chloride are removed from the filtrate by rotary evaporation to give approximately 100 ml of crude phosphine. Distillation at reduced pressure (0.25 mm Hg) affords an initial fraction (40–103°, 5–10 g), which is discarded, and the product (104–110°, 76.3 g total), which is obtained in 86% yield as a clear, colorless oil [^{31}P NMR (CDCl$_3$): $\delta = 102.8$].

Synthesis of Deoxynucleoside 3'-Phosphorothioamidites

Protected Deoxynucleosides. The protected deoxynucleosides which are currently used routinely for DNA synthesis are 5'-*O*-dimethoxytritylthymidine, 5'-*O*-dimethoxytrityl-4-*N*-benzoyldeoxycytidine, 5'-*O*-diemthoxytrityl-6-*N*-benzoyldeoxyadenosine, and 5'-*O*-dimethoxytrityl-2-*N*-isobutyryldeoxyguanosine. These intermediates are available commercially, and their synthesis has been described previously in this series.[8] There are, however, several recently described amidine[28] and amide[24] protecting groups that should be considered as substitutes for these protected deoxynucleosides. The reasons have been outlined earlier in this chapter. Immediately before use in preparing deoxynucleoside phosphoramidites or deoxynucleoside phosphorothioamidites, protected deoxynucleosides should be freed of trace amounts of water. This can most easily be accomplished by dissolving each protected deoxynucleoside in anhydrous pyridine, reconcentrating several times to a thick gum, dissolving in anhydrous pyridine, precipitating into hexane, and drying the precipitate under reduced pressure over a desiccant such as Drierite.

Synthesis of Deoxynucleoside 3'-Phosphorothioamidites. The same general procedure is used to prepare any appropriately protected deoxynucleoside 3'-phosphorothioamidite. Synthesis of the thymidine derivative is described. 5'-*O*-Dimethoxytritylthymidine (3.05 mmol, 1.66 g) in dry, freshly distilled dichloromethane (40 ml) is added to a mixture of tris(pyrrolidino)phosphine (3.60 mmol, 0.78 ml) and tetrazole (3.10 ml of a 0.50 *M* solution in dry acetonitrile). The reaction is followed to completion (5–10 min) by thin-layer chromatography (ethyl acetate–dichloromethane–triethylamine, 45:45:10). If protected deoxynucleoside remains, the addition of more phosphine is necessary. After the reaction is complete, 2,4-dichlorobenzylmercaptan (7.59 mmol, 1.22 ml) is added. After an additional 30 min, the reaction mixture is diluted with dichloromethane (150 ml containing 1% triethylamine) and extracted sequentially with saturated aqueous sodium bicarbonate (150 ml), 10% aqueous sodium carbonate (3 × 150 ml), and saturated aqueous NaCl (150 ml). Following drying of the organic solution over anhydrous sodium sulfate, removal of solids by filtration, and addition of toluene (20 ml), the product is isolated by concentration of the solution to 20 ml with the aid of a rotary evaporator. The product is then precipitated with stirring into degassed heptane (500 ml) containing 1% triethylamine (v/v). The resulting suspension is filtered and dried under reduced pressure in a desiccator containing Drierite to afford the product in 85–90% yield by weight. As measured by ^{31}P NMR, purity is usually in excess of 85%. [^{31}P NMR (CH_2Cl_2): $\delta = 165$, 162]. The synthons are stored without decomposition as dry powders under an inert gas atmosphere.

This procedure is applicable to all four protected deoxynucleosides. However because 5'-*O*-dimethoxytrityl-6-*N*-benzoyldeoxyadenosine is less soluble, it should be dissolved in 55 ml dichloromethane when a synthesis is completed at this scale. Should trace impurities persist, products can be dissolved in toluene–triethylamine (1:1, 20 ml), reprecipitated from degassed heptane containing 1% triethylamine, and isolated by filtration as described previously. The ^{31}P NMR spectra δ values (ppm) for 6-*N*-benzoyladenine, 4-*N*-benzoylcytosine, and 2-*N*-isobutyrylguanine derivatives are 164, 163; 166, 163; and 164, 161, respectively.

Synthesis of Dithioate DNA

Deoxyoligonucleotides containing any combination of phosphorodithioate and phosphate internucleotide linkages are prepared on an Applied Biosystems (Foster City, CA) Model 380A DNA Synthesizer or a comparable machine from other vendors. Deoxynucleoside 3'-phosphorothioamidite solutions are prepared by first flushing the solids with argon for 15 min and then adding solvent. Pyrimidine derivatives are dissolved in degassed acetonitrile, whereas the purine phosphorothioamidites require a degassed solution of acetonitrile containing 10% dichloromethane (v/v). Appropriate deoxynucleoside 3'-phosphorothioamidites or phosphoramidites, sulfur oxidation solution, and the other reagents required for DNA synthesis are added to the machine. The synthesis cycle can then be activated after the appropriate chemical steps are incorporated into the program.

The cycle for synthesizing one phosphorodithioate internucleotide linkage is presented in Table I. This cycle requires more time than the conventional phosphoramidite cycle due to longer coupling and sulfurization steps. Other differences involve inversion of the sulfur oxidation and capping steps and flushing to waste any residual mixture of monomer and tetrazole that remains in the lines. The latter change is a precaution to minimize side reactions of the phosphorothioamidite monomer with tetrazole which might lead to the formation of unwanted internucleotide linkages. Finally carbon disulfide is used to wash sulfur from the column because of the insolubility of sulfur in most other solvents. These changes, however, can easily be programmed into most DNA synthesizers and are also compatible with the synthesis programs and reagents for preparing natural deoxyoligonucleotides. As a result, DNA having any combination of dithioate and natural internucleotide linkages can be synthesized.

Coupling yields are estimated from trityl cation released by standard methods. Although this provides an indication of how well a synthesis is proceeding, the trityl yields may vary. Trityl yields in dry dichloromethane of 97–99% have been obtained with these monomers. Purity of the modified linkages can be assessed by ^{31}P NMR of the crude synthesis mixture obtained after deprotection. Usually this is greater than 98%.

Deprotection of DNA having phosphorodithioate linkages is a two-step process. The first step involves treatment of the deoxyoligonucleotide bound to the solid support with triethylammonium thiophenolate (thiophenol–triethylamine–dioxane, 1:2:2, v/v/v; 2 hr) to remove the 2,4-dichlorobenzyl group which protects the phosphorodithioate linkage. The second step is a standard treatment with aqueous ammonium hydroxide, which removes the β-cyanoethyl group from natural internucleotide triester linkages, cleaves the deoxyoligonucleotide from the support, and removes purine and pyrimidine protecting groups. The resulting deoxyoligonucleotides are then purified according to standard procedures. For deoxyoligonucleotides with multiple phosphorodithioate internucleotide linkages (greater than 50%), the recommended approach is reversed-phase column chromatography on a PRP-1 column (Hamilton, Reno, NV) TRITYL-ON, removal of the dimethoxytrityl group with acid, and then purification by PAGE under denaturing conditions.

These procedures have been used to synthesize numerous deoxyoligonucleotides having from one to all phosphorodithioate linkages. The sequence synthesized for the labeling experiments reported in the next section is d(GAAGATCTxTACGGCCGGCA) where x corresponds to the single phosphorodithioate linkage (DNA-1). DNA segments having phosphorothioate (DNA-2) and phosphate (DNA-3) at the site marked x are synthesized as controls. Concentrations of this segment are determined from an estimated molar absorptivity of 20.96×10^4 liter/mol.

Labeling Experiments with Dithioate DNA

All reactions are analyzed by HPLC and use an increasing gradient of acetonitrile with a triethylammonium acetate solution (50 mM) as the aqueous buffer. With acetonitrile as buffer B, alkylation reactions are assayed using the following gradient: 0% B for 2 min, 10% B for 6 min, 10% B for 10 min, and 30% B for 70 min. DNA-1, DNA-2, and DNA-3 elute at approximately 30 min in this gradient. The precise elution position is dependent on the type of Lichrosorb column used. Labeled products vary in position of elution. This position is determined by examining elution profiles of reaction mixtures relative to the deoxyoligonucleotide, the reporter group, and the reporter group after incubation with water (these reporter groups hydrolyze in buffer). If the reporter group interferes with the HPLC trace (e.g., fluorescein and *aedaens*), a sample of the reaction mixture is prepurified by precipitation of DNA prior to HPLC analysis.

Labeling with Monobromobimane. A stock solution of DNA-1 is diluted in water to a concentration of $5 \mu M$. To this solution is added acetonitrile and the required volume of stock dye solution such that the reaction mixture is 20% in acetonitrile, 4 mM dye, and 4 μM DNA. In a typical experiment, DNA-1 (12.5 μl of a 32 mM stock solution) is mixed with water (67.5 μl) and acetonitrile (18 μl). To this mixture is added 2 μl of a monobromobimane solution in acetonitrile (200 mM) and the reaction incubated at room temperature in the dark. At intervals, aliquots are injected into the HPLC column to determine the reaction parameters. For DNA-1 and DNA-2, labeling is complete (at least 90%) after 1.5 and 12 hr, respectively. No reaction is observed for DNA-3. Retention times for product elution on HPLC are as follows: DNA-1, 34 min; DNA-2, 32 min; dye, 59 min; dye plus water, 24, 36, 47, and 59 min.

Labeling with 3-(2-Iodoacetamido)-PROXYL. The same general procedure is followed (see Fig. 5) as with monobromobimane except that the reaction is in 20% N,N-dimethylformamide (DMF). This is because the dye is dissolved in DMF. Under the reaction conditions described for monobromobimane, the time for greater than 90% reaction with DNA-1 is 24 hr. The labeling can be forced to completion more rapidly (2.5 hr) by mixing DNA-1 (30 μl of a 0.98 mM stock solution), water (15 μl), DMF (15 μl), and dye (15 μl of a 200 mM solution in DMF) and allowing the labeling reaction to proceed at room temperature in the dark. Retention times on HPLC are as follows: DNA-1, 37 min; dye, 40–41 min; dye plus water, 25 and 40–41 min.

Labeling with 5-Iodoacetamidofluorescein. DNA-1 (30 μl of a 0.98 mM stock solution) is diluted with water (15 μl) and 5-IAF (30 μl of a 100 mM stock solution in DMF). An immediate orange precipitate forms. The reaction mixture is stored at room temperature in the dark and aliquots (12.5 μl) removed for analysis. Each aliquot is diluted with water (240 μl), adjusted to pH 5.5 with sodium acetate (0.3 M), and the DNA precipitated at 0° by dilution with 4 volumes of cold ethanol (30 min). The DNA is isolated by centrifugation, washed with 70% ethanol (v/v), reprecipitated, suspended in water (30 μl), and analyzed by HPLC. Labeling is complete in 1 hr, and the fluorescein-tagged DNA has a retention time of 37–38 min. Phosphorodithioate DNA labeled with these reporter groups is isolated by HPLC, concentrated via lyophilization, dissolved in water, and stored frozen.

Acknowledgments

This research was supported by U.S. Public Health Service Grants GM 25680 and GM 21120. The early contributions of J. Nielsen, W. K.-D. Brill, Y.-X. Ma, J.-Y. Tang, E. K. Yau, and A. Grandas toward the development of dithioate DNA in this laboratory are gratefully acknowledged.

[2] Purification of Synthetic DNA

By TOM BROWN and DORCAS J. S. BROWN

Introduction

In recent years structural and thermodynamic studies on synthetic oligonucleotides have led to important advances in our understanding of DNA conformation,[1] duplex stability,[2] and the effects of abnormal base pairs on the DNA duplex.[3] Such studies are possible because of the accessibility of highly pure synthetic deoxyoligonucleotides in milligram quantities. High-performance liquid chromatography (HPLC) is the most convenient and versatile method for the purification of synthetic DNA.[4]

HPLC purification is facilitated by the favorable physical properties of nucleic acids.[5] DNA is freely soluble in aqueous buffers because of the hydrophilicity of the sugar–phosphate backbone, and reversed-phase HPLC separation of impure oligonucleotides is extremely effective due to the hydrophobic nature of the heterocyclic bases. These bases provide strong ultraviolet chromophores which are of great value in the detection of small quantities of DNA. For most bases the maximum absorbance occurs between 260 and 280 nm and tails off at longer wavelengths, reach-

[1] R. E. Dickerson, H. R. Drew, B. N. Conner, R. M. Wing, A. V. Fratini, and M. L. Kopka, *Science* **216**, 475 (1982).

[2] K. J. Breslauer, R. Frank, H. Blöcker, and L. A. Marky, *Proc. Natl. Acad. Sci. U.S.A.* **83**, 3746 (1986).

[3] T. Brown, E. D. Booth, and G. A. Leonard, *in* "Molecular Mechanisms in Bio-Organic Processes" (C. Bleasdale and B. T. Golding, eds.), Royal Society of Chemistry Press, London, 1990.

[4] G. Zon and J. A. Thompson, *BioChromatography* **1**(No. 1), 22 (1986).

[5] W. Saenger, "Principles of Nucleic Acid Structure." Springer-Verlag, Berlin, 1984.

ing zero above 300 nm. A variable wavelength UV detector can be used to great effect in the HPLC purification of oligonucleotides, allowing a suitable wavelength to be selected to guarantee a reasonable peak size. Caution should be exercised when choosing the wavelength at which to monitor the HPLC chromatogram, as at relatively long wavelengths, namely, above 290 nm, AT-rich DNA gives rise to much smaller peaks than GC-rich DNA and poly(dA) absorbs very weakly. Unless the operator is aware of this, valuable samples can be lost. Hence it is important to be aware of the ultraviolet spectra of the individual bases which make up the oligonucleotide. Certain modified bases have quite characteristic ultraviolet spectra. For example, in the case of O^6-methylguanine and O^4-methylthymine there is an absorption *minimum* around 250 nm, and hypoxanthine absorbs very weakly above 280 nm.

The phosphate (phosphodiester) groups of the DNA backbone are acidic and exist as salts at physiological pH. As the number of acid groups is proportional to the length of the oligonucleotide chain, DNA strands of different lengths can be readily separated by anion-exchange HPLC. Chemical stability of DNA is not a problem, as oligonucleotides are quite stable in aqueous solution from pH 4.5 to 12 for short periods of time. However, nucleic acids are biological substances and are prone to enzymatic degradation. For this reason, care should be taken to use sterile water throughout HPLC procedures and subsequent purification. In addition, prolonged exposure to sunlight may lead to UV damage.

The basic hardware requirement for HPLC purification of oligonucleotides consists of a gradient controller, two HPLC pumps, a mixer unit, injector, a suitable column, a variable wavelength UV detector, and a chart recorder or printer. Standard safety devices such as a pressure gauge linked to a cutoff switch should also be incorporated. If a fixed wavelength ultraviolet detector is to be used, the most useful wavelengths are 254 and 280 nm. The choice of HPLC columns and eluants depends on the specific application, and this is discussed below. The following practical points should be noted.

Maintenance of Chromatography Equipment

The key to successful HPLC purification of synthetic DNA is the careful maintenance of HPLC systems. It is essential to wash out the entire system (pumps, solvent lines, loop, column, and detector) with aqueous methanol thoroughly after use. This will prevent the crystallization of buffers and will greatly increase the lifetime and reliability of the equipment. All HPLC buffers should be filtered through a Millipore (Bedford, MA) filter daily before use and should be discarded after a few days. All methanol and acetonitrile must be of HPLC grade, and deionized double-distilled water should be used throughout. For preparative HPLC the volume of the sample injector loop should not be less than 1 ml. Do not assume that the loop has been manufactured with precision, but check the exact loop volume. This can be done for a Rheodyne (Berkeley, CA) injector as follows.

Determination of Precise Volume of Chromatographic Injection Loop

With the injector in the LOAD position, inject 1.0 ml of water to wash the loop. Take a graduated syringe fitted with a Rheodyne needle and fill this with a solution of colored ink up to (say) the 1.5-ml mark. With the injector still in the LOAD position slowly inject the ink through the Rheodyne valve. When the colored ink begins to emerge from the loop overflow tube, observe the reading on the syringe. Subtract this from the first reading and hence determine the loop volume. Repeat the process to obtain a reproducible figure, then thoroughly wash the loop with distilled water. Clearly label the loop with the true volume, which may be considerably less than 1.0 ml. Failure to carry out this simple check may result in the loss of a significant proportion of all samples injected. In practice there should be no need to inject a volume of more than 0.8 ml into a 1.0-ml loop at any time.

Checking Configuration of Chromatography Pumps

Some HPLC pumps have the facility to deliver a rapid short pulse at the beginning of each compression stroke in order to pressurize the solvent and counteract the vaporization of volatile solvents which occurs during the upstroke. Unfortunately most ion-exchange resins used in HPLC columns are extremely fragile and will rapidly fragment under such harsh conditions. This will produce a large dead volume and cracks at the top of the column, resulting in high back-pressure and very broad, asymmetric peaks. Aqueous solvents are not prone to vaporization, and consequently the precompression is not necessary during oligonucleotide purification. Therefore the rapid pulse should be removed by changing the software or hardware setup according to the manufacturer's instructions.

Oligonucleotide Synthesis

The solid-phase phosphoramidite method of DNA synthesis is currently the most suitable, and deoxyoligonucleotides up to about 150 nucleotides in length can be made on modern automatic synthesizers with this chemistry.[6] The oligonucleotides described in this chapter were all prepared on an Applied Biosystems AB1 380B DNA synthesizer by the phosphoramidite method.[7] Most automatic DNA synthesizers offer a choice of three synthesis scales, typically 0.2, 1.0, and 10.0 μmol. The 0.2 μmol scale has superceded the 1.0 μmol scale in the production of small quantities of oligonucleotides for biological experiments, and the 10.0 μmol scale has been specifically developed for the preparation of large quantities of oligonucleotides (> 10 mg). We have found that for sequences in excess of 20 nucleotides the 1.0 μmol scale produces purer oligonucleotides than the 10.0 μmol scale. This is probably due to more efficient washing of the solid support which is possible in the shorter column. Six 1.0 μmol scale syntheses or one 10.0 μmol synthesis will yield more than 10 mg of fully purified product, and this is the scale we normally use for the preparation of DNA for X-ray crystallography and NMR spectroscopy.

General Strategy for Purification of Oligonucleotides

To obtain very pure oligonucleotides for physical studies it is necessary to combine two orthogonal methods of purification: (1) either anion-exchange HPLC or trityl-on reversed-phase HPLC and (2) trityl-off reversed-phase HPLC. This two-stage procedure is effective because the first method and the second method are orthogonal (i.e., they are based on different criteria for the separation of the desired product from impurities). The first method (anion-exchange HPLC or trityl-on reversed-phase HPLC) separates the full-length product from failure sequences, and the second method (trityl-off reversed-phase HPLC) separates the desired product from sequences containing base modifications.

The full protocol for the HPLC purification of synthetic oligonucleotides is described in detail below. Prior to purification, the required oligonucleotide is prepared (either six 1.0 μmol syntheses or one 10.0 μmol synthesis as outlined above) and then deprotected in the normal way. This should produce approximately 450 OD units of crude product (measured at 254 nm).

[6] M. H. Caruthers, A. D. Barone, S. L. Beaucage, D. R. Dodds, E. F. Fisher, L. J. McBride, M. D. Matteucci, Z. Stabinsky, and J. Y. Tang, this series, Vol. 154, p. 287.
[7] T. Brown and D. Brown, in "Oligonucleotides and Analogues, a Practical Approach" (F. Eckstein, ed.), Chap. 1. IRL Press at Oxford Univ. Press, Oxford, 1991.

Trityl-on Reversed-Phase Chromatographic Purification

At each step during solid-phase synthesis a small percentage of the total number of growing oligonucleotide chains ($\sim 1.5\%$) will fail to react with the activated monomer in solution. A capping step is incorporated into the synthesis cycle to acetylate these "failure" sequences and render them inert to subsequent monomer additions. Without this step, a series of one-base deletion mutations would contaminate the product, but with the step the contaminants are truncated sequences of varying length, forming a mixture much easier to purify. At the end of the synthesis the 4,4′-dimethoxytrityl protecting group may be left attached to the 5′-oxygen of the terminal deoxyribose sugar. If this is done, the product will be a mixture of the full-length tritylated product and a number of untritylated failure sequences. As the 4,4′-dimethoxytrityl group is lipophilic, it is easy to separate the desired product from untritylated impurities by reversed-phase HPLC. Hence, the trityl protecting group may be used as an aid to HPLC purification.

There are, however, a number of potential drawbacks to the technique. if the capping reaction is not efficient, the desired product will coelute with a number of trityl-on impurities. In addition, full-length impurities resulting from base modification will not be removed. Trityl-on reversed-phase purification may give low yields of purified oligonucleotides if it is carried out carelessly as some detritylation of the required product may occur during handling prior to reversed-phase purification. The trityl group is acid labile, so it is important to dissolve the oligonucleotide in aqueous buffer between pH 7.0 and 7.5 before and during trityl-on reversed-phase HPLC. If a trityl-on oligonucleotide is dissolved in distilled water, some detritylation may occur. Despite these qualifications, trityl-on HPLC is an excellent purification method, particularly for modified oligonucleotides such as phosphorothioates. However, it should always be carried out with great care.

Experimental Procedure

Equipment and Reagents

Column: Brownlee Aquapore C_8 or equivalent reversed-phase column, 25 cm $\times$ 10 mm

Buffer A: 0.1 M Ammonium acetate, 20% acetonitrile, pH 7.0

Buffer B: 0.1 M Ammonium acetate, 50% acetonitrile, pH 7.0

Flow rate: 3 ml/min

Gradient:

Time (min)	% Buffer A	% Buffer B
0	100	0
3	100	0
4	80	20
24	25	75
25	0	100
26	0	100
27	100	0
30	100	0

Protocol. Dissolve the fully deprotected crude product from trityl-on oligonucleotide synthesis (six 1 μmol syntheses or one 10.0 μmol synthesis; 450 OD units) in 3 ml of 0.1 M ammonium acetate buffer and filter through a Pasteur pipette with a glass wool plug. If necessary, add a few drops of concentrated aqueous ammonia to give a clear solution. Equilibrate the HPLC column with one blank run (i.e., run through the gradient without injecting a sample). Inject an analytical sample of the oligonucleotide, and run through the gradient, monitoring at 254 nm.

If the chromatogram is satisfactory, begin the purification procedure by injecting a 0.5-ml sample, monitoring at 295 nm to avoid saturating the UV detector. If a fixed wavelength detector is used, it should be fitted with short path length preparative flow cell to avoid huge peaks which disappear off-scale. Failure sequences (untritylated) will emerge first followed much later by the desired trityl-on product (Fig. 1). Collect the major peak, purify the remainder of the crude sample in the same way (a total of six preparative runs), and pool the product fractions. A diluted sample of the purified product should then be reinjected and examined to check the effectiveness of the purification. Finally, remove the solvent on a rotary evaporator and dissolve the residue in acetic acid—water (80%, v/v) to remove the 4,4′-dimethoxytrityl group. After 30 min, remove the solvent by evaporation on a rotary evaporator and dissolve the residue in distilled water (20 ml). Extract with ether (2 × 10 ml), discard the ether layer (upper layer), and evaporate the aqueous solution (~300 OD units) to dryness in readiness for the second stage of purification, trityl-off reversed-phase HPLC.

Strong Anion-Exchange Chromatographic Purification as Alternative to Trityl-on Chromatographic Purification

Strong anion-exchange HPLC separates oligonucleotides according to length (more correctly, the number of acidic phosphate groups). The shortest sequences elute first, and the desired product is the last major

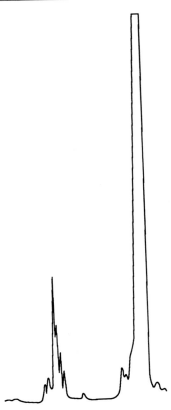

FIG. 1. Preparative trityl-on reversed-phase HPLC chromatogram of the crude self-complementary decanucleotide d(CGCAATTGCG), one 1 μmol synthesis, immediately after oligonucleotide synthesis. The small multiplet of peaks contains the untritylated failure sequences, and the main peak is the tritylated product. The sequence was synthesized on an automatic solid-phase DNA synthesizer with fresh reagents and phosphoramidite monomers. The coupling efficiency was greater than 99%. The chromatogram was monitored at 280 nm with a fixed wavelength ultraviolet detector.

peak. The anion-exchange HPLC chromatogram should show a series of small, evenly dispersed peaks arising from failure sequences which have been terminated in the capping steps, followed by the major product. Anion-exchange HPLC should be carried out on the products of trityl-off oligonucleotide syntheses. It is possible to purify short trityl-on oligonucleotides by this technique, but they elute very late and consequently an extremely long gradient is necessary.

Experimental Procedure

Equipment and Reagents

Column: Partisil P 10 SAX (strong anion-exchanger), 25 cm × 8 mm
Buffer A: 40 mM Potassium phosphate, 20% acetonitrile, pH 6.4
Buffer B: 0.67 M Potassium phosphate, 20% acetonitrile, pH 6.4
Flow rate: 3 ml/min
Gradient for sequences from 6 to 15 nucleotides in length:

Time (min)	% Buffer A	% Buffer B
0	95	5
2	95	5
25	40	60
27	40	60
30	95	5
35	95	5

Gradient for sequences from 15 to 40 nucleotides in length:

Time (min)	% Buffer A	% Buffer B
0	90	10
2	90	10
25	0	100
27	0	100
30	90	10
35	90	10

Gradient for sequences longer than 40 nucleotides:

Time (min)	% Buffer A	% Buffer B
0	80	20
2	80	20
35	0	100
37	0	100
40	80	20
45	80	20

Protocol. Dissolve the fully deprotected crude product from trityl-off oligonucleotide synthesis (six 1 μmol syntheses or one 10.0 μmol synthesis; ~450 OD units) in 3 ml of water and filter through a Pasteur pipette with a glass wool plug. If necessary, add a few drops of concentrated aqueous ammonia to give a clear solution. Equilibrate the HPLC column with two blank runs. Note that the gradient produces a sloping baseline when monitoring at 254 nm owing to the absorption of the phosphate buffer. Inject an analytical sample of the oligonucleotide, monitor at 254 nm, and run through the ion-exchange HPLC gradient.

If the gradient is satisfactory, begin the purification procedure by injecting a 0.5-ml sample, monitoring at 295 nm to avoid saturating the UV detector. A peak for benzamide (protecting groups) will emerge first followed by the shortest failure sequences and eventually the main product (Fig. 2). Collect the major peak and purify the remainder of the crude sample in this way (a total of six preparative runs). Pool the product fractions and reinject a diluted sample of the purified product to check on the effectiveness of the ion-exchange purification. If the purified sample is not diluted before reinjection it will elute very early owing to the presence of high salt buffer. Finally, remove the solvent from the purified oligonucleotide using a rotary evaporator in readiness for trityl-off reversed-phase purification (300 OD units).

Notes. Potassium phosphate buffer has a tendency to crystallize, so when the purification process is complete it is very important to purge thoroughly all HPLC columns, lines, and pumps with 5% aqueous methanol followed by 50% aqueous methanol to remove potassium phosphate. On no account should the pumps be switched off during the period of the purification procedure as this will inevitably result in the crystallization of potassium phosphate buffer B, causing severe damage to pumps, seals, and columns. The use of an SAX guard column will greatly increase the useful lifetime of the main column, which should be stored in methanol in the long term. Ion-exchange columns have a shorter lifetime than reversed-phase columns as they are chemically and physically less stable.

Acetonitrile is a convenient denaturant in anion-exchange HPLC buffers as it is volatile and therefore can be easily removed on a rotary evaporator. Buffers containing 60% formamide are more effective in extreme cases,[8] but removal of formamide is more difficult than removal of acetonitrile. Formamide can be removed from a purified aqueous solution of the oligonucleotide either by ether extraction or dialysis.

---

[8] A. Pingoud, A. Fleiss, and V. Pingoud, *in* "HPLC of Macromolecules" (R.W.A. Oliver, ed.), Chap. 7. IRL Press at Oxford Univ. Press, Oxford, 1989.

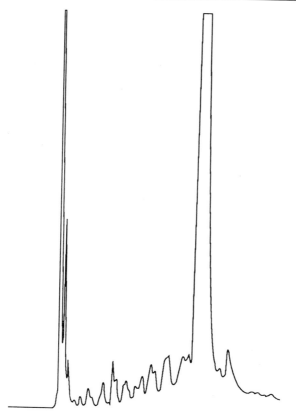

FIG. 2. Preparative anion-exchange HPLC chromatogram of the crude self-complementary dodecamer d(CGCGAATTCGCG), one 1 μmol synthesis. The sharp peak is benzamide, followed by failure sequences, and the final major peak is the desired product. The chromatogram was monitored at 280 nm with a fixed wavelength ultraviolet detector.

Trityl-off Reversed-Phase Chromatographic Purification

Trityl-off reversed-phase HPLC is the final HPLC purification step and follows either anion-exchange or trityl-on reversed-phase HPLC.

Experimental Procedure

Equipment and Reagents

Column: Brownlee Aquapore reversed-phase octyl (C_8), dimensions 25 cm $\times$ 10 mm (C_{18} columns give similar results, and C_4 columns can also be used).

Guard column: C_8 (not essential)
Buffer A: 0.1 M Ammonium acetate, pH 7.0
Buffer B: 0.1 M Ammonium acetate with 20% acetonitrile, pH 7.0
Flow rate: 3 ml/min
Gradient:

Time (min)	% Buffer A	% Buffer B
0	100	0
3	100	0
4	85	15
25	30	70
27	0	100
30	0	100
32	100	0
35	100	0

Protocol. Dissolve the product (300 OD units) from the anion-exchange or trityl-on reversed-phase HPLC purification (after detritylation) in 3 ml of distilled water. If the initial purification has been carried out by anion-exchange chromatography, it may be necessary to use a larger volume of water and to warm the solution to dissolve the residual potassium phosphate. Potassium phosphate will not otherwise present a problem as it will be removed during the reversed-phase purification. Equilibrate the reversed-phase column with a blank run and inject an analytical sample of the oligonucleotide, monitoring at 254 nm. Adjust the chromatographic conditions slightly if the product does not emerge between 15 and 22 min. (Increasing the percentage of buffer B at the 25-min step will cause the sample to elute earlier).

When a satisfactory analytical chromatogram has been obtained, carry out six preparative runs, monitoring at 295 nm and injecting 0.5 ml of the sample on each occasion. The product peak should be collected, and any shoulders should be discarded in order to remove traces of impurities (Fig. 3). Pool the purified fractions and inject a small analytical sample, monitoring at 254 nm (Fig. 4). If any impurity remains, repeat the entire reversed-phase purification procedure. Finally, remove the solvent from the purified oligonucleotide on a rotary evaporator or Speed-Vac concentrator (Savant Instruments, Hicksville, NY), dissolve the residue in 3 ml of water, and lyophilize to remove excess ammonium acetate. There should now be a total of approximately 200 OD units of purified oligonucleotide.

Notes. Trityl-off reversed-phase purification relies on the hydrophobic effect and will reveal impurities which are not apparent in anion-exchange

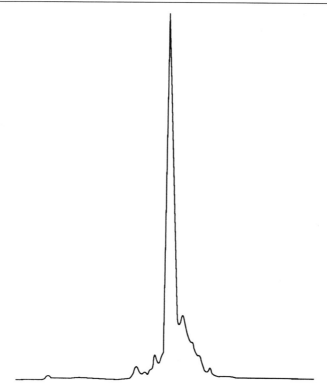

FIG. 3. Preparative trityl-off reversed-phase HPLC chromatogram of the dodecamer d(CGCGAATTCGCG), one 1 μmol synthesis, after anion-exchange HPLC purification. The chromatogram was monitored at 300 nm on a variable wavelength ultraviolet detector. A number of impurities are apparent which were not removed by trityl-on purification.

or trityl-on reversed-phase chromatography. Reversed-phase HPLC is of great value in removing base-modified impurities of the same length as the desired product. In general, longer sequences will elute after shorter ones as they are more hydrophobic. Hence, a reversed-phase chromatogram sometimes resembles an ion-exchange chromatogram.

Self-complementary (palindromic) sequences occasionally elute as double peaks on analytical reversed-phase HPLC owing to hairpin loop formation (Fig. 5). The ratio of the two peaks will depend on the concentration of the injected sample. This is because high oligonucleotide concentrations favor the duplex form, whereas hairpin loop formation is concentration independent. Certain noncomplementary G-rich sequences

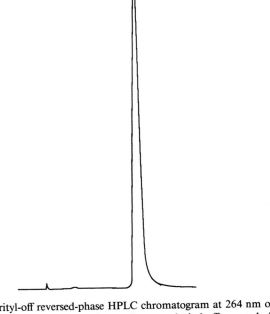

FIG. 4. Analytical trityl-off reversed-phase HPLC chromatogram at 264 nm of the dode-camer d(CGCGAATTCGCG), after both anion-exchange and trityl-off reversed-phase HPLC purification. No further HPLC purification is necessary.

tend to aggregate[9] (Fig. 6) and may bind irreversibly to reversed-phase HPLC columns. This seems to be a property of sequences with unin-terrupted runs of G residues and is most troublesome when these extend from one end of the oligonucleotide. Sequences rich in T residues require higher than normal concentrations of acetonitrile for elution but otherwise create no difficulties.

Preparation of Chromatography Buffers

Anion-Exchange Buffers

Buffer A (40 mM potassium phosphate, 20% acetonitrile): Dissolve po-tassium dihydrogen orthophosphate (4 g) in water (700 ml). Add

[9] A. Klug and W. I. Sundquist, *Nature (London)* **342,** 825 (1989).

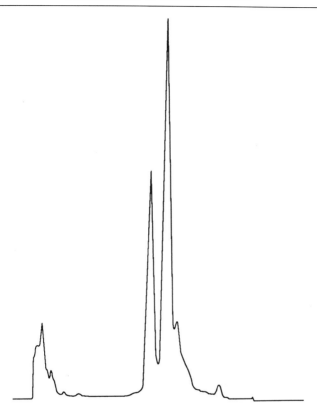

Fig. 5. Preparative trityl-off reversed-phase HPLC chromatogram of a single 1 μmol synthesis of a 21-mer oligonucleotide, d(GTC AGG TCA CAG TGA CCT GAT), a sequence compatible with hairpin loop formation. The sequence had previously been purified by anion-exchange HPLC. The small multiplet of peaks is due to phosphate buffer, and the two major peaks are stable conformations of the oligonucleotide. The ratio of the two peaks changes as a function of the DNA concentration of the injected sample. The chromatogram was monitored at 300 nm with a variable wavelength ultraviolet detector.

2.0 M potassium hydroxide dropwise to adjust the pH to 6.4. Then add acetonitrile (200 ml). Adjust the volume to 1.0 liter with water and filter through a Millipore filter.

Buffer B (0.67 M potassium phosphate, 20% acetonitrile): Repeat the above procedure using 90 g of potassium dihydrogen orthophosphate. Addition of acetonitrile should be carried out slowly with rapid stirring to avoid crystallization of the buffer. Anion-exchange HPLC buffers should be filtered daily and discarded after a few days.

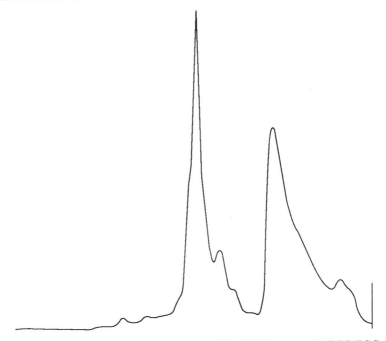

FIG. 6. Reversed-phase HPLC chromatogram of the G-rich sequence d(GGG TGG GAA GCC CTC AAA TAG G). The oligonucleotide aggregates in aqueous solution and produces a broad multiplet of peaks on HPLC.

Reversed-Phase Buffers

Stock solution of 1 M ammonium acetate: Dissolve Analar ammonium acetate (77 g) in water and make up to 1.0 liter in a volumetric flask. Stopper the solution and store in a cold room.

0.1 M Ammonium acetate, pH 7.0 (trityl-off HPLC buffer A): Add 100 ml of 1.0 M ammonium acetate to 850 ml of water. Adjust to pH 7.0 with 1.0 M aqueous ammonia or glacial acetic acid as necessary (the pH may not require adjustment). Make up to 1.0 liter with water and filter through a Millipore filter.

0.1 M Ammonium acetate, 50% acetonitrile, pH 7.0 (trityl-on HPLC buffer B): Add 100 ml of 1.0 M ammonium acetate to 300 ml of water. Check the pH, adjust to pH 7.0 if necessary, then add 500 ml of acetonitrile. Make up to 1.0 liter with water and filter through a Millipore filter.

0.1 M Ammonium acetate, 20% acetonitrile, pH 7.0 (trityl-off HPLC buffer B, trityl-on HPLC buffer A): Add 100 ml of 1.0 M ammonium acetate to 650 ml of water. Adjust to pH 7.0 if necessary and add 200 ml of acetonitrile. Make up to 1.0 liter with water and filter through a Millipore filter.

The above ammonium acetate buffers have no great tendency to crystallize. However, they must be filtered daily before use and must not be stored for more than 1 week.

Final Stages of Purification

Removal of Inorganic Salts by Sephadex Gel Filtration

Dissolve the lyophilized oligonucleotide from trityl-off reversed-phase purification ($\sim$ 200 OD units) in 2 ml of water and load onto a column of Sephadex G-10 or G-25 (column volume 1.6×50 cm). If a peristaltic pump is used, the flow rate should be 2 ml/min. Elute with water and monitor the eluant at 254 nm. Collect the entire peak, evaporate to about 1 ml, and lyophilize to give a fluffy white solid ($\sim$ 10 mg, 200 OD units). Prepacked disposable Sephadex G-25 columns can be purchased from Pharmacia.

Dialysis may be used as an alternative to gel filtration to remove traces of salt, but it is more time-consuming and less reliable for short oligonucleotides.

Appendix: Suppliers of Chromatography Columns and Other Materials

Partisil P10 SAX ion-exchange HPLC column, 10 μm particle size, column dimensions 25 cm $\times$ 8 mm (or 25 cm $\times$ 10 mm, semipreparative) and 25 cm $\times$ 4.5 mm (analytical): Hichrom Ltd., 6 Chiltern Enterprise Centre, Station Road, Theale, Reading, Berkshire RG7 4AA, U.K.

SAX ion-exchange guard column kit: Capital HPLC Specialists, Unit 13, Waverley Industrial Units, Waverley Street, Bathgate, Edinburgh EH48 4JA, U.K.

Brownlee Aquapore reversed-phase HPLC columns, octyl (C_8), 20 μm particle size, column dimensions 25 cm $\times$ 10 mm: Applied Biosystems Ltd. Birchwood Science Park North, Warrington, Cheshire WA3 7PB, U.K.

C_{18} Pellicular guard column kit for reversed-phase HPLC: Perkin Elmer Ltd. Post Office Lane, Beaconsfield, Bucks HP9 1QA, U.K.

Disposable Sephadex columns: Pharmacia NAP (nucleic acid purification) columns, Pharmacia Ltd., Milton Keynes, Bucks, U.K.

Acknowledgments

The work described in this chapter was funded in part by the S.E.R.C. Molecular Recognition Initiative.

[3] Synthetic Oligodeoxynucleotides Containing Modified Bases

By B. A. CONNOLLY

Introduction

This chapter describes the synthesis, incorporation into oligodeoxynucleotides, and characterization of some modified bases that are routinely used in our laboratory. Organic synthesis is often a daunting task for biological scientists and indeed even trained chemists working in biological laboratories. The synthetic methods used are therefore described in detail. Although some of the pathways leading to modified bases are rather long, none of the reactions are particularly tricky. In our hands all the reactions described are reliable and proceed well and in acceptable yields.

Monitoring of reactions should be performed by thin-layer chromatography. For thin-layer chromatography (TLC) we use silica gel $60F_{254}$ aluminum-backed sheets (5×7.5 cm, layer thickness 0.2 mm) purchased from Merck (Darmstadt, Germany, #5549). This material contains a fluorescent indicator, and nucleosides and their derivatives can be detected as purple spots under short-wavelength UV light. Many of the compounds are purified by flash chromatography,[1] and we always use silica gel 60 (particle size 0.040–0.063 mm, 230–400 mesh ASTM) from Merck (#9385). It is essential to master both these techniques in order to perform the organic preparations successfully. Good practical accounts of both methods have been given.[2] Many of the syntheses require dry conditions (both dry starting materials and anhydrous organic solvents, which often contain water as supplied). Once again excellent methods for drying all the solvents and reagents required are available.[2,3]

Finally many base analogs are toxic or carcinogenic. As a matter of safety it would be assumed that the nucleosides mentioned in this chapter have these properties. Therefore, always handle them in a fume hood wearing a laboratory coat and disposable gloves. When weighing out powders of the nucleosides wear a disposable gauze face mask.

[1] W. C. Still, M. Kahn, and A. Mitra, *J. Org. Chem.* **43**, 2923 (1978).
[2] B. S. Sproat and M. J. Gait, *in* "Oligonucleotide Synthesis: A Practical Approach" (M. J. Gait, ed.), p. 199. IRL Press, Oxford and Washington, D.C., 1985.
[3] D. D. Perrin and W. L. F. Armarego, "Purification of Laboratory Chemicals," 3rd Ed. Pergamon, New York, 1988.

Synthesis of Modified Bases

This section describes the preparation of the modified bases illustrated in Fig. 1. In our laboratory we use the phosphoramidite[4] approach to prepare oligodeoxynucleotides. Thus the final synthetic target is not the free deoxynucleoside itself but rather a base derivative that contains a 5'-dimethoxytrityl group[5] (procedure 8, see below), a 3'-diisopropylaminocyanoethylphosphoramidite[6] (procedure 9, see below), and protection of the exocylic groups of the base as appropriate.

Procedures 1, 2, and 3 (see below) and Fig. 2 outline the preparation of two dG derivatives, 2-aminopurine deoxyriboside ($d^{2am}P$) and 6-thiodeoxyguanosine ($d^{6S}G$). In the first derivative the 6-keto oxygen atom of dG is replaced with hydrogen and in the second with sulfur. The first two steps in the preparation of $d^{2am}P$ and $d^{6S}G$ are the same and consist of protection of the sugar 3'- and 5'-hydroxyl groups and the 2-amino group of dG by perbenzoylation followed by activation of the 6 position of dG to nucleophilic attack by sulfonylation with the 2,4,6-triisopropylbenzenesulfonyl group.[7-13] $d^{2am}P$ derivatives can be synthesized by nucleophilic displacement of the benzenesulfonyl group with hydrazine followed by oxidation and displacement of the hydrazino group with Ag_2O to give the tribenzoyl

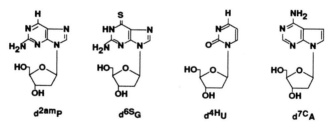

FIG. 1. Modified bases discussed in this chapter. $d^{2am}P$, 2-Aminopurine deoxyriboside; $d^{6S}G$, 6-thiodeoxyguanosine; $d^{4H}U$, 2-pyrimidinone deoxyriboside; $d^{7C}A$, 7-deazadeoxyadenosine.

[4] M. H. Caruthers, A. D. Barone, S. L. Beaucage, D. R. Dodds, E. F. Fisher, L. J. McBride, M. Matteucci, Z. Stabinsky, and J. Y. Tang, this series, Vol. 154, p. 287.
[5] H. Schaller, G. Weiman, B. Lerch, and H. G. Khorana, *J. Am. Chem. Soc.* **85**, 3821 (1963).
[6] N. D. Sinha, J. Biernat, J. McManus, and H. Koster, *Nucleic Acids Res.* **12**, 2261 (1984).
[7] C. B. Reese and A. Ubasawa, *Tetrahedron Lett.* **21**, 2256 (1980).
[8] C. B. Reese and A. Ubasawa, *Nucleic Acids Res. Symp. Ser.* **7**, 5 (1980).
[9] H. P. Daskalov, M. Sekine, and T. Hata, *Tetrahedron Lett.* **21**, 3899 (1980).
[10] B. L. Gaffney and R. A. Jones, *Tetrahedron Lett.* **23**, 2253 (1982).
[11] B. L. Gaffney and R. A. Jones, *Tetrahedron Lett.* **23**, 2257 (1982).
[12] B. L. Gaffney, L. A. Marky, and R. A. Jones, *Tetrahedron* **40**, 3 (1984).
[13] L. W. McLaughlin, T. Leong, F. Benseler, and N. Piel, *Nucleic Acids Res.* **16**, 5631 (1988).

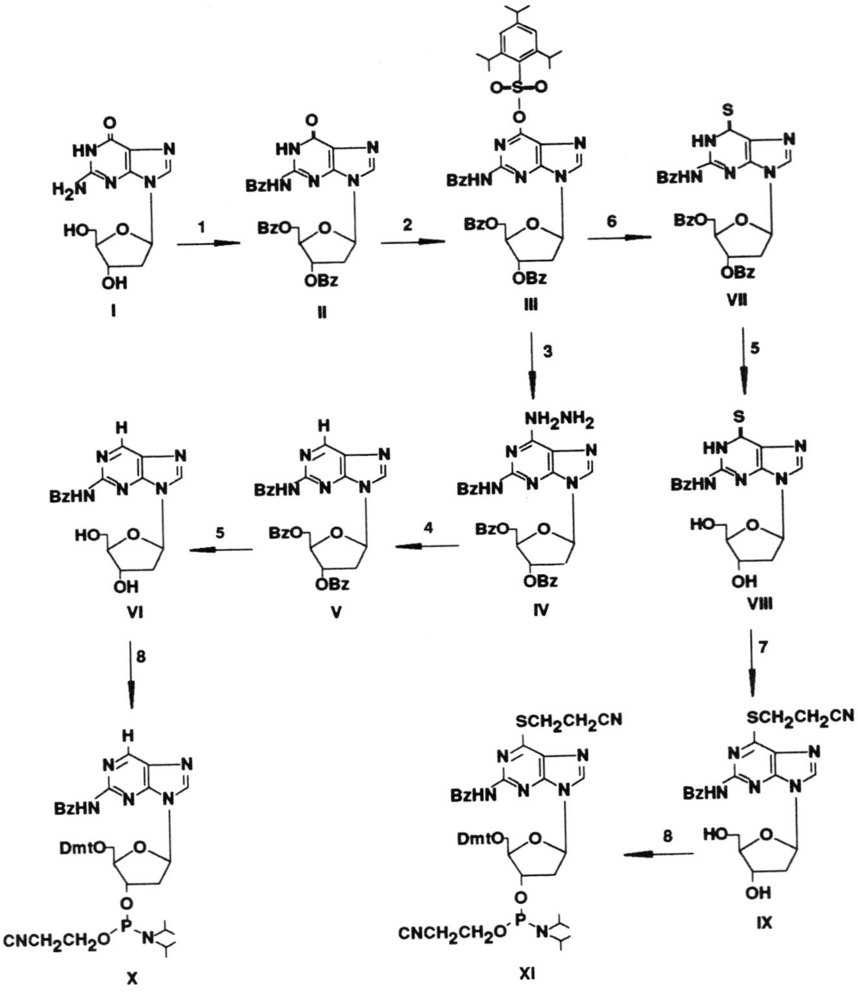

FIG. 2. Preparation of derivatives of $d^{2am}P$ and $d^{6s}G$ suitable for oligodeoxynucleotide synthesis. This figure should be used in conjunction with procedures 1, 2, and 3. Reagents: **1**, benzoic anhydride; **2**, 2,4,6-triisopropylbenzenesulfonyl chloride; **3**, hydrazine; **4**, Ag$_2$O; **5**, NaOH (brief); **6**, Li$_2$S; **7**, bromopropionitrile; **8**, dimethoxytrityl chloride, then cyanoethyl-N,N-diisopropylchlorophosphoramidite.

derivative of d^{2am}P.[13] d^{6S}G is easily prepared by direct displacement of the sulfonyl group with lithium sulfide, yielding tribenzoyl-d^{6S}G.[14] The tribenzoyl derivatives of both d^{2am}P and d^{6S}G can be converted to the N-monobenzoyl derivatives, suitably protected synthons for oligonucleotide synthesis, by brief NaOH hydrolysis.[5] With d^{2am}P, standard 5'-dimethyoxytritylation and 3'-phosphitylation yield the compound required for oligonucleotide synthesis. The sulfur atom in d^{6S}G requires further protection during oligonucleotide preparation, and this is achieved by using the cyanoethyl group introduced via reaction of N-benzoyl-d^{6S}G with bromopropionitrile.[14] Once again dimethoxytritylation and phosphitylation give the final d^{6S}G derivative.

Procedure 4 (below) and Fig. 3 detail the synthesis of 2-pyrimidinone (d^{4H}U), a dU-derived compound in which the 4-keto oxygen is replaced with hydrogen. As with the dG analogs, preparation begins with the protection of the sugar hydroxyl groups of dU by benzoylation. Treatment with Lawesson's reagent converts the 4-keto oxygen to sulfur.[15–18] The sulfur atom in d^{4S}U is susceptible to displacement by nucleophiles in an identical manner to the benzenesulfonyl group in dG derivatives. Thus d^{4H}U can be prepared from d^{4S}U by nucleophilic displacement of the sulfur with hydrazine followed by Ag$_2$O oxidation in an analogous fashion to the conversion of benzenesulfonyl-dG derivatives to d^{2am}P.[15,16,18,19] We have, however, found it more convenient to introduce the 5'-dimethoxytrityl group onto d^{4S}U prior to the hydrazine and Ag$_2$O steps. The 5'-dimethoxytrityl group confers organic solubility on deoxynucleosides and so enables them to be purified by flash chromatography. Therefore the reaction sequence represented in Fig. 3 is optimum. Alternative strategies have retained the benzoyl groups throughout the synthesis, removing them when the base derivatization is complete.[19]

Finally, Procedures 5, 6, and 7 (below) and Fig. 4 outline the preparation of 7-deazadeoxyadenosine (d^{7C}A) derivatives. These compounds have the 7 ring nitrogen of dA replaced by the isosteric CH group. As a ring nitrogen rather than an exocyclic function (as with d^{2am}P, d^{6S}G, and d^{4H}U) is changed, the synthesis is rather longer and more complicated than those described above. d^{7C}A cannot be prepared from its parent base dA (as was

[14] T. R. Waters and B. A. Connolly, *Nucleosides Nucleotides,* in press (1992).
[15] D. Cech and A. Holy, *Collect Czech. Chem. Commun.* **42,** 2246 (1977).
[16] R. Wightman and A. Holy, *Collect. Czech. Chem. Commun.* **38,** 1381 (1973).
[17] J. J. Fox, D. Van Praag, I. Wempen, I. L. Doerr, L. Cheong, J. E. Knoll, M. L. Eidinoff, A. Bendich, and G. B. Brown, *J. Am. Chem. Soc.* **81,** 178 (1959).
[18] B. A. Connolly and P. C. Newman, *Nucleic Acids Res.* **17,** 4957 (1989).
[19] B. Gildea and L. W. McLaughlin, *Nucleic Acids Res.* **17,** 2261 (1989).

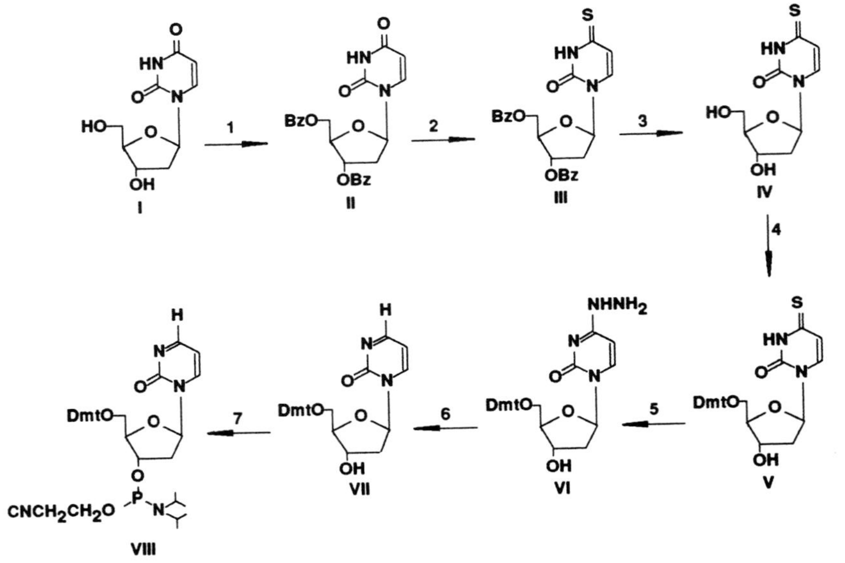

FIG. 3. Preparation of a d⁴ᴴU derivative suitable for oligodeoxynucleotide synthesis. This figure should be used in conjunction with procedure 4. Reagents: **1**, benzoyl chloride; **2**, Lawesson's reagent; **3**, sodium methoxide; **4**, dimethoxytrityl chloride; **5**, hydrazine; **6**, Ag$_2$O; **7**, cyanoethyl-*N,N*-diisopropylchlorophosphoramidite.

done with the previous modified bases), but rather the heterocycle must be synthesized *de novo* and attached to an appropriate deoxyribose derivative. This preparation is mainly given to illustrate this condensation of a heterocycle with deoxyribose by the sodium salt glycosidation procedure.[20] This is an extremely useful method that has been used to prepare many modified deoxynucleosides. The actual heterocycle required for the glycosidation reaction is 4-chloropyrrolo[2,3-*d*]pyrimidine, prepared in five steps from ethyl 2,2-diethoxyethyl cyanoacetate as described in procedure 6.[21]

Each step in the synthesis of d⁷ᶜA proceeds well and in good yield, and the intermediate products can be obtained by simple precipitations or recrystallizations. The most important points here are to use freshly prepared Raney nickel and newly redistilled POCl$_3$ in the appropriate reactions. The use of *N,N*-dimethylaniline in the POCl$_3$ reaction[22] improves the yields over the original method using POCl$_3$ alone.[21] The deoxyribose

[20] Z. Kazimierczuk, H. B. Cottam, G. R. Revankar, and R. K. Robbins, *J. Am. Chem. Soc.* **106,** 6379 (1984).

[21] J. Davol, *J. Chem. Soc.,* 131 (1960).

[22] M. J. Robins and B. Uznanski, *Can. J. Chem.* **59,** 2601 (1981).

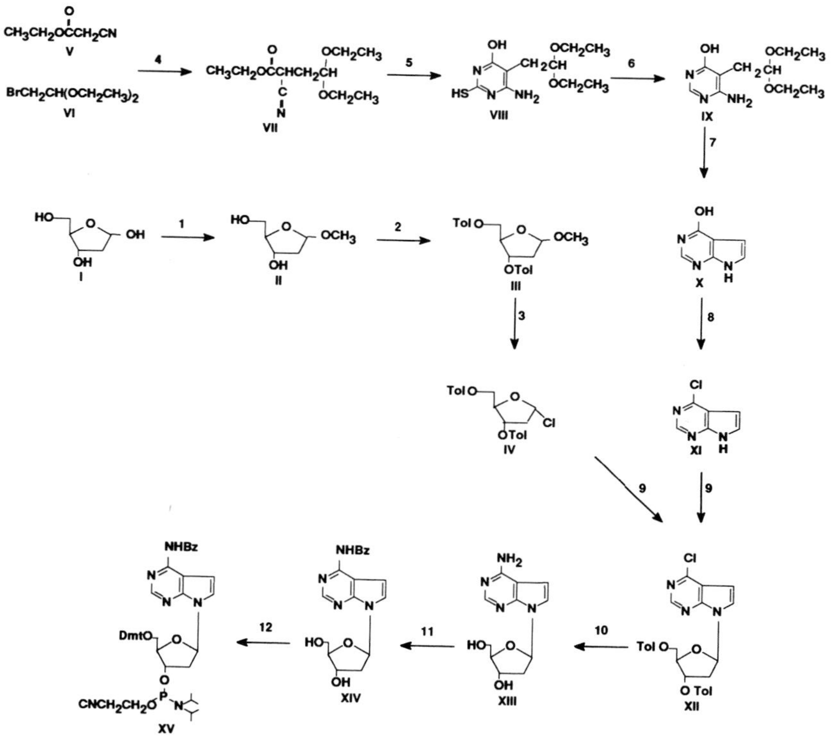

FIG. 4. Preparation of a d^{7C}A derivative suitable for oligodeoxynucleotide synthesis. This figure should be used in conjunction with procedures 5, 6, and 7. Reagents: **1**, HCl/CH$_3$OH; **2**, *p*-toluoyl chloride; **3**, HCl/CH$_3$OOH; **4**, K$_2$CO$_3$/NaI/heat; **5**, thiourea; **6**, Raney nickel; **7**, H$^+$; **8**, POCl$_3$; **9**, mix **IV** and **XI**, sodium hydride; **10**, NH$_3$/heat; **11**, trimethylsilyl chloride/ benzoyl chloride; **12**, dimethoxytrityl chloride, then cyanoethyl-*N,N*-diisopropylchlorophos- phoramidite.

derivative required is 3,5-di-*p*-toluoyl-1-*α*-chloro-2-deoxyribose synthe- sized using procedure 5.[23] The final chloro sugar is not indefinitely stable. We store it in a vacuum desiccator over KOH and use it within a few days of preparation. It is most convenient to prepare large amounts of the 3,5-di-*p*-toluoyl-1-*O*-methyl-2-deoxyriboside, which can be stored indefi- nitely at 4° (as a sticky oil) and use this to prepare the chloro sugar as needed and in the amounts required. The sodium salt glycosidation and further elaboration of d^{7C}A are very reliable, proceeding in high yields with

[23] M. Hoffer, *Chem. Ber.* **93**, 2777 (1960).

exclusive β-anomer formation.[20] Treatment of the 3',5'-di-p-toluoyl-6-chloro-7-deazapurine deoxyriboside with NH_3 gives $d^{7C}A$. This can be converted to a derivative suitable for oligodeoxynucleotide synthesis by the standard methods outlined in procedure 7. Alternative syntheses of a $d^{7C}A$ derivative suitable for oligonucleotide preparation have used phase-transfer glycosidations[24,25]; however, the sodium salt method is superior.

Oligonucleotide Synthesis, Purification, and Characterization

In our laboratory oligodeoxynucleotides containing modified bases are prepared using an Applied Biosystems (Foster City, CA) 381 A DNA synthesizer, usually on a 1 μmol scale. The modified base phosphoramidite should be dissolved in anhydrous DNA synthesis grade acetonitrile at the usual concentration of 0.1 M. The solutions should be filtered through 25-mm 0.5-μm PTFE Millex-SR filters (Millipore, Bedford, MA, SL5R 025 NB) to remove particulate material. Modified bases are introduced into oligonucleotides using the X position on the machine. No changes to the standard synthesis cycles are required to accommodate the modified bases, all of which should couple in near 100% yield as monitored by trityl cation release. Syntheses should be performed "trityl-on" to facilitate purification.[26,27]

After synthesis most of the protecting groups are removed by treatment with concentrated aqueous NH_3 at 50° overnight. These conditions are compatible with unreactive modified bases, namely, $d^{2am}P$ and $d^{7C}A$, and can be used to remove the protecting groups completely with no destruction of the bases. With $d^{6S}G$ a 4-hr treatment with NH_3 at 50° is best. This completely removes the N-benzoyl and S-cyanoethyl protecting groups.[14] Longer NH_3 treatment leads to some displacement of the sulfur atom and 2,6-diaminopurine formation, although even after overnight this side reaction is minor. Unfortunately $d^{4H}U$ is very sensitive to NH_3.[28] Best results are obtained using normal bases containing NH_3 labile groups[29] [these are available from Applied Biosystems under the name F.O.D. (fast oligonucleotide deblock) bases]. Here the deblocking can be performed using concentrated NH_3 at room temperature overnight.

[24] F. Seela and A. Kehne, *Liebigs Ann. Chem.*, 876 (1983).

[25] F. Seela and A. Kehne, *Tetrahedron* **41**, 5387 (1985).

[26] H. J. Fritz, R. Belagaje, E. L. Brown, R. H. Fritz, R. A. Jones, R. F. Lees, and H. G. Khorana, *Biochemistry* **17**, 1257 (1978).

[27] L. W. McLaughlin and N. Piel, *in* "Oligonucleotide Synthesis: A Practical Approach" (M. J. Gait, ed.), p. 117. IRL Press, Oxford and Washington, D.C., 1985.

[28] B. A. Connolly, unpublished observations (1990).

[29] J. C. Schulhof, D. Molko, and R. Teoule, *Nucleic Acids Res.* **15**, 397 (1987).

After NH$_3$ treatment oligonucleotides can be purified by reversed-phase "trityl-specific" purification[26,27] and the final dimethoxytrityl group removed by treatment with 80% acetic acid at room temperature for 1 hr. For oligonucleotides containing the four usual bases and oligomers that contain the unreactive d^{2am}P and d^{7C}A, the trityl-specific purification gives products that are pure enough for most uses. When chemically reactive bases (d^{6S}G and d^{4H}U) are present, further post-trityl-on purification is usually required. This probably arises owing to side reactions with these chemically reactive bases during synthesis and deblocking. In these cases purification by reversed-phase high-performance liquid chromatography (HPLC) using shallow acetonitrile gradients usually suffices to purify the required oligomer fully.[18]

The simplest way to characterize oligonucleotides containing modified bases is to hydrolyze them completely to their constituent mononucleosides using snake venom phosphodiesterase and alkaline phosphatase.[27,30] This method is described in procedure 10. The mononucleosides formed can be analyzed by reversed-phase HPLC. This procedure not only shows the presence of a modified base but by integration of peaks can be made quantitative and used to demonstrate that the correct mole fraction of the modified base (and the four normal bases) is present. An example of such a chromatogram is illustrated in Fig. 5.

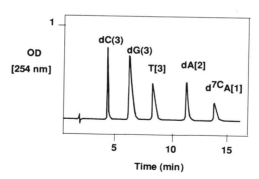

FIG. 5. Base composition analysis of d(GACGAT[^{7C}A]TCGTC). The identities of each of the peaks are indicated, and the numbers refer to the mole fraction of each deoxynucleoside present. This figure should be used in conjunction with procedure 10.

[30] J. S. Eadie, J. McBride, J. Lincoln, W. Efcavitch, L. B. Hoff, and R. Cathcart, *Anal. Biochem.* **165**, 442 (1987).

Uses of Oligonucleotides Containing Modified Bases

So far the main uses of oligonucleotides with modified bases has been to study details of specific protein–nucleic acid interactions, especially with repressor proteins, restriction endonucleases, and modification methylases.[31-38] The rationale here is to replace a group on the base capable of interacting with a protein with one that cannot form this interaction.[39] For example, the 6-keto oxygen atom of dG has the potential to form a hydrogen bond with DNA-binding proteins. The use of $d^{2am}P$, in which this 6-keto oxygen is replaced with hydrogen, clearly abolishes hydrogen bonding potential between protein and oligonucleotide. Therefore oligonucleotides in which dG is replaced by $d^{2am}P$ can be used to determine whether this oxygen atom hydrogen bonds to the protein under study. Appropriate kinetic and binding experiments can give some idea of the strength of the interaction and how much binding energy it contributes to specificity and catalysis. The uses and pitfalls of this method have been discussed.[40]

Other potential uses of modified bases arise because of their physical or chemical properties. Thus $d^{2am}P$ and $d^{4H}U$ are fluorescent, whereas $d^{6S}G$ has UV[41] and circular dichroic (CD)[14] absorptions well removed from those of the four normal bases. It has long been realized that $d^{6S}G$ is photoreactive (by 350 nm light)[42] and recently we have demonstrated that oligonucleotides containing this base may be covalently crosslinked to the *Eco*RV restriction endonuclease and modification methylase by irradiating enzyme oligodeoxynucleotide complexes with 350 nm light.[43] Finally we have shown that oligonucleotides that contain $d^{4H}U$ form stable covalent adducts with methylases that add a CH_3 group to the 5-position of dC

---

[31] E. F. Fisher and M. H. Caruthers, *Nucleic Acids Res.* **7**, 401 (1979).

[32] C. A. Brennan, M. D. van Cleve, and R. I. Gumport, *J. Biol. Chem.* **261**, 7270 (1986).

[33] C. A. Brennan, M. D. van Cleve, and R. I. Gumport, *J. Biol. Chem.* **261**, 7279 (1980).

[34] L. W. McLaughlin, F. Benseler, E. Graeser, N. Piel, and S. Scholtissek, *Biochemistry* **26**, 7238 (1987).

[35] F. Seela and A. Kehne, *Biochemistry* **28**, 2232 (1987).

[36] J. Mazzarelli, S. Scholtissek, and L. W. McLaughlin, *Biochemistry* **28**, 4616 (1989).

[37] P. C. Newman, V. U. Nwosu, D. M. Williams, R. Cosstick, F. Seela, and B. A. Connolly, *Biochemistry* **29**, 9891 (1990).

[38] P. C. Newman, D. M. Williams, R. Cosstick, F. Seela, and B. A. Connolly, *Biochemistry* **29**, 9902 (1990).

[39] N. C. Seaman, I. M. Rosenberg, and A. Rich, *Proc. Natl. Acad. Sci. U.S.A.* **73**, 804 (1976).

[40] C. R. Aiken and R. I. Gumport, this series, Vol. 208 [21].

[41] J. J. Fox, I. Wempen, A. Hampton, and I. L. Doerr, *J. Am. Chem. Soc.* **80**, 1669 (1958).

[42] H. R. Rackwitz and K. H. Scheit, *Chem. Ber.* **107**, 2284 (1974).

[43] T. T. Nikiforov and B. A. Connolly, *Nucleic Acids Res.,* submitted (1992).

residues in DNA or oligodeoxynucleotides.[44] The mechanism of these enzymes usually involves a transient covalent link between a cysteine residue and the 6-position of the dC base to be methylated. Following methylation using S-adenosylmethionine at the 5-position the temporary covalent link is broken.[45] However, when d^{4H}U replaces dC the bond between the cysteine and the 6-position becomes stable resulting in irreversible covalent attachment.

Experimental Procedures

Procedure 1: Preparation of 3',5'-N-Tribenzoyl-6-(2,4,6-triisopropyl)benzenesulfonyl-dG (III)

The following method should be used in conjunction with Fig. 2. The numbers of the compounds refer to Fig. 2.

3',5'-N-Tribenzoyl-dG (II). A 2.85 g (10 mmol) quantity of dG (monohydrate) (I) is dried by coevaporation from 2 × 50 ml dry pyridine. The dried dG is suspended in 40 ml dry pyridine, and 9.05 g (40 mmol) of benzoic anhydride together with 0.35 g of dimethylaminopyridine are added. The mixture is refluxed overnight with exclusion of moisture, after which TLC (CHCl$_3$, 90; CH$_3$OH, 10) should show only II (R_f 0.6) with no trace of I (R_f 0). Then 20 ml of 5% NaHCO$_3$ is added and the mixture evaporated to dryness and worked up. [Dissolve in 200 ml CH$_2$Cl$_2$ and extract with 2 × 200 ml of 5% NaHCO$_3$ followed by 100 ml saturated NaCl. Retain the organic (lower) layer at each stage, and finally dry the organic layer by swirling with 5 g of anhydrous Na$_2$SO$_4$. Filter to remove the Na$_2$SO$_4$ and evaporate the organic layer to dryness.] The product is purified by flash chromatography, eluting in turn with 800 ml CHCl$_3$ containing first 2% and second 4% CH$_3$OH. Compound II elutes with the second solvent in yields of about 90%.

3',5'-N-Tribenzoyl-6-(2,4,6-triisopropyl)benzenesulfonyl-dG (III). Compound II, 4.6 g (8 mmol), is dissolved in 50 ml dry CH$_2$Cl$_2$, and dimethylaminopyridine (0.5 g), dry diisopropylethylamine (5.6 ml) and 2,4,6-triisopropylbenzenesulfonyl chloride (4.8 g, 16 mmol) are added. After stirring at room temperature for 1 hr (with protection from moisture) TLC (CHCl$_3$, 97; CH$_3$OH, 3) should show complete conversion of II (R_f 0.5) to III (R_f 0.6). The mixture is evaporated to dryness and III purified by flash chromatography, eluting with CHCl$_3$ (500 ml) followed by CHCl$_3$ con-

[44] C. Taylor, K. Ford, B. A. Connolly, and D. P. Hornby, unpublished observations.
[45] L. Chen, A. M. MacMillan, W. Chang, K. Ezaz-Nikpay, W. S. Lane, and G. L. Verdine, *Biochemistry* **30,** 11018 (1991).

taining 2% CH_3OH (500 ml). Compound **III** elutes with the second solvent in yields of 60%.

Procedure 2: Preparation of N-Benzoyl-d²ᵃᵐP (VI)

The following method should be used in conjunction with Fig. 2. The numbers of the compounds refer to Fig. 2.

3',5'-N-Tribenzoyl-6-hydrazino-2-aminopurine Deoxyriboside (IV). A 3.14 g (3.7 mmol) quantity of **III** is dissolved in 70 ml dry tetrahydrofuran and cooled to $0-5°$. Anhydrous hydrazine (0.45 ml, 14.5 mmol; *caution:* extremely toxic) is added and the mixture left on ice with exclusion of moisture for 2 hr. TLC ($CHCl_3$, 92; CH_3OH, 8) should show complete conversion of **III** (R_f 0.9) to **IV** (R_f 0.1–0.2, streaks). After solvent evaporation and workup as described above, **IV** is pure enough to be used directly in the next reaction.

3',5'-N-tribenzoyl-2-aminopurine Deoxyriboside (V). Compound **IV** obtained above is dissolved in 150 ml tetrahydrofuran containing 5% water and 2.5 g of Ag_2O added. The mixture is refluxed for 2 hr and evaporated to dryness. The black solid obtained is suspended in a mixture of 100 ml ethyl acetate and 100 ml 10% potassium iodide and filtered through Celite to remove insoluble material. The filtrate is placed in a separating funnel, and the lower aqueous layer is run off and discarded. The ethyl acetate layer is extracted in turn with 100 ml of 10% potassium iodide, 2×100 ml of 10% sodium thiosulfate, and 2×100 ml of water. In each case the lower aqueous layer is discarded. The ethyl acetate layer is dried using Na_2SO_4 and evaporated to dryness. Compound **V** is purified by flash chromatography, eluting with $CHCl_3$ containing 1% CH_3OH (500 ml) followed by $CHCl_3$ containing 2.5% CH_3OH (500 ml). About 1.5 g of **V** (~70% yield from **III**) should be obtained.

N-Benzoyl-d²ᵃᵐP (VI). One and one-half grams of **V** is dissolved in 30 ml pyridine/4 ml CH_3OH and chilled to $-20°$ using an ice–salt bath. Then 3.6 ml of ice-cold 2 *M* NaOH is added and the mixture held at $-20°$ for 30 min. Dowex 50W-X8 (pyridinium cycle) resin is added in small batches until the pH of a small sample diluted with water is 7. The mixture is filtered to remove the resin, and evaporation gives **VI** in a pure enough state to be dimethoxytritylated as outlined in procedure 8 and phosphitylated as per procedure 9 to yield the required derivative **X**.

Procedure 3: Preparation of N-Benzoyl-S-cyanoethyl-d⁶ˢG (IX)

The following method should be used in conjunction with Fig. 2. The numbers of the compounds refer to Fig. 2.

3',5'-N-tribenzoyl-d⁶ˢG (VII). Four grams (4.7 mmol) of **III** is dissolved in 60 ml dry tetrahydrofuran, and 0.25 g (5.4 mmol) of Li_2S (*caution:* stench) is added. The mixture is stirred at room temperature for 2.5 hr, after which TLC should show complete conversion of **III** (R_f 0.6) to **VII** (R_f 0.1). After evaporation to dryness and standard workup, **VII** can be purified by flash chromatography using $CHCl_3$ containing 1% CH_3OH as eluent. Yields are high, usually about 90%.

N-Benzoyl-d⁶ˢG (VIII). Compound **VII** can be converted to **VIII** using a brief NaOH hydrolysis exactly as for the conversion of **V** to **VI**.

N-Benzoyl-S-cyanoethyl-d⁶ˢG (IX). The crude **VIII** produced above is suspended in 40 ml dry dimethylformamide and added dropwise (over 10 min) to a suspension of 5 ml of 3-bromopropionitrile and 6 g anhydrous K_2CO_3 in 100 ml dry dimethylformamide. The resulting mixture is stirred overnight, after which TLC ($CHCl_3$, 85; CH_3OH, 15) should show conversion of **VIII** (R_f 0.4) to **IX** (R_f 0.5). The mixture is filtered to remove insoluble material and evaporated to give a yellow oil. This is dissolved in 160 ml $CHCl_3$/40 ml CH_3OH and any insoluble material removed by filtration. Evaporation gives **IX** in a form pure enough to be dimethoxytritylated and phosphitylated as outlined in procedures 8 and 9 to furnish the required derivative **XI**.

Procedure 4: Preparation of a 2-Pyrimidinone Derivative *(VIII)* Suitable for Oligodeoxynucleotide Synthesis

The following method should be used in conjunction with Fig. 3. The numbers of the compounds refer to Fig. 3.

3′,5′-Dibenzoyl-dU (II). Deoxyuridine **(I)**, 2.28 g (10 mmol), is dried by coevaporation from 2 × 50 ml dry pyridine. Benzoyl chloride (2.55 ml, 22 mmol) is added and the mixture stirred overnight with exclusion of moisture. TLC ($CHCl_3$, 95; CH_3OH, 5) should show complete conversion of **I** (R_f 0.05) to **II** (R_f 0.6). The pyridine solution is poured into 2.5 liters of ice-cold, vigorously stirring water to precipitate the product. The precipitate is collected by filtration, washed with ice-cold water, and dried in a vacuum desiccator over silica gel. Yields should be near quantitative.

3′,5′-Dibenzoyl-d⁴ˢU (III). Compound **II**, 4.14 g (9.5 mmol), is dissolved in 50 ml of dry dioxane, and 4.85 g (12 mmol) of Lawesson's reagent [2,4-bis(4-methoxyphenyl)-1,3-dithia-2,4-diphosphetane-2,4-disulfide] *(caution:* stench) is added. The mixture is refluxed for 2 hr with moisture exclusion, when TLC ($CHCl_3$, 97; CH_3OH, 3) should show complete conversion of **II** (R_f 0.4) to **III** (R_f 0.7). Five milliliters of 5% $NaHCO_3$ is added and the mixture evaporated and worked up in the usual manner. The **III** obtained after workup is pure enough for the next stage.

d⁴ˢU (IV). Compound **III** produced above (assumed to be 9.5 mmol) is dissolved in 50 ml of dry CH_3OH, and sodium methoxide (25 mmol) is added. After 30 min, TLC (CHCl$_3$, 97; CH$_3$OH, 3) should show conversion of **III** (R_f 0.7) to **IV** (R_f 0.05). Acetic acid (1.43 ml, 25 mmol) is added and the mixture evaporated to dryness. The product is dissolved in water (50 ml) and extracted with 3 × 50 ml CHCl$_3$. The aqueous (upper) layer is retained each time and finally evaporated to dryness. The d^{4S}U **(IV)** is extracted from the contaminating inorganic salts by the addition of 100 ml acetone and stirring. Compound **IV** dissolves in the acetone, leaving the insoluble salts (which are removed by filtration) behind. The acetone is removed by evaporation to give **IV** of purity sufficient for the next reaction.

Dimethoxytrityl-d^{4S}U (V). Compound **IV** obtained above (assumed to be 9.5 mmol) is converted to its 5′-dimethoxytrityl (5′-dmt) derivative as described in procedure 8. The yield of purified **V** from **I** (four steps) should be about 50%.

Dimethoxytrityl-4-Hydrazino-2-pyrimidinone deoxyriboside (VI). A 2.6 g (5 mmol) quantity of **V** is dissolved in 100 ml ethanol containing 5 ml water, and 1.4 ml (25 mmol) hydrazine hydrate (*caution:* extremely toxic) is added. After refluxing for 1 hr, TLC (CHCl$_3$, 90; CH$_3$OH, 10) should show complete conversion of **V** (R_f 0.85) to **VI** (R_f 0.1–0.3, streaks). Evaporation and standard workup give **VI** pure enough to be used directly below.

Dimethoxytrityl-^{4H}dU (VII). Compound **VI** obtained above is dissolved in 50 ml ethanol containing 0.5 ml triethylamine. Ag$_2$O (2.3 g, 10 mmol) is added and the mixture refluxed for 2 hr. TLC (CHCl$_3$, 90; CH$_3$OH, 10) should show disappearance of **VI** and the appearance of **VII** (R_f 0.6). This reaction is never very clean, and several other spots will be visible. The mixture is evaporated to dryness, suspended in 100 ml CHCl$_3$/100 ml 10% NaI, and filtered through Celite to remove insoluble material. The filtrate is placed in a separating funnel and the CHCl$_3$ (lower) layer run off and retained. This treatment with 10% NaI is repeated twice. The CHCl$_3$ fraction is then extracted with 2 × 100 ml 5% Na$_2$S$_2$O$_3$, 100 ml water, and finally dried with Na$_2$SO$_4$. The CHCl$_3$ is removed by evaporation, and **VII** is purified by flash chromatography, eluting with 500 ml each of ethyl acetate containing 0.5% triethylamine and in turn 1, 3, and 5% CH$_3$OH. Compound **VII** elutes with 5% CH$_3$OH. Yields from **V** (two steps) are 50%. Compound **VII** can be converted to its 3′-phosphoramidite **(VIII)** as outlined in procedure 9.

Procedure 5: Preparation of 3,5-Di-p-toluoyl-1-α-chloro-2-deoxyribose IV)

The following method should be used in conjunction with Fig. 4. The numbers of the compounds refer to Fig. 4.

1-O-Methyl-2-deoxyribose (II). Dry 2-deoxyribose **(I)**, 13.6 g (100 mmol), is dissolved in 100 ml of dry CH_3OH. Then 27 ml of a 1% solution of anhydrous HCl in dry CH_3OH (100 mmol HCl) is added and the solution stirred with exclusion of moisture for 15 min. Ag_2CO_3 (3–5 g) is added and the solution filtered. Evaporation of solvents gives **II** as a thick oil, which is used immediately in the next reaction.

3,5-Di-p-toluoyl-1-O-methyl-2-deoxyribose (III). Compound **II** prepared above is dissolved in 100 ml of dry pyridine, and redistilled *p*-toluoyl chloride (34 g, 220 mmol) is added. The mixture is stirred overnight with exclusion of moisture, during which time a dense white precipitate forms. Ten milliliters of water is added and solvents removed by evaporation. The product is dissolved in 200 ml diethyl ether and extracted in turn with 2 × 200 ml of 1% H_2SO_4, 2 × 200 ml of 5% $NaHCO_3$, and 2 × 200 ml of water. The ether (upper layer) is retained at each stage and finally dried with Na_2SO_4. Evaporation of solvents gives **III** as a dense oil, which can be stored for months at 4°.

3,5-Di-p-toluoyl-1-α-chloro-2-deoxyribose (IV). Compound **III** produced above is dissolved in 80 ml of ice-cold acetic acid saturated with HCl. The mixture is cooled to 0–3°, and HCl is bubbled through the solution for 15 min, maintaining the temperature below 5°. Copious white crystals of **IV** should form. These are collected by filtration and washed thoroughly with dry diethyl ether. Compound **IV** should be stored in a desiccator over KOH and used within a few weeks of preparation. The yield of **IV** from **I** (three steps) should be 70%. This step can be performed with portions of **III** rather than the entire amount by appropriate scale down in the amount of acetic acid used.

Procedure 6: Preparation of 4-chloropyrrolo[2,3-d]pyrimidine (XI)

The following method should be used in conjunction with Fig. 4. The numbers of the compounds refer to Fig. 4.

Ethyl 2,2-diethoxyethyl cyanoacetate (VII). A mixture of ethyl cyano-acetate (**V**) (225 g, 2 mol), bromoacetaldehyde diethyl acetal (**VI**) (80 g, 400 mmol), K_2CO_3 (60 g), and NaI (4 g) is heated at 110–120° with vigorous stirring and exclusion of moisture overnight. The mixture is cooled, and 1 liter each of diethyl ether and water are added. This mixture is placed in a separating funnel and the aqueous (lower) layer run off and discarded. The ether layer is further extracted with 2 × 1 liter of water and finally dried with Na_2SO_4. Ether is removed by evaporation. Compound **VII** is purified by distillation at 0.7–1.2 mm Hg using a 6-inch column filled with glass spirals. Residual **V** distills at 67° and the product **VII** at 117°. About 50 g (55% yield) should be obtained.

2-Mercapto-4-oxo-5-(2,2-diethoxyethyl)-6-aminopyrimidine   (VIII). Dry thiourea (3.8 g, 50 mmol) (*caution:* extremely toxic) is dissolved in

40 ml of dry ethanol, and sodium ethoxide (3.4 g, 50 mmol) is added. This mixture is added to 9.16 g (40 mmol) of **VII** and the mixture refluxed with exclusion of moisture for 5 hr. The mixture is evaporated to one-third volume, 20 ml of water is added, and the pH is adjusted to 7 with acetic acid. Compound **VIII** precipitates and is collected by filtration. Recrystallization from CHCl$_3$/diethyl ether gives 8 g (70%).

4-Oxo-5-(2,2-diethoxyethyl)-6-aminopyrimidine (IX). The 8 g of **VIII** produced above is dissolved in 150 ml of water containing 25 ml of a 35% aqueous ammonia solution, and 10 g of freshly prepared Raney nickel is added. [NaOH (20 g) is dissolved in water (40 ml), cooled on ice, and nickel–aluminum alloy (10 g; 1 : 1, w/w) is gradually added to maintain the temperature between 70 and 75°. Five minutes after the last addition of alloy the aqueous layer is decanted off and the Raney nickel produced washed rapidly with water (5 × 50 ml) and ethanol (3 × 50 ml.[46]]. After refluxing for 4 hr, the mixture is filtered and solvents removed by freeze-drying (evaporation is extremely difficult owing to severe frothing). A total of 67 g (78%) of pure **IX** can be obtained by recrystallisation from methanol/diethyl ether.

4-Oxopyrrolo[2,3-d]pyrimidine (X). The 6.7 g of **IX** produced above is suspended in 67 ml of water, and 4.7 ml of 10 *M* HCl is added (final HCl concentration 0.65 *M*, with the excess of HCl over **IX** being 1.5). The mixture is stirred overnight, during which **X** precipitates and is collected by filtration. A total of 2.7 g (68%) is obtained of purity sufficient for the next reaction.

4-Chloropyrrolo[2,3-d]pyrimidine (XI). Two grams of dried (vacuum desiccator over silica gel) **X** is added to 200 ml of freshly distilled POCl$_3$ containing 3 ml freshly distilled *N,N*-dimethylaniline. The mixture is refluxed with exclusion of moisture for 6 hr. Excess POCl$_3$ is removed by evaporation using an oil pump, and 50 ml of ice-cold water is added carefully to the remaining material. The pH is adjusted to 8–9 using NaOH, and **XI** is extracted into 4 × 50 ml CHCl$_3$. The CHCl$_3$ layers are pooled, dried with Na$_2$SO$_4$, and evaporated to about 20 ml. On standing at 4° **XI** crystallizes, and about 1.4 g (63%) of **XI** is obtained.

Procedure 7: Preparation of N-Benzoyl-7-deaza-dA (XIV)

The following method should be used in conjunction with Fig. 4. The numbers of the compounds refer to Fig. 4.

3′,5′-Di-p-toluoyl-6-chloro-7-deazapurine Deoxyriboside (XII). A

[46] E. Caspi, P. J. Ramm, and R. E. Gain, *J. Am. Chem. Soc.* **91**, 4012 (1969).

1.3 g (8 mmol) quantity of 4-chloropyrrolo[2,3-d]pyrimidine (**XI**) (prepared in procedure 6) and 0.4 g (9.7 mmol) sodium hydride (60% dispersion in oil) are mixed in 25 ml dry (DNA synthesis grade) acetonitrile for 30 min. To this solution is added 3.1 g (8 mmol) of 3,5-di-*p*-toluoyl-1-α-chloro-2-deoxyribose (**IV**) (prepared in procedure 5) as a solution in 25 ml dry acetonitrile. The reaction mixture is stirred overnight under N_2 with exclusion of moisture. After filtration to remove insoluble material, the filtrate is evaporated to dryness and worked up in the usual way. Compound **XII** can be purified by flash chromatography, eluting with 500 ml $CHCl_3$ containing 2.5% CH_3OH. Pure **XII**, 3.5 g (80%), is obtained.

7-Deaza-dA (XIII). Three grams of **XII** is dissolved in 50 ml ethanol and 100 ml of 35% aqueous ammonia in a thick-walled glass vessel fitted with a Teflon screw-cap stopper. The mixture is heated at 80° for 72 hr and then cooled on ice for 30 min before opening. Solvents are removed by evaporation and 200 ml water added. The solution is extracted with 3 × 100 ml $CHCl_3$, with the aqueous (upper) layer being retained. Removal of water by evaporation gives **XIII**, which cannot be recrystallized or easily purified and so is used directly in the next step. The yield of this reaction is assumed to be 100% for the purpose of determining reagent stoichiometries in the next reaction.

N-Benzoyl-7-deaza-dA (XIV). Compound **XIII** produced above was dried by coevaporation from 2 × 50 ml dry pyridine, suspended in 100 ml dry pyridine, and cooled to 5° on ice. Freshly distilled trimethylsilyl chloride (7.7 ml, 5 Eq) is added in about 5 aliquots and the mixture held on ice with exclusion of moisture for 30 min.[47,48] Benzoyl chloride (7.3 ml, 5 Eq) is added in 5 aliquots, and the mixture is removed from the ice bath and left at room temperature for 2 hr. The mixture is then cooled to 5° and 20 ml of ice-cold water slowly added. Solvents are removed by evaporation and the oil obtained suspended in 200 ml water. The aqueous solution is extracted with 3 × 200 ml diethyl ether, with the aqueous (lower) layer being retained. The aqueous solution is concentrated to about 40 ml and left at 4° whereon **XIV** crystallizes. About 1.3 g (65%) of **XIV** from **XII** (two steps) should be obtained. Compound **XIV** can be dimethoxytritylated as described in procedure 8 and then converted to its 3′-phosphoramidite as outlined in procedure 9 to give the required derivative **XV**.

[47] G. S. Ti, B. L. Gaffney, and R. A. Jones, *J. Am. Chem. Soc.* **104**, 1316 (1982).
[48] R. A. Jones, *in* "Oligonucleotide Synthesis: A Practical Approach" (M. J. Gait, ed.), p. 23. IRL Press, Oxford and Washington, D.C., 1985.

Procedure 8: Preparation of 5'-Dimethoxytrityldeoxynucleosides

The deoxynucleoside to be dimethoxytritylated[5,48] (usually between 0.5 and 5 g) is dried by coevaporation from 2×50 ml dry pyridine and suspended in dry pyridine (between 10 and 50 ml, depending on amount). Dimethoxytrityl chloride (1.2 Eq) is added and the mixture left at room temperature for 3 hr. TLC ($CHCl_3$, 95; CH_3OH, 5 is a useful solvent) should show complete conversion of starting material (low R_f) to product (higher R_f). Side products derived from the dimethoxytrityl chloride will be present at the solvent front. Compounds containing the dmt group will stain orange when spotted with 3% trichloroacetic acid in dichloromethane, and this is often helpful in identification. If the reaction is not complete it should be left overnight or more dimethoxytrityl chloride added (0.2–0.4 Eq). On completion CH_3OH (2- to 3-fold excess over dimethoxytrityl chloride) is added and the solvents removed by evaporation. The oil produced is worked up in the normal manner.

The product should be purified by flash chromatography, using $CHCl_3$ containing 0.5% triethylamine and between 0 and 5% CH_3OH as appropriate. Many of the modified deoxynucleosides to be dimethoxytritylated are not completely pure prior to reaction. It is usually much easier to purify the dmt derivative by flash chromatography. As it is important to obtain extremely pure dmt-deoxynucleosides, only the purest fractions should be pooled. Solid products can be obtained by removing solvents first with a water pump and then with an oil pump. Yields should be between 60 and 80%.

*Procedure 9: Preparation of Deoxynucleoside-3'-O-(N,N-
diisopropylamino)cyanoethylphosphoramidites*

Between 0.5 and 2.5 g of the appropriate deoxynucleoside is dissolved in about 25 ml dry tetrahydrofuran and 5 Eq of dry diisopropylethylamine added. The mixture is protected from moisture with a rubber septum and cooled on ice. One and one-half equivalents of cyanoethyl-*N,N*-diisopropylchlorophosphoramidite[6] is added over 5 min using a syringe, after which the mixture is removed from the ice and left at room temperature for 30 min. TLC (ethyl acetate, 45; $CHCl_3$, 45; triethylamine, 10) should indicate disappearance of the starting material (low R_f, 0.2) and the presence of the two diastereomers of the product (high R_f, between 0.6 and 0.8). Traces of deoxynucleoside 3'-phosphonate (a side product, R_f usually between 0.3 and 0.5) may be present but should be minor. CH_3OH (1–2 ml) is added and solvents removed by evaporation. The product should be worked up but using ethyl acetate rather than $CHCl_3$ as the organic solvent. Retain the upper organic layer after each aqueous extraction. The products can be purified by flash chromatography, eluting with ethyl acetate (45), $CHCl_3$ (45), triethylamine (10). The products elute very quickly and can be obtained in a solid form either by evaporation of solvents with an oil pump or after evaporation by dissolving in a small volume (5–10 ml) of ethyl acetate and adding dropwise to vigorously stirred petroleum ether (500 ml). In the latter procedure the products precipitate and are collected by filtration. The products should be dried *in vacuo* over silica gel and stored desiccated at −20°. Yields of this reaction vary between 60 and 90%.

Procedure 10: Deoxynucleoside Composition Analysis of Oligonucleotides Containing Modified Bases

About 0.5 OD_{254} units of the oligonucleotide is dissolved in 0.1 ml of 10 mM KH_2PO_4, pH 7, containing 10 mM $MgCl_2$. Ten microliters each of snake venom phosphodiesterase (Boehringer-Mannheim, Germany, #108260, 1 mg/0.5 ml) and alkaline phosphatase (Boehringer, #713023, 1 unit/μl, special quality for molecular biology) are added and the mixture incubated at 37° for 5 hr. An aliquot is injected onto an analytical C_{18} reversed-phase column which is developed with 0.1 M triethylammonium acetate, pH 6.5, containing 5% CH_3CN. The bases elute in the following order (extinction coefficients, M^{-1} cm^{-1}, in parentheses): dC (6×10^3), $d^{4H}U$ (6.47×10^3), dG (13.5×10^3), T (7×10^3), $d^{6S}G$ (21.0×10^3), $d^{2am}P$ (3×10^3), dA (14.3×10^3), $d^{7C}A$ (7.8×10^3). The elution times from dC to $d^{7C}A$ occur between 4 and 14 min.

dC, dG, T, $d^{2am}P$, dA, and $d^{7C}A$ are detected at 254 nm, and the extinction coefficients refer to this wavelength. $d^{4H}U$ and $d^{6S}G$ are detected at 305 and 340 nm, respectively, and the extinction coefficients for these two bases refer to this wavelength. With these bases two aliquots should be injected, with an initial detection at 254 nm being used for the four usual bases and a second run at 305 or 340 nm for picking up the modified bases. An example is shown in Fig. 5. Quantitation is possible by integration of the areas under each peak and division of these areas by the appropriate extinction coefficient.

Acknowledgments

I would like to thank the following members of my group for contributions in the area of oligonucleotides containing modified bases: Dr. P. C. Newman, Dr. V. U. Nwosu, Mr. T. R. Waters, Ms. H. Jones, and Dr. T. Nikiforov. The work undertaken in my laboratory was supported by the U.K. S.E.R.C., the U.K. M.R.C., and the E.E.C. (Science Plan). The author is a Fellow of the Lister Institute of Preventive Medicine. Thanks are due to Mrs. S. Broomfield for typing the manuscript.

[4] Preparation of Psoralen-Derivatized Oligodeoxyribonucleoside Methylphosphonates

By PAUL S. MILLER

Introduction

Oligodeoxyribonucleoside methylphosphonates are nonionic nucleic acid analogs which contain nuclease-resistant methylphosphonate linkages in place of the negatively charged phosphodiester internucleotide bonds found in naturally occurring nucleic acids. Because these analogs are lipophilic and are taken up intact by mammalian cells in culture, they have been found to be effective "antisense" reagents that are capable of selectively controlling viral or cellular gene expression at the messenger RNA level.[1-3] Oligonucleoside methylphosphonates form duplex hybrids with complementary nucleic acids via standard Watson–Crick base-pairing interactions. The formation of these hybrids is reversible, and their stability is dependent on the length and base composition of the oligomer.

Duplex formation can be rendered irreversible by derivatizing oligonucleoside methylphosphonates with the photoreactive, cross-linking group, 4'-{[N-(aminoethyl)amino]methyl}-4,5',8-trimethylpsoralen   [(ae)AMT]. On irradiation with long-wavelength ultraviolet light (356 nm), the psoralen functionality undergoes a 2 + 2 cycloaddition reaction with an appropriately placed pyrimidine residue in the target strand, thus cross-linking the oligomer to its target. Psoralen-derivatized oligonucleoside methylphosphonates have been shown to cross-link in a sequence-specific manner to single-stranded DNA and to mRNA *in vitro*.[4-7] These oligomers are taken up intact by mammalian cells and have been shown to have light-dependent antiviral activity in cultured cells infected with herpes simplex virus type-1 (HSV-1).[8] This chapter describes the preparation of

[1] P. S. Miller and P. O. P. Ts'o, *Annu. Rep. Med. Chem.* **23**, 295 (1988).

[2] P. S. Miller, *in* "Oligodeoxynucleotides: Antisense Inhibitors of Gene Expression" (J. S. Cohen, ed.), p. 79. Macmillan, London, 1989.

[3] P. S. Miller, *in* "Herpesviruses, the Immune System and AIDS" (L. Aurelian, ed.), p. 343. Kluwer Academic Publishers, Boston, Massachusetts, 1990.

[4] B. L. Lee, A. Murakami, K. R. Blake, S.-B. Lin, and P. S. Miller, *Biochemistry* **27**, 3197 (1988).

[5] J. M. Kean, A. Murakami, K. R. Blake, C. D. Cushman, and P. S. Miller, *Biochemistry* **27**, 9113 (1988).

[6] B. L. Lee, K. R. Blake, and P. S. Miller, *Nucleic Acids Res.* **16**, 10681 (1988).

[7] P. Bhan and P. S. Miller, *Bioconjugate Chem.* **1**, 82 (1990).

[8] M. Kulka, C. Smith, L. Aurelian, R. Fishelvich, K. Meade, P. Miller, and P. O. P. Ts'o, *Proc. Natl. Acad. Sci. U.S.A.* **86**, 6868 (1989).

(ae)AMT)-derivatized oligodeoxyribonucleoside methylphosphonates and general procedures for cross-linking the oligomers to complementary nucleic acids.

Structure of Psoralen-Derivatized Oligodeoxyribonucleoside Methylphosphonates

The general structure of a psoralen-derivatized oligodeoxyribonucleoside methylphosphonate is shown in Fig. 1. The (ae)AMT group is attached to the 5′ end of the oligomer via a phosphoramidate linkage. The phosphoramidate linkage and the methylphosphonate linkages of the oligomer are completely resistant to hydrolysis by both endo- and exonucleases, whereas the 5′-phosphodiester internucleotide bond can be slowly hydrolyzed by S1 nuclease. The methylphosphonate linkages are stable at acidic and neutral pH but can be cleaved by bases or organic amines, such as 1 M aqueous piperidine. The phosphoramidate linkage, on the other hand, is resistant to hydrolysis under basic or neutral pH conditions but can be cleaved by acid, such as 0.1 N hydrochloric acid.

Choosing Target Site for Psoralen-Derivatized Oligodeoxyribonucleoside Methylphosphonates

The sequence of the methylphosphonate oligomer is complementary to some region of the targeted nucleic acid. The target site is chosen such that

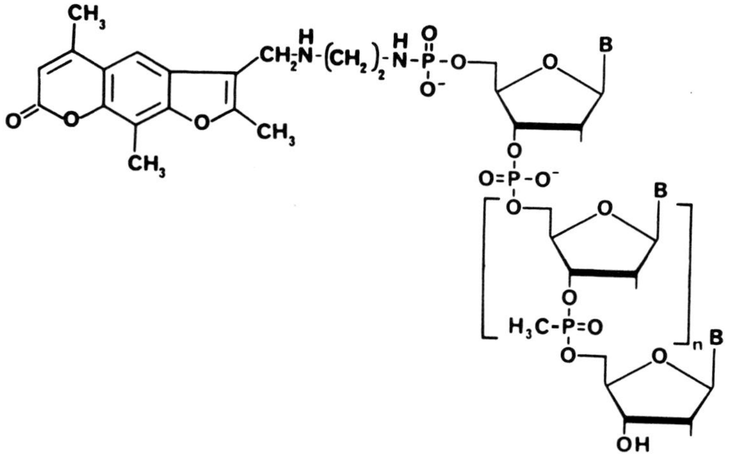

Fig. 1. General structure of a 4′-{[N-(aminoethyl)amino]methyl}-4,5′,8-trimethylpsoralen-derivatized oligodeoxyribonucleoside methylphosphonate.

the nucleoside at the psoralen cross-linking site of the target strand is a pyrimidine. This nucleoside occurs 3' to the last base pair formed between the target and the 5'-nucleotide of the oligomer. Cross-linking occurs most efficiently when the psoralen cross-linking site is a thymidine or uridine.[5] In the case of mRNA, it appears that the most efficient cross-linking occurs when the binding site for the oligomer is in a single-stranded loop region of the mRNA.

Synthesis of Psoralen-Derivatized Oligodeoxyribonucleoside Methylphosphonates

Psoralen-derivatized oligonucleoside methylphosphonates are synthesized by a three-step procedure (Fig. 2), which can be carried out in a single reaction vessel. The three steps shown in Fig. 2 are (1) conversion of the oligonucleoside methylphosphonate to its 5'-phosphorylated derivative using polynucleotide kinase and ATP; (2) conversion of the 5'-phosphorylated oligomer to its imidazolide derivative by reaction with imidazole in the presence of a water-soluble carbodiimide; and (3) conversion to the (ae)AMT-oligomer by reaction of the imidazolide with 4'-{[N-(aminoethyl)amino]methyl}-4,5',8-trimethylpsoralen. The following procedure describes the synthesis of d-(ae)AMTpApATAACTCTCTC. The underline indicates the position of the methylphosphonate linkages, and p represents the phosphoramidate or phosphodiester linkage. The synthesis is carried out on a 0.2 μmol (20 A_{260} units) scale.

Preparation of d-ApATAACTCTCTC. Oligodeoxyribonucleoside methylphosphonates are prepared in an automated DNA synthesizer on controlled-pore glass supports using commercially available 5'-O-dimethoxytrityldeoxyribonucleoside 3'-O-(N,N-bisdiisopropylamino)methylphosphonamidite synthons. The methods of synthesis, deprotection, and purification of these oligomers have recently been described in detail.[9] After synthesis and deprotection, the oligomer is purified by diethylaminoethyl (DEAE)-cellulose chromatography and, if necessary, by preparative reversed-phase high-performance liquid chromatography (HPLC). The purified d-ApATAACTCTCTC appears as a single peak on reversed-phase HPLC as shown in Fig. 3A.

Preparation of d-pApATAACTCTCTC. A solution containing 20 A_{260} units (~0.2 μmol) of d-ApATAACTCTCTC in 500 μl of acetonitrile–water (1 : 1, v/v) is placed in a 1.1 × 10 cm glass test tube and the solvents are evaporated using a Speed-Vac vacuum concentrator (Savant Instru-

---

[9] P. S. Miller, *in* "Oligonucleotides and Analogues: A Practical Approach" (F. Eckstein, ed.), p. 137. IRL Press at Oxford Univ. Press, Oxford, 1991.

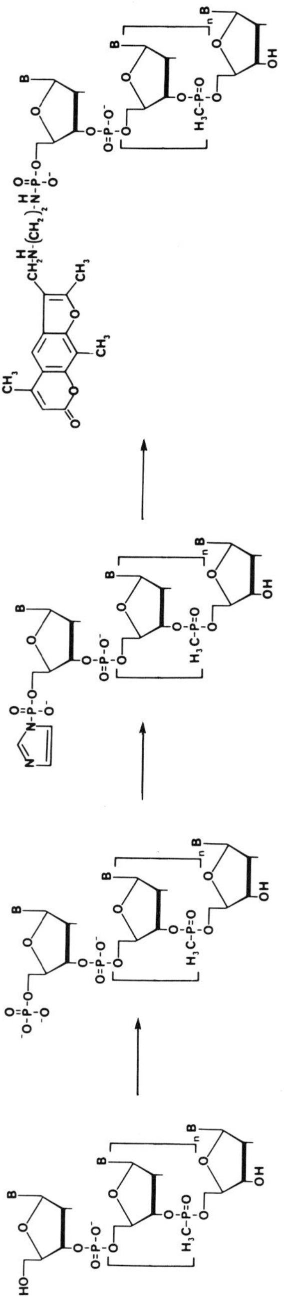

Fig. 2. Synthetic scheme for preparing psoralen-conjugated oligonucleoside methylphosphonates.

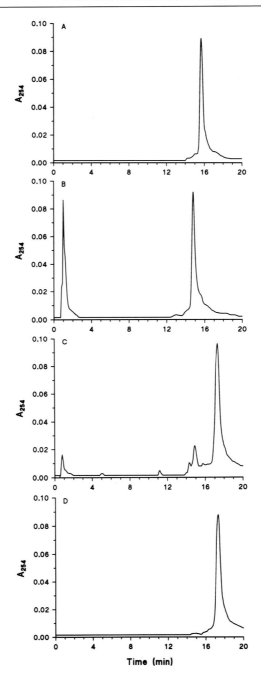

ments, Hicksville, NY). The residue is dissolved in 310 μl of acetonitrile–
water (1 : 9, v/v), and the following components are added to the solution:
10 μl of acetonitrile, 40 μl if 50 mM dithiothreitol, 11 μl of 0.1 M adenosine triphosphate, and 40 μl of a solution containing 0.5 M Tris-HCl, pH
7.6, and 50 mM magnesium chloride. The reaction is initiated by adding
450 units of polynucleotide kinase in 15 μl of 50 mM Tris-HCl, pH 7.5,
1 mM dithiothreitol, 0.1 mM EDTA, and 50% glycerol. The reaction mixture is incubated at 37° for 24 hr. The reaction can be checked for com-
pleteness by reversed-phase HPLC. The phosphorylated oligomer has a
shorter retention time on the column than does the starting oligomer (Fig.
3B). The first peak in the chromatogram is ATP. The solvents are evaporated from the reaction mixture using a Speed-Vac concentrator. The dried
reaction mixture can be stored at −20°.

Preparation of Imidazolide Derivative of d-pApATAACTCTCTC. The
dry residue obtained from the reaction described above is dissolved in a
solution which contains 350 μl of 0.1 M imidazole hydrochloride, pH 6.0,
60 μl of acetonitrile, and 40 μl of 1 M 1-ethyl-3-[3-(dimethylamino)
propyl]carbodiimide hydrochloride (CDI). The reaction mixture is incubated at room temperature for 4 hr. The solvents are evaporated using a
Speed-Vac concentrator. The residue is extracted with 0.6 ml of a solution
containing 2 ml of methanol and 25 μl of triethylamine. A precipitate
forms on addition of the methanolic solution. This is centrifuged, and the
supernatant, which contains imidazole and CDI, is removed using a Pasteur pipette. Care is taken not to remove the precipitate, which contains
the derivatized oligomer. The extraction process is repeated two more
times. The test tube containing the precipitate is placed under reduced
pressure in a vacuum desiccator attached to an oil pump in order to ensure
complete removal of solvents. It is important to remove all traces of
triethylamine from the residue before proceeding to the next step.

Preparation of d-(ae)AMTpApATAACTCTCTC. The residue obtained
from the reaction described above is dissolved in a solution containing
280 μl of 0.25 M 2,6-lutidine hydrochloride, pH 7.5, 70 μl of 50 mM
4'-{[N-(aminoethyl)amino]methyl}-4,5',8-trimethylpsoralen,[4] and 66 μl of
acetonitrile. The test tube is wrapped with aluminum foil to exclude light,
and the reaction mixture is incubated at room temperature for 48 hr. The

---

Fig. 3. Reversed-phase HPLC profiles of (A) d-ApATAACTCTCTC, (B) d-
pApATAACTCTCTC reaction mixture, (C) d-(ae)AMTpApATAACTCTCTC reaction mixture, and (D) purified d-(ae)AMTpApATAACTCTCTC. Chromatography was performed on
a C_{18} ODS3 RAC II column (Whatman, Clifton, NJ) using a 30 ml linear gradient of 2 to
30% acetonitrile in 0.1 M sodium phosphate, pH 5.8, at a flow rate of 1.5 ml/min.

solvents are evaporated on a Speed-Vac concentrator and the residue extracted with three 0.6-ml aliquots of methanol–triethylamine solution as described above. The extract contains unreacted 4'-{[N-(amino-ethyl)amino]methyl}-4,5',8-trimethylpsoralen, and the solid residue contains the derivatized oligomer. After the last extraction, any solution remaining in the test tube is removed by evaporation using an oil pump. The residue is dissolved in 1.0 ml of acetonitrile–water (1 : 1, v/v). An aliquot is removed and subjected to analytical reversed-phase HPLC. As shown in Fig. 3C, the (ae)AMT-derivatized oligomer has a retention time which is longer than that of either the phosphorylated oligomer or the starting oligomer.

Preparation of d-(ae)AMT[^{32}P]pApATAACTCTCTC. Psoralen-derivatized oligomers which contain a ^{32}P-labeled phosphoramidate linkage may be also be prepared. In this case the oligomer is phosphorylated using [γ-^{32}P]ATP. A typical reaction mixture contains the following components: 2 A_{260} units of oligomer, 18 μl of 10% aqueous acetonitrile, 2 μl acetonitrile, 4 μl of 50 mM dithiothreitol, 1.1 μl of 0.1 M ATP, 11 μCi [γ-^{32}P]ATP, 4 μl of 0.5 M Tris-HCl (pH 7.6)/50 mM magnesium chloride, and 45 units of polynucleotide kinase. The specific activity of the phosphorylated oligomer is 1 Ci/mmol. The remaining steps in the derivatization are carried out as described above, except the amounts of reagents are reduced by a factor of 10. After the last extraction with the methanol–triethylamine solution, the residue is dissolved in 100 μl of acetonitrile–water (1 : 1, v/v).

Purification of d-(ae)AMTpApATAACTCTCTC. The solution containing the derivatized oligomer is diluted with 0.1 ml of 1 M sodium phosphate and 4 ml of 0.1 M sodium phosphate, pH 5.8. The solution is pumped onto a preparative reversed-phase HPLC column (Whatman, Clifton, NJ, ODS3, 0.9 × 10 cm). The column is washed with 15 ml of 2% acetonitrile in 0.1 M sodium phosphate, pH 5.8. The column is eluted with a 50 ml linear gradient of 2 to 35% acetonitrile in 0.1 M sodium phosphate, pH 5.8, at a flow rate of 2.5 ml/min. The column is monitored at 254 or 280 nm, and the peak corresponding to the product is collected. The solution containing the product is diluted with sufficient 0.1 M sodium phosphate to make the final acetonitrile concentration approximately 10%.

Purification of d-(ae)AMT[^{32}P]pApATAACTCTCTC. The solution containing the ^{32}P-labeled, derivatized oligomer is transferred to a 1-ml Eppendorf tube, and the solvents are evaporated using a Speed-Vac concentrator. The residue is dissolved in 10 μl of gel loading buffer, which consists of 90% formamide, 0.2% xylene cyanol, and 0.2% bromphenol blue in 1 × Tris–borate–EDTA (TBE) buffer.[10] The solution is loaded onto a 20 × 20 cm, 15% polyacrylamide gel which contains 7 M urea in TBE buffer. A sample of [^{32}P]pApATAACTCTCTC, which serves as a marker, is loaded onto an adjacent lane of the gel. The gel is run at a constant voltage of 800 V until the bromphenol blue tracking dye is approximately 2 cm from the bottom of the gel. The wet gel is wrapped in plastic wrap, and small filter paper squares spotted with radioactive ink are placed at three corners of the wrapped gel. The wet gel is then subjected to autoradiography at room temperature. The (ae)AMT-derivatized oligomer has a lower mobility on the gel than does the phosphorylated oligomer. The area of the gel containing the derivatized oligomer is excised using the markers at the corners of the gel to align the gel with the film. The gel slice is placed in a 1-ml Eppendorf tube and crushed to a paste using a pestle. The crushed gel is extracted four times with 1 M aqueous triethylammonium bicarbonate (TEAB). For each extraction, 0.5 ml of TEAB is added to the gel paste and the suspension vortexed. After 15 min, the gel suspension is centrifuged in a microcentrifuge, and the supernatant is transferred to a glass test tube which is kept on ice. The combined supernatant solution is diluted with 8 ml of water.

Desalting d-(ae)AMTpApATAACTCTCTC or d-(ae)AMT[^{32}P]-ApATAACTCTCTC. The (ae)AMT-derivatized oligomers obtained as described above are desalted using a Sep-Pak C_{18} reversed-phase cartridge (Waters, Milford, MA). The cartridge is attached to a three-way nylon valve (Bio-Rad, Richmond, CA) fitted to a 10-ml plastic disposable syringe. The three-way valve allows air to be drawn into the syringe without disturbing the bed of the Sep-Pak cartridge. The cartridge is preequilibrated with the following solutions: 10 ml of acetonitrile; 10 ml of acetonitrile–water (1:1, v/v); and 10 ml of 0.1 M sodium phosphate, pH 5.8. The solution containing the (ae)AMT-derivatized oligomer is slowly loaded onto the Sep-Pak cartridge. The cartridge is washed with 10 ml of 0.1 M sodium phosphate, pH 5.8, followed by 10 ml of water. The (ae)AMT-derivatized oligomer is eluted with four 3-ml aliquots of acetonitrile–water (1:1, v/v), which are collected in separate test tubes. Each tube is checked for the presence of the oligomer. Most of the oligomer is usually present in the first two fractions. Fractions containing the oligomer are combined, and the solvents are evaporated using a Speed-Vac concentrator. The residue is dissolved in acetonitrile–water (1:1, v/v) and the oligomer is checked for purity by reversed-phase HPLC or by electrophoresis on a 15%

---

[10] T. Maniatis, E. F. Fritsch, and J. Sambrook, "Molecular Cloning: A Laboratory Manual," p. 454. Cold Spring Harbor Laboratory, Cold Spring Harbor, New York, 1982.

denaturing polyacrylamide gel. The oligomer should appear as a single peak on HPLC (Fig. 3D) or as a single band on the polyacrylamide gel.

Characterization of d-(ae)AMTpApATAACTCTCTC or d-(ae)AMT-[³²P]pApATAACTCTCTC. The structure of the (ae)AMT-derivatized oligomer can be confirmed by treating the oligomer with acid under conditions which do not result in hydrolysis of purine nucleosides within the oligomer. Acid hydrolysis of the phosphoramidate linkage occurs at the N–P bond and results in the regeneration of the 5'-phosphorylated oligomer. An aliquot of the solution containing 0.2 A_{260} units of unlabeled or 20,000 disintegrations/min (dpm) of labeled d-(ae)AMT-pApATAACTCTCTC is placed in a 1-ml Eppendorf tube and the solvents are removed using a Speed-Vac concentrator. The residue is dissolved in 10 μl of 0.1 N hydrochloric acid. The solution is incubated at 37° for 30 min. The solution is then frozen and lyophilized. The residue is dissolved in 10 μl of water and lyophilized. This process is repeated two more times. The residue is then subjected to reversed-phase HPLC in the case of the unlabeled oligomer or to polyacrylamide gel electrophoresis in the case of the labeled oligomer. The identity of the product is confirmed by the observation of d-pApATAACTCTCTC on the HPLC chromatogram or on the gel.

Cross-linking (ae)AMT-Derivatized Oligonucleoside Methylphosphonates to Complementary Nucleic Acids

Psoralen-derivatized oligonucleoside methylphosphonates cross-link to single-stranded oligo DNA or RNA targets and to mRNA *in vitro.* In the case of single-stranded oligomer targets, the target molecule carries a 5'-[³²P]phosphate group and the psoralen-derivatized oligomer is unlabeled. After irradiation, the reaction mixture is subjected to electrophoresis on a polyacrylamide gel under denaturing conditions. Under these conditions, target which is cross-linked with the oligomer will have a reduced electrophoretic mobility relative to non-cross-linked target. The extent of cross-linking can be easily determined by counting the radioactivity contained in the gel bands corresponding to the starting target and the cross-linked target.

In the case of mRNA, the (ae)AMT-oligomer is ³²P-labeled as described above. After irradiation, the reaction mixture is subjected to polyacrylamide or agarose gel electrophoresis under denaturing conditions. The presence of cross-linked material is signaled by a radioactive band that moves coincidently with the target mRNA. The following generalized procedures describe methods for studying cross-linking to a single-stranded nucleic acid target or to mRNA. The precise electrophoretic conditions used to detect cross-linking will depend on the target being studied.

Cross-linking to Single-Stranded Nucleic Acid Targets. The single-stranded nucleic acid target molecule is 5′-end-labeled with [^{32}P]phosphate using [γ-^{32}P]ATP and polynucleotide kinase and purified by preparative polyacrylamide gel electrophoresis.[11] A solution containing 0.15 μM ^{32}P-labeled target oligomer and 10 μM (ae)AMT-derivatized oligonucleoside methylphosphonate in 5 μl of 10 mM Tris-HCl, 0.1 mM EDTA, pH 7.5, is incubated at 37° for 5 min in a borosilicate glass test tube. Depending on the nature of the experiment, other target/oligomer concentrations and other buffer conditions can be used. The solution is then incubated for 10 min at the temperature at which the irradiation is to be carried out. The solution is then irradiated at 365 nm using a Ultra-Violet Products (San Gabriel, CA) long-wavelength lamp. The irradiation is carried out for 0 to 30 min at an intensity of 0.83 J cm^{-2} min^{-1} in a thermostatted water bath. The reaction mixture is lyophilized and the residue dissolved in gel loading buffer. The reaction mixture is loaded onto a polyacrylamide gel which contains 7 M urea. After electrophoresis the wet gel is autoradiographed at −80°. The autoradiogram is scanned by a densitometer to determine the extent of cross-linking. Alternatively, the gel bands corresponding to the cross-linked target and the unreacted target can be excised and counted directly.

Cross-linking to mRNA. The following example describes conditions for cross-linking sequence-specific (ae)AMT-derivatized oligonucleoside methylphosphonates to rabbit globin mRNA.[5] A solution containing 0.33 μM rabbit globin mRNA and 0–5 μM ^{32}P-labeled (ae)AMT-oligomer in 12 μl of water or a solution containing 30 mM Tris-HCl, 20 mM magnesium chloride, 0.3 M potassium chloride, pH 6.5, is preincubated at 37° for 2 min in a borosilicate glass test tube. The reaction mixture is then further incubated for 10 min at the temperature at which the irradiation will be carried out. The reaction mixture is irradiated for 0 to 15 min with 365 nm light at an intensity of 0.83 J cm^{-2} min^{-1}. The reaction mixture is lyophilized and dissolved in 10 μl of gel loading buffer. The reaction mixture is subjected to electrophoresis on a 7% polyacrylamide/0.25% agarose gel containing 7 M urea. The gel is run in TBE buffer at 500 V for 4 hr. Under these conditions α- and β-globin mRNAs are separated. The wet gel is then subjected to autoradiography at −80°.

[11] A. Murakami, K. R. Blake, and P. S. Miller, *Biochemistry* **24,** 4041 (1985).

Use of Psoralen-Conjugated Oligonucleoside Methylphosphonates in Cell Culture

Psoralen-conjugated oligonucleoside methylphosphonates are resistant to hydrolysis by nucleases and may thus be added directly to serum-containing cell culture medium. The following example outlines how a psoralen-conjugated 12-mer, d-(ae)AMTpApTCTTCTTTTAAG, is used to inhibit vesicular stomatitis virus (VSV) protein synthesis in VSV-infected mouse L cells.[12] This oligomer is complementary to nucleotides 50–61 near the initiation codon of VSV M-protein mRNA.

The oligomer is dissolved in modified Eagle's minimal essential medium supplemented with 2% (v/v) fetal bovine serum to a final concentration of 5 μM. The oligomer solution is added to a subconfluent culture of mouse L cells growing in a 2.4-cm culture chamber. The cells are infected with VSV at a multiplicity of 5. The cells are then incubated at 37° for 6 hr, after which the culture medium is removed and replaced with fresh methionine-free medium. These cells are incubated for 15 min, and [35S]methionine is then added to the medium. The cells are immediately irradiated at 365 nm for 5 min at room temperature at a flux of 0.83 J cm^{-2} min^{-1}. A sheet of cellulose acetate is placed between the cell culture dish and the lamp to prevent irradiation with light below 300 nm. The cells are incubated another 5 min after irradiation, the medium is removed, and the cells are washed with a solution containing 0.14 M sodium chloride and 20 mM Tris-HCl, pH 7.4.

The cells are then lysed on ice with a solution containing 0.14 M sodium chloride, 20 mM Tris-HCl, and 0.75% Nonidet P-40 (NP-40), pH 7.4. Aliquots of the lysate are added to loading buffer containing 0.125 M Tris-HCl, pH 6.8, 0.4% sodium dodecyl sulfate, 20% glycerol, 10% 2-mercaptoethanol, and 2% bromphenol blue. The solution is heated in a boiling water bath for 2 min and then subjected to polyacrylamide gel electrophoresis on a Laemmli gel under conditions which separate the five VSV proteins.[13] The gel is dried and subjected to autoradiography. The amounts of protein in the gel are determined by scanning densitometry. Under these conditions the synthesis of M-protein is inhibited 37% relative to a control in which VSV-infected cells were irradiated at 365 nm for 5 min in the absence of the oligomer.[12] Separate experiments showed that the oligomer alone had no effect on VSV protein synthesis.

---

[12] P. S. Miller, P. Bhan, C. D. Cushman, J. M. Kean, and J. T. Levis, *Nucleosides Nucleotides* **10**, 37 (1991).

[13] C. H. Agris, K. R. Blake, P. S. Miller, M. P. Reddy, and P. O. P. Ts'o, *Biochemistry* **25**, 6268 (1986).

Section II

Nonstandard DNA Structures and Their Analysis

[5] DNA Structure from A to Z

By RICHARD E. DICKERSON

Introduction

The man who generally is credited with having coined the term "molecular biology" is British X-ray crystallographer W. T. Astbury at Leeds University, England. In the decade before World War II, he studied the diffraction patterns from many protein fibers such as hair, wool, silk, and collagen, identifying two especially characteristic patterns that were labeled α and β.[1] For a decade and a half following the war, molecular biology was "the British Science," centering on the X-ray crystal structure group at Cambridge University that had been initiated by J. D. Bernal in the 1930s and continued by W. L. Bragg. An illustrious body of scientists congregated there, including Bernal's former graduate student Max F. Perutz, Perutz' own student John C. Kendrew, Francis Crick, and postdoctoral fellow James D. Watson. They set out to do nothing less than to establish the three-dimensional structures of biological macromolecules, both by finding explanations for Astbury's fiber patterns, and by examining the crystallizable proteins hemoglobin and myoglobin.

The Cambridge hegemony was severely challenged in 1951, when the α fiber pattern was explained successfully, not by a member of the club, but by a rank outsider in the colonies, Linus Pauling at the California Institute of Technology in Pasadena. Watson and Crick, as is well known, turned to the diffraction patterns of DNA fibers, which Maurice Wilkins and Rosalind Franklin at Kings' College, London, had shown to exist in two principal families, the low-humidity A form and the high-humidity B. After a worrying near-miss by Pauling (again),[2] the Cambridge group came through in 1953 with a structure for B-DNA.[3]

Kendrew and Perutz triumphed in 1959 with the first single-crystal X-ray structure analyses of any protein — myoglobin and then hemoglo-

---

[1] Astbury also is responsible for the preservation, in a 1949 review, of the semi-immortal lines by Lindo Patterson, whose more lasting contribution to crystal structure analysis was the Patterson Function:

> "Amino acids in chains,
> Are the cause, so the X-ray explains,
> Of the stretching in wool,
> And its strength when you pull,
> And show why it shrinks when it rains."

[2] L. Pauling and R. B. Corey, *Proc. Natl. Acad. Sci. U.S.A.* **39,** 84 (1953).
[3] J. D. Watson and F. H. C. Crick, *Nature (London)* **171,** 737 (1953).

bin.[4] They (and Pauling) were gratified to find that both proteins were constructed from folded α helices. It is interesting to speculate on probable reactions if the first protein crystal structure to be solved had been that of an immunoglobulin, which has no α helices at all. The single-crystal analyses of the globins revealed a wealth of molecular detail that was impossible to obtain by fiber methods. It must have been obvious immediately that a similar dividend of increased knowledge would result from single-crystal structure analyses of DNA. Yet 20 years were to go by before a DNA crystal structure would be solved, and this on the other side of the Atlantic, at the Massachusetts Institute of Technology (MIT) in "New Cambridge." What took so long?

The 20-year delay occurred because of an inability to obtain pure, homogeneous DNA of defined length and sequence for crystallization. Natural genomic DNA is far too long a molecule to fold and crystallize as globular proteins do, and sheared fragments were heterogeneous in both size and base sequence. Single-crystal DNA X-ray structure analysis had to wait on improvements in methods for synthesizing short, uniform deoxyoligonucleotides. By the mid-1970s this became practical, and three research groups independently embarked on single-crystal DNA studies: our own group (then at Caltech), that of Alexander Rich at MIT, and a collaboration between Olga Kennard at Cambridge University, Zippora Shakked and Dov Rabinovich at the Weizmann Institute (Rehovat, Israel), and Viswamitra at the Indian Institute of Science in Bangalore. Both the Caltech and the MIT laboratories focused on alternating oligomers of C and G, because of the observations of Pohl and Jovin that poly(dC-dG) underwent a mysterious salt- or alcohol-induced phase transition in solution.[5]

In the intervening two decades between 1959 and 1979, Wilkins and co-workers at Kings' College had established helical structures for both A- and B-DNA and had demonstrated that virtually all of the observed fiber diffraction patterns could be assigned to one of these two families. As the single-crystal analyses progressed, the natural question was raised: "Would the first crystalline deoxyoligonucleotide double helix be in the A or the B form?" As so often happens, nature took us by surprise with the answer: "No." Andrew H.-J. Wang and co-workers at MIT found in 1979 that the hexamer C-G-C-G-C-G adopted a new and totally unexpected left-handed form, which they called Z-DNA.[6,T2] Horace R. Drew and co-workers at

---

[4] R. E. Dickerson, in "The Proteins" (H. Neurath, ed.), 2nd Ed., Vol. 2, p. 603. Academic Press, New York, 1964

[5] F. M. Pohl and T. M. Jovin, *J. Mol. Biol.* **67**, 375 (1972).

[6] A citation preceded by "T" refers to the literature accompanying Tables I–IV. These references do not appear at the foot of the page on which they are cited; rather they are given at the end of Table IV. All tables (Tables I–VII) are placed at the end of this chapter.

Caltech confirmed this structure with the tetramer C-G-C-G, and a third family of DNA helices was born.[T4]

More than 100 deoxyoligonucleotide crystal structures have been solved since 1979. For reference, these are listed in Tables I–IV,* classified by helix type and then by their manner of packing in the crystal. Tables I–IV form the basis for most of the discussion of this chapter. Figures 1–3 show sixteen-base-pair lengths of A-, B-, and Z-DNA on a uniform scale, as a comparison and an illustration of their distinguishing features. Mean helix parameters[7,8,T15] for A-, B-, and Z-DNA as observed in single-crystal structure analyses of oligonucleotides are compared in Table V.* The most striking immediate difference from these stereo drawings (aside from the right-handed helix sense of A and B versus the left-handedness of Z) is that the A helix appears to be broader and more compressed along its axis relative to B and Z. In fact a critical distinction is the displacement of base pairs from the helix axis.[9] In B-DNA the base pairs sit directly on the axis so that the major and minor grooves are of equal depth. In A-DNA the pairs are displaced off-axis in a direction that makes the major groove cavernously deep, and the minor groove extremely shallow. In Z-DNA the base pairs are displaced in the opposite direction, yielding a very deep minor groove and a major groove that is little more than a shallow depression winding around the helix cylinder. These differences are measured by the X displacement in Table V.2: typically $+3$ to $+4$ Å for Z, near zero for B, and -4 Å for A. This parameter seems to mark a true distinction between the three helix families; intermediate values of X displacement have not been found in single-crystal studies.

The mean twist angle in B-DNA oligomer structures is $36°$, corresponding to the expected 10 base pairs per turn, although individual steps range between $24°$ and $51°$. The mean twist in A-DNA oligomers is $31°$, again with considerable local variation. Twelve base pairs constitute one turn of a Z helix. Because of the alternation of purines and pyrimidines in Z-DNA, the helical repeat is two successive base pairs, so the mean helical twist is $60°$. As Table V.1 shows, the two kinds of steps in an alternating Z helix are unequal. Each G-C step turns the helix by $50°$, and each C-G step, by $10°$.

Two other distinguishing characteristics of the three helix families are the pucker or conformation of the sugar ring, and the geometry of the

---

[7] R. E. Dickerson, in "Biological Macromolecules and Assemblies" (F. A. Jurnak and A. McPherson, eds.), Vol. 2, Appendix, p. 471. Wiley (Interscience), New York, 1985.

[8] R. E. Dickerson et al., EMBO J. **8**, 1 (1989); J. Biomol. Struct. Dyn. **6**, 627 (1989); J. Mol. Biol. **205**, 787 (1989); Nucleic Acids Res. **17**, 1797 (1989).

[9] R. E. Dickerson, Sci. Am. **249**, (No. 6), 94 (December, 1983).

* Tables I–VII appear at the end of this chapter.

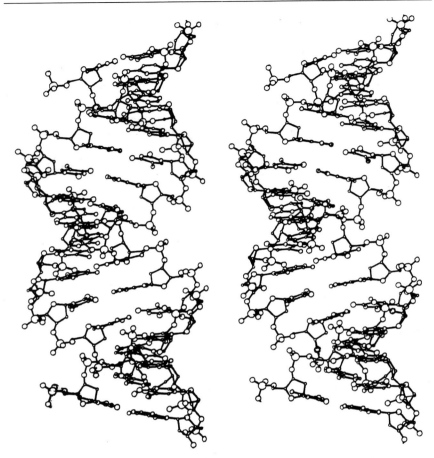

Fig. 1. Sixteen base pairs of A-DNA, generated from the crystal structure of the octamer G-G-T-A-T-A-C-C.[T12,T22] The shallow minor groove faces the viewer at top, and the very deep major groove at bottom. Variation in the width of the opening of this major groove is a significant part of the polymorphism observed in A-DNA crystal structure analyses.[T82,T87,T102]

glycosyl bond, χ, connecting base to sugar. Sugar pucker[9] is measured by the pseudorotation angle, P,[10,11] in Table V.3, and by main-chain torsion angle δ (or $C5'-C4'-C3'-O3'$). In practice these two measures correlate extremely well, as shown in the figures in Ref. 12. The C3'-*endo* sugar

[10] C. Altona and M. Sundaralingam, *J. Am. Chem. Soc.* **94**, 8205 (1972).

[11] S. T. Rao, E. Westhof, and M. Sundaralingam, *Acta Crystallogr. Sect.* A: *Cryst. Phys. Diffr. Theor. Gen. Crystallogr.* **36**, 432 (1981).

[12] R. E. Dickerson, *in* "Structure and Methods, Volume 3: DNA and RNA" (R. H. Sarma and M. H. Sarma, eds.), p. 1. Adenine Press, Albany, New York, 1990.

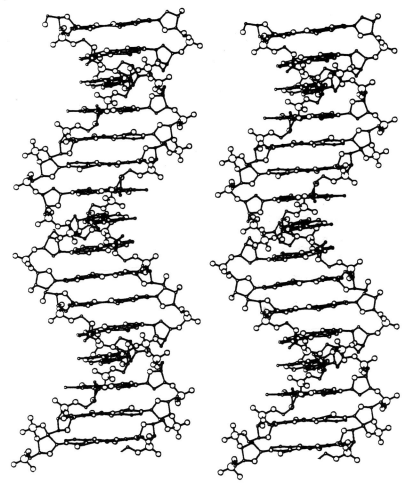

FIG. 2. Sixteen base pairs of B-DNA, excerpted from stacked decamers of C-C-A-A-C-G-T-T-G-G as they occur in the crystal structure.[T110,T111] Minor groove facing viewer at top, major groove below. The stacking is so perfect that the break from one helix to the next is very difficult to detect. It only reveals itself by the absence of connecting $-PO_2-$ groups, six steps from the bottom of the stereo diagram.

conformation that generally is associated with the A helix has a pseudorotation angle P of approximately $18°$ and torsion angle δ around $83°$. The C2'-endo conformation associated with the B helix has conformations centered around $P = 162°$ and $\delta = 144°$, although in practice the distributions usually are considerably broader in B helices than in A. These distri-

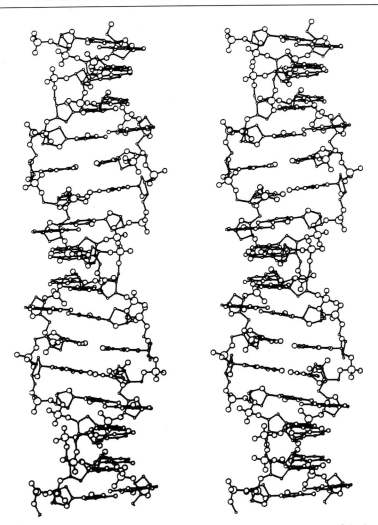

FIG. 3. Sixteen base pairs of Z-DNA, generated from the crystal structure of the hexamer C-G-C-G-C-G.[T2,T17] Minor groove at top, major groove below. Note the close approach of guanosine phosphate groups across the minor groove.

butions are illustrated quite clearly in Ref. 12. In both A- and B-DNA, the glycosyl bond connecting base to sugar is in the anti conformation, that is, the planes of the sugar rings are turned away from the minor groove as in Fig. 4a. An inevitable consequence of this anti conformation is that the backbone chain on the right of the minor groove runs downward, whereas that on the left rises upward, as indicated by the arrows in Fig. 4a.

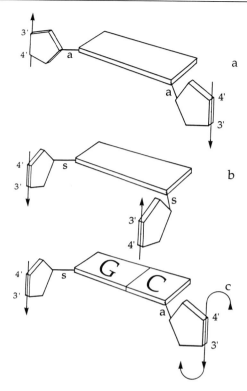

FIG. 4. Schematic views toward the minor groove edge of a base pair, illustrating the connection between glycosyl bond conformation and backbone chain direction. (a) Anti glycosyl bonds (labels a) with sugars turned away from the minor groove. The conventional "forward" chain direction from C-4′ to C-3′ on the sugar ring points down at right and up at left. This is the conformation in A- and B-DNA. (b) Syn glycosyl bonds (labels s), with sugars rotated 180° inward around their glycosyl bonds toward the minor groove. The backbone chain directions now are reversed: down at left and up at right. This is the chain sense of Z-DNA but is not quite the correct sugar geometry. (c) A syn glycosyl bond is too cramped for cytosine, so each cytosine sugar in Z-DNA must rotate back to the anti conformation. This introduces a local chain reversal, marked by the doubly curved arrow, and gives rise to a zigzag backbone pathway. The overall chain direction, however, remains as with (b): down at left and up at right.

The Z helix represents a radical departure from A and B. Its structure imposes severe inherent limitations on the base sequences that can adopt a Z conformation. The "type structure" for Z-DNA is an alternating (dG-dC)$_n$ polymer. The two backbone strands run in opposite directions from those of A and B: downward at the left of the minor groove and upward at the right, as in Fig. 4b. In a perfectly regular helix these inverted chain directions would require both sugar rings to rotate 180° about their glycosyl bonds to the bases, and to adopt a syn conformation, as drawn in Fig. 4b. But the conformation at cytosine cannot be syn because of steric clash between the sugar and the six-membered cytosine ring. (The five-membered ring connected to the sugar in guanosine provides just enough extra latitude to make a syn conformation possible.) Hence each cytosine sugar must rotate back to the anti conformation, as in Fig. 4c. This causes a local chain reversal that generates the zigzag backbone path and produces a helical repeat consisting of two successive bases—purine plus pyrimidine—instead of one base as with A and B. The overall chain sense in Z-DNA, however, remains opposed to that of A and B: down to the left of the minor groove and up to the right.

The syn conformation in Z-DNA is represented in Table V.3 by glycosyl angles χ in the neighborhood of 60° rather than 210°. The sugar pucker is C2′-endo at cytosines and C3′-endo at guanines. The Z helix apparently is compatible with poly(dG-dC)·poly(dG-dC) and poly(dG-dT)·poly(dA-dC), but not with poly(dA-dT)·poly(dA-dT) or with any sequence that does not alternate purines and pyrimidines. The requirement of purine/pyrimidine alternation is easily understood, but the preference of G and C over A and T is not. Isolated deviations from this pattern have been induced in synthetic hexamers such as [br]C-G-A-T-[br]C-G, but only at a cost in free energy that forbids multiple violations of the "standard" sequences in longer runs of DNA. Hence, although the suggestion is frequently made that left-handed DNA might be involved in some aspect of genetic *control* because of the possibility of changing the supercoiling state via right-to-left helix transitions, the Z-DNA helix imposes too many sequence restrictions to be involved in genetic *coding*.

Chain-Sense Paradox in B-to-Z Helix Transitions

Because the backbone chains of B- and Z-DNA run in opposite directions, B-DNA cannot be converted to Z-DNA by simply twisting on the two ends of the helix; a more elaborate intermediate mechanism must be invoked. This chain sense paradox was noted in the first crystal structure paper on Z-DNA,[T2] but it has received relatively little attention since. The crux of the dilemma is represented schematically in Fig. 5. Parts (a) and (e) depict unwound Z and B helices, viewed into what had been their minor groove prior to unwinding. The two unwound structures are quite different, for their backbone chain directions are reversed. (See Fig. 6 for the standard convention about chain direction.) It is a sometime practice to identify the two strands of a B-DNA double helix by labeling one strand "Watson" and the other strand "Crick." Since neither of these two men ever conceived of anything as bizarre as a left-handed helix with a zigzag

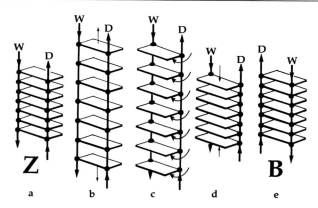

FIG. 5. The chain-sense paradox in the B-to-Z helix transition. If the Z and B helices are unrolled to zero twist and then viewed into their minor grooves, phosphate backbone chains are seen to run in opposite directions as in (a) and (e). Before one helix can be converted to the other, this chain sense must be reversed. One way of doing this involves (b) pulling the "ladder" out to double its normal length and (c) rotating each base pair about its long axis to the other side of the ladder uprights. The helix then can be compacted again (d), with the direction of its chains reversed when viewed from the minor groove (e). Without this or an equivalent transition between (a) and (e), a Z-to-B transition is impossible.

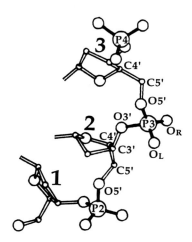

FIG. 6. Standard nomenclature conventions in a DNA sugar–phosphate backbone chain. The forward direction is along the sequence $P \rightarrow O\text{-}5' \rightarrow C\text{-}5' \rightarrow C\text{-}4' \rightarrow C\text{-}3' \rightarrow O\text{-}3' \rightarrow P$. Each phosphate is numbered and associated with the sugar and base that follows it in the forward direction, as P2 and sugar 2. As one travels in a forward direction and rises over the hump of a phosphate group via atoms $O\text{-}3 \rightarrow P \rightarrow O\text{-}5'$, oxygen atom O_L is to the left and O_R to the right. In A- and B-DNA, O_L points toward the major groove and O_R toward the minor. (Alternate oxygen nomenclature: O_L, O1P; O_R, O2P).

backbone, it seems more fitting in this context to identify the descending and ascending backbone strands as "Wang" and "Drew," respectively, after the people who independently solved the first two examples of Z-DNA.[T2,T4] In the unwound Z structure of Fig. 5a, the descending Wang strand is at left and the rising Drew strand is at right when viewed into the former minor groove. By contrast, in an unwound B helix (Fig. 5e), Drew is at left and Wang at right. These are distinct topological entities, and it is not easy to think of a way of transforming one into the other.

One solution, which was proposed in principle by Wang and co-workers in the first Z-DNA structure paper,[T2] is diagrammed in Fig. 5b–d. This model requires that the backbone chains first be stretched (b) until each base pair has sufficient room to rotate about its glycosyl bonds between the uprights of the backbone "ladder," to bring it to the other side (c). After this is done, the bases can be restacked to their normal separation (d), and a rotation about the vertical axis (e) will reveal that the Wang and Drew strands have changed places relative to (a).

But this is a major and barely tolerable transition process. Harvey[13] and Olson et al.[14] have both considered the details of such an "expand–rotate–collapse" transition. Harvey's experiments with physical models have convinced him that sufficient room for rotation of the base pair can only be achieved with difficulty, and at a single base step at a time, so that the B-to-Z transition would have to ripple down the helix in a cooperative manner. One would expect that attaching additional bulky chemical groups to the edges of the base pairs would immobilize this rotation entirely, and thus prevent a B-to-Z transition. To the contrary, certain substituents such as N-acetoxy-N-acetyl-2-aminofluorene (AAF) actually facilitate the transition.[15-17] An alternative solution to the chain-sense paradox would be for each base pair to break apart and to re-form at the back of the helix, like a person clasping his hands behind his back. Yet NMR spectroscopy,[18-21] single-strand nuclease sensitivity, and circular

[13] S. C. Harvey, *Nucleic Acids Res.* **11**, 4867 (1983).

[14] W. Olson, A. R. Srinivasan, N. L. Markey, and V. N. Balaji, *Cold Spring Harbor Symp. Quant. Biol.* **47**, 229 (1983).

[15] A. Rich, A. Nordheim, and A. H.-J. Wang, *Annu. Rev. Biochem.* **53**, 791 (1984).

[16] T. M. Jovin, L. P. McIntosh, D. J. Arndt-Jovin, D. A. Zarling, M. Robert-Nicoud, J. H. van de Sande, K. F. Jorgenson, and F. Eckstein, *J. Biomol. Struct. Dyn.* **1**, 21 (1983).

[17] T. M. Jovin, D. M. Soumpasis, and L. P. McIntosh, *Annu. Rev. Phys. Chem.* **38**, 521 (1987).

[18] M. H. Sarma, G. Gupta, M. M. Dhingra, and R. H. Sarma, *J. Biomol. Struct. Dyn.* **1**, 59 (1983).

[19] P. Mirau and D. R. Kearns, *Proc. Natl. Acad. Sci. U.S.A.* **82**, 1594 (1985).

[20] S. Tran-Dihn, J. Taboury, J.-M. Neumann, T. Huynh-Dinh, B. Genissel, B. L. d'Estaintot, and J. Igolen, *Biochemistry* **23**, 1362 (1984).

[21] M. Kochoyan, J. L. Leroy, and M. Guéron, *Biochemistry* **29**, 4799 (1989).

dichroism (CD) and absorption spectroscopy[13] have all failed to associate breakage of the imino hydrogen bonds with the B-to-Z transition. There seems to be no escape from the base pair "expand–rotate–collapse" model, in its sequential if not its concerted form. (A recent attempt to find a B-to-Z pathway that avoided unstacking of bases foundered on the issue of backbone chain sense.[22] The final "Z" form of that study still had the same chain directions as B-DNA, or the opposite of true Z.)

Other difficulties exist for Z-DNA. The zigzag backbone and alternating syn–anti glycosyl bond conformations confine the Z helix to an alternation of purines and pyrimidines, with at most a small number of isolated local anomalies. But no one has ever been able to explain why the Z form should be possible with poly(dG-dC)·poly(dG-dC) and poly(dG-dT)·poly(dA-dC), but not with poly(dA-dT)·poly(dA-dT). Nor is there an explanation of why hydrogen exchange from guanine N-2 amine groups should be 10 times slower in the left-handed form than in the right-handed state of the same sequence.[19,23,24] It also is unclear why adding a methyl group to the O_R oxygen (defined in Fig. 6) of the phosphate group *preceding* guanine in short alternating C-G oligomers should block the B-to-Z transition, whereas forming a methylphosphonate at the O_L oxygen of the same phosphate, or at either oxygen of the phosphate *following* guanine, should still permit formation of Z-DNA.[25]

Ansevin and Wang[26] have recently proposed an alternative model for a left-handed zigzag double helix that answers all of these difficulties and avoids the chain-sense paradox by the simple expedient of having the *same* chain directions as B-DNA. They call this the Watson–Crick form of Z-DNA, or Z[WC]-DNA. But their structure, depicted in Fig. 7, is much more different from Z than the A and B helices are from one another at the other end of the alphabet. W-DNA is a convenient abbreviation of Z[WC]-DNA that has the triple advantages of reminding one that it is distinct from Z-DNA, that it lies closer to B-DNA structurally than does the Z helix, and that it possesses a Watson–Crick chain sense.

W-DNA has many features in common with Z (Table V). The glycosyl bond geometry is similar to that of Z (Fig. 4c). However, the reversed overall backbone directions require that the local chain reversal that generates a zigzag backbone must occur at the syn guanine instead of the anti cytosine. Sugar puckers also are reversed in W: C3′-*endo* at cytosines and

[22] W. Saenger and U. Heinemann, *FEBS Lett.* **257**, 223 (1989).

[23] J. Ramstein and M. Leng, *Nature (London)* **288**, 413 (1980).

[24] B. Hartmann, J. Ramstein, and M. Leng, *FEBS Lett.* **225**, 11 (1987).

[25] L. Callahan, F. S. Han, W. Watt, D. Duchamp, F. J. Kezdy, and K. Agarwal, *Proc. Natl. Acad. Sci. U.S.A.* **83**, 1617 (1986).

[26] A. T. Ansevin and A. H. Wang, *Nucleic Acids Res.* **18**, 6119 (1990).

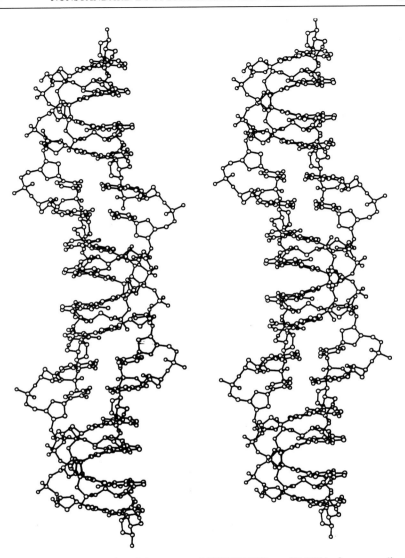

FIG. 7. Sixteen base pairs of the proposed Z[WC]-DNA, or W-DNA, from coordinates deposited in the Brookhaven Protein Data Bank by Allen T. Ansevin and Amy H. Wang. Minor groove facing viewer at top, major groove below. The chain sense in W-DNA is reversed from that in Z-DNA but is the same as in B-DNA. Hence the B form can be converted to the W by simple twisting of the ends of the helix, without the elaborate unstacking and flipping of base pairs seen in Fig 5. Both Z and W helices have zigzag backbone chains with phosphates in close opposition across a narrow but deep minor groove, and a broad, shallow major groove.

C2'-*endo* at guanines. In both W and Z the minor groove is deep and narrow, whereas the major groove is broad and shallow. Phosphate groups preceding guanines come into close juxtaposition across the minor groove in both helices, with a minimum P–P distance of 8 Å for Z and 6 Å for W. The X displacement of base pairs is much less for W-DNA than for Z (−1.6 versus +3.6 Å); its base pairs sit nearly on the helix axis as with the B form. In Z-DNA the 60° rotation every two base pairs is realized by a 50° rotation at the G-C step and a 10° rotation at the C-G step. Rotations in W-DNA are more extreme: an overrotation of 68° at the C-G step and a back-rotation of 8° at G-C.

One key structural feature of W-DNA answers simultaneously (1) the incompatibility of poly(dA-dT)·poly(dA-DT) with a left-handed state, (2) the abnormally slow exchange of guanine N-2 amine hydrogens in the left-handed helical conformation, and (3) the incompatibility of the left-handed helix with methylphosphonation of guanine O_R oxygens. The W helix is stabilized by an intrachain hydrogen bond between each guanine N-2 amine and the O_R oxygen on the preceding phosphate group (Fig. 8). The O_R–N-2 distance is 2.70 Å, and the O_R–N-2–C-2 angle is close to 120°, demonstrating almost ideal hydrogen bond geometry. If the left-handed helix that is produced with long DNA polymers in solution actually is the W helix rather than the Z, then the slow hydrogen exchange would be explained because the amines are involved in intrachain as well as interchain hydrogen bonding. Methylation of the guanine O_R oxygen would eliminate this hydrogen bond and hence would destabilize the

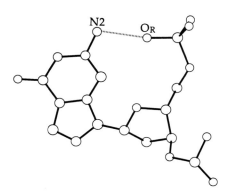

FIG. 8. View of one guanine nucleotide of W-DNA, down the helix axis. The dotted line indicates an intrachain hydrogen bond between the guanine N-2 amine and the preceding phosphate O_R oxygen, which is proposed as an important means of stabilizing the left-handed W helix. Each guanine N-2 is 2.7 Å from its own O_R atom.

left-handed helix. The observed sequence preference in alternating purine–pyrimidine polymers also would be accounted for. Poly(dG-dC)·poly(dG-dC) has $NH-O_R$ intrachain hydrogen bonds stabilizing both strands of the helix. The absence of an N-2 amine on adenine means that only one strand of poly(dG-dT)·poly(dA-dC) has such bonds, but this evidently still is sufficient to stabilize the helix. But when the bonds are completely lacking in poly(dA-dT)·poly(dA-dT), the left-handed helix becomes unstable relative to the right-handed B form, which is accessible to it merely by a simple twist.

At present the W-DNA helix is only a model structure, refined by molecular modeling programs to eliminate bad contacts and geometry, but without direct X-ray support. Indeed, Ansevin and Wang do not suggest that the Z-DNA structures obtained with single crystals of oligomers are wrong. They propose that the Z structure represents a global minimum of stability for left-handed double helices, but one that only is accessible to short strands that can come apart and re-form with an inverted chain sense. The W helix constitutes no more than a local energy minimum in conformational space, but it is the only one that is accessible to a B helix by simple untwisting. Hence short oligomers would be expected to adopt the Z form, whereas long stretches of DNA in closed circles or with restricted ends could only interconvert between right-handed B and left-handed W (not Z). We have been led astray for 10 years, Ansevin and Wang maintain, because we assumed that the left-handed form adopted by short oligomers must necessarily be the left-handed form that is used by longer stretches of DNA in solution. *Se non è vero, è ben trovato.* We now need to think about finding some way of restricting the ends of short DNA oligomers, to see whether they can be induced to crystallize in the W form.

Evaluation of Crystal Structure Analysis

Elegant computer graphics, stereo pairs, and color plates all combine to make a macromolecular structure paper attractive and convincing. This is one of the advantages of the field. But how can an outside observer see behind the graphics to judge the value of the structure analysis itself? Observable local features in an oligonucleotide crystal structure can have three origins: *(1)* effects of base sequence, *(2)* effects of intermolecular contacts within the crystal; and *(3)* inadequacies in data and refinement, including a loss of detail because of low resolution of the analysis. The goal is to study *(1)*, to monitor and control *(2)*, and as much as possible to eliminate or at least minimize *(3)*. How does one assess how well this has been done?

With oligonucleotides, a few simple calculations can be of considerable help, using the quantities listed in Tables I–IV. Rather than being chronological, or classifying structure determinations according to base sequence, Tables I–IV list the analyses according to their mode of crystallization or space group, with cell dimensions and the number of unique base pairs in one asymmetric unit of the crystal. All structures that share these features, for example, the 14 A-DNA structures in Table I.a, are said to be *isomorphous.* They essentially are variants of one and the same structure, with atoms in equivalent places, and with the very same intermolecular contacts. If one were given atomic coordinates of any member of this group of 14, and the F_o values or X-ray intensity data from another, it would be a relatively straightforward process to solve the second structure by a refinement technique known as "molecular replacement."

Given only a set of isomorphous structures such as the first 14 in Table I, one could learn much about the effect on local helix structure that is produced by substituting one base for another [*(1)* in our list above]. But one could learn nothing at all about the perturbations arising from crystal packing [*(2)* in our list], because all 14 structures experience the very same crystal environment. For this purpose we would need to add structures that crystallize differently, for example, the 11 listed in Table I.b.

Notice that the sequence G-G-G-C-G-C-C-C is to be found under both Table I.a and I.b, and at two different temperatures in each. This is especially useful, in that one can begin to unravel the relative effects on local helix structure that arise from crystal packing, base sequence, and temperature. Considerable environmental polymorphism has been found for the A-DNA helix recently, especially in terms of an opening or closing of the deep major groove that is visible in Fig. 1.[T82,T87,T102]

The quantity Z given in parentheses after each space group symbol is the number of equivalent points, or the number of identical units to be found within each cell of that space group. This "asymmetric unit" of the cell frequently is an entire double helix. But it often happens that a crystallographic 2-fold axis passes through the duplex perpendicular to the helix axis, relating one end of the helix to the other. In such a case the asymmetric unit is only half the double helix, or alternatively one strand of the helix. "Ubp" in Tables I–IV specifies the number of unique base pairs in the crystallographically independent asymmetric unit. Note that for all of the A-helical octamer crystal structures in tetragonal space group $P4_32_12$, Ubp is 4, indicating that the asymmetric unit is half the helix, and hence that the two ends of the helix are crystallographically identical. In contrast, for octamers in hexagonal space group $P6_1$, Ubp is 8. Here the two ends of the octamer helix are crystallographically distinct. The two ends can be compared in order to learn something about possible effects of crystal packing or of errors in the structure analysis.

The product of Ubp and Z yields the total number of base pairs in the entire cell. The volume per base pair, V/bp, is simply the total cell volume divided by this product: V/bp = (cell volume)/(Ubp $\times Z$). This volume per base pair is a measure of how tightly the helices are packed within the crystal. Note that it is a fairly uniform 1400 cubic angstroms (Å^3) per base pair for the tetragonal A-DNA structures and slightly larger for the hexagonal form. B-DNA packs with roughly similar density (Table II), but Z-DNA crystals are significantly more densely packed, with only 1100 Å^3 per base pair (Table III). Not surprisingly, some of the DNA – drug complexes in Table IV require an even greater volume per base pair, because of the additional presence of the drug molecule. A densely packed crystal is potentially advantageous because its molecules may be held in a more orderly array, leading to a higher resolution diffraction pattern. But at the same time such a dense packing could produce local contact deformations in the helices, and so the issue of specific volume becomes a trade-off.

The resolution in angstrom units is a measure of how far out the diffraction pattern has been measured, and it is a rough guide to the quality of the resulting structure. The farther out the data are collected, the finer is the resolution. In principle, an electron density map at 3.0 Å should only allow one to discriminate features in the structure that are 3.0 Å or more apart, but a higher resolution 1.5 Å map should almost permit one to see resolved atoms. Because a crystal is three-dimensional, the amount of data that must be collected increases as the inverse cube of the resolution; that is, a 1.5 Å resolution map differs from a 3.0 Å map by a factor of 2 in linear resolution, but by a factor of $2^3 = 8$ in data required for its calculation. In return, however, it should yield 8 times as much information as the lower resolution (3.0 Å) map. Protein crystal structures frequently are solved in stages of increasing resolution, but for a small molecule such as an oligonucleotide, one usually collects the maximum possible data from the outset. A lower resolution analysis then usually means that the degree of order within the crystal was not good enough to permit collection of data to a higher resolution.

The nominal resolution has the disadvantage that, although it tells one how far out the X-ray data were collected, it provides no information about what percentage of data within a given resolution limit actually were strong enough to be measured. In a weak crystal, less than half the possible reflections might be measurable, and the resulting map could be inferior to one at lower nominal resolution but from a more strongly diffracting crystal. Hence crystallographers usually quote the number of reflections that actually were observed to be greater than some multiple of σ, the standard error in data measurement. A σ cutoff of 2.0 means that all reflections weaker than 2.0 times the standard error level have been deleted from the analysis. A 2σ or 3σ cutoff is customary. Lower σ levels dilute the signal in the data with experimental error, but a too-high σ cutoff can eliminate meaningful information. As representative examples, the [i]C-C-G-G analysis at the beginning of Table I had 1486 X-ray reflections above the 2σ level, whereas the C-C-C-C-G-G-G-G analysis below it had 1018 reflections measured in total, without deleting any of the weak F_o values.

An excellent practical indication of our level of knowledge about an oligonucleotide crystal structure is the number of observed data per unique base pair; in Tables I–IV (No. data)/Ubp. In Table II.b, the C-C-A-A-C-G-T-T-G-G decamer analysis used 4402 X-ray intensities to define 5 base pairs, or 880 data points per base pair. In contrast, the nicked strand C-G-C-G-A-A-A-A-C-G-C-G dodecamer in Table II.a.2 had only 553 intensities with which to define 12 unique base pairs, or just 46 data points per base pair. Hence the decamer structure analysis contains $880/46 = 19$ times as much information per base pair as the dodecamer. The respective nominal resolutions of 1.4 versus 3.0 Å would have predicted a smaller factor of only $(3.0/1.4)^3 = 10$ in relative information content. The extra detail in the decamer structure evidently arises because nearly twice as many of the possible reflections within the nominal resolution limit actually were strong enough to be measured.

Of course such a comparison can only be made fairly for data sets with the same σ cutoff level, or threshold for rejection of weak data. Nevertheless, because 2σ or 3σ cutoffs are the custom, the simple number of reflections used per base pair usually is a better assessment of information content of a map than is the nominal resolution alone. The inverse cube of the nominal resolution specifies only the total volume of diffraction pattern space ("reciprocal space") that should or could have been measured; the number of data per base pair tells how many data actually were observed, and hence includes information about the quality of the diffraction pattern. By this criterion the most accurate oligonucleotide structure of all is the Z-helical C-G-C-G-C-G with spermine in Table III.b, which is defined by no fewer than 2500 reflections per base pair. Fig. 9 illustrates the effect of increased amounts of data and compares data per base pair versus nominal resolution as measures of information content. Note the better definition of the adenine ring in Fig. 9b, the tetrahedral phosphate at lower right, and the well-defined water molecules above and below the thymine oxygens.

The percent residual error or crystallographic "R-factor" is a quantitative measure of how well a proposed molecular structure matches the observed X-ray data, but it must be interpreted with caution. If F_o is the observed structure factor (roughly the square root of intensity) for a given X-ray reflection, if F_c is the theoretical value based on a given structure

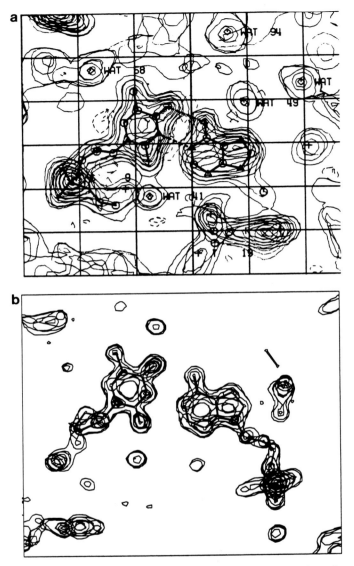

FIG. 9. Improvement in quality of an electron density map by inclusion of more X-ray data per base pair. (a) Base pair T8·A17 from the analysis of the B-DNA dodecamer C-G-C-G-A-A-T-T-C-G-C-G, which has a nominal resolution of 1.9 Å and uses 227 X-ray reflections per base pair. (b) Base pair T8·A13 from analysis of the B-DNA decamer C-C-A-A-C-G-T-T-G-G, with a resolution of 1.4 Å and 880 independent reflections per base pair. Nominal resolution alone would predict an increase in information content for the decamer map of $(1.9/1.4)^3 = 2.5$. By contrast, the numbers of data involved indicate a higher factor of 4. Not only is the decamer structure at a higher nominal resolution; a larger percentage of the potentially observable X-ray reflections for the resolution in question was in fact measured and used in the analysis. Compare relative definition of phosphate groups at lower left of (a) and lower right of (b). (From Ref. T110.)

model, and if Σ indicates summation over all the X-ray data, then $R = \Sigma||F_o| - |F_c||/\Sigma|F_o|$, where the vertical bars indicate the taking of magnitudes only. R-Factors are monitored during a crystal structure analysis, as one measure of how well the refinement is progressing. Starting R-factors with a reasonable structure model at the beginning of an analysis typically are in the 50–60% range. An R in the neighborhood of 30% during refinement usually is an indication that refinement is moving toward the true structure. Most macromolecular structure analyses end with an R-factor between 20 and 15%.

It is dangerous, however, to accept the R-factor uncritically as the only measure of the quality of the final structure. Archimedes once commented, "Give me a lever long enough, and a fulcrum, and I can move the Earth." Crystallographer William N. Lipscomb paraphrased this, "Give me enough terms in a power series, and I can fit anything." Indeed, too many degrees of freedom in a structure refinement can depress the residual R-factor to a meaninglessly low figure. One way of introducing excess degrees of freedom is to add too many "water" molecules to the analysis. Although a certain number of water molecules definitely are part of an oligonucleotide structure, it would be extremely useful for a reader to know by how much the R-factor would rise, were all of these water molecules to be deleted at the end of the analysis, leaving only the DNA helix. Such data seldom are quoted in the final paper.

The interactions of solvent molecules around a DNA double helix can be important in stabilizing helix structure,[T9,T21,T41,T46,T110,T113] so much so that bound and ordered water molecules are sometimes described as the fourth component of DNA structure, after bases, sugars, and phosphates. Solvent peaks usually are found late in the analysis process, by searching electron density maps and difference maps after the structures of the DNA helix and any bound drug molecules have been established and refined. Overcaution in accepting solvent peaks may obscure useful structure information, but, as has been mentioned, excessive zeal in calling every bump in the map a water molecule will lead to a meaningless lowering of the residual error or R-factor and a collection of water sites of no significance.

The number of water molecules located and accepted per base pair is a useful measure of the scepticism vs. credulity of the investigator, yet one would expect this number to be greater at high resolution than at low. A more judicious measure is the number of observed X-ray reflections per water molecule accepted. Consider as examples three dodecamers and one decamer from Table II: C-G-C-G-A-A-T-T-C-G-C-G, C-G-C-A-T-A-T-A-T-G-C-G, C-G-C-A-A-A-A-A-A-G-C-G, and C-C-A-A-C-G-T-T-G-G, identifying them for reference as the Drew, Yoon, Nelson, and Privé helices, respectively. Their parameters are shown below. Although the Drew analysis located twice as many waters per base pair as the Yoon, this is justified by its better resolution. The number of X-ray intensities per water molecule is roughly comparable for the two studies (34 versus 44). The Nelson structure has many fewer waters than the Drew (27 versus 80), which is justified in part because of its lower resolution. But the Nelson analysis may have erred on the side of caution, because it has twice as many X-ray data per water molecule accepted as in the case of the Drew or Yoon structures (74 versus 44 and 34). In contrast to these three dodecamers, the Privé decamer has what at first appears to be a surprisingly large number of water molecules per base pair: 14 instead of 2 to 7. But this is permissible because of its increased resolution and greater number of X-ray data. The number of X-ray intensities per water molecule is still a respectable 62.

Helix	Ubp	Resolution	No. data	No. waters	Waters/bp	Data/Water
Drew	12	1.9 Å	2725	80	6.7	34
Yoon	12	2.2 Å	1915	43	3.6	44
Nelson	12	2.5 Å	2003	27	2.2	74
Privé	5	1.4 Å	4402	71	14.2	62

A few simple calculations using quantities corresponding to those listed in Tables I–IV can be of great help in evaluating the quality of a new crystal structure analysis. The two key issues are those of the information content of a map and of the ratio of information fed into analysis to structural features deduced from it. The positive counterpart to the gloomy acronym of the computer programmer, GIGO, "Garbage in, garbage out," is DIKO, "Data in, knowledge out." One needs always to be aware of the ratio of input to output.

Local Variation in Helix Parameters

One surprise from single-crystal structure analyses of synthetic oligomers is the breadth of variation of local helix parameters about mean values that generally follow expectations from fiber diffraction. Among eight B-DNA dodecamer structures and four decamers, the mean value of the helical twist angle between base pairs is 36.1°, but the standard deviation in this quantity is 5.9° and the range in values extends from 24° to 51°. The mean rise per base pair is 3.36 Å, but the standard deviation is 0.46 Å with a range of 2.5 to 4.4 Å. (Rise is measured between C-1' atoms at the ends of the base pairs, and it can be smaller than the thickness of a base pair if the ends of the two base pairs bow toward one another. Such bowing is defined as positive cup.[T111]) Roll angles between successive base pairs average $+0.6°$ (slight opening toward the minor groove), but with a standard deviation of $6.0°$ and a range of $-18°$ to $+16°$. None of this was expected from fiber diffraction studies, which can only look at average helix values. Such variations in twist and roll would have the effect of substantially reorienting potential hydrogen bond acceptors and donors on base edges along the floors of the grooves, and may be an important part of the recognition process by drugs and proteins. Hence it becomes a matter of considerable interest to unravel any dependence of local helix parameters on base sequence.

(All local helix parameters discussed here were calculated using the NEWHEL91 program, which is available directly from the author or from the Brookhaven Protein Data Bank.[T110,T113] This program has been thoroughly user-tested, and it fits the best straight helix axis to a crystal structure of a short double helix. Other helix parameter calculating programs also are available that vary the mode of calculation of parameters in minor details or are designed to follow a curved helix.[27–29])

Standard rotational helix parameters are depicted in Fig. 10, and analogous translational parameters are given in Ref. 8. Rotational parameters such as twist, roll, propeller, buckle, and inclination seem to be important in DNA helix structure; other parameters such as tip and opening have found no utility. Strong correlations exist between various *base step* parameters from one base pair to the next, such as twist, rise, roll, and cup, with linear correlation coefficients greater than $r = 0.90$.[T111,T113] In marked contrast, *base pair* parameters, such as propeller, buckle, and inclination, seem to be mutually uncorrelated, with few r values above 0.25. The important consideration in B-DNA structure appears to be how one base pair is oriented relative to its neighbors, and not how a base pair is positioned relative to some overall helix axis.

Twist, rise, cup, and roll are so closely correlated in B-DNA that they scarcely deserve to be described as independent variables.[T111] That low twist should be associated generally with high rise, and high twist with low rise, follows from the finite extensibility of the backbone chain connecting one nucleotide to the next. Most base steps follow one of two profiles:

High twist profile (HTP): High twist, low rise, positive cup, and negative roll

[27] D. M. Soumpasis and C. S. Tung, *J. Biomol. Struct. Dyn.* **6,** 397 (1988).
[28] D. Bhattacharyya and M. Bansal, *J. Biomol. Struct. Dyn.* **6,** 635 (1989).
[29] R. Lavery and H. Sklenar, *J. Biomol. Struct. Dyn.* **6,** 655 (1989).

Low twist profile (LTP):   Low twist, high rise, negative cup, and
positive roll

A few steps exhibit intermediate twist profiles (ITP), and a few others such as C-G and C-A are variable (VTP), depending strongly on the nature of the steps to either side. Large twists to either side will restrict a VTP step to a small twist, and small flanking twists allow the central step to open up.

From an empirical examination of parameters in eight dodecamers and five decamers, Grzeskowiak *et al.*[T113] deduce the following general twist profile behavior for base steps in B-DNA, where Y is pyrimidine and R is purine:

Low twist profile: All R-R (and Y-Y) except for G-A, which
is HTP
Intermediate twist profile: All R-Y except for G-C, which is HTP
Variable twist profile: All Y-R except for T-A, which is HTP

If these are plotted on a 4×4 base matrix, the result is the intriguing pattern of Fig. 11. Base steps along the diagonal from lower left to upper right are self-complementary pairs, as C-G or T-A. Base steps disposed symmetrically about this diagonal are complementary pairs, as A-G and C-T, and hence do not represent distinct double helical entities. The unique set of steps therefore consists of the diagonal, plus all the entries on one side of the diagonal, or 10 steps in all. Low, intermediate, and variable profile steps each have a quadrant of the matrix, with the upper right entry in each quadrant being a high twist profile step instead.

The structural basis for this curiously ordered matrix is not yet clear. Stereo pair drawings of the overlap of two successive base pairs, given in Refs. T110 and T113, have not provided easy explanations. Base pair stacking energy calculations are needed badly, employing the actual stacking from observed crystal structure analyses. An important principle surely must be embodied in Fig. 11, but one that remains to be discovered.

The fact that the VTP steps C-G and C-A are influenced by the nature of their neighboring steps suggests that the best sequence–structure correlation would be obtained by examining each step in its flanking context: three successive base steps, or a tetrad of four successive base pairs. This principle indeed was enunciated by Calladine in 1982.[30,31] Although Calladine's algorithms based on propeller-induced purine–purine clash in major or minor grooves were not sufficiently comprehensive, the unit of study was defined correctly: a tetrad. How many unique tetrads can there be in a DNA double helix? The answer, as with dimer steps, is: All of the self-complementary sequences, plus half the number of non-self-comple-

[30] C. R. Calladine, *J. Mol. Biol.* **161,** 343 (1982).
[31] R. E. Dickerson, *J. Mol. Biol.* **166,** 419 (1983).

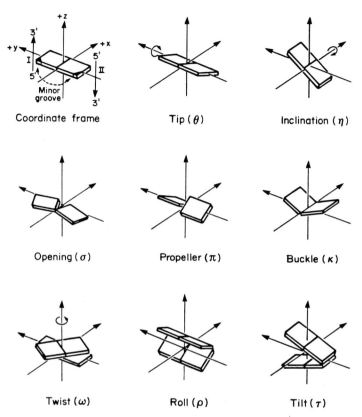

FIG. 10. Definitions of local rotational helix parameters, from the Cambridge DNA nomenclature accord.[8] Rotations within one vertical column are about x, y, and z axes, from right to left. Top row: Bases of a pair moving in concert. Middle row: Bases of a pair moving in opposition. Bottom row: Two-base pair step. Comparable definitions of translational parameters are given in Ref. 8.

	G	A	T	C
G	L	H	i	H
A	L	L	i	i
T	v	H	L	H
C	v	v	L	L

FIG. 11. Matrix of empirical observations about twist profile behavior at base steps. First base of a step at left; second base along top. Steps symmetrically disposed about the diagonal from lower left to upper right are complementary, as G-A and T-C. H, High twist profile; L, low twist profile; i, intermediate twist profile; v, variable twist profile.

GGGG	GGGA	GGGT	GGGC	GGAG	GGAA	GGAT	GGAC	GGTG	GGTA	GGTT	GGTC	GGCG	GGCA	GGCT	GGCC
GGGG	GGGA	GGGT	GGGC	GGAG	GGAA	GGAT	GGAC	GGTG	GGTA	GGTT	GGTC	GGCG	GGCA	GGCT	**GGCC**
AGGG	AGGA	AGGT	**AGGC**	AGAG	AGAA	AGAT	AGAC	AGTG	AGTA	AGTT	AGTC	AGCG	AGCA	**AGCT**	
TGGG	TGGA	TGGT	**TGGC**	TGAG	**TGAA**	TGAT	TGAC	TGTG	TGTA	TGTT	TGTC	**TGCG**	TGCA		
CGGG	CGGA	CGGT	CGGC	CGAG	**CGAA**	**CGAT**	CGAC	**CGTG**	CGTA	**CGTT**	CGTC	**CGCG**			
GAGG	GAGA	GAGT	GAGC	GAAG	GAAA	**GAAT**	GAAC	GATG	**GATA**	**GATT**	**GATC**				
AAGG	AAGA	AAGT	**AAGC**	**AAAG**	**AAAA**	**AAAT**	AAAC	AATG	AATA	**AATT**					
TAGG	TAGA	TAGT	TAGC	TAAG	TAAA	**TAAT**	TAAC	TATG	**TATA**						
CAGG	CAGA	CAGT	CAGC	CAAG	**CAAA**	CAAT	**CAAC**	**CATG**							
GTGG	**GTGA**	GTGT	GTGC	GTAG	GTAA	GTAT	**GTAC**								
ATGG	ATGA	ATGT	**ATGC**	ATAG	ATAA	**ATAT**									
TTGG	TTGA	TTGT	**TTGC**	TTAG	**TTAA**										
CTGG	CTGA	CTGT	CTGC	**CTAG**											
GCGG	**GCGA**	GCGT	**GCGC**												
ACGG	ACGA	**ACGT**													
TCGG	**TCGA**														
CCGG															

FIG. 12. Matrix of the 136 unique tetrads or four-base pair sequences in a double helix. Tetrads along the diagonal are self-complementary. Those above the diagonal are each complementary to another tetrad below the diagonal, not shown because it represents the same double helix, as AGGC and GCCT. Decamer and dodecamer B-DNA crystal structure analyses to date have provided examples only of the 35 tetrads given in boldface.

mentary sequences because they occur in pairs.[T111] The 136 unique tetrads are shown in Fig. 12. Our sampling of this set to date has been severely limited: only the 35 examples given in bold type. More examples will be needed before a dependable model of sequence–structure correlations can be derived.

Another striking local variation in B-DNA crystal structures is the width of the minor groove and the nature of ordered water down that groove. Drew et al.[T9] and Kopka et al.[T21] observed that the A-A-T-T region of the minor groove in C-G-C-G-A-A-T-T-C-G-C-G was especially narrow, barely wide enough to admit a 3.5 Å thick planar organic ring of a drug molecule,[T33,T34] and in the absence of a groove-binding drug was filled with a highly ordered zigzag spine of hydration. Similar water spines were observed for those other dodecamer structures in Table II having a resolution of approximately 2.3 Å or better. The minor groove of C-G-C-G-A-A-T-T-C-G-C-G widened in the C-G-C-G ends, but details of hydration in these regions were obscured by overlapping contacts between molecules in the crystal, a feature common to all of the isomorphous dodecamers in Table II. The more recent and higher resolution B-DNA decamers[T58,T80,T110–T113] show an unexpected variability in minor groove width, sometimes with a periodicity as short as five base pairs. Narrow regions exhibit a spine of hydration exactly as with the dodecamers, whereas wider regions of minor groove have two ribbons of waters down the two walls of the groove, bridging from nitrogen or oxygen atoms on the base edges to O-4' sugar atoms.

The width of the minor groove in these B-DNA decamers appears to be influenced by base sequence. The minor groove is widest whenever phosphates opposed across the groove have the B_{II} conformation in which main chain torsion angles ϵ (C-4'–C-3'–O-3'–P) and ζ (C-3'–O-3'–P–O-5') are (gauche⁻, trans) rather than the more common (trans, gauche⁻). The phosphate conformation in turn is related to base stacking and overlap, and this has a sequence component. B_{II} phosphates are observed only at steps of the type Y-R and R-R, where the second base is a purine.[T113] It is too early to generalize with only five high-resolution B-DNA decamer structures to examine, but a sequence dependence of minor groove width probably can be found as more examples become known.

Symmetry Check Plots and Resolution of Trilemma

Earlier in this chapter we noted that observed local features in a DNA crystal structure could have three origins: (1) effects of base sequence; (2) effects of intermolecular contacts within the crystal; and (3) inadequacies in data and refinement. The desired goals were to study (1), monitor and control *(2)*, and minimize *(3)*. Several methods were given by which the success of this strategy in a given crystal structure analysis might be evaluated. The most powerful diagnostic test, however, remains to be discussed: the symmetry check plot. This test is applicable to any double helix structure in which the Ubp in Tables I–IV is equal to the number of base pairs in the helix, and whose base sequence is self-complementary. It can be used, for example, with any of the A-DNA octamers of Table I.b for which Ubp = 8, but not with those of Table I.a for which Ubp = 4. It is applicable to most of the isomorphous B-DNA decamers in Table II.a, all of which have Ubp = 12, but not to C-G-C-A-A-A-A-A-A-G-C-G because that sequence is non-self-complementary.

The test, although powerful, is simplicity itself: Examine the symmetry or asymmetry of the two ends of the double helix, comparing parameters whose values would be identical if the helix possessed a 2-fold axis relating the ends. Calibrate these differences by comparing them with the mean and standard deviation for that parameter along the helix. The test is of limited value in considering one structure in isolation but becomes much more significant when a set of isomorphous structures is analyzed.

Two operating principles allow one to discriminate among base sequence, crystal packing, and refinement error effects. First, if differences between ends of the helix are found, they cannot arise from base sequence, since the strands of the helix are self-complementary. The two ends of the helix presumably would be truly identical in solution. The asymmetry must arise from some combination of crystal packing forces and/or errors in data or refinement. Second, if a series of isomorphous structures is compared, then crystal packing effects presumably are common to all. In that case, the *smallest* mean asymmetry of any member of the series places an upper bound on that which can be attributed to crystal packing. Greater asymmetry in other structures of the isomorphous series must arise from data or analysis errors.

Table VI* analyzes eight local helix parameters in the KK decamer (C-G-A-T-C-G-A-T-C-G) and in nine isomorphous dodecamers (code letters explained in the footnote to Table VI). *AV* is the average value of a parameter down the length of the helix, and *SD* is the standard deviation. *DI* is the mean of the differences between parameter values that would be identical if the two ends of the helix were truly symmetrical. For base step parameters such as roll, cup, twist, and rise, this is the mean of differences equidistant from the two ends of the helix, as steps 1 and 9 for a decamer, 2 and 8, 3 and 7, etc. For parameters that pertain only to one strand, such as main chain torsion angles, it is the mean of differences of that angle at

---

* Tables I–VII appear at the end of this chapter.

equivalent positions along strands 1 and 2, both measured from their respective 5' ends. An analogous table of noncrystallographic symmetry in main chain torsion angles is to be found in Ref. 32, where symmetry check plots are treated in more detail. If DI is small by comparison with SD, then the helix is quite symmetrical, so crystal contacts and refinement errors can be said to be small or negligible. But if DI is equal to or larger than SD, then the end-for-end asymmetry of the helix is as great as the variation along the helix itself. Crystal packing and refinement errors then are fully as significant as are base sequence effects, and one must treat the structure with caution.

As an example, the KK decamer has a mean helix twist of 37.1° with a standard deviation of 4.1°, but its end-for-end symmetry differences average only 1.4°. The structure is well determined where twist is concerned, with little indication of packing deformation or of refinement problems. The best of the dodecamers, MD, has a mean twist of 35.8° and a standard deviation of 4.8°. Its two ends have a mean symmetry difference of only 2.3°. By contrast, the worst twist asymmetry in these nine dodecamers is with helix N1, which has $DI = 5.2°$. Because all of the dodecamers are isomorphous and have the same crystal packing interactions, no more than 2.3° of the symmetry difference, at most, can be ascribed to packing deformations. A better, although rough, rule of thumb might be to assume that the mean symmetry difference in the best of a set of structures was evenly divided between packing distortion and refinement errors—with twist in our dodecamer examples, roughly 1.1° each. Then only 1.1° of the 5.2° of asymmetry in helix N1 can be explained by packing distortions, and the other 4.1° must indicate the error level of the structure analysis. The end-for-end uncertainty of this analysis is nearly as great as the sequence-induced variation of twist along the helix, and so one would not really be justified in making strong statements about relationships between base sequence and helix twist from this analysis alone.

As another example, the KK decamer has a mean end-for-end asymmetry of 2.2° in roll, and the best of the dodecamers, LA, has 3.5° mean symmetry. The worst of this set of dodecamer structures, YO, has a mean asymmetry of 10.8° in roll, a quantity that is greater than the standard deviation in roll along the helix. The two lowest and two highest DI values in each parameter column in Table VI are marked. A crude "scoring" of each structure analysis can be obtained by assigning + 1 for each of the two analyses with the smallest DI, and − 1 for the two highest. The 10 helix entries in Table VI have been arranged in order of this decreasing score.

A graphic display of the same information in the form of symmetry check plots is given for twist in Fig. 13 and for roll in Fig. 14. The individual parameters are simply plotted in the forward and reverse direc-

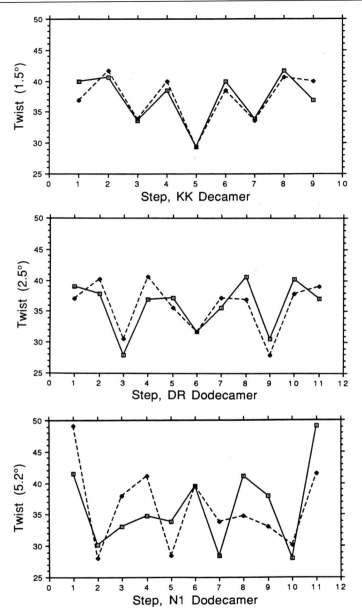

FIG. 13. Symmetry check plots of twist angle in a forward (solid line) and reverse (dashed line) direction down the helix, for the KK decamer and the DR and N1 dodecamers, as defined in Table VI. The mean difference in twist angle between solid and dashed curves is given in parentheses along the left axis and is listed as *DI* in Table VI. (From Ref. 32.)

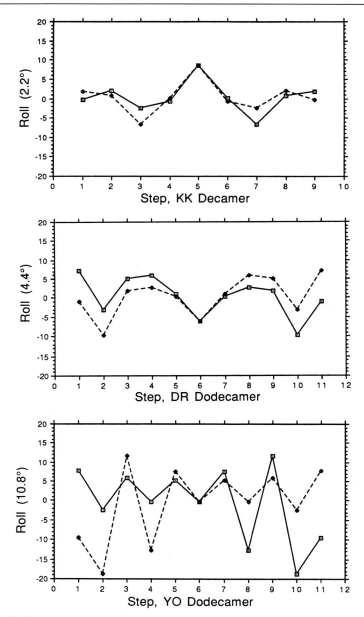

FIG. 14. Symmetry check plots of roll angles between base pairs. The KK decamer is quite symmetric end-for-end. The downward drift to the right in the DR helix is a real effect and is a consequence of a bend in the helix axis. The regular oscillation of roll in the YO helix with its strict alternation of pyrimidine (T) and purine (A) probably is real, but the wild swings at the right end are more problematical. (From Ref. 32.)

tions, and the two plots are examined for congruence. The twist plot for the KK decamer shows it to be quite symmetric, and the DR dodecamer is nearly as good. But twist in the N1 dodecamer undergoes wild swings from one end to the other. These cannot be ascribed to intermolecular packing, or they would be seen in the DR plot also. Hence one must simply say that the N1 structure is less well determined, a conclusion that would follow equally well from nominal resolution (1.9 Å for DR versus 2.4 Å for N1) or from the number of observed reflections per base pair (227 for DR versus 154 for N1).

Some features of the symmetry check plots can be identified as arising from crystal packing. All of the isomorphous dodecamers in Table II.a except for those with ^{5br}C at the ninth position have a pronounced curvature at the upper third of the helix. This means that roll angles are stretched in the first part of the helix and compressed in the second part, an effect that leads to the solid line in the DR roll plot in Fig. 14 having an overall downward slope from left to right. But the YO helix should show this same effect to the same degree. The added oscillations of roll in the second half of the YO helix can arise neither from base sequence nor from crystal packing, and it must represent simple error in the analysis. The alternating oscillations of roll probably are correct, but the radical excursions at the right of the plot cannot be real.

Symmetry check plots isolate crystal packing effects because they are shared by a set of isomorphous structures. Another powerful technique is to examine the same sequence in radically different crystal forms, whenever this can be accomplished, so that the base sequence is the same but the crystal contacts are different. As mentioned earlier, this has been done with great success for the A-DNA sequence G-G-G-C-G-C-C-C.[T62,T68,T87,T102] It also has succeeded with the B-DNA decamer C-C-A-G-G-C-C-T-G-G[T80] and its ^{5me}C derivative.[T112] Heinemann and Alings have demonstrated convincingly that the large twist angle seen at C-A steps in their unmethylated sequence[T80] and in two other decamers with the same monoclinic space group[T58,T110] is not an artifact of crystal packing. The methylated decamer adopts a hexagonal space group with quite different intermolecular contacts, but it still shows a large twist at the C-A step, which therefore must have its origin in the base sequence.

Reference 32 has symmetry check plots for inclination, propeller, backbone angle γ, and the difference between angles $(\epsilon - \zeta)$, and a fuller discussion of their uses. The test has validity, of course, only so long as no attempt has been made to symmetrize the two ends of the helix in the

[32] R. E. Dickerson, K. Grzeskowiak, M. Grzeskowiak, M. L. Kopka, T. Larsen, A. Lipanov, G. G. Privé, J. Quintana, P. Schulze, K. Yanagi, H. Yuan, and H.-C. Yoon, *Nucleosides Nucleotides* **10**, 3 (1991).

course of the refinement process. If the two ends of the helix were symmetrized as part of the boundary conditions of refinement, then their similarity would mean nothing, other than that the refinement constraints worked. Such artificial symmetrization is a highly questionable procedure in any event, because it gets rid of not only experimental errors, but also crystal packing distortions. As data from many structures accumulate, knowing what parts of a DNA double helix are intrinsically *deformable* may be important in understanding the recognition process between DNA and proteins.

The symmetry test also only works when the examiner has access to the coordinates resulting from an X-ray structure analysis. One might think that this would be automatic — of what merit is a structure analysis if the investigator refuses to divulge the results to the general scientific public? Yet the record of public disclosure of atomic coordinates from X-ray crystal structure analyses of oligonucleotides has only been fair until recently, and the record of disclosure of the actual intensity data is much worse. Table VII* shows the yearly cumulative record of public disclosure since the first oligonucleotide structure analysis in 1978. The numbers in Table VII have been obtained by counting entries under "F_o" and "XYZ" to the right of Tables I–IV. No effort has been made in Table VII to identify the pattern of contributions from various laboratories, but patterns do exist. Strenuous lobbying in recent years by scientific societies, some journals, and the National Institutes of Health has led to increasing disclosure of atomic coordinates, but release of the data on which these coordinates were based has lagged abysmally. At present, one-half of the announced and published oligonucleotide structures have coordinates available from the Brookhaven Protein Data Bank, the traditional repository of macromolecular structure data and results. Another one-quarter are currently on deposit in the Cambridge Structural Data Base, and are available on request. The Cambridge holdings as of February 1991 are listed in Tables I–IV in anticipation of the new policy. But Brookhaven possesses X-ray intensity data for only one-quarter of the published oligonucleotide structures, and the Cambridge Structural Data Base does not store X-ray data at all.

It has been remarked that macromolecular crystallography must be the only field of science where one is obliged to make public neither data nor results, and one can get by with publishing nothing more than illustrated commentaries on the results. This situation is beginning to change where results are concerned and, it is hoped, will improve with respect to the data as well.

---

* Tables I–VII appear at the end of this chapter.

TABLE I

SINGLE-CRYSTAL X-RAY STRUCTURE ANALYSES OF SYNTHETIC A-HELICAL OLIGONUCLEOTIDES[a]

Sequence	Temp.[b]	Unit cell dimensions[c] (Å) a	b	c	Ubp[d]	V/bp[e] (Å³)	Res.[f] d (Å)	R[g] (%)	σ level[h]	No. data[i]	No. solv.[j]	Publ. date[k]	Research group	Deposited in public domain[l] F_o	XYZ	Refs.[m]
a. Space group $P4_32_12$ ($Z = 8$)[r]																
[5i]CCGG	2°	41.10	41.10	26.70	4	1409	2.0	16.5	2	1486	86	1981	UCLA (CIT)	R1ANASF	1ANA/C	8,13,23
CCCCGGGG	15°	43.36	43.36	24.83	4	1459	2.25	15	0	1018	41	1987	Weiz/MIT	No	No	52
GCCCGGGC	RT	43.25	43.25	24.61	4	1439	1.8	17.1	3	1359	34	1987	Berlin	R9DNASF	9DNA/C	53
GGCCGGCC	−8°	42.06	42.06	25.17	4	1391	2.25	17.4	3	920	75	1982	MIT	No	No	17,19
GGCGGGCC	−18°	40.51	40.51	24.67	4	1265	2.25	15.8	3	874	84	1982	MIT	No	No	17,19
GGCGCGCC	RT	43.28	43.28	24.66	4	1444	1.7	15.9	2	2082	88	1988	Weizmann	No	No	68,102
GGGCGCCC	115 K	42.74	42.74	24.57	4	1403	1.7	21	2	1694	62	1988	Weizmann	No	No	62,102
GG[5mC]CCGGCC	15°	42.51	42.51	24.91	4	1407	2.25	14.8	2	802	41	1987	MIT	No	1DNS	51
GTGTACAC/spermine	4°	42.43	42.43	24.75	4	1392	2.0	11.5	0	1214	43	1987	Wisconsin	No	1DNS	55,83
CTCTAGAG	RT	42.53	42.53	24.33	4	1375	2.15	14.7	1	931	35	1989	Cambridge	No	Cam	81
GTACGTAC	−10°	42.32	42.32	25.04	4	1401	2.25	18.4	0	1233	53	1990	Kansas	No	5ANA	103
GTACGTAC	RT	42.50	42.50	24.79	4	1399	2.4	17.0	1	894	56	1990	Bordeaux	R18DNSF	18DN	94
ATGCGCAT	?	42.41	42.41	24.90	4	1400	1.5	17.7	?	?	110	1990	InstCanRes	No	No	93
ATGCGCAT/spermine	?	42.53	42.53	24.92	4	1409	1.7	18.0	?	?	113	1990	InstCanRes	No	No	93
b. Space group $P6_1$ ($Z = 6$)[n]																
GG[5br]UA[5br]UACC	RT	45.05	45.05	41.72	8	1528	1.7	14	2	4669	84	1981	Weiz/Camb	No	Cam	12,22,46
GGTATACC	RT	45.01	45.01	41.55	8	1519	1.8	19.8	2	3251	66	1981	Weiz/Camb	No	No	12,22
GGGATCCC	RT	46.83	46.83	44.49	8	1525	2.5	16.6	3	1428	9	1988	Berlin	R3ANASF	3ANA	65
GGGTACCC	RT	46.80	46.80	44.52	8	1759	2.5	11.9	1	1582	89	1990	Weizmann	No	No	96
GGGTACCC	100 K	46.06	46.06	44.09	8	1688	2.1	16	2	2008	76	1990	Weizmann	No	No	96
GGGGCCCC	RT	46.96	46.96	44.31	8	1763	2.9	12	2	1068	45	1989	Weizmann	No	No	87,102
GGGCGCCC	110 K	43.62	43.62	41.49	8	1424	1.9	16	2	2827	100	1989	Weizmann	No	No	87,102
GGGGCCCC	18°	45.32	45.32	42.25	8	1566	2.5	14	2	1283	106	1985	Cambridge	R2ANASF	2ANA/C	35
GGGGCTCC Mis	2°	45.20	45.20	42.97	8	1584	2.25	13.6	2	1924	52	1985	Camb/Weiz	No	Cam	28,43
GGGGTCCC Mis	2°	44.71	44.71	42.40	8	1529	2.1	14.5	2	2486	104	1985	Camb/Weiz	No	Cam	32
GGGTGCCC Mis	RT	45.62	45.62	40.99	8	1539	2.5	15.2	2	1362	63	1988	Weizmann	No	No	68

c. Space group $P4_3$ ($Z = 4$)[n]

GGATGGGAG	RT	45.29	45.29	24.73	9	1409	3.0	33	0	1032	0	1986	Cambridge	No	1DN6	47

d. Space group $P6_122$ ($Z = 12$)[n]

GTGTACAC		32.4	32.4	79.5	4	1505	2.0	12.5	?	1561	50	1989	*Wisconsin*	No	No	82
ACCGGCCGGT		39.23	39.23	78.00	5	1733	2.0	18.0	2	1434	36	1989	MIT	No	*1D13/C*	78
CpsCGTACGTACGG		46.2	46.2	71.5	6	1836	2.5	13.6	0	2415	71	1989	Wisconsin	No	No	71

e. Space group $P2_12_12_1$ ($Z = 4$)[n]

r(GCG)d(TATACGC)	0°	24.20	43.46	49.40	10	1299	2.0	16	1.5	2521	179	1982	MIT	No	Cam	17,20
r[U(UA)$_6$A]	RT	34.11	44.61	49.11	14	1334	2.25	13.1	2	2492	91	1988	Strasbourg	No	*Cam*	61,77

[a] DNA is assumed unless specified otherwise by r( ). RNA. p, Leading phosphate group; ps, phosphorothioate; Mis, base pair mismatch (mismatch bases underlined); Adr, Adriamycin; Ber, berenil; Cis, cisplatin; Dau, daunomycin; Dis, distamycin; Ech, echinomycin; Hoe, Hoechst 33258; Net, netropsin; Nog, nogalomycin; Tri, triostin A; U-5, U-58872 analog of nogalomycin.

[b] Temperature at which data were collected (not at which crystals were grown). RT, Room temperature.

[c] All angles 90° except $\gamma = 120°$ for hexagonal cells, and β as indicated for monoclinic cells.

[d] Unique base pairs, number of base pairs of DNA within one independent, asymmetric unit of the cell.

[e] Volume per base pair, unit cell volume/($Z \times$ Ubp).

[f] Resolution of the crystal structure analysis.

[g] Final residual error or R-factor.

[h] Statistical level of cutoff of weak data.

[i] Number of observed reflections in the data set to the resolution and σ level specified, not the total number of possible reflections.

[j] Number of water or other solvent molecules per asymmetric unit (or per set of unique base pairs), not per double helix.

[k] Date of first publication.

[l] F_o, Observed X-ray intensity data available to the general public; XYZ, final refined atomic coordinates available to the general public; Paper, coordinates printed in journal article. Serial numbers are Brookhaven Protein Data Bank identification codes of coordinates or structure factors. Those deposited but not yet released to the public are italicized. /C, Also obtainable from the Cambridge Structure Data Base; Cam, obtainable only from the Cambridge Structure Data Base.

[m] References are listed at the end of Table IV.

[n] Number of equivalent points or asymmetric units per cell for the indicated space group.

TABLE II
Single-Crystal X-Ray Structure Analyses of Synthetic B-Helical Oligonucleotides[a]

Sequence	Temp.[b]	Unit cell dimensions[c] (Å)			Ubp[d]	V/bp[e] (Å³)	Res.[f]		σ level[h]	No. data[i]	No. solv.[j]	Publ. date[k]	Research group	Deposited in public domain[l]		Refs.[m]
		a	b	c			d (Å)	R[g] (%)						F_o	XYZ	
a. Space group $P2_12_12_1$ ($Z = 4$)[n]																
1. "Drew" CGCGAATTCGCG sequence and variants																
CGCGAATTCGCG	16 K	23.44	39.31	65.26	12	1253	2.7	15.1	2	1051	83	1982	UCLA (CIT)	R2BNASF	2BNA/C	14
CGCGAATTCGCG	RT	24.87	40.39	66.20	12	1385	1.9	17.8	2	2725	80	1980	UCLA (CIT)	R1BNASF	1BNA/C	6–9,11
CGCGAATTCGCG	RT	24.87	40.39	66.20	12	1385	1.9	18.8	?	3979	64	1987	Strasbourg	No	9BNA	59
CGCGAATTCGCG	RT	24.87	40.39	66.20	12	1385	1.9	14.9	2	2728	80	1985	Berkeley	R7BNASF	7BNA	30
CGCGAATT⁵ᵇʳCGCG	20°	24.71	40.56	65.62	12	1370	3.0	17.3	2	1515	44	1982	UCLA	R3BNASF	3BNA/C	15,21
CGCGAATT⁵ᵇʳCGCG	7°	24.20	40.09	63.95	12	1293	2.3	17.3	2	2919	115	1982	UCLA	R4BNASF	4BNA/C	15,21
CGCGA⁶ᵐᵉATTCGCG	RT	25.62	40.70	67.32	12	1462	2.0	16.9	2	2379	86	1988	MIT	No	4DNB	63
CGCGAATTTGCG Mis	4°	25.53	41.22	65.63	12	1433	2.5	13	2	2000	71	1985	Cambridge	No	Cam	54
CGCGAATTAGCG Mis	5°	25.69	41.96	65.19	12	1464	2.5	17.0	2	2028	83	1986	Cambridge	No	Cam	39,44
CGCAAATTCGCG Mis	4°	25.37	41.44	65.20	12	1428	2.5	19.0	2	2029	82	1986	Cambridge	No	Cam	45
With drugs																
CGCGAATTCGCG/Cis	4°	24.16	39.93	66.12	12	1329	2.6	11.2	2	2088	128	1984	UCLA	R5BNASF	5BNA/C	26
CGCGAATT⁵ᵇʳCGCG/Net	4°	24.27	39.62	63.57	12	1273	2.2	21.1	2	2528	75	1985	UCLA	R6BNASF	6BNA	33,34
CGCGAATTCGCG/Hoe	RT	25.04	40.33	65.85	12	1385	2.2	14.0	2	1569	175	1987	UCLA	R8BNASF	8BNA/C	57
CGCGAATTCGCG/Hoe	15°	25.23	40.58	66.08	12	1409	2.25	15.7	2	2000	126	1988	MIT	R1DNHSF	1DNH/C	69
CGCGAATTCGCG/DAPI	RT	25.35	40.71	66.53	12	1430	2.4	21.5	2	2428	25	1989	UCLA	R1D30SF	1D30	84
CGCGAATTCGCG/Ber	7°	24.51	39.98	66.23	12	1352	2.5	17.7	2	1759	49	1990	InstCanRes	No	2DBE	91
2. Other isomorphous dodecamer sequences																
CGCAAAAAGCG	4°	25.4	40.7	65.8	12	1417	2.5	20	0.5	2003	27	1987	Cambridge	No	Cam	56
CGCAAAATGCG	12°	24.54	40.32	65.86	12	1358	2.6	20.1	2	1975	23	1989	Yale	*R1BDN*	*IBDN*	76
CGCATATATGCG	6°	23.54	38.86	66.57	12	1269	2.2	18.7	1	1915	43	1988	UCLA	*R1DN9SF*	*IDN9*	70
CGTGAATTCACG	0°	24.94	40.78	66.13	12	1401	2.5	15.8	2	1475	36	1991	UCLA	*R1D29SF*	*ID29*	114
CGTGAATTCACG	−90°	24.78	40.85	65.67	12	1385	2.7	17.0	2	1532	85	1991	Rutgers	*R1D28SF*	*ID28*	115
CGCAAATTGGCG Mis	4°	25.23	41.16	65.01	12	1406	2.25	15.8	2	2262	94	1989	Edinburgh	No	Cam	72,98
CGCAAGCTGGCG Mis	7°	25.29	41.78	64.76	12	1426	2.5	19.3	3	1710	47	1990	InstCanRes	No	*IDNM*	104
CGCGAAAACGCG/CGCGTT/ TTCGCG (nicked strand)	25°	25.99	44.03	66.62	12	1588	3.0	18.2	2	553	79	1990	MIT	No	*INDN*	90
With drugs																
CGCAAATTTGCG/Dis	15°	25.20	41.07	64.65	12	1393	2.2	19.6	2	2369	70	1987	MIT	No	2DND/C	50
CGCGATATCGCG/Net	10°	25.48	41.26	66.88	12	1464	2.4	20.1	2	1848	60	1989	MIT	No	1DNE/C	74
CGCGATATCGCG/Hoe	15°	25.59	40.56	67.10	12	1451	2.3	13.7	2	1573	112	1989	MIT	No	Yes??	73

3. Other oligonucleotide lengths

Sequence	Temp[b]	a[c]	b	c	β	[d]	Vol[e]	Res[f]	R[g]	σ[h]	Obs[i]	Water[j]	Date[k]	Location	[l]	Cam	Ref[m]
GpsCGpsCGpsC	18°	34.90	39.15	20.64		6	1175	2.17	14.5	2	1327	72	1987	Cambridge	No	Cam	42
CGATCGATCG	0°	38.93	39.63	33.30		10	1284	1.5	16.1	2	5107	142	1991	UCLA	*R1D23SF*	*1D23*	113

b. Space group C2 (Z = 4)[n]

Sequence	Temp[b]	a[c]	b	c	β	[d]	Vol[e]	Res[f]	R[g]	σ[h]	Obs[i]	Water[j]	Date[k]	Location	[l]	Cam	Ref[m]
CAAGATTGG Mis	5°	32.52	26.17	34.30	β = 118.9°	5	1278	1.3	16.4	2	4016	68	1987	UCLA	R3DNBSF	3DNB	58,67
CCAACGTTGG	4°	32.25	25.49	34.65	β = 114.0°	5	1301	1.4	17.9	1	4402	71	1991	UCLA	*R5DNBSF*	*5DNB*	110,111
CCAGGCCTGG	RT	32.15	25.49	34.82	β = 116.71°	5	1275	1.6	16.9	1	2420	49	1989	Berlin	R1BD1SF	1BD1/C	80
CGCAGAATTCGCG A bulge	123 K	78.48	42.85	25.16	β = 99.36°	12	1739	2.8	15	2	?	50	1988	Weizmann	No	1D3II	64

c. Space group P6 (Z = 6)[n]

Sequence	Temp[b]	a[c]	b	c	[d]	Vol[e]	Res[f]	R[g]	σ[h]	Obs[i]	Water[j]	Date[k]	Location	[l]	Cam	Ref[m]
CCAGGC^{5me}CTGG	RT	53.77	53.77	34.35	10	1433	2.25	17.2	1	2407	72	1991	Berlin	Yes	1D25	112

d. Space group I222 (Z = 8)[n]

Sequence	Temp[b]	a[c]	b	c	[d]	Vol[e]	Res[f]	R[g]	σ[h]	Obs[i]	Water[j]	Date[k]	Location	[l]	Cam	Ref[m]
CGCGAAATTTACGCG A bulge	4°	37.0	53.7	101.6	14	1802	3.0	26.3	1.5	1943	?	1988	NCI/Weiz	No	No	66

e. Space group R3 (Z = 9)[n]

Sequence	Temp[b]	a[c]	b	c	[d]	Vol[e]	Res[f]	R[g]	σ[h]	Obs[i]	Water[j]	Date[k]	Location	[l]	Cam	Ref[m]
ACCGGCGCCACA	—	65.9	65.9	47.1	12	1640	2.8	15.3	3	1101	20	1989	Strasbourg	No	Cam	89

[a] DNA is assumed unless specified otherwise by r( ), RNA. p, Leading phosphate group; ps, phosphorothioate; Mis, base pair mismatch (mismatch bases underlined); Adr, Adriamycin; Ber, berenil; Cis, cisplatin; Dau, daunomycin; Dis, distamycin; Ech, echinomycin; Hoe, Hoechst 33258; Net, netropsin; Nog, nogalomycin; Tri, triostin A; U-5, U-58872 analog of nogalomycin.

[b] Temperature at which data were collected (not at which crystals were grown). RT, Room temperature.

[c] All angles 90° except γ = 120° for hexagonal cells, and β as indicated for monoclinic cells.

[d] Unique base pairs, number of base pairs of DNA within one independent, asymmetric unit of the cell.

[e] Volume per base pair, unit cell volume/(Z × Ubp).

[f] Resolution of the crystal structure analysis.

[g] Final residual error or R-factor.

[h] Statistical level of cutoff of weak data.

[i] Number of observed reflections in the data set to the resolution and σ level specified, not the total number of possible reflections.

[j] Number of water or other solvent molecules per asymmetric unit (or per set of unique base pairs), not per double helix.

[k] Date of first publication.

[l] F_o, Observed X-ray intensity data available to the general public; *XYZ*, final refined atomic coordinates available to the general public; Paper, coordinates printed in journal article. Serial numbers are Brookhaven Protein Data Bank identification codes of coordinates or structure factors. Those deposited but not yet released to the public are italicized. /C, Also obtainable from the Cambridge Structure Data Base; Cam, obtainable only from the Cambridge Structure Data Base.

[m] References are listed at the end of Table IV.

[n] Number of equivalent points or asymmetric units per cell for the indicated space group.

TABLE III
SINGLE-CRYSTAL X-RAY STRUCTURE ANALYSES OF SYNTHETIC Z-HELICAL OLIGONUCLEOTIDES[a]

Sequence	Temp.[b]	Unit cell dimensions[c] (Å) a	b	c	Ubp[d]	V/bp[e] (Å³)	Res.[f] d (Å)	R[g] (%)	σ level[h]	No. data[i]	No. solv[j]	Publ. date[k]	Research group	Deposited in public domain[l] F_o	XYZ	Refs.[m]
a. Space group C222₁ (Z = 8)[n]																
CGCGCG	RT	19.50	31.27	64.67	4	1232	1.5	21	2	1900	84	1980	UCLA (CIT)	R1ZNASF	1ZNA	4,8,10
b. Space group P2₁2₁2₁ (Z = 4)[n]																
CGCGCG/spermine	RT	17.88	31.55	44.58	6	1048	0.9	13	2	15000	68	1979	MIT	No	2DCG/C	2,17,79
CGCGCG	RT	18.01	31.03	44.80	6	1043	1.0	17.5	2	10893	88	1989	MIT	No	1DCG	79
5meCG⁵meCG⁵meCG	−8°	17.76	30.57	45.42	6	1027	1.3	15.6	1.5	4208	98	1982	MIT	No	Paper	16,17
5brCG⁵brCG⁵meCG	18°	18.01	30.88	44.76	6	1037	1.4	13.3	2	2919	61	1986	Strasbourg	R1DN4SF	1DN4	41,59
5brCG⁵brCG⁵brCG	37°	17.93	30.83	44.73	6	1030	1.4	12.5	2	3765	83	1986	Strasbourg	R1DN5SF	1DN5	41,59
5meCGTA⁵meCG	−10°	17.91	30.43	44.96	6	1021	1.2	16	1.5	5412	98	1984	MIT	No	Paper/C	24
5meCGUA⁵meCG	RT	17.82	30.44	44.52	6	1006	1.3	20.8	1.5	2870	62	1990	Oregon State	No	No	109
5brCGAT⁵brCG	?	18.30	31.10	44.10	6	1046	1.54	19.3	1.5	2386	88	1985	MIT	No	No	37
d(CG)t(CG)d(CG)	20°	18.25	30.93	43.05	6	1013	1.5	20.4	2	1796	68	1989	MIT	No	No	88
d(CG)(araC)dGCG)	20°	18.57	30.68	42.85	6	1017	1.5	16.7	2	2720	77	1989	MIT	No	No	88
CGC⁶moCGCG	−90°	17.85	30.87	43.98	6	1010	1.9	19.0	0	1217	60	1990	Rutgers	Yes	1D24	99
CGCG⁴moCG	4°	18.17	30.36	43.93	6	1010	1.7	18.1	0	2559	68	1990	Cambridge	No	No	105
CGCGTG Mis	15°	17.45	31.63	45.56	6	1048	1.0	19.5	1	10964	91	1985	MIT	No	Cam	31
CGCG5ˢᵘUG Mis	10°	17.38	31.06	45.39	6	1021	1.5	20.0	2	3008	64	1989	MIT	No	1DNF	75
5brUGCGCG Mis	4°	17.94	30.85	49.94	6	1152	2.25	15.6	2	1105	64	1986	Cambridge	No	Cam	40

c. Space group P6₅ (Z = 6)[n]

CGCG	Disordered	31.25	31.25	44.06	6	1035	1.5	19.3	?	?	85	1980	MIT	No	No	No	3
CGCGCGCG	Disordered	31.27	31.27	43.56	6	1025	1.6	19	?	?	85	1985	MIT	No	No	No	29
CGCATGCG	Disordered	30.90	30.90	43.14	6	991	2.5	16	?	?	?	1985	MIT	No	No	No	29
CGTACGTACG	Disordered	17.93	17.93	43.41	2	1007	1.5	25.5	2	506	7	1985	Wis/Stras.	No	No	1DN8	27,38

d. Space group C2 (Z = 4)[n]

CGCGCGTTTTCGCGCG Hairpin	7 C	57.18	21.63 $\beta = 95.22°$	36.40	8	1400	2.1	20.0	2	1509	70	1988	UCLA	R1D16SF	1D16	60,92

[a] DNA is assumed unless specified otherwise by r( ); RNA. p, Leading phosphate group; ps, phosphorothioate; Mis, base pair mismatch (mismatch bases underlined); Adr, Adriamycin; Ber, berenil; Cis, cisplatin; Dau, daunomycin; Dis, distamycin; Ech, echinomycin; Hoe, Hoechst 33258; Net, netropsin; Nog, nogalomycin; Tri, triostin A; U-5, U-58872 analog of nogalomycin.

[b] Temperature at which data were collected (not at which crystals were grown). RT, Room temperature.

[c] All angles 90° except γ = 120° for hexagonal cells, and β as indicated for monoclinic cells.

[d] Unique base pairs, number of base pairs of DNA within one independent, asymmetric unit of the cell.

[e] Volume per base pair, unit cell volume /(Z × Ubp).

[f] Resolution of the crystal structure analysis.

[g] Final residual error or R-factor.

[h] Statistical level of cutoff of weak data.

[i] Number of observed reflections in the data set to the resolution and σ level specified, not the total number of possible reflections.

[j] Number of water or other solvent molecules per asymmetric unit (or per set of unique base pairs), not per double helix.

[k] Date of first publication.

[l] F_o, Observed X-ray intensity data available to the general public; XYZ, final refined atomic coordinates available to the general public; Paper, coordinates printed in journal article. Serial numbers are Brookhaven Protein Data Bank identification codes of coordinates or structure factors. Those deposited but not yet released to the public are italicized. /C, Also obtainable from the Cambridge Structure Data Base; Cam, obtainable only from the Cambridge Structure Data Base.

[m] References are listed at the end of Table IV.

[n] Number of equivalent points or asymmetric units per cell for the indicated space group.

TABLE IV
SINGLE-CRYSTAL X-RAY STRUCTURE ANALYSES OF IRREGULAR HELICAL OLIGONUCLEOTIDE STRUCTURES[a]

Sequence	Temp.[b]	Space group	Z[c]	Unit cell dimensions[d] (Å)			Ubp[e]	V/bp[f] (Å³)	Res.[g] d (Å)	R[h] (%)	σ level[i]	No. data[j]	No. solv.[k]	Publ. date[l]	Res. group	Deposited in public domain[m] F_o	XYZ	Refs.[n]
				a	b	c												
Cyclic r(GG)	RT	$I4_1$	8	20.06	20.06	39.40	2*	991	0.9	9.4	2	3877	27	1990	MIT	No	No	95,101
							(*Two independent cyclic dimers)											
pATAT	?	$P2_1$	2	21.12	21.29	8.77	2	986	1.0	15.2	1.5	2717	43	1978	Cam/Wz	No	No	1,18
With drugs																		
CGTACG/Dau	?	$P4_32_12$	8	27.92	27.92	52.89	3	1718	1.5	20	2	2108	40	1980	MIT	No	1D11/C	5
CGATCG/Dau	4°	$P4_32_12$	8	27.98	27.98	52.87	3	1724	1.5	25	5(!)	1845	40	1989	Camb	No	Cam	86
CGATCG/Dau	15°	$P4_32_12$	8	28.11	28.11	53.08	3	1748	1.5	17.5	2	?	47	1990	MIT	No	1D10/C	97
CGATCG/Adr	15°	$P4_32_12$	8	28.21	28.21	53.19	3	1764	1.7	17.7	2	?	53	1990	MIT	No	1D12/C	97
CGATCG/4-epiAdr	RT	$P4_32_12$	8	28.04	28.04	53.15	3	1741	1.5	19.6	1.5	1714	3	1990	MIT	No	1D15	107
CGTpsACG/11-deoxyDau	10°	$P2$ $\beta = 98.0°$	2	17.73	19.13	26.36	3	1476	1.5	22.4	2	2129	36.5	1990	MIT	No	1D14/C	106
CGTACG/Ech	10°	$F222$	16	30.71	62.56	61.70	3	2470	1.8	19.9	1	2015	88	1985	MIT	No	No	36
CGTACG/Tri	−16°	$F222$	16	31.35	62.38	61.26	3	2496	1.67	18	1	2995	135	1984	MIT	No	No	25,36
GCGTACGC/Tri	15°	$P6_522$	12	40.9	40.9	80.7	4	2436	2.25	20	1.5	1130	91	1986	III	No	Cam	48,49
CGTpsACGC/Tri	25°	$C222_1$	8	22.98	47.27	64.44	6	1458	1.3	19.2	3	5983	113	1989	III	No	No	85
5meCGTpsA5meCG/Nog	25°	$P6_1$	6	26.31	26.31	100.3	6	1669	1.7	20.8	2	3386	?	1989	III	No	1D21	85
5meCGTpsA5meCG/U-5	25°	$P6_1$	6	26.27	26.27	100.2	6	1663	1.8	19.6	2	2143	?	1990	III	No	1D22	100
5meCGTpsA5meCG/Nog	RT	$P6_122$	12	26.30	26.30	100.0	3	1664	2.0	20.6	2	809	39	1990	MIT	No	1D17/C	108

[a] DNA is assumed unless specified otherwise by r( ), RNA. p, Leading phosphate group; ps, phosphorothioate; Mis, base pair mismatch (mismatch bases underlined); Adr, Adriamycin; Ber, berenil; Cis, cisplatin; Dau, daunomycin; Dis, distamycin; Ech, echinomycin; Hoe, Hoechst 33258; Net, netropsin; Nog, nogalomycin; Tri, Triostin A; U-5, U-58872 analog of nogalomycin.

[b] Temperature at which data were collected (not at which crystals were grown). RT, Room temperature.

[c] Number of equivalent points or asymmetric units per cell for the indicated space group.

[d] All angles 90° except γ = 120° for hexagonal cells, and β as indicated for monoclinic cells.

[e] Unique base pairs, number of base pairs of DNA within one independent, asymmetric unit of the cell.

[f] Volume per base pair, unit cell volume /(Z × Ubp).

[g] Resolution of the crystal structure analysis.

[h] Final residual error or R-factor.

[i] Statistical level of cutoff of weak data.

j Number of observed reflections in the data set to the resolution and σ level specified, not the total number of possible reflections.

k Number of water or other solvent molecules per asymmetric unit (or per set of unique base pairs), not per double helix.

l Date of first publication.

m F_o, Observed X-ray intensity data available to the general public; Paper, coordinates printed in journal article. Serial numbers are Brookhaven Protein Data Bank identification codes of coordinates or structure factors. Those deposited but not yet released to the public are italicized. /C, Also obtainable from the Cambridge Structure Data Base; Cam, obtainable only from the Cambridge Structure Data Base.

n This list contains the first and the principal references to all known single-crystal X-ray structure analyses of synthetic deoxyoligonucleotides. References are numbered in chronological order by year, but alphabetically by first author within each year. In the text these table references are cited with a leading "T," as T2, T76.

1978

1. M. A. Viswamitra, O. Kennard, P. G. Jones, G. M. Sheldrick, S. Salisbury, L. Falvello, and Z. Shakked, *Nature (London)* **273**, 687 (1978).

1979

2. A. H.-J. Wang, G. J. Quigley, F. K. Kolpak, J. L. Crawford, J. H. van Boom, G. van der Marel, and A. Rich, *Nature (London)* **282**, 680 (1979).

1980

3. J. L. Crawford, F. J. Kolpak, A. H.-J. Wang, G. J. Quigley, J. H. van Boom, G. van der Marel, and A. Rich, *Proc. Natl. Acad. Sci. U.S.A.* **77**, 4016 (1980).

4. H. R. Drew, T. Takano, S. Tanaka, K. Itakura, and R. E. Dickerson, *Nature (London)* **286**, 567 (1980).

5. G. J. Quigley, A. H.-J. Wang, G. Ughetto, G. van der Marel, J. H. van Boom, and A. Rich, *Proc. Natl. Acad. Sci. U.S.A.* **77**, 7204 (1980).

6. R. M. Wing, H. R. Drew, T. Takano, C. Broka, S. Tanaka, K. Itakura, and R. E. Dickerson, *Nature (London)* **287**, 755 (1980).

1981

7. R. E. Dickerson and H. R. Drew, *J. Mol. Biol.* **149**, 761 (1981).

8. R. E. Dickerson, H. R. Drew, and B. N. Conner, *in* "Biomolecular Stereodynamics" (R. H. Sarma, ed.), Vol. 1, p. 1. Adenine Press, Albany, New York, 1981.

9. H. R. Drew and R. E. Dickerson, *J. Mol. Biol.* **151**, 535 (1981).

10. H. R. Drew and R. E. Dickerson, *J. Mol. Biol.* **152**, 723 (1981).

11. H. R. Drew, R. M. Wing, T. Takano, C. Broka, S. Tanaka, K. Itakura, and R. E. Dickerson, *Proc. Natl. Acad. Sci. U.S.A.* **78**, 2179 (1981).

12. Z. Shakked, D. Rabinovich, W. B. T. Cruse, E. Egert, O. Kennard, G. Sala, S. A. Salisbury, and M. A. Viswamitra, *Proc. R. Soc. London, B* **213**, 479 (1981).

1982

13. B. N. Conner, T. Takano, S. Tanaka, K. Itakura, and R. E. Dickerson, *Nature (London)* **295**, 294 (1982).

14. H. R. Drew, S. Samson, and R. E. Dickerson, *Proc. Natl. Acad. Sci. U.S.A.* **79**, 4040 (1982).

15. A. V. Fratini, M. L. Kopka, H. R. Drew, and R. E. Dickerson, *J. Biol. Chem.* **257**, 14686 (1982).

16. S. Fujii, A. H.-J. Wang, G. van der Marel, J. H. van Boom, and A. Rich, *Nucleic Acids Res.* **10**, 7879 (1982).

17. S. Fujii, A. H.-J. Wang, J. van Boom, and A. Rich, *Nucleic Acids Res. Symp. Ser.* **11**, 109 (1982).

18. M. A. Viswamitra, Z. Shakked, P. G. Jones, G. M. Sheldrick, S. A. Salisbury, and O. Kennard, *Biopolymers* **21**, 513 (1982).

19. A. H.-J. Wang, S. Fujii, J. H. van Boom, and A. Rich, *Proc. Natl. Acad. Sci. U.S.A.* **79**, 3968 (1982).

20. A. H.-J. Wang, S. Fujii, J. H. van Boom, G. A. van der Marel, S. A. A. van Boeckel, and A. Rich, *Nature (London)* **299**, 601 (1982).

1983

21. M. L. Kopka, A. V. Fratini, H. R. Drew, and R. E. Dickerson, *J. Mol. Biol.* **163**, 129 (1983).

22. Z. Shakked, D. Rabinovich, O. Kennard, W. B. T. Cruse, S. A. Salisbury, and M. A. Viswamitra, *J. Mol. Biol.* **166**, 183 (1983).

(Continued)

105

TABLE IV *(Continued)*

1984

23. B. N. Conner, C. Yoon, J. L. Dickerson, and R. E. Dickerson, *J. Mol. Biol.* **174**, 663 (1984).
24. A. H.-J. Wang, T. Hakoshima, G. van der Marel, J. H. van Boom, and A. Rich, *Cell (Cambridge, Mass.)* **37**, 321 (1984).
25. A. H.-J. Wang, G. Ughetto, G. J. Quigley, T. Hakoshima, G. A. van der Marel, J. H. van Boom, and A. Rich, *Science* **225**, 1115 (1984).
26. R. M. Wing, P. Pjura, H. R. Drew, and R. E. Dickerson, *EMBO J.* **3**, 1201 (1984).

1985

27. R. G. Brennan and M. Sundaralingam, *J. Mol. Biol.* **181**, 561 (1985).
28. T. Brown, O. Kennard, G. Kneale, and D. Rabinovich, *Nature (London)* **315**, 604 (1985).
29. S. Fujii, A. H.-J. Wang, G. J. Quigley, H. Westerink, G. van der Marel, J. H. van Boom, and A. Rich, *Biopolymers* **24**, 243 (1985).
30. S. R. Holbrook, R. E. Dickerson, and S.-H. Kim, *Acta Crystallogr. Sect. B: Struct. Sci.* **41**, 255 (1985).
31. P. S. Ho, C. A. Frederick, G. J. Quigley, G. A. van der Marel, J. H. van Boom, A. H.-J. Wang, and A. Rich, *EMBO J.* **4**, 3617 (1985).
32. G. Kneale, T. Brown, O. Kennard, and D. Rabinovich, *J. Mol. Biol.* **186**, 805 (1985).
33. M. L. Kopka, C. Yoon, D. Goodsell, P. Pjura, and R. E. Dickerson, *Proc. Natl. Acad. Sci. U.S.A.* **82**, 1376 (1985).
34. M. L. Kopka, C. Yoon, D. Goodsell, P. Pjura, and R. Dickerson, *J. Mol. Biol.* **183**, 553 (1985).
35. M. McCall, T. Brown, and O. Kennard, *J. Mol. Biol.* **183**, 385 (1985).
36. G. Ughetto, A. H.-J. Wang, G. J. Quigley, G. A. van der Marel, J. H. van Boom, and A. Rich, *Nucleic Acids Res.* **13**, 2305 (1985).
37. A. H.-J. Wang, R. V. Gessner, G. A. van der Marel, J. H. van Boom, and A. Rich, *Proc. Natl. Acad. Sci. U.S.A.* **82**, 3611 (1985).

1986

38. R. G. Brennan, E. Westhof, and M. Sundaralingam, *J. Biomol. Struct. Dyn.* **3**, 649 (1986).
39. T. Brown, W. N. Hunter, G. Kneale, and O. Kennard, *Proc. Natl. Acad. Sci. U.S.A.* **83**, 2402 (1986).
40. T. Brown, G. Kneale, W. N. Hunter, and O. Kennard, *Nucleic Acids Res.* **14**, 1801 (1986).
41. B. Chevrier, A. C. Dock, B. Hartmann, M. Leng, D. Moras, M. T. Thuong, and E. Westhof, *J. Mol. Biol.* **188**, 707 (1986).
42. W. B. T. Cruse, S. A. Salisbury, T. Brown, R. Cosstick, F. Eckstein, and O. Kennard, *J. Mol. Biol.* **192**, 891 (1986).
43. W. N. Hunter, T. Brown, G. Kneale, N. N. Anand, D. Rabinovich, and O. Kennard, *J. Mol. Biol.* **190**, 605 (1986).
44. W. N. Hunter, T. Brown, and O. Kennard, *J. Biomol. Struct. Dyn.* **4**, 173 (1986).
45. W. N. Hunter, T. Brown, N. N. Anand, and O. Kennard, *Nature (London)* **320**, 552 (1986).
46. O. Kennard, W. B. T. Cruse, J. Nachman, T. Prange, Z. Shakked, and D. Rabinovich, *J. Biomol. Struct. Dyn.* **3**, 623 (1986).
47. M. McCall, T. Brown, W. N. Hunter, and O. Kennard, *Nature (London)* **322**, 661 (1986).
48. G. J. Quigley, G. Ughetto, G. A. van der Marel, J. H. van Boom, A. H.-J. Wang, and A. Rich, *Science* **232**, 1255 (1986).
49. A. H.-J. Wang, G. Ughetto, G. J. Quigley, and A. Rich, *J. Biomol. Struct. Dyn.* **4**, 319 (1986).

1987

50. M. Coll, C. A. Frederick, A. H.-J. Wang, and A. Rich, *Proc. Natl. Acad. Sci. U.S.A.* **84**, 8385 (1987).
51. C. A. Frederick, D. Saal, G. A. van der Marel, J. H. van Boom, A. H.-J. Wang, and A. Rich, *Biopolymers* **26**, S145 (1987).
52. T. E. Haran, Z. Shakked, A. H.-J. Wang, and A. Rich, *J. Biomol. Struct. Dyn.* **5**, 199 (1987).
53. U. Heinemann, H. Lauble, R. Frank, and H. Blöcker, *Nucleic Acids Res.* **15**, 9531 (1987).
54. W. N. Hunter, T. Brown, G. Kneale, N. N. Anand, D. Rabinovich, and O. Kennard, *J. Biol. Chem.* **21**, 9962 (1987).
55. S. Jain, G. Zon, and M. Sundaralingam, *J. Mol. Biol.* **197**, 141 (1987).
56. H. C. M. Nelson, J. T. Finch, B. F. Luisi, and A. Klug, *Nature (London)* **330**, 221 (1987).
57. P. E. Pjura, K. Grzeskowiak, and R. E. Dickerson, *J. Mol. Biol.* **197**, 257 (1987).

58. G. G. Privé, U. Heinemann, S. Chandrasegaran, L.-S. Kan, M. L. Kopka, and R. E. Dickerson, *Science* **238**, 498 (1987).

59. E. Westhof, *J. Biomol. Struct. Dyn.* **5**, 581 (1987).

1988

60. R. Chattopadhyaya, S. Ikuta, K. Grzeskowiak, and R. E. Dickerson, *Nature (London)* **334**, 175 (1988).

61. A. C. Dock-Bregeon, B. Chevrier, A. Podjarny, D. Moras, J. S. deBear, G. R. Gough, P. T. Gilham, and J. E. Johnson, *Nature (London)* **335**, 375 (1988).

62. M. Eisenstein, H. Hope, T. E. Haran, F. Frolow, Z. Shakked, and D. Rabinovich, *Acta Crystallogr. Sect. B: Struct. Sci.* **44**, 625 (1988).

63. C. A. Frederick, G. J. Quigley, G. A. van der Marel, J. H. van Boom, A. H.-J. Wang, and A. Rich, *J. Biol. Chem.* **263**, 17872 (1988).

64. L. Joshua-Tor, D. Rabinovich, H. Hope, F. Frolow, E. Appella, and J. L. Sussman, *Nature (London)* **334**, 82 (1988).

65. H. Lauble, R. Frank, H. Blöcker, and U. Heinemann, *Nucleic Acids Res.* **16**, 7799 (1988).

66. M. Miller, R. W. Harrison, A. Wlodawer, E. Appella, and J. L. Sussman, *Nature (London)* **334**, 85 (1988).

67. G. G. Privé, U. Heinemann, S. Chandrasegaran, L.-S. Kan, M. L. Kopka, and R. E. Dickerson, *in* "Structure and Expression, Volume 2: DNA and Its Drug Complexes" (R. H. Sarma and M. H. Sarma, eds.), p. 27. Adenine Press, Albany, New York, (1988).

68. D. Rabinovich, T. Haran, M. Eisenstein, and Z. Shakked, *J. Mol. Biol.* **200**, 151 (1988).

69. M. Teng, N. Usman, C. A. Frederick, and A. H.-J. Wang, *Nucleic Acids Res.* **16**, 2671 (1988).

70. C. Yoon, G. G. Privé, D. S. Goodsell, and R. E. Dickerson, *Proc. Natl. Acad. Sci. U.S.A.* **85**, 6332 (1988).

1989

71. C. A. Bingman, S. Jain, D. Jebaratnam, and M. Sundaralingam, Abstracts, Sixth Conversation in Biomolecular Stereodynamics, Albany, New York, p. 28 (1989).

72. T. Brown, G. A. Leonard, E. D. Booth, and J. Chambers, *J. Mol. Biol.* **207**, 455 (1989).

73. M. A. A. F. de C. T. Carrondo, M. Coll, J. Aymami, A. H.-J. Wang, G. A. van der Marel, J. H. van Boom, and A. Rich, *Biochemistry* **28**, 7849 (1989).

74. M. Coll, J. Aymami, G. A. van der Marel, J. H. van Boom, A. Rich, and A. H.-J. Wang, *Biochemistry* **28**, 310 (1989).

75. M. Coll, D. Saal, C. A. Frederick, J. Aymami, A. Rich, and A. H.-J. Wang, *Nucleic Acids Res.* **17**, 911 (1989).

76. A. D. DiGabriele, M. R. Sanderson, and T. A. Steitz, *Proc. Natl. Acad. Sci. U.S.A.* **86**, 1816 (1989).

77. A. C. Dock-Bregeon, B. Chevrier, A. Podjarny, J. Johnson, J. S. de Bear, G. R. Gough, P. T. Gilham, and D. Moras, *J. Mol. Biol.* **209**, 459 (1989).

78. C. A. Frederick, G. J. Quigley, M.-K. Teng, M. Coll, G. A. van der Marel, J. H. van Boom, A. Rich, and A. H.-J. Wang, *Eur. J. Biochem.* **181**, 295 (1989).

79. R. V. Gessner, C. A. Frederick, G. J. Quigley, A. Rich, and A. H.-J. Wang, *J. Biol. Chem.* **264**, 7921 (1989).

80. U. Heinemann and C. Alings, *J. Mol. Biol.* **210**, 369 (1989).

81. W. N. Hunter, B. L. D'Estaintot, and O. Kennard, *Biochemistry* **28**, 2444 (1989).

82. S. Jain and M. Sundaralingam, *J. Biol. Chem.* **264**, 12780 (1989).

83. S. Jain, G. Zon, and M. Sundaralingam, *Biochemistry* **28**, 2360 (1989).

84. T. A. Larsen, D. S. Goodsell, D. Cascio, K. Grzeskowiak, and R. E. Dickerson, *J. Biomol. Struct. Dyn.* **7**, 477 (1989).

85. Y.-D. Liaw, Y.-G. Gao, H. Robinson, G. A. van der Marel, J. H. van Boom, and A. H.-J. Wang, *Biochemistry* **28**, 9913 (1989).

86. M. H. Moore, W. N. Hunter, B. L. D'Estaintot, and O. Kennard, *J. Mol. Biol.* **206**, 693 (1989).

87. Z. Shakked, G. Guerstein-Guzikevich, F. Frolow, and D. Rabinovich, *Nature (London)* **342**, 456 (1989).

88. M.-K. Teng, Y.-C. Liaw, G. A. van der Marel, J. H. van Boom, and A. H.-J. Wang, *Biochemistry* **28**, 4923 (1989).

89. Y. Timsit, E. Westhof, R. P. P. Fuchs, and D. Moras, *Nature (London)* **341**, 459 (1989).

1990

90. J. Aymami, M. Coll, G. A. van der Marel, J. H. van Boom, A. H.-J. Wang, and A. Rich, *Proc. Natl. Acad. Sci. U.S.A.* **87**, 2526 (1990).

91. D. G. Brown, M. R. Sanderson, J. V. Skelly, T. C. Jenkins, T. Brown, E. Garman, D. I. Stuart, and S. Neidle, *EMBO J.* **9**, 1329 (1990).

92. R. Chattopadhyaya, K. Grzeskowiak, and R. E. Dickerson, *J. Mol. Biol.* **211**, 189 (1990).

(Continued)

107

TABLE IV *(Continued)*

93. G. R. Clark, D. G. Brown, M. R. Sanderson, T. Chwalinksi, S. Neidle, J. M. Veal, R. L. Jones, W. D. Wilson, G. Zon, E. Garman, and D. I. Stuart, *Nucleic Acids Res.* **18**, 5521 (1990).

94. C. Courseille, A. Dautant, M. Hospital, B. Langlois d'Estaintot, G. Precigoux, D. Molko, and R. Teoule, *Acta Crystallogr. Sect A: Found. Crystallogr.* **46**, FC9 (1990).

95. M. Egli, R. V. Gessner, L. D. Williams, G. J. Quigley, G. A. van der Marel, J. H. van Boom, A. Rich, and C. A. Frederick, *Proc. Natl. Acad. Sci. U.S.A.* **87**, 3235 (1990).

96. M. Eisenstein, F. Frolow, L. D. Williams, Z. Shakked, and D. Rabinovich, *Nucleic Acids Res.* **18**, 3185 (1990).

97. C. A. Frederick, L. D. Williams, G. Ughetto, G. A. van der Marel, J. H. van Boom, A. Rich, and A. H.-J. Wang, *Biochemistry* **29**, 2538 (1990).

98. G. A. Leonard, E. D. Booth, and T. Brown, *Nucleic Acids Res.* **18**, 5617 (1990).

99. S. L. Ginell, S. Kuzmich, R. A. Jones, and H. M. Berman, *Biochemistry* **29**, 10461 (1990).

100. Y.-G. Gao, Y.-C. Liaw, H. Robinson, and A. H.-J. Wang, *Biochemistry* **29**, 10307 (1990).

101. Y.-C. Liaw, Y.-G. Gao, H. Robinson, G. M. Sheldrick, L. A. J. M. Sliedregt, G. A. van der Marel, J. H. van Boom, and A. H.-J. Wang, *FEBS Lett.* **264**, 223 (1990).

102. Z. Shakked, G. Guerstein-Guzikevich, A. Zaytzev, M. Eisenstein, F. Frolow, and D. Rabinovich, *in* "Structure and Methods, Volume 3: DNA and RNA" (R. H. Sarma and M. H. Sarma, eds.), p. 55. Adenine Press, Albany, New York, 1990.

103. F. Takusagawa, *J. Biomol. Struct. Dyn.* **7**, 795 (1990).

104. G. D. Webster, M. R. Sanderson, J. V. Skelly, S. Neidle, P. F. Swann, B. F. Li, and I. J. Tickle, *Proc. Natl. Acad. Sci. U.S.A.* **87**, 6693 (1990).

105. L. Van Meervelt, M. H. Moore, P. K. T. Lin, D. M. Brown, and O. Kennard, *J. Mol. Biol.* **216**, 773 (1990).

106. L. D. Williams, M. Egli, G. Ughetto, G. A. van der Marel, J. H. van Boom, G. J. Quigley, A. H.-J. Wang, A. Rich, and C. A. Frederick, *J. Mol. Biol.* **215**, 313 (1990).

107. L. D. Williams, C. A. Frederick, G. Ughetto, and A. Rich, *Nucleic Acids Res.* **18**, 5533 (1990).

108. L. D. Williams, M. Egli, Q. Gao, P. Bash, G. A. van der Marel, J. H. van Boom, A. Rich, and C. A. Frederick, *Proc. Natl. Acad. Sci. U.S.A.* **87**, 2225 (1990).

109. G. Zhou and P. S. Ho, *Biochemistry* **29**, 7229 (1990).

1991

110. G. G. Privé, K. Yanagi, and R. E. Dickerson, *J. Mol. Biol.* **217**, 177 (1991).

111. K. Yanagi, G. G. Privé, and R. E. Dickerson, *J. Mol. Biol.* **217**, 201 (1991).

112. U. Heinemann and C. Alings, *EMBO J.* **10**, 35 (1991).

113. K. Grzeskowiak, K. Yanagi, G. G. Privé, and R. E. Dickerson, *J. Biol. Chem.* **266**, 3861 (1991).

114. T. A. Larsen, M. L. Kopka, and R. E. Dickerson, *Biochemistry* **30**, 4443 (1990).

115. N. Narendra, S. L. Ginell, I. M. Russu, and H. M. Berman, *Biochemistry* **30**, 4450 (1990).

TABLE V
COMPARISON OF HELIX PARAMETERS FOR A- AND B-DNA CRYSTAL STRUCTURES AND THREE MODEL LEFT-HANDED HELICES[a]

1. Base step parameters

Helix	Step	Roll (°)	Tilt (°)	Cup (°)	Slide (Å)	Twist (°)	Rise (Å)	D_{xy} (Å)	Rad_p (Å)
B	All	0.6	0.0	0.0	0.4	36.1	3.36	3.5	9.4
A	All	6.3	—	—	−1.6	31.1	2.9	—	9.5
Z_I	C-G	−5.8	0.0	12.5	5.4	−9.4	3.92	5.0	6.3
Z_{II}	C-G	−5.4	0.0	12.3	5.2	−10.2	3.88	4.8	6.1
W	C-G	0.2	0.0	7.8	−0.7	−68.4	3.93	6.0	5.6
Z_I	G-C	5.8	0.0	−12.5	−1.1	−50.6	3.51	6.0	7.3
Z_{II}	G-C	5.4	0.0	−12.3	−0.4	−49.8	3.55	5.5	8.0
W	G-C	−0.2	0.0	−7.8	−0.0	+8.4	3.50	1.3	9.1

2. Base pair parameters

Helix	Base	Tip (°)	Incl (°)	Prop (°)	Buck (°)	X Dsp (Å)	Y Dsp (Å)	P–P (Å)
B	All	0.0	2.4	−11.1	−0.2	0.8	0.1	8.8–14.0
A	All	11.0	12.0	−8.3	−2.4	−4.1	—	11.5–15.9
Z_I	C	2.9	−6.2	−1.3	−6.2	3.0	−2.3	13.7
Z_{II}	C	2.7	−6.2	−0.4	−6.2	3.6	−2.1	18.1
W	C	−0.1	−5.8	−22.7	−3.9	−1.6	−0.3	13.3
Z_I	G	−2.9	−6.2	−1.3	6.2	3.0	2.3	7.7
Z_{II}	G	−2.7	−6.2	−0.4	6.2	3.6	2.1	8.6
W	G	0.1	−5.8	−22.7	3.9	−1.6	0.3	5.7

3. Main chain torsion angles (°)

Helix	Base	α	β	γ	δ	Sugar	Pseu	ϵ	ζ	χ	Glyc
B	All	295	167	51	129	2′en	146	203	240	257	anti
A	All	287	173	64	78	3′en	20	209	283	195	anti
Z_I	C	223	221	55	138	2′en	152	266	80	201	anti
Z_{II}	C	146	163	67	147	2′en	163	259	74	213	anti
W	C	316	313	198	82	3′en	35	188	64	207	anti
Z_I	G	48	179	190	100	3′en	−1	256	291	67	syn
Z_{II}	G	93	193	157	93	3′en	51	182	55	57	syn
W	G	270	248	223	160	2′en	173	279	160	12	syn

[a] Parameters are calculated with the NEWHEL91 helix analysis program, a copy of which can be obtained from the author on request. Helix variables are defined in the NEWHEL91 writeup, and in Refs. 7, 8, and T15. B, Average over four B-DNA decamer single-crystal structure analyses (see Refs. T110 and T113); A, averages over six A-DNA octamer single-crystal structure analyses (see Ref. T96); Z and W, model coordinates from Brookhaven Protein Data Bank. P–P, Shortest P–P distance across minor groove. For a helix with N base pairs, shortest oppositions are at phosphates as follows: B helices, n versus $(2N + 5) - n$, as P5/P20 in a decamer; Z helices, n versus $(2N + 4) - n$, as P2/P14 in a hexamer; W helix, n versus $(2N) - n$, as P3/P29 in a 16-mer; and A helix, *Major* groove width estimated from terminal 5′ separations in octamers. Pseu, Sugar pseudorotation angle; Sugar, sugar pucker (2′en, C2′-endo; 3′en, C3′-endo; etc.); Glyc, glycosyl bond geometry (syn or anti).

TABLE VI

NONCRYSTALLOGRAPHIC SYMMETRY IN LOCAL HELIX PARAMETERS[a]

Helix	Res (Å)		Roll (°)	Cup (°)	Twist (°)	Rise (Å)	D_{xy} (Å)	Incl (°)	Prop (°)	Buckle (°)
KK	1.5	AV	0.32	1.65	37.14	3.25	3.70	0.70	−11.48	−0.62
		SD	4.04	8.69	4.11	0.47	0.51	1.24	5.24	6.97
+6		DI	2.20	3.77	1.45	0.19	0.20	1.08	3.55	3.17
M7	2.3	AV	−2.29	−1.80	36.01	3.37	3.55	−2.31	−15.00	−2.95
		SD	5.77	14.26	4.67	0.55	0.66	3.73	7.85	10.90
+3		DI	5.02	6.47	3.36	0.11	0.28	4.31	4.79	6.37
LA	2.5	AV	0.61	0.12	35.98	3.33	3.53	0.68	−14.16	−0.42
		SD	4.37	6.75	4.28	0.41	0.61	5.70	4.82	4.90
+1		DI	3.47	5.56	3.96	0.20	0.54	9.27	5.14	4.81
MD	2.0	AV	0.01	−0.52	35.80	3.39	3.53	0.00	−10.79	−2.70
		SD	5.71	7.97	4.80	0.43	0.70	6.05	5.92	4.83
+1		DI	4.06	5.20	2.26	0.18	0.39	8.46	7.47	5.92
DR	1.9	AV	0.32	−0.37	35.86	3.34	3.56	0.72	−12.14	−0.35
		SD	5.13	7.58	4.09	0.45	0.64	5.86	5.72	5.27
0		DI	4.36	5.57	2.46	0.16	0.44	9.36	4.12	5.23
DW	1.9	AV	0.81	−0.42	35.74	3.35	3.56	0.69	−11.65	−0.72
		SD	5.72	8.53	3.29	0.46	0.61	6.44	4.96	5.56
0		DI	4.12	9.47	3.42	0.18	0.48	10.56	2.44	4.88
NE	2.5	AV	0.48	1.66	36.72	3.30	3.59	0.12	−14.83	−0.61
		SD	5.79	9.28	3.71	0.40	0.74	4.67	7.12	7.12
0		DI	6.07	6.78	2.42	0.27	0.50	6.85	6.26	5.14
N1	2.4	AV	1.24	0.83	36.16	3.35	3.58	1.35	−9.43	−0.29
		SD	6.25	14.14	6.45	0.63	0.71	4.42	4.87	8.17
−1		DI	5.79	6.70	5.24	0.27	0.63	6.58	3.25	5.30
CO	2.2	AV	0.18	0.85	36.60	3.30	3.67	0.32	−14.72	−1.28
		SD	5.56	13.40	5.19	0.45	0.63	4.64	5.64	9.65
−5		DI	6.37	12.43	4.50	0.34	0.49	6.87	5.72	8.24
YO	2.2	AV	−0.54	1.12	36.10	3.32	3.51	−0.26	−13.75	0.26
		SD	9.57	11.07	4.29	0.36	0.55	6.15	6.60	8.59
−6		DI	10.77	11.28	2.76	0.33	0.54	9.03	10.34	9.20

[a] *AV*, Average value of parameter along the helix; *SD*, standard deviation of parameter; DI, mean difference between parameter values at opposite ends of helix. The smallest two *DI* values per column are double underlined, and the largest two are italicized. Sequences are listed in descending order of net score (lower left) = (# lowest two *DI*) − (# highest two *DI*). Oligonucleotide sequences are as follows:

		Ref.
KK	C-G-A-T-C-G-A-T-C-G	T113
M7	C-G-C-G-A-A-T-T-[5me]C-G-C-G	T15,T21
LA	C-G-T-G-A-A-T-T-C-A-C-G	T114,T115
MD	C-G-C-G-A-[6me]A-T-T-C-G-C-G	T63
DR	C-G-C-G-A-A-T-T-C-G-C-G	T6
DW	C-G-C-G-A-A-T-T-C-G-C-G	T59
NE	C-G-C-A-A-A-A-A-A-G-C-G	T56
N1	C-G-C-G-A-T-A-T-C-G-C-G (+ Net)	T74
CO	C-G-C-A-A-A-T-T-T-G-C-G (+ Dis)	T50
YO	C-G-C-A-T-A-T-A-T-G-C-G	T70

TABLE VII
CUMULATIVE TALLY OF PUBLIC DEPOSITION OF DNA CRYSTAL
STRUCTURE DATA (F_o) AND RESULTS (XYZ), FROM ALL
LABORATORIES

Year published	XYZ and F_o	XYZ only	XYZ total	Nothing released	All structures
1978	0	0	0	1	1
1979	0	1	1	1	2
1980	2	2	4	2	6
1981	3	3	6	3	9
1982	6	5	11	5	16
1983	6	5	11	5	16
1984	7	6	13	6	19
1985	10	11	21	10	31
1986	12	16	28	10	38
1987	15	21	36	12	48
1988	19	23	42	17	59
1989	22	33	55	24	79
1990	24	44	68	30	98
1991	27	45	72	31	103

[6] A-DNA in Solution as Studied by Diverse Approaches

By VALERY I. IVANOV and DMITRY YU. KRYLOV

Introduction

A-DNA is a high-energy conformation of the double helix under physiological conditions. However, it can be stabilized by a decrease in water activity in drying fibers or by the presence of alcohols in aqueous solutions. This conformation may be biologically interesting.[1] First, it is a stable form of double-helical fragments of RNA and thereby DNA–RNA hybrid duplexes. (This is due to an extra hydroxyl group on the ribose that cannot fit the tight space allotted to it in the B form.) Second, the local B-to-A transition of DNA has been repeatedly suggested to occur during transcription. Third, proteins exist (e.g., a protein from sporulating bacteria) that transform DNA into the A form when complexed.[2]

[1] V. I. Ivanov, L. E. Minchenkova, E. E. Minyat, M. D. Frank-Kamenetskii, and A. K. Schyolkina, *J. Mol. Biol.* **87,** 817 (1974).
[2] S. C. Mohr, N. V. H. A. Sokolov, Ch. He, and P. Setlow, *Proc. Natl. Acad. Sci. U.S.A.* **88,** 77 (1991).

METHODS IN ENZYMOLOGY, VOL. 211

Preparation of A-Form DNA in Solution

DNA Samples

Various natural and synthetic DNAs or oligodeoxyduplexes may be used. In the cases of high molecular weight natural DNAs, they should be sheared down to approximately 10^5 Da by repeated passage through a syringe (to prevent aggregation of DNA samples in the media containing a high alcohol content used in the experiments with A-DNA). To remove heavy metal and divalent cations the DNA is subjected to ethanol precipitation in the presence of 10 μM NaEDTA and 0.1 M NaCl. Then the DNA is dissolved in deionized water containing 1 mM NaCl to a final concentration of about 1 mg/ml. (To estimate the DNA concentration, the molar absorption coefficient of 6600 at 259 nm is used.) From this stock solution the sample solutions having the desired DNA concentration (no more than 100 μg/ml) are prepared.

Other Components

To avoid the complications due to the effects of extra ions on the B–A transition, unbuffered solutions are recommended. Use double-distilled and deionized water and alcohols of spectroscopic grade.

Procedure

To prepare A-DNA the following procedure is used.[1,3] To a test tube add the calculated quantity of DNA from the stock solution, then NaCl, and water. Finally, add the required quantity of alcohol stepwise with careful stirring after each addition. In this way DNA can be dissolved in 80% ethanol without precipitation provided that sodium concentration does not exceed 1 mM.

Check the occurrence of A-DNA by dichroism; the A form must have a molar circular dichroism per nucleotide, $\Delta\varepsilon$, at the long-wavelength maximum (~270 nm) no less than 8 (compared to $\Delta\varepsilon = 2-3$ for the long-wavelength maximum for B-DNA). It should be stressed that in a water–organic solution of DNA even traces of the foreign cationic contaminations can perceptibly modify the data obtained.

When presenting the results, the corresponding values of water activity are preferable to the percentage of alcohol in many cases; we have shown that the free energy difference between the B and A forms is a linear

[3] V. I. Ivanov, L. E. Minchenkova, A. K. Schyolkina, and A. I. Poletayev, *Biopolymers* **12**, 89 (1973).

TABLE 1
WATER ACTIVITY AT DIFFERENT ETHANOL CONCENTRATIONS

Ethanol (%)	Water activity (r.h. %)	Ethanol (%)	Water activity (r.h. %)	Ethanol (%)	Water activity (r.h. %)
1	99.69	35	89.96	68	80.46
2	99.38	36	89.70	69	80.02
3	99.07	37	89.44	70	79.56
4	98.77	38	89.18	71	79.06
5	98.46	39	88.92	72	78.54
6	98.16	40	88.66	73	77.98
7	97.86	41	88.40	74	77.37
8	97.56	42	88.15	75	76.73
9	97.26	43	87.89	76	76.02
10	96.96	44	87.63	77	75.27
11	96.66	45	87.38	78	74.44
12	96.37	46	87.12	79	73.54
13	96.08	47	86.86	80	72.56
14	95.78	48	86.61	81	71.49
15	95.49	49	86.35	82	70.31
16	95.20	50	86.09	83	69.02
17	94.91	51	85.82	84	67.59
18	94.63	52	85.56	85	66.01
19	94.34	53	85.29	86	64.26
20	94.06	54	85.02	87	62.32
21	93.77	55	84.75	88	60.16
22	93.49	56	84.47	89	57.75
23	93.21	57	84.19	90	55.06
24	92.93	58	83.90	91	52.05
25	92.66	59	83.61	92	48.67
26	92.38	60	83.31	93	44.88
27	92.11	61	83.00	94	40.61
28	91.83	62	82.67	95	35.78
29	91.56	63	82.34	96	30.33
30	91.29	64	82.00	97	24.14
31	91.02	65	81.64	98	17.12
32	90.76	66	81.26	99	9.12
33	90.49	67	80.87	100	0.00
34	90.23				

function of water activity.[4] Table I lists the corresponding values for the binary ethanol–water mixtures according to data compiled and interpolated by Malenkov and Minasyan.[5] We have presented experimental evi-

[4] L. E. Minchenkova, A. K. Schyolkina, B. K. Chernov, and V. I. Ivanov, *J. Biomol. Struct. Dyn.* **4**, 463 (1986).
[5] G. G. Malenkov and K. A. Minasyan, *Mol. Biol. (U.S.S.R.)* **11**, 352 (1977).

dence that using another alcohol, trifluoroethanol, to study the B–A transition does not alter these figures significantly.[4]

There are six methods (approaches) which allow one to study various characteristics of the B–A transition of DNA in solution: (1) circular dicroism (CD), which is the most convenient technique for obtaining the fraction of the A form in a DNA sample[1,6]; (2) A-DNA-specific ligands, which can be used either to reveal a preexisting A form[7] or to make the DNA acquire the A conformation; (3) three-state phase diagrams (A form, B form, coil), which allow one to "extract" thermodynamic parameters of the B–A transition[8]; (4) orientation of DNA in flow, which testifies to the emergence of the B–A junctions within a long DNA molecule undergoing the B-to-A shift[9]; (5) relaxation of superhelical DNA due to B–A change, which makes it possible to estimate directly the change in winding angle[10]; and (6) photofootprinting, which provides the opportunity to study the B–A transitions along a DNA molecule at nucleotide level resolution.[11]

Application of Circular Dichroism to Studying the B–A Transition

Values Relevant to Circular Dichroism

A parameter, measured by circular dichroism, is the difference in absorptions, ΔA, of left- and right-circularly polarized light:

$$A_L - A_R = \Delta A \tag{1}$$

Dividing this value by the optical path of light within the cell, l (in cm), and the concentration of the sample, d (in M), the molar CD, $\Delta\varepsilon$, is obtained:

$$\Delta\varepsilon = \Delta D/dl \tag{2}$$

The CD parameters are sometimes given in terms of ellipticity, θ, in grad:

$$\theta = 3300\ \Delta\varepsilon \tag{3}$$

The dependence of ΔD or $\Delta\varepsilon$ on wavelength, λ (in nm), is called a CD spectrum. (For details of CD methods and CD spectra of various conformations of DNA, see [19] in this volume.)

[6] J. Brahms and M. F. Mommaerts, *J. Mol. Biol.* **10**, 73 (1964).
[7] H.-Y. Mei and J. K. Barton, *Proc. Natl. Acad. Sci. U.S.A.* **85**, 1339 (1988).
[8] V. I. Ivanov, D. Yu. Krylov, and E. E. Minyat, *J. Biomol. Struct. Dyn.* **3**, 43 (1985).
[9] S. K. Zavriev, L. E. Minchenkova, M. D. Frank-Kamenetskii, and V. I. Ivanov, *Nucleic Acids Res.* **5**, 2657 (1978).
[10] D. Yu. Krylov, V. L. Makarov, and V. I. Ivanov, *Nucleic Acids Res.* **18**, 759 (1989).
[11] M. M. Becker and Zh. Wang, *J. Biol. Chem.* **264**, 4163 (1989).

Obtaining B–A Transition Curve with CD Method

The curve for the B–A transition is the change in the A-form fraction, [A]/[A + B], as a function of an effective factor, such as alcohol volume percent or the corresponding value of water activity (relative humidity) (see Table I). Record the CD spectrum (or its magnitude at a selected wavelength, e.g., at 270 nm) for the A-DNA, prepared as discussed above. Typically, the absorbancy within the spectral interval of interest should not exceed a value of about 1.0 to avoid distortion of the CD spectra. Note that ordinary free DNA is present totally as the A-form in a 78% (v/v) ethanol–water or trifluoroethanol (TFE)–water solution. Then dilute this initial sample stepwise with small volumes of a water–NaCl solution, each time recording the CD spectrum or its magnitude. The water aliquots should be of such a volume as to reduce the alcohol concentration by approximately 1%, as the B–A transition interval is rather narrow, approximately 6%.

In this way a set of CD spectra, like that in Fig. 1, is obtained. The existence of an isodichroic point usually means that just two species are in equilibrium. Plotting the value of the A-form fraction,

$$\theta = (\Delta\varepsilon - \Delta\varepsilon_{\mathbf{B}})/(\Delta\varepsilon_{\mathbf{A}} - \Delta\varepsilon_{\mathbf{B}}) \tag{4}$$

the B–A transition curve can be built. The curve is characterized by two

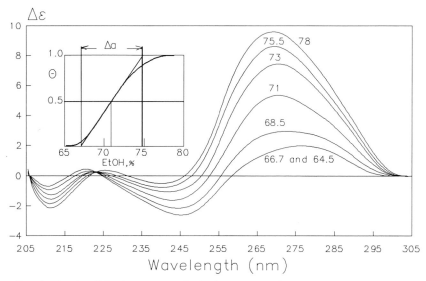

FIG. 1. Circular dichroism spectra of calf thymus DNA in water–ethanol solutions. The spectra are labeled with the volume percent of ethanol. (Inset) Curve of the B–A transition based on the change in the CD magnitude at 270 nm. (For conditions, see text.)

parameters: transition point, a_o, and width, Δa. These parameters are self-explanatory (see fig. 1).

A-DNA-Specific Ligands

An example has been given of a chiral metal–organic complex that specifically binds and cleaves A-form polynucleotides.[7] It can probably be used to reveal the preexisting A conformation or one easily transformed into the A form. Other examples of obvious biological interest are the small acid-soluble spore proteins (SASP) from bacteria that were recently shown by CD and IR spectroscopy to transform DNAs of different origin from the B to the A form. This protein protects DNA of spores from UV damage, since the A form has been known to be much more resistant to UV than the B form.

It is probable, however, that most of the A-DNA-specific ligands will not be able to make DNA acquire the A form in an aqueous solution; rather, they will only shift the B–A equilibrium to the side of A form, thereby facilitating the B–A transition, induced by alcohol. The polyamines spermine and spermidine are an example of this equilibrium shift.[12] We recommend the following simple procedure for testing a substance for this transition ability.

Procedure

Prepare a DNA solution with an ethanol content near the B–A transition point, approximately 72% (v/v); $\Delta\varepsilon$ (270 nm) must be approximately 4. The appropriate conditions include a concentration of DNA of about 15 μg/ml; a 1-cm cell; a NaCl concentration of about 0.3 mM, the nonbuffer solution; and room temperature (B–A equilibrium does not depend on temperature). Set the wavelength at 270 nm in the spectrometer. Add small aliquots of the ligand solution to the cell, with the resulting concentration being 1 μM for each aliquot in the cell. Record the CD spectrum after each addition.

The CD magnitude will rise if a ligand is A-DNA specific, and the CD will drop if the ligand is B-DNA specific. The method is extremely sensitive, since at the transition point the free energies of the B and A forms are equal, and even very small preferences of a ligand for the A (or B) form will be seen at once. Figure 2 illustrates this procedure. The method has limitations; it does not work when a high alcohol content interferes with ligand

[12] E. E. Minyat, V. I. Ivanov, A. M. Krytzsyn, L. E. Minchenkova, and A. K. Shyolkina, *J. Mol. Biol.* **128**, 397 (1978).

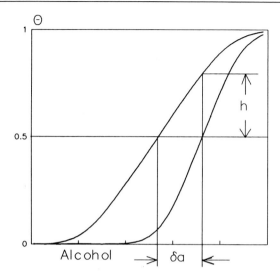

FIG. 2. Estimating the parameters of cooperativity with the use of small ligands. The righthand curve is for the B-to-A transition in free DNA, and the lefthand one is for the DNA solution in the presence of the ligand, specific for A-DNA.

binding to DNA. Note that the method is not restricted to the B–A transition; it can be applied to other cooperative transitions as well (e.g., B–Z, helix–coil).

Quantitative Parameters Relevant to B–A Transition

DNA participating in a cooperative transition is schematically depicted in Fig. 3 (for the model having two states, A and B). In fact, an additional state must also arise, the B–A junction. At the transition point the B and A forms are isoenergetic and the number of junctions is a maximum. Their linear density on DNA, j, is maximum and determined by the free energy, F_J:

$$j = \exp(-F_J/RT) \tag{5}$$

A reciprocal value, that is, mean interjunctional distance or mean length of the A (or B) fragments at the transition point, is called cooperativity length, v:

$$1/j = v = \exp(F_J/RT) \tag{6}$$

The greater the junction energy, the more cooperative the transition is, and the transition curve will have less width. The width of the transition, Δa, is

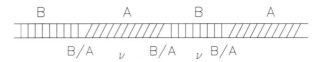

FIG. 3. Scheme of DNA structure within the B–A transition interval (see text).

also controlled by a parameter Q, which relates the change in free energy between the A and B forms, $F_A - F_B$, in the vicinity of the transition point, a_o, as a function of water activity, a:

$$F_A - F_B = \frac{RT}{Q} (a_o - a)$$
(7)

Thus, Δa approximately equals Q/v; the statisticomechanical theory for the B–A transition gives the following equation (e.g., Ivanov et al.[1]):

$$\Delta a = 4Q/v$$
(8)

Determining Cooperativity Length, v

The cooperativity length may be determined by the use of ligands specific for A-DNA (or B-DNA). As shown in Fig. 2, in the presence of the ligand concentration, c (moles/mole of DNA base pairs), the B–A transition curve shifts relative to that of the free DNA by a value δa_c and widens by a value $\delta \Delta a$. As shown in Ref. 1,

$$\delta a_c = 2Qc$$
(9)

$$\delta \Delta a = 4Qc$$
(10)

provided that the ligand-binding constant of one of the DNA forms is at least 10 times larger than that of the other. Vice versa, if the widening between the two curves is 2 times larger than the shift of the transition point, then the requirement for the constants is fulfilled. (This condition is not burdensome. In most real cases it is fulfilled.) For low ligand concentration, namely, for $c \ll 1$, the v value can be estimated[12]:

$$v = 2\delta a_c/c\Delta a$$
(11)

Here Δa stands for the width of the B–A transition curve in the absence of a ligand.

It follows from Fig. 2 that Eq. (11) may be rewritten as

$$v \approx 2h/c$$
(12)

for low h values. (In practice, it is sufficient to have $h < 0.1$). Using various ligands and DNAs, the v value for the B–A transition was found to be within the limits of 10–30 base pairs[1,12] that corresponds to a junction energy, F_J, of 1.4 to 2 kcal/mol [Eq. (6)]. The accuracy of v and Q estimation by this method is not high.

Determining Q Values with Greater Precision

An unavoidable defect of the ligand method for determining the v and Q parameters is due to the involvement of a foreign molecule, the ligand, whose size is comparable to the cooperativity length of the B–A transition. This obstacle can be overcome by an approach that involves no ligand at all, but uses the short DNA fragments. If the size of the fragment were equal to v or less, the duplex would change into the A form according to an all-or-none principle, that is, there could be no B–A junctions. In this case a simple equation can be derived (e.g., see Ref. 4):

$$\Delta a = 4Q/(n - 1) \tag{13}$$

where n is the number of base pairs of the fragment used and Δa is the width of the B–A transition curve of this fragment. Figure 4 shows the data we obtained with several synthetic decanucleotide duplexes having different nucleotide sequences.[4] All the transitions have the same width: about 6% in terms of relative humidity, giving Q a value of 13% relative humid-

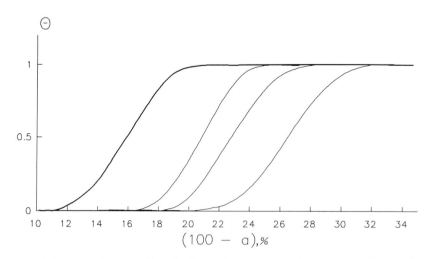

FIG. 4. B–A transition profiles of self-complementary decadeoxyduplexes. From left to right: CCCCCGGGGG, ATACCGGTAT, GCTACGTAGC, CCAAGCTTGG (only one strand is listed). Note that the first duplex (marked by a heavy line) has the lowest B–A transition point and has the highest number of CC dinucleotide steps (contacts).

ity. Taking into account that for a typical DNA (calf thymus) Δa is about 3% relative humidity, one obtains v equal to 16 base pairs [Eq. (8)].

We have thus considered two independent methods for determining the two general energetic parameters that describe the B–A transition (in the model of two states), F_J ($=RT \ln v$) and RT/Q. Now we turn to the third method.

Energetics of B–A Change as Revealed by Three-State Diagrams

The double helix, whatever its form, A or B, melts with an increase in temperature, T. Hence, a point (more correctly, a small region) should exist in the plane (T, a) where all three phases, A, B, and coil, are in equilibrium with each other. So, a phase diagram (A, B, coil) can be built.

Obtaining a Diagram of (A, B, Coil) in Experiments

As an example we shall consider the case of an alternating polynucleotide, poly[d(A-T)].[8,13] Although the diagram pattern for natural DNAs is very similar to that for poly[d(A-T)], this polynucleotide has two advantages: first, it is a homopolymer, so that no questions arise regarding complications due to sequence effects; second, because of easy renaturation (unlike the DNAs) a whole diagram may be obtained with one and the same sample. This reduces the random experimental error in the melting point to $0.2°$, whereas it is $1°$ for the DNAs.

Experimental Conditions

Poly[d(A-T)] from P-L Biochemicals (Madison, WI) without additional purification, doubly distilled water, trifluoroethanol (TFE) from Sigma (St. Louis, MO), and EDTA were used. Concentrations in the sample are as follows: NaCl, 1 mM; EDTA, 0.5 mM; and poly[d(A-T)] within the limits of about 20 to 10 μg/ml during experiment, because of dilution due to the addition of water or TFE aliquots. The use of TFE instead of ethanol and EDTA in a concentration of 0.5 mM is essential; otherwise the branch for A–coil equilibrium is distorted, probably because of aggregation. Because TFE content varies during the experiment, two samples of poly[d(A-T)] should be prepared: one for the TFE interval from 0 to 50% (v/v) and the other for the interval from 50 to 80%. To estimate the polynucleotide concentration, use a molar absorbancy of 6700 at 260 nm.

[13] V. I. Ivanov, D. Yu. Krylov, E. E. Minyat, and L. E. Minchenkova, *J. Biomol. Struct. Dyn.* **1,** 453 (1983).

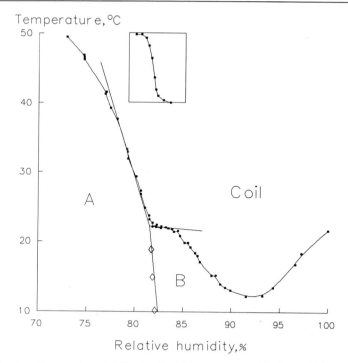

FIG. 5. Phase diagram for poly[d(A-T)]. (Inset) Corresponding B–A transition curve. Note that the B–A transition point, measured by CD, coincides with the break point in the helix–coil branch. For explanation, see text.

The samples are melted in a 0.5 cm cell at a rate of 0.7°/min in a spectrophotometer, set at a wavelength 260 nm. After each melting, the sample is allowed to cool down for 1 hr to 8–10° to get complete renaturation. Then the next aliquot of TFE (or water solution) is added, and a new melting cycle begins. The branch for B–A equilibrium is obtained by the circular dichroism method as described above. As a result the three-state phase diagram is obtained (Fig. 5).

Obtaining Q from Three-State Diagrams

From Fig. 5, the A-helix, B-helix, and coil are stable within the three provinces. The boundaries between these states are determined by Eq. (14)–(16):

$$F_A(a, T) - F_{coil}(a, T) = 0 \tag{14}$$

$$F_B(a, T) - F_{coil}(a, T) = 0 \tag{15}$$

$$F_A(a, T) - F_B(a, T) = 0 \tag{16}$$

Equation (16) is the difference between Eqs. (14) and (15) and connects thermodynamic parameters of the B–A transition with those of the helix–coil.

In the vicinity of the triple point, specified by the condition $F_A(a_o, T_o) = F_B(a_o, T_o) = F_{coil}(a_o, T_o)$, the free energy differences can be converted to a series, preserving only the linear terms[14]:

$$F_A - F_{coil} = RT[\beta_A(a_o - a) + \alpha_A(T_o - T)] \tag{17}$$

$$F_B - F_{coil} = RT[\beta_B(a_o - a) + \alpha_B(T_o - T)] \tag{18}$$

Here α_A, α_B, β_A, and β_B are constants.

Because the B–A equilibrium depends only slightly (if at all) on temperature (see the nearly vertical orientation of the B–A branch in Fig. 5), then $\alpha_A = \alpha_B = \alpha$ with good accuracy. On the other hand, $F_{helix} - F_{coil}$ is expressed through the enthalpy of polynucleotide melting, ΔH:

$$F_{helix} - F_{coil} = \Delta H(T - T_o)/T_o \tag{19}$$

Comparing Eqs. (17), (18), and (19), one obtains

$$\alpha = -\Delta H/RT_o^2 \qquad (\Delta H > 0) \tag{20}$$

Hence

$$\beta_A = [(T_o - T)/(a_o - a)][\Delta H/RT_o^2] \qquad (a > a_o) \tag{21}$$

$$\beta_B = [T_o - T)/(a_o - a)][\Delta H/RT_o^2] \qquad (a < a_o) \tag{22}$$

The free energy difference between the A and B form can be written as:

$$F_A - F_B = RT(\beta_A - \beta_B)(a_o - a) = RT/Q(a_o - a) \tag{23}$$

Because the coefficients β_A and β_B are proportional to the slopes of the branches A–coil and B–coil, the Q value can be estimated from the slopes of the branches in the vicinity of the triple point of the experimentally obtained phase diagram, if the ΔH value is known. According to calorimetric determinations, $\Delta H = 7-8$ kcal/mol.[13] For $Q = 1/\beta_A - \beta_B$, one obtains a value of about 8% relative humidity, and for $v = 4Q/\Delta a$, the value of about 30 base pairs, since $\Delta a = 1.1\%$ relative humidity for poly[d(A-T)].

[14] M. D. Frank-Kamenetskii and G. I. Chogovadze, J. Biomol. Struct. Dyn. 1, 1517 (1984).

The two basic parameters of the cooperative B–A transition in the homopolynucleotide are thus similar to those for natural DNA.

To complete this section, consider the general case where a transition, for example, B–Z or, to some extent, B–A for poly[d(A-T)], depends on temperature. Then, one more equation will be needed to connect the parameters found with the inclination, for example, the B–A branch, if one considers the B–A transition. Now, $\alpha_A \neq \alpha_B$, and one obtains

$$\delta a/\delta T \text{ (for B–A branch)} = RT(\alpha_A - \alpha_B)/(\beta_A - \beta_B) \tag{24}$$

The Q value is determined from Eq. (25):

$$Q = \alpha_B^{-1}[(1 - \delta a/\delta T_{B-A})(\delta T/\delta a_{A-coil})]/(\delta T/\delta a_{A-coil} - \delta T/\delta a_{B-coil}) \tag{25}$$

Revelation of B–A Junctions by Orienting DNA in Flow

The flow method for studying the B–A transition is based on the change in the hydrodynamic properties of DNA. The B–A junctions (whose existence is a direct consequence of the cooperativity) appear as a distortion of the regular helix and have an increased energy. Hence DNA is expected to have increased flexibility at these sites and, probably, a bend. Experimentally this can be studied by observing changes in specific viscosity, sedimentation, gel electrophoresis, etc. We shall present the orientation method that is most convenient.[9]

Procedure

To orient DNA, a cell, similar to that described by Chung and Holzwarth,[15] may be used. Glass capillaries 33 mm in length with an internal diameter of 0.5 mm are placed along the optical path; the total length of the optical path is 35 mm. The solution is pumped back and forth through the capillaries using a syringe, connected by a silicon tube to the input of the cell. The cell is set in a spectrophotometer, and the degree of DNA orientation, b, is measured by the increase in the absorption at 260 nm, δA_f,

$$b = 2\delta A_f/A \tag{26}$$

A sample of A-DNA, prepared as described above, in 82% ethanol (v/v), is diluted stepwise with the water–NaCl solution to achieve the A-to-B transition. At each step the orientation value, b, and circular dichroism magnitude are measured.

[15] S.-Y. Chung and G. Holzwarth, *J. Mol. Biol.* **92**, 449 (1975).

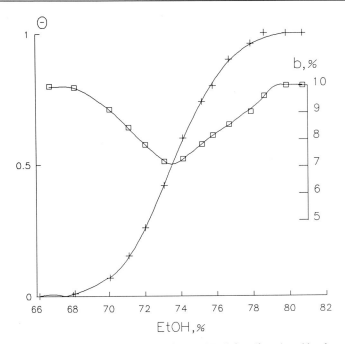

FIG. 6. The B–A transition of DNA as studied by CD (left ordinate) and by the method of DNA orientation in flow (V-shaped curve, right ordinate).

DNA Orientation within B–A Transition Interval

As a result, two curves for one sample are obtained (Fig. 6): the B–A transition curve (left ordinate) and the V-shaped curve for the orientation value (right ordinate). Clearly, the orientation minimum coincides with the B–A transition point, where the number of the B–A junctions should be maximum. We thus have an independent method to trace the B–A conformational change. As illustrated clearly in Fig. 7, the shift of the transition curve in the presence of a stabilizer of the B form (guanidinium cation) is accompanied by a corresponding shift in the orientation minimum position.[16]

[16] V. I. Ivanov, L. E. Minchenkova, E. E. Minyat, and A. K. Schyokina, *Cold Spring Harbor Symp. Quant. Biol.* **47**, 243 (1983).

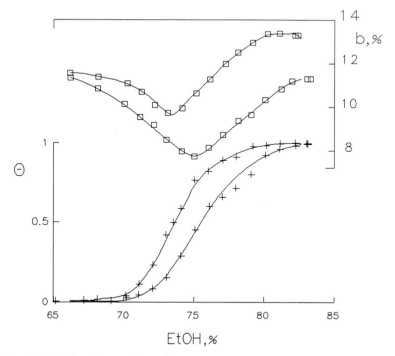

FIG. 7. Inhibiting the transition of DNA to the A form in the presence of guanidine hydrochloride. Note that guanidine (at a concentration of 0.1 mM) shifts the CD B–A transition curve to the right from that of free DNA, thereby stabilizing the B form. This is also seen in the B–A transitions measured by the orientation method (two upper V-shaped curves).

To complete discussion of the orientation method, a recent important improvement should be mentioned (Krylov *et al.*[10]). A linear dicroism (LD) attachment to the dichrograph is used, and a flow cell rather than a multicapillary cell is also used, which greatly increases sensitivity. A continuous change in the fraction of trifluoroethanol is provided by the continuous administration of the water component into the cell for measuring the LD of the DNA oriented in flow, so that smooth (rather than punctuated) curves of the B–A transition are obtained. When applied to the study of superhelical DNA relaxation in the course of the B-to-A shift, these innovations allow for the estimation of the average winding angle of A-DNA in solution. This angle was shown to coincide with that in fibers and crystals (32.6°).[10]

B−A transition point, r.h.%

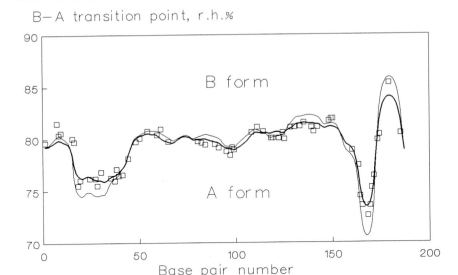

FIG. 8. Positions of the B–A transition point (in r.h. %) along the DNA molecule at nucleotide level resolution, estimated by Becker and Wang[11] (squares), and the theoretical line, calculated by us with the use of the following parameters: the transition point for AA/TT step is at 66%; that for CC/GG step is 88.3%; and that for the remaining eight steps is supposed to be one and the same, and equals 81%. $v = 20$ (thin line) and 30 (heavy line) base pairs; $Q = 10\%$. (See text for more information.)

Ultraviolet Footprinting as Radical Breakthrough in B–A Studies

Becker and Wang,[11] using the UV footprinting technique, have been able to "visualize" the B–A transition at nucleotide level resolution. The procedure they developed[11] is based on the fact that A-DNA is about 10 times less sensitive to UV damage than B-DNA. By irradiating a definite DNA fragment in the presence of different TFE concentrations with subsequent chemical cleavage of the photoproducts and gel electrophoresis, one obtains the B–A transition curve for each nucleotide in the DNA fragment under investigation.

Figure 8 shows the results of this work for a fragment of the 5 S rRNA gene, presented in such a way as to demonstrate in an explicit form the sequence effect in the B–A transition. The right end of this fragment contains a string of 8 T residues in one strand, then, after a 2-base pair interval, a string of 12 C residues. The first string is highly B-philic, whereas the second one is highly A-philic.

In this chapter, we have only touched on the problem of sequence effect on the B–A transition. Additional study is needed. For the current status of this problem see [17] in this volume and Minchenkova *et al.*[4]

Acknowledgment

This work was supported by the Institute of New Technologies in Moscow.

[7] Generation and Detection of Z-DNA

By Brian H. Johnston

Introduction

Z-DNA is the only known helical conformation of DNA that is left-handed. The Z-DNA helix is longer and thinner than B-DNA, and the bases lie relatively farther from the helix axis, creating a single deep, narrow groove. The backbone follows a zigzag pattern, and the conformation of the glycosidic bond alternates between syn and anti. In addition to dictating what factors promote Z-DNA formation, these structural properties of the helix also permit its detection: (1) The left-handed helical sense means that any torsional stress directed opposite the right-handed helical sense of B-DNA (i.e., negative supercoiling) will tend to favor the formation of Z-DNA. (2) The zigzag pattern of the backbone results in some phosphate–phosphate distances smaller than those in B-DNA; consequently, high salt concentrations tend to stabilize Z-DNA. (3) The syn–anti alternation dictates that alternating purine–pyrimidine sequences most easily form Z-DNA because pyrimidines favor the anti orientation, whereas purines can accommodate either the syn or the anti conformation. (4) As a result of the displacement of the bases from the helix axis, Z-DNA can be stabilized by adding bulky groups at sites that are sterically hindered in B-DNA but become more accessible in the Z conformation; similarly, certain chemicals that react at those accessible sites can be used to recognize Z-DNA. (5) The overall conformational differences between B-DNA and Z-DNA permit the two structures to be distinguished by many nucleases and DNA-binding proteins, which generally recognize right-handed DNA, and by antibodies specifically directed against Z-DNA.

The objective of this chapter is to provide an overview of the techniques available for inducing the formation of Z-DNA and detecting it. This chapter is not intended to be a review of the literature on Z-DNA; because this literature is so extensive, citations are generally limited to papers that were among the first to report the phenomenon or technique or that provide detailed procedures.

Factors Influencing B–Z Equilibrium

Supercoiling is the most important inducer of the Z conformation in natural DNA. Nonsupercoiled, natural DNA contains virtually no Z-DNA, but the same DNA under extreme negative supercoiling, as in "form V" DNA (see Detection of Z-DNA in Homopolymers, below), may have as much as 35–40% of its sequences in the Z form.[1] However, several factors beside supercoiling also influence the B–Z equilibrium, including both sequence effects and environmental factors, and they are discussed in turn.

Nucleotide Sequence

As mentioned above, in Z-DNA the orientation of the bases about the glycosidic bond alternates between anti and syn,[2] whereas in B-DNA the orientation is always anti. Because purines are more stable in syn than are pyrimidines,[3] Z-DNA is favored by alternating purine–pyrimidine sequences. The base sequence that most favors Z-DNA is alternating CG $[d(C-G)_n \cdot d(C-G)_n]$.[2] Among the other simple repeating sequences, (C-A)$_n \cdot$(T-G)$_n$ sequences also easily form Z-DNA,[4,5] but (A-T)$_n \cdot$(A-T)$_n$ is quite resistant to adopting the Z conformation.[6]

Nonalternating Sequences. The requirement for purine–pyrimidine alternation is not absolute. When plasmids are subjected to increasing superhelical density, regions of Z-DNA can extend outward from alternating sequences to include adjacent nonalternating regions.[7,8] Recently, we found that some sequences which have one base pair of every four out of alternation, such as $(CAGG)_n$ and $(CAGA)_n$, can also form Z-DNA under moderately high superhelical density[9]; these particular sequences are asso-

[1] S. Brahms, J. Vergine, J. G. Brahms, E. DiCapua, P. Bucher, and T. Koller, *Cold Spring Harbor Symp. Quant. Biol.* **47,** 119 (1983).

[2] A. H.-J. Wang, G. J. Quigley, F. J. Kolpak, J. L. Crawford, J. H. van Boom, G. van der Marel, and A. Rich, *Nature (London)* **282,** 680 (1979).

[3] A. E. V. Haschemeyer and A. Rich, *J. Mol. Biol.* **27,** 369 (1967).

[4] B. D. Haniford and D. E. Pulleyblank, *Nature (London)* **302,** 632 (1983).

[5] A. Nordheim and A. Rich, *Proc. Natl. Acad. Sci. U.S.A.* **80,** 1821 (1983).

[6] M. J. Ellison, J. Feigon, R. J. Kelleher, A. H.-J. Wang, J. F. Habener, and A. Rich, *Biochemistry* **25,** 3648 (1986).

[7] B. H. Johnston and A. Rich, *Cell (Cambridge, Mass.)* **42,** 713 (1985).

[8] B. H. Johnston, *J. Biomol. Struct. Dyn.* **6,** 153 (1988).

[9] B. H. Johnston, L. Carter, and A. Rich, *Proc. Natl. Acad. Sci. U.S.A.,* in press.

ciated with recombinational hot spots in the mouse major histocompatibility locus.[10,11] Measurements based on two-dimensional gel electrophoresis[12] indicate that the energy cost of a base out of alternation is only about 2.5 kcal.[13] Thus, the order of ease forming Z-DNA among some repeated sequences is $(CG)_n > (CA)_n > (CAGG)_n > (CAGA)_n > (AT)_n$.

Z–Z Junctions. Another irregularity in purine–pyrimidine alteration that affects the energetics of the B–Z transition is a change in the phase of alternation, which creates a dislocation in the helix called a Z–Z junction.[7,14–17] An example of a Z–Z junction that has been examined in some detail[17] is the (GG)·(CC) step in the sequence $(CG)_8G(CG)_8] \cdot [(CG)_8C(CG)_8$. This junction destabilizes the Z helix by about 3.5 kcal relative to the purely alternating sequence $(CG)_{16}$, and it possesses a distinctive reactivity signature (see Hydroxylamine, below). A model for this Z–Z junction based on the energy minimization of two juxtaposed Z-DNA helices[17] has been proposed.

Solution Environment

The conformation that was eventually recognized as Z-DNA was first observed for poly(dG-dC) in high concentrations of sodium chloride by Pohl and Jovin,[18] who noticed that it showed a circular dichroism (CD) spectrum that was nearly the inverse of the low-salt form. Several years later, the first Z-DNA crystal structure was determined[2] and identified as a novel left-handed helix. The Z-DNA backbone follows a zigzag pattern, with the result that every other phosphate group is closer to another phosphate than is the case with B-DNA. The repulsion of the negatively charged phosphates destabilizes the Z conformation unless the charges are shielded by cations; this repulsion is thought to be the reason why Z-DNA is stabilized by high salt.

Multivalent cations are more effective in this stabilization than monovalent cations; the midpoint of the B–Z transition for poly(dG-dC) occurs

[10] M. Steinmetz, D. Stephan, and K. F. Lindahl, *Cell (Cambridge, Mass.)* **44**, 895 (1986).
[11] Y. Uematsu, H. Kiefer, R. Schulze, K. Fischer-Lindahl, and M. Steinmetz, *EMBO J.* **5**, 2123 (1986).
[12] R. Bowater, F. Aboul-Ela, and D. M. J. Lilley, this series, Vol. 212, [5].
[13] M. J. Ellison, R. J. Kelleher, A. H.-J. Wang, J. F. Habener, and A. Rich, *Proc. Natl. Acad. Sci. U.S.A.* **82**, 8320 (1985).
[14] M. J. McClean and R. W. Wells, *J. Biol. Chem.* **263**, 7370 (1988).
[15] M. J. McClean, J. W. Lee, and R. W. Wells, *J. Biol. Chem.* **263**, 7378 (1988).
[16] S. M. Mirkin, V. I. Lyamichev, V. P. Kumarev, V. F. Kobzev, V. V. Nosikov, and A. V. Vologodskii, *J. Biomol. Struct. Dyn.* **5**, 79 (1987).
[17] B. H. Johnston, G. J. Quigley, M. J. Ellison, and A. Rich, *Biochemistry* **30**, 5257 (1991).
[18] F. M. Pohl and T. M. Jovin, *J. Mol. Biol.* **67**, 375 (1972)

TABLE I

CONCENTRATIONS OF CATIONS OR ETHANOL AT MIDPOINT OF B–Z TRANSITION[a,b]

Ion	poly(dG-dC)·poly(dG-dC)	poly(dG-m⁵dC)·poly(dG-m⁵dC)
Na^+	2500	700
Mg^{2+}	700	0.6
Ca^{2+}	100[c]	0.6
Ba^{2+}	40[c]	0.6
$NH_3 3(CH_2)_2 NH_3^{2+}$	d	1.0
$NH_3(CH_2)_3 NH_3^{2+}$	d	1.0
$NH_3(CH_2)_4 NH_3^{2+}$	d	2.0
$NH_3(CH_2)_5 NH_3^{2+}$	d	30[c]
$Co(NH_3)_6^{3+}$	0.02	0.005
Spermidine[3+]	e	0.05
Spermine[4+]	f	0.002
Ethanol	60%	20%

[a] From Behe and Felsenfeld.[19]

[b] Ion concentrations are given as millimolar; ethanol concentration is volume/volume. Values were determined for chloride as the counterion and in the presence of 50 mM NaCl and 5 mM Tris-HCl, pH 8.0.

[c] Transition produces anomalous CD spectrum.

[d] No change in optical properties is observed up to 100 mM.

[e] Produces B-form absorption spectrum up to 1 mM spermidine and then aggregates.

[f] Produces B-form absorption spectrum up to 50 μM spermine and then aggregates.

at 2.5 M NaCl, 700 mM MgCl$_2$, and 20 mM Co(NH$_3$)$_6$Cl$_3$ (Table I[19]). The relative effectiveness of multivalent ions is even greater for methylated polymers (see Methylation at Cytosine C-5, below). One class of multivalent cations, polyamines, are found in substantial concentrations in the cell and are as effective as Co(NH$_3$)$_6$Cl$_3$ at inducing the B-to-Z transition in poly(dG-m⁵dC) (Table I[19]). For this reason, polyamines, together with supercoiling and possibly Z-DNA-binding proteins, are one of the intracellular factors most likely to stabilize any Z-DNA existing *in vivo*.

An anomaly in the effects of sodium ion on the B–Z equilibrium has been noted for sequences in supercoiled plasmids. High salt concentrations (e.g., 4 M NaCl) stabilize Z-DNA in plasmids,[20] but intermediate salt

[19] M. Behe and G. Felsenfeld, *Proc. Natl. Acad. Sci. U.S.A.* **78**, 1619 (1981).

[20] J. Klysik, S. M. Stirdivant, J. E. Larson, P. A. Hart, and R. D. Wells, *Nature (London)* **290**, 672 (1981).

concentrations (350 mM) destabilize Z-DNA relative to either high or low (50 mM) salt, as assayed by binding of Z-DNA-specific antibodies.[21,22]

Agents that lower the dielectric constant of water, such as ethanol, methanol, and ethylene glycol, also tend to stabilize Z-DNA. This stabilization may be due to a closer clustering of counterions around the DNA because ionic effects are felt more strongly, thus providing more effective shielding of the mutually repelling phosphate groups. For example, as shown in Table I, poly(dG-dC) converts to Z-DNA in 60% ethanol, and poly(dG-m⁵dC) converts at 20% ethanol.[19]

Chemical Modification

Bromination. Chemical bromination was one of the first techniques discovered for stabilizing poly(dG-dC) in the Z conformation.[23] If the polymer is treated with aqueous bromine under high-salt conditions until approximately one-third of the guanine residues and one-fifth of the cytosine residues are brominated, it remains in the Z conformation even at low salt concentrations (150 mM NaCl or less). The effectiveness of bromination is attributed to steric effects. Bromine substitutes for the hydrogens on the ring carbons C-8 of guanine and (to a lesser extent) C-5 of cytosine. In B-DNA, the proton at the C-8 position of guanine is in van der Waals contact with the sugar–phosphate backbone, leaving insufficient room for substitution by an atom the size of bromine. In contrast, the proton at the 8-position of guanine in Z-DNA is on the outside of the helix where there is essentially no steric constraint. Thus, bromination probably stabilizes the Z form by raising the energy of or even prohibiting the B form.

Stabilization can also be achieved by specifically incorporating 5-bromocytosine in place of cytosine; studies on poly(dG-br⁵dC) have shown that this molecule exists in the Z conformation in 5 mM Tris.[24] Poly(dG-br⁵dC) can be synthesized using the Klenow fragment of DNA polymerase I and poly(dI-dC) as a template,[24] or, alternatively, br⁵dC residues may be introduced during the chemical synthesis of DNA. Brominated poly(dG-dC) is an effective immunogen for the development of polyclonal[25] and

[21] C. K. Singleton, J. Klysik, S. M. Stirdivant, and R. D. Wells, *Nature (London)* **299**, 312 (1982).

[22] F. Azorin, A. Nordheim, and A. Rich, *EMBO J.* **2**, 649 (1983).

[23] A. Möller, A. Nordheim, S. A. Kozlowski, D. J. Patel, and A. Rich, *Biochemistry* **23**, 54 (1984).

[24] B. Malfoy, N. Rousseau, and M. Leng, *Biochemistry* **21**, 5463 (1982).

[25] E. M. Lafer, A. Möller, A. Nordheim, B. D. Stollar, and A. Rich, *Proc. Natl. Acad. Sci. U.S.A.* **78**, 3546 (1981).

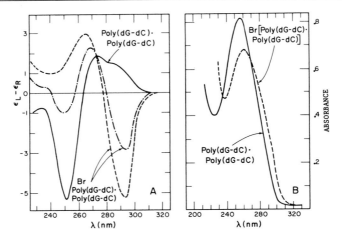

FIG. 1. Circular dichroism spectra (A) of poly(dG-dC) in the B form (—) and brominated polymers with $\theta = 0.6$ (- · -) and $\theta = 1$ (---) in 10 mM sodium phosphate (pH 7.0), 150 mM NaCl, and 1 mM EDTA. Absorption spectra (B) of poly(dG-dC) (B form) (—) and brominated poly(dG-dC) (Z form, $\theta = 1$) (---) under the same ionic conditions as listed for (A). (From Möller et al.[23])

monoclonal[26] antibodies, which have been widely used in determining the presence of Z-DNA in a variety of biological systems (see Detection Using Antibodies, below).

For homopolymers, circular dichroism is a particularly sensitive indicator of the extent of the B–Z transition θ, which can be operationally defined as $\theta = (\alpha_B - \alpha)/(\alpha_B - \alpha_Z)$. Here α is the specific ellipticity of the solution, and α_B and α_Z are the specific ellipticities for the B and Z forms, respectively.[18] Figure 1A shows the CD spectra of poly(dG-dC), unmodified and modified by bromine at two different levels. The solid curve represents B-DNA, whereas the interrupted lines are those typical of partially ($\theta = 0.6$) and fully converted ($\theta = 1.0$) Z-DNA. The dashed line ($\theta = 1.0$) was produced by a polymer in which 38% of the guanine residues and 18% of the cytosine residues are brominated and shows a near inversion of the CD spectrum. Further bromination does not change the CD spectrum.

The procedure for bromination described below was developed[23] for stabilizing poly(dG-dC) in the Z form. In the procedure, the polymer is converted to the Z form by raising the NaCl concentration to 3.8 M and is then treated with aqueous bromine to "lock in" that conformation. The

[26] A. Möller, J. E. Gabriels, E. M. Lafer, A. Nordheim, A. Rich, and B. D. Stollar, J. Biol. Chem. 257, 12081 (1982).

extent of bromination is monitored by following the A_{295}/A_{260} ratio in the UV spectrum. In the absence of bromination, conversion of poly(dG-dC) to the Z form by adding NaCl to 3.8 M results in a small decrease in A_{260} and a considerable increase in A_{295}, which, in turn, result in an increase in the A_{295}/A_{260} ratio. Bromine-stabilized poly(dG-dC) shows an additionally increased A_{295}/A_{260} ratio because the extinction coefficient at 260 nm of brominated nucleosides is reduced (Fig. 1B).

A higher level of bromination is required to stabilize short chains of poly(dG-dC) than long chains. Figure 2 shows the extent of the B–Z

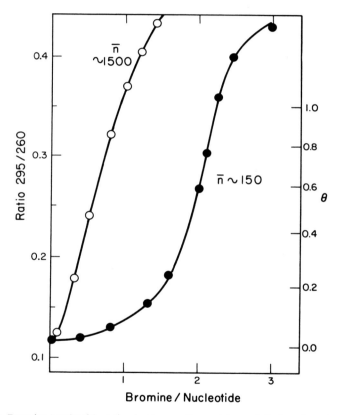

FIG. 2. Bromine/nucleotide ratios in the reaction mixture plotted against the A_{295}/A_{260} ratios and the corresponding degree of transition (θ) values, determined from CD spectra, measured under low-salt buffer conditions. Two different polymers were used with average chain lengths $\bar{n}$ of approximately 1500 bp (O) and 150 bp (●). Note that the degree of transition (θ) scale is not linear, and the absorbance continued to change with added bromine even after the transition is complete ($\theta = 1$). (From Möller et al.[23])

transition (determined from CD spectra) and the A_{295}/A_{260} ratios in low salt as a function of the extent of bromination for polymers of average length 1500 and 150 base pairs (bp).

Protocol for Brominating Poly(dG-dC). The following procedure is adapted from Möller *et al.*[23] and produces a large amount of brominated polymer suitable for a Z-DNA affinity column; it can be scaled down. A slightly different procedure is given by Fishel *et al.*[27]

1. Dissolve 100 A_{260} units of poly(dG-dC) (available from Pharmacia, Piscataway, NJ) in 1 ml of water.
2. Remove 50 A_{260} units and add 5 M NaCl to a final concentration of 3.8 M. The volume at this point should be about 4 ml for convenience in later steps. Heat to 60° for 10 min to ensure full conversion to Z-DNA.
3. Dilute 0.1 ml of a saturated solution of Br_2 in water (shaken well to ensure saturation) to 2 ml with water (approximate final concentration, 11 mM). All operations with bromine should be performed in a fume hood.
4. Add 1/11 volume of 0.22 M sodium citrate (pH 7.2) to the DNA solution (the pH is important; it drops to about 6.4 on addition of salt and bromine).
5. Add 1/11 volume of the 5% saturated bromine water, and incubate at room temperature for 3 min. Immediately bubble a gentle stream of air through the solution for 3 min using a 2-inch syringe needle to remove the bromine.
6. Dilute 40 μl to 500 μl with water and determine the A_{295}/A_{260} ratio.

For B-DNA the A_{295}/A_{260} ratio in low salt is 0.12. For DNA to be in the Z-form, the ratio must be at least 0.3; 0.33–0.38 is desirable. If the ratio is too low, repeat Steps 5 and 6, shortening the incubation time. *Note:* The A_{260} drops by 30% on bromination (see Fig. 1B), so to convert A_{260} to DNA concentration, multiply it by 65 μg/ml. Over time the DNA will gradually depurinate, and the A_{295}/A_{260} ratio can increase to as high as 0.6.

7. Dialyze the DNA overnight against 10 mM Tris, 1 mM EDTA, 50 mM NaCl. A final A_{295}/A_{260} ratio of 0.33–0.36 is ideal.

Note: Poly(dG-dC) chemically brominated by this procedure is unstable; an affinity column made with this material is effective for about 2 months. An alternative procedure for making more stable Z-DNA affinity columns is to use poly(dG-m⁵dC) (available from Pharmacia; see next section) in

---

[27] R. Fishel, P. Anziano, and A. Rich, this series, Vol. 184, p. 328.

the presence of 10–15 mM MgCl$_2$ and no more than 0.3 M monovalent cation.[27] See Fishel *et al.*[27] for detailed procedures for making Z-DNA affinity columns.

Methylation at Cytosine C-5. Behe and Felsenfeld[19] found that methylation of cytosine at the 5-position helps to stabilize poly(dG-dC) in the Z form. Only millimolar amounts of Mg^{2+} and micromolar amounts of trivalent cations are required to convert poly d(dG-m^{5}dC) to Z-DNA, concentrations orders of magnitude lower than those needed by unmodified poly(dG-dC). For example, the midpoint of the B–Z transition occurs at 700 mM MgCl$_2$ for poly(dG-dC) and at 0.6 mM for poly(dG-m^{5}dC). Table I shows the effectiveness of m^{5}dC methylation in stabilizing the Z form of poly(dG-dC) for a number of cations. Low salt has been reported to be stabilizing relative to intermediate salt concentrations for poly(m^{5}dC-dG),[28] just as for (dC-dG)$_n$ in supercoiled plasmids (see below).

Several Z-DNA crystal structures have been solved for sequences containing m^{5}dC residues. Methylation at this position fills a small depression in the surface of Z-DNA, as discerned from the three-dimensional crystal structure of (m^{5}dC-dG)$_3$.[29] In the case of m^{5}CGATm^{5}CG,[30] the methyl groups provide enough stabilization to overcome the unfavorable syn conformation of the thymines.

Preparation of methylated DNAs. Poly(dG-m^{5}dC) can be synthesized from poly(dG-dC) using the Klenow fragment of DNA polymerase I from *Escherichia coli*,[19] and it is available commercially (Pharmacia). Methylated oligonucleotides may be chemically synthesized using methylated phosphoramidites.

Other Modifications. Certain other covalent modifications also promote the B-to-Z transition. Methylation at the N-7 position of guanine stabilizes Z-DNA; when fully methylated at this point, poly(dG-dC) exists in the Z form at physiological salt concentrations.[31] The carcinogen N-acetoxy-N-acetyl-2-aminofluorene (AAF) reacts primarily with the C-8 position of guanines and stabilizes the Z form of poly(dG-dC),[32,33] perhaps because of steric interference in the B form. Platinum complexes such as

[28] B. G. Feuerstein, L. J. Marton, M. A. Keniry, D. L. Wade, and R. H. Shafer, *Nucleic Acids Res.* **13**, 4133 (1985).

[29] S. Fujii, A. H.-J. Wang, G. A. van der Marel, J. H. Van Boom, and A. Rich, *Nucleic Acids Res.* **10**, 7879 (1982).

[30] J. Feigon, A. H.-J. Wang, G. A. van der Marel, J. H. van Boom, and A. Rich, *Science* **230**, 82 (1985).

[31] A. Möller, A. Nordheim, S. R. Nichols, and A. Rich, *Proc. Natl. Acad. Sci. U.S.A.* **78**, 4777 (1981).

[32] E. Sage and M. Leng, *Proc. Natl. Acad. Sci. U.S.A.* **77**, 4597 (1980).

[33] E. Sage and M. Leng, *Nucleic Acids Res.* **9**, 1241 (1981).

chloro(diethylenetriamine)platinum(II) chloride also tend to stabilize Z-DNA.[34,35]

Generation of Z-DNA through Negative Supercoiling

Because the free energy of supercoiled DNA is higher than that of relaxed DNA, supercoiling facilitates any process that reduces the superhelical tension. The conversion of right-handed helical turns of B-DNA to left-handed helical turns reduces the superhelical tension of negatively supercoiled DNA; consequently, negative supercoiling is a very effective way of inducing Z-DNA in susceptible sequences. For plasmids with alternating CG inserts ranging from 24 to 42 bp, the level of supercoiling required to induce the B-to-Z transition in a medium of low ionic strength is well below that of typical DNA from natural sources ("native" superhelical density; $\sigma = -0.055$ when isolated from E. coli).[36] Other sequences that form Z-DNA at native superhelical density include alternating CA/TG sequences, mixed CA/CG sequences, and sequences such as CACGGGTGCGCATG, found in pBR322, which, despite a disruption in purine–pyrimidine alternation, largely converts to Z-DNA at $\sigma = 0.053$.[8,37] For sequences that require a higher superhelical density to undergo the B–Z transition, plasmids can be adjusted to the required superhelicity by incubating them with topoisomerase I in the presence of an unwinding agent (usually an intercalator such as ethidium bromide). The intercalator reduces the helical twist so that when the intercalated plasmid is relaxed by the enzyme, the linking number (the number of times the two DNA strands cross each other) is reduced. When the DNA is then purified away from the intercalator and the enzyme, it is negatively supercoiled. The reduction in linking number can be accurately determined by electrophoresing the DNA in the presence of different concentrations of chloroquine, alongside samples of the same plasmid having known linking number, and counting topoisomer bands.[38]

Protocol for Preparation of Negatively Supercoiled Plasmids. Plasmid DNA is added to different concentrations of ethidium bromide in 10 mM Tris (pH 8.0), 200 mM NaCl, 0.1 mM EDTA. The final volume is 10 μl for every microgram of DNA. Calf thymus topoisomerase I is added at a level of 1–3 units per microgram of DNA, with the higher levels used in the

[34] B. Malfoy, B. Hartmann, and M. Leng, *Nucleic Acids Res.* **9**, 5659 (1981).
[35] H. M. Ushay, R. M. Santella, J. P. Caradonna, D. Grunberger, and S. J. Lippard, *Nucleic Acids Res.* **10**, 3573 (1982).
[36] L. J. Peck, A. Nordheim, A. Rich, and J. C. Wang, *Proc. Natl. Acad. Sci. U.S.A.* **79**, 4560 (1982).
[37] W. Herr, *Proc. Natl. Acad. Sci. U.S.A.* **82**, 8009 (1985).
[38] W. Keller, *Proc. Natl. Acad. Sci. U.S.A.* **72**, 4876 (1975).

presence of higher concentrations of ethidium bromide ($> 15\ \mu M$; these levels inhibit the enzyme somewhat), and the mixture is incubated overnight in the dark at room temperature. Under subdued light, samples are extracted successively with equal volumes of phenol and phenol–chloroform ($1:1$) and twice with n-butanol (saturated with buffer, to complete extraction of ethidium bromide), then ethanol-precipitated. Pellets are taken up in 10 mM Tris (pH 7.4), 1 mM EDTA, and aliquots are electrophoresed on separate agarose gels containing two or three concentrations of chloroquine so that overlapping series of topoisomers are seen. Bands that run as relaxed circles on gels lacking chloroquine are assigned a linking number of zero; the linking numbers of all bands of other samples can be assigned by counting bands from that position if there is enough overlap between different samples and if chloroquine concentrations are chosen so that all bands are resolved.

The relationships between ethidium bromide concentration, linking number, and the required chloroquine concentrations depend on the size of the plasmid, and some trial and error may be required to obtain and measure the desired level of supercoiling. For pUC18, ethidium bromide in the range of $1-20\ \mu M$ gives topoisomers approximately in the range $0.01 - \sigma \leq 0.12$; a gel containing 1 μg/ml chloroquine sulfate is sufficient to resolve topoisomers in the range of $-\sigma < 0.05$, and 4 μg/ml resolves $0.03 < -\sigma < 0.1$. For higher levels of supercoiling and certain plasmids, as much as 40 μg/ml chloroquine may be needed. This method is described in more detail by Keller.[38]

Stabilization of Z-DNA through Protein Binding: Antibodies

The free energy of association of a specific Z-DNA-binding protein can contribute to the stabilization of Z-DNA. Perturbing effects of anti-Z-DNA antibodies on the salt-induced B–Z transition of poly(dG-dC) have been observed.[34,39] Similarly, anti-Z-DNA antibodies can shift the B–Z equilibrium in a supercoiled plasmid so that the B-to-Z transition occurs at a lower superhelical density than is required to drive the transition in the absence of antibody binding. The magnitude of the perturbation in the equilibrium depends on the antibody concentration and the affinity of the antibody for the particular Z-DNA sequence. High-affinity antibodies can even stabilize the Z conformation of $(dG-dC)_n$ sequences in plasmids in the absence of supercoiling. Revet et al.[40] observed that if anti-Z-DNA anti-

[39] T. M. Jovin, L. P. McIntosh, D. J. Arndt-Jovin, D. A. Zarling, M. Robert-Nicoud, J. H. van de Sande, K. F. Jorgenson, and F. Eckstein, J. Biomol. Struct. Dyn. 1, 21 (1983).
[40] B. Revet, D. A. Zarling, T. M. Jovin, and E. Delain, EMBO J. 3, 3353 (1984).

bodies were bound to supercoiled plasmids in very low ionic strength and the plasmids were then cut with a restriction enzyme, some antibodies could be visualized on the linear plasmids in the electron microscope, suggesting that Z-DNA had been stabilized in the plasmid through its interaction with the anti-Z-DNA antibody. Lafer *et al.*[41] further showed that, after incubation with antibodies for a prolonged time (hours or days), linear plasmids would bind the antibodies, locking their $(C-G)_n$ or $(C-A)_n$ sequences in the Z conformation. Because natural Z-DNA-binding proteins have been isolated from several sources (e.g., see Refs. 42 and 43), they may contribute to the stabilization of Z-DNA *in vivo*.

Techniques for Detecting Z-DNA

The ultimate tool for studying Z-DNA is X-ray crystallography[2] (reviewed in Ref. 44). The structural information gained from this technique is of the highest resolution available; however, the technique requires a specialist and is applicable only to the study of oligonucleotides. The following discussion therefore focuses on techniques of more general availability that are useful for identifying Z-DNA in homopolymers and "natural" DNA, particularly plasmids.

Detection of Z-DNA in Homopolymers: Spectroscopic Techniques

Z-DNA in homopolymers is easily detected by spectroscopic techniques such as absorption, NMR, and Raman spectroscopies, and circular dichroism. As described above for poly(dG-dC), CD spectra are dramatically different in right- and left-handed DNA, one spectrum being almost an inversion of the other. The transition from right- to left-handed DNA can easily be followed by observing the ellipticity at any wavelength that changes sharply with the transition. The differences in the CD spectrum between B- and Z-DNA are particularly pronounced in the vacuum ultraviolet; differences seen at wavelengths between 180 and 200 nm are about 10 times greater than those seen between 230 and 300 nm.[45] Where only part of a DNA sequence is undergoing the transition, as in supercoiled plasmids, CD is of marginal utility unless the transition involves a substan-

[41] E. M. Lafer, R. Sousa, and A. Rich, *EMBO J.* **4**, 3655 (1985).
[42] A. Nordheim, P. Tesser, F. Azorin, Y. H. Kwon, A. Möller, and A. Rich, *Proc. Natl. Acad. Sci. U.S.A.* **79**, 7729 (1982).
[43] E. M. Lafer, R. Sousa, B. Rosen, A. Hsu, and A. Rich, *Biochemistry* **24**, 5070 (1985).
[44] A. Rich, A. Nordheim, and A. H.-J. Wang, *Annu. Rev. Biochem.* **53**, 791 (1984).
[45] J. C. Sutherland, K. P. Griffin, P. C. Keck, and P. Z. Takacs, *Proc. Natl. Acad. Sci. U.S.A.* **78**, 4801 (1981).

tial proportion of the molecule. In the case of form V DNA, the extremely highly supercoiled DNA that is formed by annealing complementary single-stranded circles, the CD spectrum is intermediate between that of B-DNA and Z-DNA; assuming that 10% of the bases were unpaired (partly on the basis of hyperchromicity and Raman measurements), Brahms et al.[1] estimated from CD spectra that form V DNA contains some 35–40% Z-DNA.

Ultraviolet absorption spectra, which change only modestly in the B-to-Z transition, are useful only for rough monitoring of the transition in polymers, as described for brominated poly(dG-dC) above.[23] Raman spectra of Z-form sequences, however, which reflect stretching and bending frequencies of the chemical bonds, show some distinct bands not present in B-DNA.[46-48] A particular advantage of Raman spectroscopy is that signals can be obtained from both solutions and crystals; this capability allowed Thamann et al.[49] to demonstrate in an early study that the high-salt conformation of poly(dG-dC) is the same as the Z-DNA crystal structure.

NMR spectroscopy can provide definitive information on DNA conformation in homopolymers and relatively short oligonucleotides (reviewed in Ref. 44). ^{31}P NMR shows two distinct peaks for the nonequivalent phosphorus atoms of the dinucleotide repeat of poly(dG-dC), whereas in the B form only a single peak is seen. Use of the nuclear Overhauser effect (NOE) in ^{1}H NMR spectroscopy allows the determination of specific interproton distances that can distinguish between B-DNA and Z-DNA.

Detection of Z-DNA in Supercoiled DNA

Early studies of B–Z transitions in supercoiled plasmids relied largely on two techniques for detecting and crudely mapping Z-DNA: (1) gel electrophoresis to detect unwinding resulting from the transition; and (2) binding of anti-Z-DNA antibodies, together with either nitrocellulose filter binding assays or electron microscopy, to detect the Z-DNA segments directly. These techniques are still useful for detecting Z-DNA when precise information about the site of the transition is not needed. Unwinding assays are particularly useful for in vivo studies[50] or, with the use of

[46] W. L. Peticolas, W. L. Kubasek, G. A. Thomas, and M. Tsuboi, in "Biological Applications of Raman Spectroscopy" (T. Spiro, ed.), Vol. 1, p. 81. Wiley, New York, 1987.

[47] G. J. Thomas, Jr., J. M. Benevides, and B. Prescott, in "Proceedings of the Fourth Conversation in the Discipline Biomolecular Stereodynamics" (R. H. Sarma and M. H. Sarma, eds.), p. 227. Adenine Press, Guiderland, New York, 1987.

[48] W. L. Peticolas, Y. Wang, and G. A. Thomas, Proc. Natl. Acad. Sci. U.S.A. 85, 2579 (1988).

[49] T. J. Thamann, R. C. Lord, A. H.-J. Wang, and A. Rich, Nucleic Acids Res. 9, 5443 (1981).

[50] D. B. Haniford and D. E. Pulleyblank, J. Biomol. Struct. Dyn. 1, 593 (1983).

two-dimensional gel electrophoresis, for determining energetic parameters for the transition.[50,51] Filter assays, on the other hand, are useful for the rapid detection of Z-DNA somewhere in a plasmid. However, for mapping Z-DNA and for following structural transitions at specific sequences, chemical probes (see section on chemical probing, below), which allow mapping at nucleotide resolution, have become the technique of choice in recent years.

Unwinding Assays. Singleton *et al.*[21] and Peck and Wang[36] used changes in gel electrophoretic mobility to detect the formation of left-handed DNA segments in plasmids. This approach was used by Haniford and Pulleyblank[50] to detect the presence of the Z conformation in plasmids *in vivo*. The technique relies on the fact that, when *n* nucleotides of a negatively supercoiled plasmid convert from the B to the Z conformation, the net supercoiling of the plasmid is reduced by approximately $n/10.5 + n/12$ superhelical turns. If the transition occurs within the range of topoisomers that are resolved by agarose gel electrophoresis, the transition can be seen as a shift to lower electrophoretic mobility. The shift is most easily displayed by two-dimensional electrophoresis[50,51] in which the first dimension is as usual and the second is in the presence of an intercalator that, by sufficiently unwinding the helix, disallows all unwinding transitions. The reader is referred to the discussion of this technique by Lilley *et al.*[12]

Detection of Z-DNA Using Antibodies

Poly(dG-dC) stabilized in the Z form by modification with bromine or chloro(diethylenetriamine)platinum(II) chloride is highly immunogenic and, when injected into goats or rabbits, elicits the production of antibodies specific for Z-DNA. Both polyclonal[24,25] and monoclonal[26] antibodies have been isolated and characterized. Monoclonal antibodies raised against Z-DNA can recognize different parts of the Z-DNA helix; for example, methylation of the C-5 position of cytosine blocks some antibodies but not others, leading to the conclusion that some antibodies bind to the base pairs and others bind to the backbone.[26] Antibodies have proved very useful for detecting Z-DNA in a variety of ways; some of the most useful types of antibody-based assays are described in the following sections.

Nitrocellulose Filter Assays. Over a range of solution conditions, nitrocellulose filters allow double-stranded DNA to pass through while proteins are retained. If a plasmid contains a segment of Z-DNA, in the presence of an anti-Z-DNA antibody it would be retained as a DNA–protein complex on a nitrocellulose filter. If the plasmid is radioactive, the presence of

---

[51] L. J. Peck and J. C. Wang, *Proc. Natl. Acad. Sci. U.S.A.* **80**, 6206 (1983).

Z-DNA anywhere in the plasmid can then be quantitatively detected by counting the filter in a liquid scintillation counter.

The plasmid to be tested is usually prepared in radioactive form by isolating it from a plasmid-producing *E. coli* strain grown in the presence of [^{3}H]thymidine[52]; for usage over short time periods, ^{32}P- or ^{35}S-labeled plasmids can also be used. Convenient amounts of radioactivity, typically 5000–10,000 counts/min (cpm) representing about 0.2 μg DNA for each assay, are incubated with an appropriate antibody in 60 mM sodium phosphate, 30 mM EDTA (pH 8), 200 mM NaCl for 15 min (shorter time periods can usually also be used; see Measuring Rate of B–Z conversion, below) at room temperature in a volume of 150 μl. Typically a concentration series is done to identify the amount of antibody needed to give half-retention on the filter, with the antibody serially diluted over several logarithms of concentration. After the incubation period, the sample is filtered through a nitrocellulose filter (Millipore, Bedford, MA, HAWP 025 00, soaked in 0.3 M NaOH for 3 min and thoroughly rinsed with distilled water before use) using a multiple filtration apparatus (e.g., Hoefer, San Francisco, CA) with the flow rate controlled to a low level (2 ml/min) by adjusting the suction. After being washed with several milliliters of the incubation buffer, the filters are dried under a heating lamp and counted in a scintillation counter. The level of background binding is determined by including as a control a nonspecific antibody such as rabbit IgG. Nonspecific binding is strongly influenced by the ionic environment, particularly the concentration of di- and trivalent cations, and can be reduced in the presence of EDTA. This procedure has been used extensively for assaying Z-DNA-binding proteins using Z-forming polymers or Z-containing plasmids, with anti-Z-DNA antibodies as positive controls.

Antibody–DNA Cross-linking. Complexes between anti-Z-DNA antibodies and supercoiled plasmids usually dissociate if the plasmid is nicked or linearized, unless the binding affinity is strong (see section on protein binding, above). Complexes can be cross-linked by treatment with glutaraldehyde.[53,54] In this procedure, 1.5–2 μg of plasmid DNA is incubated with approximately 7–20 μg (depending on the antibody–DNA affinity) of antibody in 100 μl of 40 mM triethanolamine (pH 7.4), 50 mM NaCl for 15 min at 37°. Glutaraldehyde is added to a final concentration of

---

[52] J. Sambrook, E. F. Fritsch, and T. Maniatis, eds., "Molecular Cloning: A Laboratory Manual," 2nd Ed. Cold Spring Harbor Laboratory, Cold Spring Harbor, New York, 1989.

[53] A. Nordheim, E. M. Lafer, L. J. Peck, J. C. Wang, B. D. Stollar, and A. Rich, *Cell (Cambridge, Mass.)* **31**, 309 (1982).

[54] A. Nordheim, M. L. Pardue, L. M. Weiner, K. Lowenhaupt, P. Scholten, A. Möller, A. Rich, and B. D. Stollar, *J. Biol. Chem.* **261**, 468 (1986).

0.1%, and the mixture is allowed to incubate for 15 min at room temperature. Free glutaraldehyde and unbound antibody are then removed by passing the mixture over a small Sephacryl S-400 column (0.7 × 5 cm) equilibrated with 10 mM Tris-HCl (pH 8), 1 mM EDTA. The elution of the antibody–DNA complex is monitored by UV absorption.

Nordheim et al.[53,54] used the cross-linking approach to localize antibody-binding sites within restriction fragments by digesting the cross-linked DNA with an appropriate restriction endonuclease, passing the digest through a nitrocellulose filter (Millipore Millex-HA, 45 mm), precipitating the eluate, and analyzing the precipitate on a polyacrylamide gel. The fragment retained on the filter, and thus depleted in the eluate, contained the Z-DNA site. The retained fragment can be recovered from the filter by digesting with 50 μg/ml proteinase K in the presence of 1% sodium dodecyl sulfate (SDS) at 60° for 20 min.[54] If the Z-DNA sequence happens to overlap a restriction site, the cross-linked antibody will block digestion at that site, leading to an altered pattern of fragments on subsequent gel electrophoresis.[53] For mapping purposes, this procedure has been almost entirely supplanted by chemical probing techniques; however, fixation by glutaraldehyde (and formaldehyde) is useful for electron microscopic detection of antibody-bound Z-DNA.

Electron microscopy of Antibody-Bound DNA. Antibody–Z-DNA complexes can be visualized by electron microscopy.[55–57] After incubating the DNA with antibodies, complexes are purified by gel filtration chromatography and then visualized either directly, by shadowing, or indirectly by conjugation with ferritin via a second antibody. This approach has been used to identify Z-DNA in form V DNA[55] as well as ordinary plasmids. Approximate locations of Z-DNA regions on supercoiled plasmids have been determined by linearizing the plasmid (after fixing the antibody–DNA complex with formaldehyde and glutaraldehyde) and measuring the positions of the bound antibodies relative to an end.[56]

Antibodies Bound to Solid Supports. Antibodies bound to solid supports are useful in separating Z-DNA (or plasmids containing Z-DNA) from molecules lacking Z-DNA.[58] Examples are isolating Z-DNA-forming sequences from plasmid libraries or selecting from a family of topoisomers those having enough supercoiling to include Z-DNA. The availability of

[55] M. C. Lang, B. Malfoy, A. M. Freund, M. Daune, and M. Leng, *EMBO J.* **10**, 1149 (1982).
[56] F. J. Stockton, F. D. Miller, K. F. Jorgenson, D. A. Zarling, A. R. Morgan, J. B. Rattner, and F. H. van de Sande, *EMBO J.* **2**, 2123 (1983).
[57] E. DiCapua, A. Stasiak, T. Koller, S. Brahms, R. Thomae, and F. M. Pohl, *EMBO J.* **2**, 1531 (1983).
[58] R. Thomae, S. Beck, and F. M. Pohl, *Proc. Natl. Acad. Sci. U.S.A.* **80**, 5550 (1983).

cross-linked agarose beads conjugated to protein A of *Staphylococcus aureus* (Repligen Corp., Cambridge, MA) permits easy attachment and subsequent elution of antibodies (Fig. 3); this technique has recently been used to confirm the ability of CAGG sequences to form Z-DNA when supercoiled.[9] Alternatively, beads conjugated to a second antibody that

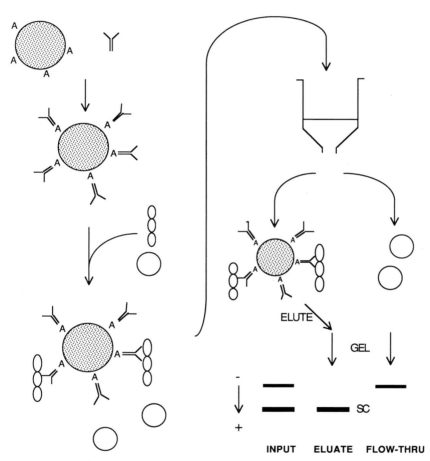

FIG. 3. Use of anti-Z-DNA antibodies immobilized on protein-A-conjugated agarose beads to isolate Z-DNA-containing plasmids.[9] Antibodies are allowed to bind to the beads, and the DNA mixture is added. After a short incubation, the beads are filtered; supercoiled DNA (SC) containing segments of Z-DNA is retained, while plasmids lacking Z-DNA because of nicking (NC) (or sequence) pass through the filter. Supercoiled DNA can be eluted from the beads and recovered by adding a detergent. Gel electrophoretic analysis of the input DNA, the filtrate, and the eluate confirms the separation.

binds the anti-Z-DNA antibodies by their F_c end can be used, but we have had better results using protein A-conjugated beads.

Measuring Rate of B–Z Conversion in Supercoiled Plasmids using Antibody Binding. Although the rate constants for the B-to-Z transition in polymers can be easily followed by circular dichroism, absorption, or Raman spectroscopy, these spectroscopic techniques are not so useful for monitoring transitions in supercoiled plasmids, where usually only a small fraction of the DNA undergoes the transition. Peck *et al.*[59] measured the forward rate of the B-to-Z transition induced by negative supercoiling using the binding of Z-DNA-specific antibodies to follow the transition. Supercoiled, radiolabeled plasmids containing a $d(CG)_{16} \cdot d(CG)_{16}$ insert were supercoiled, and enough ethidium bromide was bound to the DNAs to relax them and to keep the alternating C-G sequence in the right-handed helical form. On the rapid removal of ethidium ions by passage through a column of cation-exchange resin, the DNA became negatively supercoiled, inducing the conversion of the C-G insert to Z-DNA. After this step, which was completed in less than 40 sec, anti-Z-DNA antibodies were added and the samples were passed through nitrocellulose filters.[53] The amount of radioactivity retained on the filters gave a measure of the extent of conversion to Z-DNA. The rate of the transition was found to be strongly dependent on the superhelical density σ: the transition was complete in less than 50 sec for $\sigma = -0.09$, whereas the half-time of the transition was about 2 min at $\sigma = -0.07$ and about a factor of 10 slower at $\sigma = -0.05$. Control experiments in which the DNAs remained negatively supercoiled (i.e., no ethidium bromide was added) gave complete retention of DNA on the filter at the first time point (50 sec), indicating that the binding of antibodies to the Z-DNA sequences in the plasmids was fast relative to the time scale of the B-to-Z transition, and therefore the binding rates reflected the rate of the B-to-Z transition.

Detecting Z-DNA in Permeabilized Nuclei Using Antibodies. The ability of antibodies to enter permeabilized nuclei has been used in recent studies[60,61] to detect Z-DNA in a system having some of the properties of living cells, including the ability to perform transcription and replication. Cells were embedded in agarose, the cell membrane and cytoplasm were removed by detergent, and the resulting nuclei were incubated with biotinylated monoclonal antibodies, which can enter the nuclei. Unbound antibodies were washed out by repeatedly pelleting the cells, and bound antibodies were detected either microscopically, after labeling with streptavidin

[59] L. J. Peck, J. C. Wang, A. Nordheim, and A. Rich, *J. Mol. Biol.* **190**, 125 (1986).
[60] B. Wittig, T. Dorbic, and A. Rich, *J. Cell Biol.* **108**, 755 (1989).
[61] B. Wittig, T. Dorbic, and A. Rich, *Proc. Natl. Acad. Sci. U.S.A.* **88**, 2259 (1991).

tagged with fluorescein isothiocyanate (FITC), or by scintillation counting, after labeling the antibodies with ^{125}I-labeled streptavidin. At moderate antibody concentrations, the amount of bound antibody was found to be constant over a 200-fold range in the concentration of added antibody; this result was interpreted as being due to preexisting Z-DNA in the nucleus. At higher levels of antibody, further binding was detected that was interpreted as the result of "induction" of Z-DNA by pertubation of the B–Z equilibrium. Increased levels of Z-DNA were detected by this method when negative supercoiling was increased by adding a topoisomerase I inhibitor[60] or when transcription or replication was stimulated by adding nucleoside triphosphates.[61]

Detection of Z-DNA Using Chemical Probes

Chemical probing techniques allow the simultaneous detection and mapping of Z-DNA at nucleotide resolution, and they are generally the methods of choice. They are based on methods of DNA sequence determination and can utilize either the chemical degradation[62] or the chain-termination[63] approach. In Maxam–Gilbert sequencing, end-labeled DNA fragments are exposed to reagents that modify specific bases in a manner that allows subsequent cleavage of the DNA backbone. The fragments generated in this way are resolved as a ladder of bands on a denaturing polyacrylamide gel. For structural mapping, reagents are used whose reactivities depend not only on nucleotide sequence but also on conformation. Appropriate reagents were identified[7] by screening a number of potential sequencing reagents for their ability to generate altered band intensities with model DNA sequences known to form Z-DNA from antibody-binding or two-dimensional gel electrophoresis studies. Several reagents were found to give characteristic signals in the presence of Z-DNA, either within the Z-DNA segments themselves or at junctions between right-handed B-DNA and left-handed Z-DNA (Table II). Three of these — hydroxylamine, osmium tetroxide, and diethyl pyrocarbonate — are especially useful and are discussed in detail elsewhere in this series.[64-66] These reagents have been used to map the extent of Z-DNA structure in plasmids that contain a wide variety of sequences, including simple alternating purine–pyrimidine sequences such as $(CG)_n$ and $(CA)_n \cdot (TG)_n$,[7,37] irregular

---

[62] A. M. Maxam and W. Gilbert, this series, Vol. 65, p. 499.
[63] F. Sanger, S. Nicklen, and A. R. Coulson, *Proc. Natl. Acad. Sci. U.S.A.* **74,** 5463 (1977).
[64] B. H. Johnston, this series, Vol. 212, [10].
[65] E. Paleček, this series, Vol. 212, [8].
[66] T. Kohwi-Shigematsu and Y. Kohwi, this series, Vol. 212, [9].

TABLE II
SPECIFICITIES OF CHEMICAL PROBES FOR Z-DNA

Reagent	Target base in double-stranded DNA	Hyperreactive for
Diethyl pyrocarbonate	A (N-7, N-6); G (N-7)	Z-DNA (syn), single-stranded regions
Dimethyl sulfate	G (N-7)	Z-DNA (anti)
Hydroxylamine, methoxylamine	C (C-4, C-6)	B–Z junctions, Z–Z junctions, single-stranded regions
Osmium tetroxide	T (5,6-double bond)	B–Z junctions, single-stranded regions

natural sequences,[8,37] and nonalternating repeated sequences such as (CAGG)$_n$.[9]

Detection of Adducts by Strand Cleavage. In the Maxam–Gilbert version of chemical probing, the chemical modification is performed on intact plasmid DNA at various superhelical densities instead of on isolated end-labeled restriction fragments as with ordinary Maxam–Gilbert sequencing (Fig. 4). Supercoiling can be induced by incubating covalently closed circular plasmid DNA containing the sequence of interest with ethidium bromide and topoisomerase I.[26] After purification, the DNA is tested for the presence of appropriate topoisomers on chloroquine–agarose gels and is then reacted with the reagents to an extent of less than one modified site per several hundred base pairs. All the reagents described have in common the ability to modify DNA bases at some unusual structural feature without nicking the DNA backbone and thereby losing supercoiling. After modification, a restriction fragment radioactively labeled at one end is prepared and purified by standard procedures.[52,62] The backbone is then cleaved at each modification site, and the resulting fragments are resolved on a sequencing gel. For a particular reagent, equal amounts of radioactivity are loaded on the gel for each level of supercoiling tested. The intensities of particular bands at different levels of supercoiling are then compared. In addition, the intensity of an adenine, for example, within a Z region is compared with that of adenines outside the region but in the same lane.

Detection of Adducts by Polymerase Blockage. In the chain-termination approach to chemical probing, the chemical reactions are performed just as with chemical cleavage, except that smaller amounts of DNA can be used. The sites of reaction are detected by their ability to block polymerases. Either DNA polymerases (including reverse transcriptase) or RNA polymerases can be used for detection. In the case of some reagents, certain

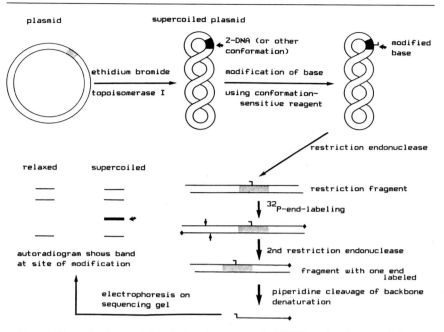

FIG. 4. General scheme of chemical probe analysis for Z-DNA using the strand cleavage approach. Plasmids are supercoiled by treatment with topoisomerase I in the presence of ethidium bromide, inducing segments of Z-DNA. The supercoiled DNA is treated with the chemical reagent to "tag" the Z-DNA sequence, then linearized at a nearby site. The ends are labeled with ^{32}P, and a second cut is made to remove the label from the unwanted end. If necessary, the resulting singly end-labeled fragment is isolated from a gel; with an appropriate choice of vector, this step can usually be eliminated by cutting off the unwanted label close enough to the end. The desired fragment is then treated with hot piperidine to cleave the backbone at the tagged sites, and the resulting fragments are analyzed on a sequencing gel alongside appropriate sequence markers.

adducts not susceptible to chemical cleavage will produce polymerase blocks (e.g., example, dimethyl sulfate adducts on unpaired cytosines). For further details on the use of this approach, see the chapters by Htun and Johnston[67] and Sasse-Dwight and Gralla[68] elsewhere in this series.

  Chemical Reagents Useful for Detecting Z-DNA. Most of the reagents useful for mapping Z-DNA are discussed individually in Volume 212 of this series[64–66]; therefore, the following discussion focuses on their applica-

---

[67] H. Htun and B. H. Johnston, this series, Vol. 212, [15].
[68] S. Sasse-Dwight and J. D. Gralla, this series, Vol. **208**, [10].

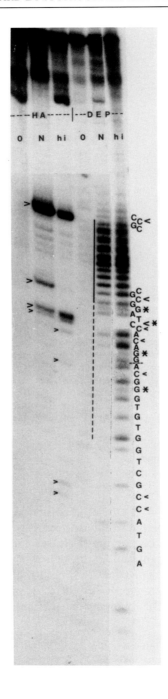

bility to Z-DNA and the patterns of reactivity that are characteristic of Z-DNA.

Diethyl pyrocarbonate. Diethyl pyrocarbonate (DEP) is hyperreactive with purines in the syn conformation in Z-DNA[7,37] and, therefore, with all purines in alternating purine–pyrimidine sequences. It reacts mainly with the N-7 position, which is more exposed in Z-DNA than in B-DNA. Because of the syn–anti repeat pattern characteristic of Z-DNA of any sequence, DEP-reacted Z-DNA segments appear on sequencing gels as a pattern of strong bands at every other nucleotide, with "holes" in the pattern at bases out of alternation (i.e., *syn*-pyrimidines). Z–Z junctions appear as a shift in the pattern by one nucleotide.[7,17] These patterns are seen in Fig. 5, which shows a (dG-dC)$_{16}$ sequence (vertical solid line) and surrounding regions from a plasmid reacted with DEP. The pattern of intense bands at every two nucleotides is clearly seen within the C-G sequence (fifth lane, DEP-N). When the plasmid is more highly supercoiled (sixth lane, DEP-hi), the pattern of hyperreactivity extends into the adjacent nonalternating sequences, particularly on one side (vertical dashed line). Where the predominant alternating purine–pyrimidine sequence pattern shifts, a Z–Z junction is evident as the alternation of light and dark bands shifts by one nucleotide (horizontal dashed line). The two purines at the Z–Z junction show less reactivity than *syn*-purines elsewhere in the Z-DNA segment.[17]

The syn–anti dinucleotide repeat characteristic of Z-DNA is illuminated by DEP reactivity even in sequences that are not alternating purines and pyrimidines. One such sequence is (CAGG)$_n$, which, as mentioned above, is associated with a hot spot for recombination in the mouse major histocompatibility (MHC) region[10] and forms Z-DNA when supercoiled.[9] As shown in Fig. 6, when this DNA, containing 17 repeats of approximately the sequence CAGG (two repeats are CATG), is analyzed for reactivity with DEP when relaxed or linearized, light bands at all three purines are seen for each repeat, with the adenine more reactive than the two guanine residues. When supercoiled, the adenine and the second

---

FIG. 5. Reactivity of the sequence (CG)$_{16}$ in plasmid pLP32[36] to hydroxylamine and diethyl pyrocarbonate. Hydroxylamine (HA) and diethyl pyrocarbonate (DEP) were used on intact plasmids in the relaxed state (O), at native ($\sigma \sim -0.05$) (N), and at high (~ -0.1) (hi) superhelical densities. Cytosines strongly or slightly reactive to hydroxylamine when supercoiled are labeled with large and small arrowheads, respectively. The latter appear mostly at high levels of supercoiling. Bases in perfect purine–pyrimidine alternation [(CG)$_{16}$] are within the solid vertical line, whereas those mainly alternating but with some out of alternation (asterisks) are along the dotted vertical line. The short horizontal dotted line in the sequence marks a Z–Z junction, where the phase of alternation changes. (From Johnston and Rich.[7])

σ: **0.032 0.047 0.074**

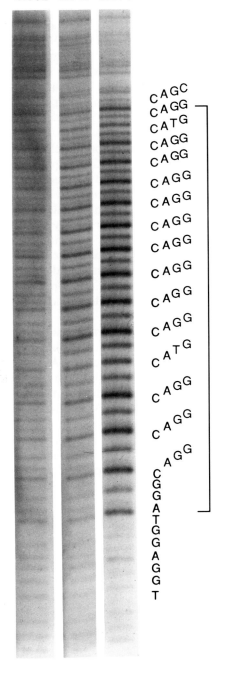

guanine become much more reactive, whereas the first guanine shows no hyperreactivity. This pattern is indicative of transition to Z-DNA, with the adenine and second guanine in the reactive syn conformation and the first guanine in the less reactive anti conformation.

Hydroxylamine. Hydroxylamine reacts very strongly with cytosines at junctions between B- and Z-DNA.[7] Thus, when a Z-DNA segment having cytosine residues at the B–Z junctions is reacted with DEP and, separately, with hydroxylamine, strong hydroxylamine bands will "bracket" the block of DEP bands corresponding to the Z-DNA segment. This result is seen very clearly for the plasmid containing $(C\text{-}G)_{16}$ at moderate supercoiling in Fig. 5 (lane HA-N). In the transition from moderate to high levels of supercoiling (lane HA-hi), one B–Z junction band at the edge of the insert is lost and other junction bands appear or are enhanced in the adjacent nonalternating region as the Z-DNA region expands.

Hydroxylamine reactivity is not always seen to bracket DEP reactivity on sequencing gels. When only part of a long alternating plasmid sequence is in the Z conformation, the entire sequence may show hyperreactivity to DEP because the Z-DNA segment can appear anywhere within the repeat. However, the junction hyperreactivities with hydroxylamine are "smeared out" to the point of invisibility in this case, becoming visible only when the entire sequence is in the Z conformation.[69]

Hydroxylamine also reacts with Z–Z junctions at the sequence 5'-Cp(C/T)-3', with only the 5'-most cytosine reactive.[17] Reactivity of hydroxylamine at both B–Z and Z–Z junctions is clearly demonstrated in the sequence $(CG)_8 G(CG)_8] \cdot [(CG)_8 C(CG)_8$, shown in Fig. 7.

Although hydroxylamine is highly specific for B–Z or Z–Z junctions, it is used at molar concentrations; thus, possible perturbing effects of the solution on DNA structure should be considered. Because it is prone to oxidation in the free amine form, hydroxylamine is generally prepared by titrating the hydrochloride salt with diethylamine. As a result, substantial levels of organic cations and chloride anions are contributed to the solution; to be sure they are not affecting the DNA structure being examined, it

[69] B. H. Johnston, W. Ohara, and A. Rich, *J. Biol. Chem.* **263**, 4512 (1988).

FIG. 6. Diethyl pyrocarbonate reactivity of a nonalternating Z-DNA-forming sequence.[9] The sequence is from a recombination hot spot found in the mouse MHC locus[10] containing 17 imperfect repeats of CAGG (bracket). The superhelical density σ is shown at the top. Note the change in the relative reactivities of the adjacent guanines in each CAGG repeat, indicative of the different conformations of their glycosidic bonds (syn is hyperreactive; anti is not) when the sequence converts to Z-DNA. Note also the hyperreactivity of the adenines, which have become syn.

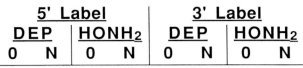

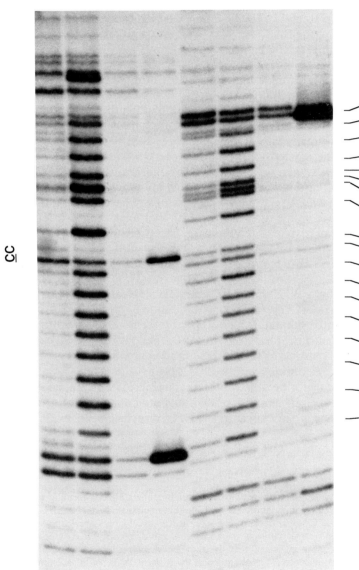

is prudent to include appropriate controls. In the case of Z-DNA in super-coiled plasmids, levels of salt of these magnitudes have not been observed to have gross effects on the B–Z equilibrium as detected by DEP reactiv-ity.[7]

Osmium tetroxide. Like hydroxylamine, OsO_4 in the presence of pyri-dine reacts with B–Z junctions,[7,70] except that OsO_4 reacts with exposed thymines. Both reagents react with ring carbon atoms and presumably must attack from outside the plane of the base; this fact may explain why they are quite unreactive with regular helical DNA, whether it be the B or Z form. Although little structural information exists for a B–Z junction, its reactivity with these reagents suggests that it involves a dislocation in the helix that permits out-of-plane attack. Osmium tetroxide has the advan-tage that its use does not contribute significantly to the electrolyte compo-sition of the DNA solution, and, with 2,2'-bipyridine as cofactor, it has been used to identify Z-DNA in *E. coli* cells.[71,72] However, it is somewhat less specific than hydroxylamine for B–Z junctions, and its high toxicity and volatility necessitate care in handling. It also unwinds DNA on forma-tion of a covalent adduct[73]; hence, when applied at high concentrations, OsO_4 can reduce supercoiling and potentially remove underwound struc-tures such as Z-DNA early in the reaction.

Dimethyl sulfate. Dimethyl sulfate, which is a standard agent for the detection of guanines in Maxam–Gilbert sequencing, has a minor but distinctive usefulness in detecting Z-DNA in sequences that lack perfect purine–pyrimidine alternation, specifically those in which guanines ap-pear in the anti conformation. Whereas *syn*-guanines in Z-DNA show essentially the same reactivity with dimethyl sulfate as *anti*-guanines in

[70] K. Nejedly, M. Kwinkowski, G. Galazka, J. Klysik, and E. Palecek, *J. Biomol. Struct. Dyn.* **3**, 467 (1985).

[71] E. Palecek, E. Rasovska, and P. Boublikova, *Biochem. Biophys. Res. Commun.* **150**, 731 (1988).

[72] A. R. Rahmouni and R. D. Wells, *Science* **246**, 358 (1989).

[73] M. Vojtiskova, J. Stokrova, and E. Palecek, *J. Biomol. Struct. Dyn.* **2**, 1013 (1985).

FIG. 7. Reactivity of the synthetic Z–Z junction $(CG)_8G(CG)_8] \cdot [(CG)_8C(CG)_8$ to diethyl pyrocarbonate (DEP) and hydroxylamine ($HONH_2$) when relaxed (0) or at native bacterial supercoiling (N), displayed on a 20% polyacrylamide sequencing gel containing 1.5% bisac-rylamide and $8.3\ M$ urea. The left-hand four lanes were 5'-end-labeled and the right-hand four lanes were 3'-labeled, all at the same *Taq*I site; hence, for both strands, bands of the same mobility correspond to approximately the same position in the sequence. The hydroxyl-amine-reactive C̲ (and the less DEP-reactive G̲G̲) residues of the Z–Z junction sequence d(CC)·d(GG) are underlined; the hyperreactive cytosines at the B–Z junctions are the darkest bands at the ends of the CG insert in the supercoiled lanes. (From Johnston *et al.*[17])

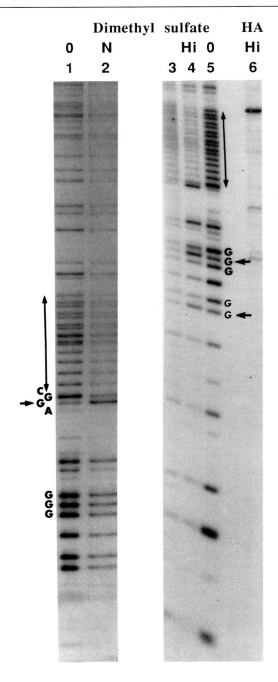

B-DNA, *anti*-guanines in Z-DNA are hyperreactive with dimethyl sulfate. This characteristic is seen in Fig. 8, which shows the dimethyl sulfate reactivity of the same sequence shown in Fig. 5. Here, with enough supercoiling, Z-DNA extends into GGA and GGG sequences; the middle guanine becomes hyperreactive to dimethyl sulfate while at the same time the outer purines become hyperreactive with DEP as seen in Fig. 5.[7]

Because the reagents described above (with the exception of dimethyl sulfate) show hyperreactivity toward Z-DNA and little or no reactivity toward B-DNA, Z-DNA segments can be detected even if a substantial proportion of plasmids lack Z-DNA, which can be the case if the DNA is supercoiled and a significant level of nicking has occurred (e.g., in the process of preparing supercoiled plasmids).

Chemical Modification Reactions

Diethyl pyrocarbonate. For modification with diethyl pyrocarbonate, dilute DNA (2–10 μg is usually convenient) into 0.2 ml of 50 mM sodium cacodylate (pH 7) and 1 mM EDTA; add 5 μl diethyl pyrocarbonate (Sigma, St. Louis, MO, stored at 4°, opened for less than 6 months). The diethyl pyrocarbonate may remain as a separate phase; thorough vortex mixing should achieve a saturated solution. Incubate for 15–30 min at 20–37°, depending on the length of the sequence to be examined, with occasional mixing to maintain saturation. Terminate the reaction by adding 50 μl of 1.5 M sodium acetate (pH 5.2), mix, and add 0.75 ml of cold ethanol. Recover the precipitated DNA by chilling in dry ice for 15 min and centrifuging for 15 min at 20,000 g at 4°. Take up the pellet in 0.25 ml of 0.3 M sodium acetate (pH 5.2) and reprecipitate by adding 0.75 ml ethanol, chilling and centrifuging, then rinse with 70% ethanol and dry the pellet in a Speed-Vac concentrator (Savant Instruments, Hicksville, NY) or similar device. (For further details, see Ref. 66.)

Hydroxylamine. For the hydroxylamine reactions,[74] prepare the re-

[74] C. M. Rubin and C. W. Schmid, *Nucleic Acids Res.* **8,** 4613 (1980).

FIG. 8. Reactivity of the (C-G)$_{16}$ plasmid sequence shown in Fig. 5 to dimethyl sulfate. Levels of supercoiling are 0, relaxed; N, native (~ -0.05); and Hi, high (> -0.1). The vertical line spans the CG repeat. Note that at native superhelical density the first G out of alternation at the bottom of the repeat becomes hypersensitive to dimethyl sulfate methylation (lane 2, horizontal arrow), whereas only at higher superhelical density is the same effect seen at out-of-alternation guanines further from the repeat, as the region of Z-DNA extends to include those sequences as well (lane 4, horizontal arrows). Lane 3, Purine ladder; lane 6, highly supercoiled plasmid reacted with hydroxylamine (HA) showing B–Z junction sites. (From Johnston and Rich.[7])

agent freshly by titrating hydroxylamine hydrochloride to pH 6 (or other desired pH between 5 and 8) with diethylamine and adjust the final volume to a concentration of 4 M. Add the DNA contained in 7 μl water to 21 μl of 4 M hydroxylamine and incubate for 20–30 min at 37°. Add 0.2 ml of 0.3 M sodium acetate, 0.1 mM EDTA and mix; then add 0.75 ml of cold ethanol and pellet, reprecipitate, rinse, and dry the DNA as described for the DEP reaction. (For further details, see Johnston.[64])

Dimethyl sulfate. Dimethyl sulfate reactions are performed according to Maxam and Gilbert,[62] except that MgCl$_2$ can be omitted from the reaction mixture. Dilute the DNA into 0.2 ml of 50 mM sodium cacodylate (pH 7), 1 mM EDTA; chill on ice, and add 1 μl of dimethyl sulfate. Incubate at 20° for 4–5 min, then terminate the reaction by adding 50 μl of a cold solution containing 1.5 M sodium acetate and 1 M 2-mercaptoethanol. Mix, then add 0.75 ml of cold ethanol. Recover the precipitated DNA, reprecipitate, rinse, and dry as described for the DEP reaction.

Osmium tetroxide. For OsO$_4$ modification, combine in a final volume of 50–100 μl DNA [in water or 10 mM Tris (pH 7.4), 1 mM EDTA], pyridine (final concentration 1% v/v), and aqueous osmium tetroxide (final concentration ~ 1 mM; chill to 0° before adding). Vortex to mix and incubate at 25° for 5–60 min, depending on the length of sequence to be examined. (2,2'-Bipyridine can also be used in place of pyridine; it forms a more stable complex and can be used at 0.5 mM.[71]) Chill the solution on ice, extract 3 times with carbon tetrachloride, and then add 0.2 ml of 0.3 M sodium acetate (pH 5.2), 0.1 mM EDTA, followed by 0.75 ml cold ethanol. Chill, pellet, rinse, and dry the DNA; a second precipitation is not necessary since the extraction removes most of the reagent. (For further discussion of OsO$_4$, see Palecek.[65]

pH effects. Most of the reagents except hydroxylamine are effective over a fairly wide pH range; hence, DNA structure can be examined at pH values other than those listed above provided that compatible buffers of the appropriate pK_a are used. Buffers that have been successfully used for OsO$_4$, dimethyl sulfate, and DEP are cacodylate for pH 6–8 and acetate for pH 4–5.5, with the pyridine cofactor titrated to the appropriate pH before addition in the case of OsO$_4$. For hydroxylamine, the pH is adjusted to the desired value by titration with diethylamine. Although the pK_a values of hydroxylamine and cytosine give rise to a broad pH optimum of 6 for this reagent, it has been used successfully over a pH range of 4.5–8.[75]

Little additional nicking of plasmids should occur as a result of any of the modification reactions. Reference purine ladders for the subsequent sequencing gel analysis are generated using standard sequencing reactions.

[75] B. H. Johnston, *Science* **241**, 1800 (1988).

Sequence Analysis. Modified DNA samples can be stored frozen for some time until used. For analysis by the chemical cleavage approach, the DNA is cut with an appropriate restriction enzyme and ^{32}P-labeled at the 3′ ends using the Klenow fragment of DNA polymerase I or at the 5′ ends using polynucleotide kinase. The label on the unwanted end is removed by a second restriction cut, and if necessary the singly end-labeled fragment of interest is isolated from a gel by electroelution or other technique. The fragment is ethanol-precipitated, thoroughly rinsed with 70% ethanol to remove salts, dried, and dissolved in water. The DNA backbone is cleaved at sites of modification by treatment with 1 M piperidine at 90° for 30 min (15 min in the case of hydroxylamine). Piperidine is removed by evaporation using a Speed-Vac concentrator followed by addition of 25–50 μl water, evaporation, addition of another 15–25 μl water, and final evaporation. The DNA is taken up in 80% formamide loading buffer,[62] and, for a particular reagent, equal amounts of radioactivity for each level of supercoiling are loaded in adjacent lanes of a sequencing gel. After running, gels are dried (except those of over 12% acrylamide) and autoradiographed on X-ray film, using an intensifying screen if needed.

Detection of Z-DNA by Restriction Endonucleases

Restriction endonucleases are highly selective for B-DNA (or whatever is the normal conformation of their recognition sequence) and have been used to monitor transitions from B-DNA to Z-DNA.[76] They have also been used[77] to probe for general regions of non-B-DNA structure in the extremely negatively supercoiled form V DNA (see Detection of Z-DNA in Homopolymers, above). Although these probes are convenient for assaying the structural status of specific sequences, care must be taken if they are used for quantitation because, in negatively supercoiled plasmids, a dynamic equilibrium exists between the B and Z conformations. For example, at a superhelicity in which 50% of the population of a plasmid containing an alternating C-G sequence exists in the Z conformation as measured by retention on nitrocellulose filters in the presence of anti-Z-DNA antibodies, *Bss*HII (which recognizes GCGCGC) can still cleave to completion given enough time.[76] This caution applies to any detection procedure, including antibody binding, that can trap either the B or the Z conformation and thus can shift the equilibrium.

[76] F. Azorin, R. Hahn, and A. Rich, *Proc. Natl. Acad. Sci. U.S.A.* **81**, 5714 (1984).

[77] S. K. Brahmachari, N. Ramesh, Y. S. Shouche, R. K. Mishra, R. Bagga, and G. Meera, *in* "Structure and Methods" (R. H. Sarma and M. H. Sarma, eds.), Vol. 2, p. 33. Adenine Press, Albany, New York, 1990.

Detection of Z-DNA by Restriction Methylases

Restriction methylases are unable to methylate their target sequence if it is in the Z conformation.[78] This observation has been used to assay for Z-DNA formation in supercoiled plasmids *in vivo*. Jaworski *et al.*[79] and Rahmouni and Wells[72] introduced a plasmid expressing a temperature-sensitive *Eco*RI methylase into *E. coli* and showed that, when grown at the permissive temperature, all available *Eco*RI sites were methylated. When cotransfected with plasmids containing alternating CG sequences of different lengths adjacent to an *Eco*RI site, those plasmids having large CG inserts were not methylated *in vivo* (as determined by susceptibility to *Eco*RI endonuclease after isolation of the plasmid) at the permissive temperature, indicating that they existed in the Z form *in vivo*.

Acknowledgments

I thank A. Rich for suggestions, and A. Rich and G. J. Quigley for critical reading of the manuscript. This work was supported in part by National Institutes of Health Grant 1R29GM41423.

[78] L. Vardimon and A. Rich, *Proc. Natl. Acad. Sci. U.S.A.* **81**, 3268 (1984).
[79] A. Jaworski, W.-T. Hsieh, J. A. Blaho, J. E. Larson, and R. D. Wells, *Science* **238**, 773 (1987).

[8] Supercoiled DNA and Cruciform Structures

By ALASTAIR I. H. MURCHIE and DAVID M. J. LILLEY

Supercoiled DNA and Local DNA Structure

The sequence-dependent rearrangement of DNA structure to adopt a new geometry usually involves a change in local DNA twist, and this is almost always negative; that is to say, perturbed DNA structures are usually underwound relative to the normal B-form double helix. For this reason, such structures will be more stable in negatively supercoiled DNA circles than their relaxed counterparts, and DNA supercoiling is well known to stabilize a number of structural polymorphs, including left-handed Z-DNA,[1,2] cruciform structures,[3-5] and H-triplex structures.[6] The

[1] L. J. Peck, A. Nordheim, A. Rich, and J. C. Wang, *Proc. Natl. Acad. Sci. U.S.A.* **79**, 4560 (1982).
[2] C. K. Singleton, J. Klysik, S. M. Stirdivant, and R. D. Wells, *Nature (London)* **299**, 312 (1982).

basis of this stabilization lies in simple topology. A circular double-stranded molecule can exist in a number of isomeric forms (topoisomers) that differ in the number times one strand is linked with the other (linking number, Lk). For an unconstrained, relaxed DNA molecule, the linking number is given by

$$Lk = Lk° = N/h \qquad (1)$$

where N is the number of base pairs in the circle, and h is the helical repeat under the conditions of the experiment. Note that although Lk is required to be an integer, N/h may not be so, and thus a completely relaxed topoisomer may not be attainable for a particular sized molecule under a certain set of conditions of temperature, ionic strength, etc. It is useful to define the linkage of the molecule relative to the relaxed state, by means of the linking difference (ΔLk):

$$\Delta Lk = Lk - Lk° \qquad (2)$$

For the reasons above, the linking difference need not necessarily be an integer. The linking difference measures the number of helical turns by which a given topoisomer differs from the relaxed state. Most natural DNA molecules are underwound, or negatively supercoiled, so that ΔLk has negative sign. To provide a measure of linkage changes that is independent of molecular size, we define the specific linking difference or superhelix density (σ):

$$\sigma = \Delta Lk/Lk° \qquad (3)$$

Many natural DNA molecules extracted from cells have a level of supercoiling corresponding to an underwinding of about 1 turn in 20, that is, a superhelix density of -0.05. Energy is required to supercoil DNA. The free energy of a supercoiled DNA molecule (ΔG_s) relative to the relaxed state is quadratically related to the linking difference,[7-9] namely,

$$\Delta G_s = (1050RT/N)\, \Delta Lk^2 \qquad (4)$$

[3] M. Gellert, K. Mizuuchi, M. H. O'Dea, H. Ohmori, and J. Tomizawa, *Cold Spring Harbor Symp. Quant. Biol.* **43**, 35 (1979).

[4] D. M. J. Lilley, *Proc. Natl. Acad. Sci. U.S.A.* **77**, 6468 (1980).

[5] N. Panayotatos and R. D. Wells, *Nature (London)* **289**, 466 (1981).

[6] V. Lyamichev, S. M. Mirkin, and M. D. Frank-Kamenetskii, *J. Biomol. Struct. Dyn.* **3**, 667 (1986).

[7] R. E. Depew and J. C. Wang, *Proc. Natl. Acad. Sci. U.S.A.* **72**, 4275 (1975).

[8] D. S. Horowitz and J. C. Wang, *J. Mol. Biol.* **173**, 75 (1984).

[9] D. E. Pulleyblank, M. Shure, D. Tang, J. Vinograd, and H.-P. Vosberg, *Proc. Natl. Acad. Sci. U.S.A.* **72**, 4280 (1975).

where R is the gas constant and T the absolute temperature. The alteration in linkage requires geometric changes to occur in the molecule, of torsional or flexural character, and this is stated mathematically by

$$\Delta Lk = \Delta Tw + Wr \tag{5}$$

where Tw is the helical twist and Wr is the writhing of the helix axis in three dimensions. Although this treatment has recently been absorbed into a much more detailed analysis of the topology of circular DNA,[10] it is adequate to understand the basis of the stabilization of certain perturbed structures, such a cruciforms, in supercoiled DNA. Because such structures are associated with a local unwinding, the negative twist change helps to compensate for the linkage deficiency by Eq. (5) and thus the molecule becomes partially relaxed by virtue of the change in topology. Thus the energetic cost of forming the new DNA structure locally (assumed to be positive) may be balanced by the negative free energy change associated with the global relaxation of the molecule, which can be calculated from Eq. (4). Because the supercoiling energy that is available for a given amount of relaxation increases quadratically with linking difference, the structure will be stable at a level of supercoiling that is greater than some critical value. Thus it is generally observed that structures such as cruciforms, which are unstable in relaxed or linear DNA, are formed cooperatively at a threshold level of supercoiling.

Cruciform Structures: Energetics and Kinetics

Cruciform structures are paired hairpin loop structures, formed by intrastrand base pairing that is possible when a sequence possesses 2-fold symmetry. An example is shown in Fig. 1. When an inverted repeat containing a total of n bases (stem plus loop) forms a cruciform structure, there is a local unwinding of the DNA (ΔTw_c) that is approximately given by

$$\Delta Tw_c = n/h \tag{6}$$

The free energy due to this relaxation for a given topoisomer can be calculated from Eq. (4), and this is offset against the free energy of cruciform formation (ΔG_c), which will be the cost of forming two hairpin loops and a four-way junction. For inverted repeats of average sequence and base

---

[10] J. H. White, N. R. Cozzarelli, and W. R. Bauer, *Science* **241**, 323 (1988).

5' AAA**GTCCTAGCAATCC**AAAAT**G****GGATTGCTAGGAC**CAA 3'
3' TTT**CAGGATCGTTAGG**TTTAC**CCTAACGATCCTG**GTT 5'

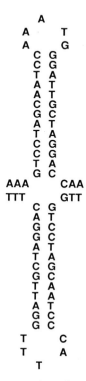

FIG. 1. Formation of a cruciform structure by an inverted repeat. The base sequence of the ColE1 inverted repeat is shown along with its cruciform conformation.[4,5]

composition ΔG_c has been measured to be 17–18 kcal mol^{-1},[11-13] although this should be regarded as a lower limit, since the introduction of longer loops or other imperfections into the 2-fold symmetry mean that there is no real upper limit. The energetic limits mean that at normal levels of superhelix densities ($-\sigma \leq 0.06$) a cruciform will not usually be detected

[11] A. J. Courey and J. C. Wang, *Cell (Cambridge, Mass.)* **33**, 817 (1983).
[12] M. Gellert, M. H. O'Dea, and K. Mizuuchi, *Proc. Natl. Acad. Sci. U.S.A.* **80**, 5545 (1983).
[13] D. M. J. Lilley and L. R. Hallam, *J. Mol. Biol.* **180**, 179 (1984).

if it has a stem shorter than $10-11$ base pairs (bp), a loop longer than $5-6$ bases, and greater than 5% mismatching in the stems. For special sequences ΔG_c can be lower; ΔG_c was measured to be around 13.5 kcal mol^{-1} for alternating adenine–thymine sequences,[14,15] whereas for a $(GATC)_{10}$ sequence a value of 16 kcal mol^{-1} was found.[16]

Significant kinetic barriers to cruciform extrusion exists,[11,12,17-19] and the extrusion can be a very slow process. We found that the kinetics were markedly dependent on local DNA sequence and base composition, and we observed that the extrusion of the majority of sequences could be divided into two kinetic classes,[19] namely, S-type and C-type extrusion. The great majority of sequences are S type. These sequences do not undergo extrusion in the absence of salt, requiring $50-60$ mM sodium for optimal extrusion rates. Under these conditions they exhibit moderate temperature dependence, with activation energies in the $30-60$ kcal mol^{-1} range. An alternative kinetic behavior was found to be associated with inverted repeats that were flanked by $(A + T)$-rich DNA sequences.[20] These sequences undergo extrusion with maximal rates at low ionic strength and are suppressed by salt. C-Type extrusion has a very high temperature dependence, with activation energies in the $100-200$ kcal mol^{-1} range.

We proposed two mechanistic schemes to account for these findings (Fig. 2). These are reviewed in detail by Lilley.[21] The S-type sequences are proposed to undergo a local opening at the center of the inverted repeat, followed by the initial formation of a junction structure to form the protocruciform. This may then extrude to the fully formed cruciform structure by a branch migration process. The smaller degree of initial opening is consistent with the relatively low activation energy and with the unwinding estimated by Courey and Wang[22]; it is also consistent with observations that sequence changes[22-24] and base modifications[25] in the center of the inverted repeat have the greatest kinetic consequences. We demonstrated that the rate of S-type extrusion is proportional to the anhy-

[14] D. R. Greaves, R. K. Patient, and D. M. J. Lilley, *J. Mol. Biol.* **185**, 461 (1985).
[15] J. A. McClellan, E. Palecek, and D. M. J. Lilley, *Nucleic Acids Res.* **14**, 9291 (1986).
[16] L. H. Naylor, D. M. J. Lilley, and H. van de Sande, *EMBO J.* **5**, 2407 (1986).
[17] R. R. Sinden and D. E. Pettijohn, *J. Biol. Chem.* **259**, 6593 (1984).
[18] I. Panyutin, V. Klishko, and V. Lyamichev, *J. Biomol. Struct. Dyn.* **1**, 1311 (1984).
[19] D. M. J. Lilley, *Nucleic Acids Res.* **13**, 1443 (1985).
[20] K. M. Sullivan and D. M. J. Lilley, *Cell (Cambridge, Mass.)* **47**, 817 (1986).
[21] D. M. J. Lilley, *Chem. Soc. Rev.* **18**, 53 (1989).
[22] A. J. Courey and J. C. Wang, *J. Mol. Biol.* **202**, 35 (1988).
[23] A. I. H. Murchie and D. M. J. Lilley, *Nucleic Acids Res.* **15**, 9641 (1987).
[24] G. Zheng and R. R. Sinden, *J. Biol. Chem.* **263**, 5356 (1988).
[25] A. I. H. Murchie and D. M. J. Lilley, *J. Mol. Biol.* **205**, 593 (1989).

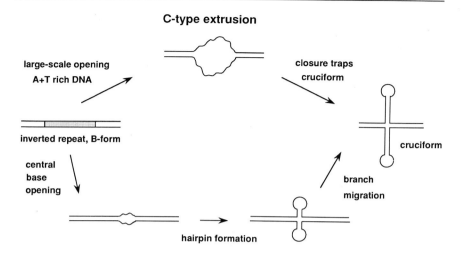

C-type extrusion

large-scale opening
A+T rich DNA

closure traps
cruciform

inverted repeat, B-form

cruciform

central
base
opening

branch
migration

hairpin formation

S-type extrusion

Fig. 2. Mechanistic schemes for cruciform extrusion. Kinetic data indicate that cruciform extrusion may proceed via two alternative mechanisms.[19,20] The majority of sequences undergo cruciform extrusion by the S-type pathway in which there is a relatively contained initial opening (probably around 10 bp) at the center of the inverted repeat, followed by intrastrand base pairing and junction formation. The structure may then undergo branch migration to generate the fully extruded cruciform. This pathway requires salt. A small number of sequences, in which the inverted repeat is contiguous with (A + T)-rich DNA, may extrude via the C-type mechanism. At low salt concentrations the (A + T)-rich DNA may initiate thermal helix opening of a relatively large region that encompasses the inverted repeat. Intrastrand pairing of the entire inverted repeat may then occur to generate the fully extruded cruciform in a single step.

drous radius of the cation present,[26] suggesting that the transition state for the reaction has partial four-way junction character.

The larger activation energy for the C-type cruciform extrusion is consistent with a greater degree of helix opening, which would be facilitated by low ionic strength. This large-scale opening is associated with the (A + T)-rich context of the inverted repeat.[20] Indeed, C-type extrusion can be closely approximated in normal DNA sequences if the helix stability is reduced by solvents such as dimethyl formamide[27]; conversely, if the helical stability of (A + T)-rich DNA is increased with distamycin, C-type cruciforms extrude with kinetic properties that are essentially similar to

[26] K. M. Sullivan and D. M. J. Lilley, *J. Mol. Biol.* **193**, 397 (1987).
[27] K. M. Sullivan and D. M. J. Lilley, *Nucleic Acids Res.* **16**, 1079 (1988).

S-type cruciforms.[27] The (A + T)-rich sequences can be relatively short; a core sequence of 30 bp from ColE1 confers normal C-type extrusion on an adjacent inverted repeat,[28] and we have observed residual effects down to a length of only 12 bp (A. I. H. Murchie and D. M. J. Lilley, unpublished data). The effect can be blocked by a short (G + C)-rich DNA sequence if placed in an intervening position.[28] The efficacy of the (A + T)-rich sequence can be considerably increased by methylation of adenine bases (A. I. H. Murchie and D. M. J. Lilley, unpublished data.)

Direct evidence for large-scale opening in the (A + T)-rich DNA has been obtained by two approaches. First, the sequences are chemically reactive above 25°[29a] to probes that react with unpaired bases (see [7] in Volume 212, this series). Second, thermal unwinding of (A + T)-rich sequences has been detected in supercoiled DNA by two-dimensional gel electrophoresis,[30,31] and the ColE1 sequences that promote C-type cruciform extrusion exhibit a series of temperature-dependent transitions (see [5] in Volume 212, this series). In addition, statistical mechanical helix–coil calculations on these sequences indicate a high propensity for cooperative melting, and they demonstrate a correlation between the calculated melting temperature and the experimental temperature of cruciform extrusion.[32]

Some sequences undergo cruciform extrusion without a measurable kinetic barrier. Alternating adenine–thymine sequences [$(AT)_n$] undergo very rapid cruciform formation even at 0°.[14] We presume this to reflect the low helix stability of the $(AT)_n$ sequences themselves, undergoing rapid, cooperative bubble formation and permitting facile extrusion of the cruciform.

Structure of the Cruciform

A cruciform structure comprises three components: the loops, the stems, and the four-way junction. The stems appear to consist of fairly normal duplex DNA, being cleavable by restriction enzymes, for example. However, one distinction from the rest of the plasmid DNA should be noted: the DNA of the stems is topologically distinct from the remaining DNA in the circular molecule. In other words, the cruciform stems are the only sections of duplex DNA in the molecule that are *not* subject to superhelical stress. Progress in understanding the structures of the loops

[28] K. M. Sullivan, A. I. H. Murchie, and D. M. J. Lilley, *J. Biol. Chem.* **263,** 13074 (1988).
[29] J. C. Furlong, K. M. Sullivan, A. I. H. Murchie, G. W. Gough, and D. M. J. Lilley, *Biochemistry* **28,** 2009 (1989).
[29a] R. Bowater, F. Aboul-ela, and D. M. J. Lilley, *Biochemistry* **30,** 11495 (1991).
[30] F. S. Lee and W. R. Bauer, *Nucleic Acids Res.* **13,** 1665 (1985).
[31] D. Kowalski, D. A. Natale, and M. J. Eddy, *Proc. Natl. Acad. Sci. U.S.A.* **85,** 9464 (1988).
[32] F. Schaeffer, E. Yeramian, and D. M. J. Lilley, *Biopolymers* **28,** 1449 (1989).

and the junctions has come from studies of these entities in isolation. It is difficult to study structures that require DNA supercoiling for their continued existence for two reasons. First, a cleavage anywhere within the DNA molecule releases the superhelicity, and the structure is instantly destabilized. Second, supercoiled circles are relatively large molecules for study, making most physical techniques difficult or even inapplicable.

There have been a number of studies of hairpin loops in DNA by spectroscopic and thermodynamic methods. Chemical probing of symmetrical loops in cruciforms suggested that the optimal size is around four nucleotides,[33] in agreement with an oligonucleotide melting study of the stability of a series of thymine loops of different sizes.[34] However, some sequences are capable of forming a loop of only two unpaired bases.[35] NMR studies of hairpin loop structures of various sequences indicate that there is considerable structure within the loop and that in general the formally unpaired bases of the loop may undergo stacking and hydrogen bonding interactions.[35-38] Hairpin structure has been reviewed by van de Ven and Hilbers.[39]

The structure of the four-way junction is of considerable interest and importance, as it is formally equivalent to the Holliday junction, the putative central intermediate of genetic recombination,[40-46] and there is good evidence for the involvement of a four-way junction in the integrase class of site-specific recombination events.[47-50] DNA junctions are sub-

[33] G. W. Gough, K. M. Sullivan, and D. M. J. Lilley, *EMBO J.* **5,** 191 (1986).

[34] C. A. G. Haasnoot, C. W. Hilbers, G. A. van der Marel, J. H. van Boom, U. C. Singh, N. Pattibiraman, and P. A. Kollman, *J. Biomol. Struct. Dyn.* **3,** 843 (1986).

[35] L. P. M. Orbons, A. A. van Beuzekom, and C. Altona, *J. Biomol. Struct. Dyn.* **4,** 965 (1987).

[36] L. P. M. Orbons, G. A. van der Marel, J. H. van Boom, and C. Altona, *Nucleic Acids Res.* **14,** 4188 (1986).

[37] D. R. Hare and B. R. Reid, *Biochemistry* **25,** 5341 (1986).

[38] M. J. J. Blommers, C. A. G. Haasnoot, C. W. Hilbers, J. H. van Boom, and G. A. van der Marel, *NATO ASI Ser. Ser. E* **133,** 78 (1987).

[39] F. J. M. van de Ven and C. W. Hilbers *Eur. J. Biochem.* **178,** 1 (1988).

[40] R. Holliday, *Genet. Res.* **5,** 282 (1964).

[41] T. R. Broker and I. R. Lehman, *J. Mol. Biol.* **60,** 131 (1971).

[42] T. L. Orr-Weaver, J. W. Szostak, and R. J. Rothstein, *Proc. Natl. Acad. Sci. U.S.A.* **78,** 6354 (1981).

[43] H. Potter and D. Dressler, *Proc. Natl. Acad. Sci. U.S.A.* **73,** 3000 (1976).

[44] H. Potter and D. Dressler, *Proc. Natl. Acad. Sci. U.S.A.* **75,** 3698 (1978).

[45] N. Sigal and B. Alberts, *J. Mol. Biol.* **71,** 789 (1972).

[46] H. M. Sobell, *Proc. Natl. Acad. Sci. U.S.A.* **69,** 2483 (1972).

[47] P. A. Kitts and H. A. Nash, *Nature (London)* **329,** 346 (1987).

[48] S. E. Nunes-Düby, L. Matsomoto, and A. Landy, *Cell (Cambridge, Mass.)* **50,** 779 (1987).

[49] R. Hoess, A. Wierzbicki, and K. Abremski, *Proc. Natl. Acad. Sci., U.S.A.* **84,** 6840 (1987).

[50] M. Jayaram, K. L. Crain, R. L. Parsons, and R. M. Harshey, *Proc. Natl. Acad. Sci. U.S.A.* **85,** 7902 (1988).

strates for a special class of resolving enzymes, which have been isolated from widespread sources including bacteriophages,[51,52] yeast,[53,54] and calf thymus.[55]

Stable DNA junctions have been constructed in linear DNA by choosing sequences that are unable to undergo branch migration.[56-58] Gel electrophoretic experiments on a pseudocruciform structure in linear DNA showed that the junction caused a considerable retardation in mobility in polyacrylamide, which was interpreted in terms of a bending or kinking of the DNA at the junction,[58] and this was shown to be dependent on the concentration of cations.[59] As in so many areas of nucleic acid structure, gel electrophoresis has proved to be immensely important in deducing the structure of the four-way DNA junction.[58,60,61] We used this technique to show that, in the presence of cations, the DNA junction folds into an X-shaped structure, in which the helical arms undergo pairwise stacking to form two quasi-continuous coaxial helices.[61] Coaxial helical stacking was consistent with the cleavage of junctions by the restriction enzyme *Mbo*II, when binding and cleavage sites were on opposite sides of the point of strand exchange.[62] These stacked arms are rotated (in the manner of opening a pair of scissors) in order to generate the stacked X-structure. Two isomers of the structure are possible depending on which arm is stacked with which, but we have found that for the majority of sequences, one isomer is favored over the other. The structure is consistent with gel electrophoretic experiments,[61] fluorescence resonance energy transfer measurements,[63] enzyme and chemical probing,[64-66] and electric birefringence measurements.[67]

[51] B. Kemper and M. Garabett, *Eur. J. Biochem.* **115**, 123 (1981).
[52] B. de Massey, F. W. Studier, L. Dorgai, F. Appelbaum, and R. A. Weisberg, *Cold Spring Harbor Symp. Quant. Biol.* **49**, 715 (1984).
[53] L. Symington and R. Kolodner, *Proc. Natl. Acad. Sci. U.S.A.* **82**, 7247 (1985).
[54] S. C. West and A. Korner, *Proc. Natl. Acad. Sci. U.S.A.* **82**, 6445 (1985).
[55] K. M. Elborough and S. C. West, *EMBO J.* **9**, 2931 (1990).
[56] L. R. Bell and B. Byers, *Proc. Natl. Acad. Sci. U.S.A.* **76**, 3445 (1979).
[57] N. R. Kallenbach, R.-I. Ma, and N. C. Seeman, *Nature (London)* **305**, 829 (1983).
[58] G. W. Gough and D. M. J. Lilley, *Nature (London)* **313**, 154 (1985).
[59] S. Diekmann and D. Lilley, *Nucleic Acids Res.* **14**, 5765 (1987).
[60] J. P. Cooper and P. J. Hagerman, *J. Mol. Biol.* **198**, 711 (1987).
[61] D. R. Duckett, A. I. H. Murchie, S. Diekmann, E. von Kitzing, B. Kemper, and D. M. J. Lilley, *Cell (Cambridge, Mass.)* **55**, 79 (1988).
[62] A. I. H. Murchie, J. Portugal, and D. M. J. Lilley, *EMBO J.* **10**, 713 (1991).
[63] A. I. H. Murchie, R. M. Clegg, E. von Kitzing, D. R. Duckett, S. Diekmann, and D. M. J. Lilley, *Nature (London)* **341**, 763 (1989).
[64] M. E. Churchill, T. D. Tullius, N. R. Kallenbach, and N. C. Seeman, *Proc. Natl. Acad. Sci. U.S.A.* **85**, 4653 (1988).
[65] M. Lu, Q. Guo, N. C. Seeman, and N. R. Kallenbach, *J. Biol. Chem.* **264**, 20851 (1989).

On folding of the junction into the stacked X-structure, the reduction of symmetry generates two different kinds of strands in the structure; the continuous strands possess a continuous helical axis, whereas the exchanging strands pass from one stack to the other. Stereochemical considerations suggest that unfavorable steric and electrostatic interactions will be minimized if the X-structure forms a right-handed cross with an antiparallel alignment of continuous strands: the exchanging strands enter and leave the point of strand exchange about the small angle of X. This generates a favorable accommodation of the continuous strands 3' to the point of strand exchange in the major groove of the other helical stack, thereby avoiding steric clash, consistent with the results of DNase I probing studies.[66] The alignment of strands and grooves will be optimal if the small angle of the X-structure is 60°, and a crystal structure showing similar packing between oligonucleotides has been presented.[68] The stacked X-structure is illustrated in Fig. 3. Analysis of the structure of a four-way DNA junction using molecular mechanics confirmed that the right-handed, antiparallel stacked X-structure was likely to be the most energetically favorable conformation.[69] The two sides of the X-structure are not equivalent, as the four base pairs at the point of strand exchange are oriented in the same direction. This generates a side that presents the major groove edges and one that presents the minor groove edges of these base pairs, and the two sides have distinct structural properties. It also appears that enzymes are able to distinguish the two sides of the junction; a number of resolving enzymes (e.g., T4 endonuclease VII) selectively cleave on the minor groove side.[70]

Detection of Cruciform Structures

There are two contrasting ways in which cruciform structures can be detected in a supercoiled circular DNA molecule.

Gel Electrophoresis

Formation of a stable cruciform may be detected by the relaxation of a topoisomer, according to Eq. (6),[13,71,72] which is revealed as a mobility shift

[66] A. I. H. Murchie, W. A. Carter, J. Portugal, and D. M. J. Lilley, *Nucleic Acids Res.* **18**, 2599 (1990).
[67] J. P. Cooper and P. J. Hagerman, *Proc. Natl. Acad. Sci. U.S.A.* **85**, 4653 (1989).
[68] Y. Timsit, E. Westhof, R. P. P. Fuchs, and D. Moras, *Nature (London)* **341**, 459 (1989).
[69] E. von Kitzing, D. M. J. Lilley, and S. Diekmann, *Nucleic Acids Res.* **18**, 2671 (1990).
[70] A. Bhattacharyya, A. I. H. Murchie, E. von Kitzing, S. Diekmann, B. Kemper, and D. M. J. Lilley, *J. Mol. Biol.* **221**, 1191 (1991).
[71] K. Mizuuchi, M. Mizuuchi, and M. Gellert, *J. Mol. Biol.* **156**, 229 (1982).

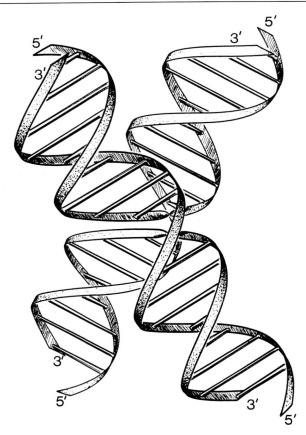

FIG. 3. Ribbon representation of the stacked X-structure for the four-way DNA junction.[63] The structure is constructed by pairwise stacking of arms to generate quasi-continuous coaxial pairs, with a right-handed rotation of approximately 60°. The 2-fold symmetric structure contains two kinds of strands; the continuous strands have unbroken axes through the junction, whereas the exchanging strands pass between the two stacked pairs at the point of strand exchange. The two continuous strands have an approximately antiparallel alignment. The arrangement leads to a favorable mutual accommodation of the continuous strands in the major groove of the opposite stacked helical pair, thereby minimizing unfavorable steric and electrostatic interactions. The folded structure of the junction requires cations for stability.

in an agarose or polyacrylamide gel. This is seen most clearly using a two-dimensional gel,[73] where the formation of the cruciform above a threshold level of supercoiling generates a discontinuous "jump." The degree of unwinding on cruciform formation may be measured from the amplitude of the mobility shift, and this coupled with the critical linking difference allows one to calculate the free energy of cruciform formation. The full method is described in [5] in Volume 212, this series. This approach is very powerful and is completely nonperturbing. The drawback of the gel electrophoresis method is that only global changes in the entire circular molecule are observed, and these cannot be interpreted directly in terms of local structure. The gel method shows that there is a structural change somewhere in the molecule, but not where. Frequently the twist change can be correlated with the sequence of the molecule in order to deduce the likely nature of the structural transition undergone.

Probing

Enzyme or chemical probes provide a radical alternative to gel electrophoresis methods, and they help to localize the structural change to a particular place in the base sequence. The first demonstration of cruciform formation in natural DNA molecules was based on the cleavage by single-strand-specific nucleases at the center of inverted repeat sequences when present in supercoiled DNA.[4,5] Probing methods rely on some feature of the structure being recognizably different from the remaining duplex DNA of the molecule, and for a cruciform structure this means either the single-stranded loops or the four-way junction (see above). A summary of some of the probes that have been employed for cruciform structures is given in Table I, and more details may be found in Volume 212 of this series.

The potential problem with probes is a kind of uncertainty principle; in order to recognize the modified structure there must be an interaction between probe and DNA, and this might modify the structure under investigation. This is of particular concern for enzyme probes, where the interaction between the substrate and the enzyme binding site might alter the DNA structure. For this reason probing methods should be applied with caution, and the results interpreted with care. However, probes provide the information that gel methods lack, namely, the location of the

---

[72] V. I. Lyamichev, I. G. Panyutin, and M. D. Frank-Kamenetskii, *FEBS Lett.* **153**, 298 (1983).

[73] J. C. Wang, L. J. Peck, and K. Becherer, *Cold Spring Harbor Symp. Quant. Biol.* **47**, 85 (1983).

TABLE I
PROBES USED FOR THE DETECTION AND ANALYSIS OF CRUCIFORM STRUCTURES IN
SUPERCOILED DNA[a]

Enzyme	Target	Special conditions or comments	Refs.
S1 nuclease	Loop	Low pH, Zn^{2+} required	4,5
Micrococcal	Loop	Ca^{2+} required	b
BAL31	Loop	Ca^{2+} required	13
P1 nuclease	Loop	Range of pH possible	c
Mung bean	Loop	Low salt possible	d
T4 endo VII	Junction	Highly specific cleavage	e, f
T7 endo I	Junction	Also has single-strand specificity	5,52,g
Restriction	Loop, junction	Cleavage inhibited by nonduplex	11
Methylases	Loop, junction	Methylation inhibits subsequent restriction	25

Reagent	Target	Product	Analysis[h]	Refs.
Haloacetaldehyde	Loop A, C	Etheno adduct	S1 or DMS	i
Osmium tetroxide	Loop T > C	cis-Diester	Piperidine, sequence	j
Diethyl pyrocarbonate	Loop A > G	Carbethoxylate	Piperidine, sequence	k
Glyoxal	Loop G	Etheno adduct	S1	l
Bisulfite	Loop C	Deamination to dU	Sequence C to T	33
Psoralen	Stem	Cross-linking	Cross-linked product	17

[a] Further details on many of these probes may be found in Volume 212, this series. The references given are the applications of the probes to the study of cruciforms.

[b] D. M. J. Lilley, *Cold Spring Harbor Symp. Quant. Biol.* **47**, 101 (1983).

[c] D. B. Haniford and D. E. Pulleyblank, *Nucleic Acids Res.* **13**, 4343 (1985).

[d] L. G. Sheflin and D. Kowalski, *Nucleic Acids Res.* **12**, 7087 (1984).

[e] K. Mizuuchi, B. Kemper, J. Hays, and R. A. Weisberg, *Cell (Cambridge, Mass.)* **29**, 357 (1982).

[f] D. M. J. Lilley and B. Kemper, *Cell (Cambridge, Mass.)* **36**, 413 (1984).

[g] P. Dickie, G. McFadden and A. R. Morgan, *J. Biol. Chem.* **262**, 14826 (1987).

[h] S1 refers to cleavage of the adduct with S1 nuclease and DMS (dimethyl sulfate) to secondary chemical modification.

[i] D. M. J. Lilley, *Nucleic Acids Res.* **11**, 3097 (1983).

[j] D. M. J. Lilley and E. Palecek, *EMBO J.* **3**, 1187 (1984).

[k] J. C. Furlong and D. M. J. Lilley, *Nucleic Acids Res.* **14**, 3995 (1986); P. M. Scholten and A. Nordheim, *Nucleic Acids Res.* **14**, 3981 (1986).

[l] D. M. J. Lilley, in "The Role of Cyclic Nucleic Acid Adducts in Carcinogenesis and Mutagenesis" (B. Singer and H. Bartsch, eds.), p. 83. IARC Publ. 70, Lyon, France, 1986.

structural alteration. For these reasons the best way to proceed is by a combination of methods, using topology along with a number of enzyme and chemical probes. Such comparisons have shown that in the case of cruciform structures the concerns about perturbation by the probes were unfounded,[13] but for other structures this might not be the case.

A number of useful enzyme probes are single-strand-specific nucleases, such as S1 or P1 nucleases, that cleave the bases of loop regions more rapidly than duplex DNA. Such cleavage should always be followed by a complete digestion by a restriction enzyme before analysis of the products by gel electrophoresis, so that only cleavage at a *specific* site is studied. This is necessary because enzymes such as S1 nuclease generate significant background levels of cleavage throughout a supercoiled DNA molecule. It is also important that "single-hit" conditions are employed, since random cleavage anywhere in the molecule releases the supercoiling. Most enzyme probes will function at reduced temperature, and we find that background cleavage elsewhere in the molecule is minimized if the digestion with the nuclease probe is performed at 15° or lower. Restriction enzymes[11] and methylases[25] can also be used as probes of cruciform structure, as the loops are refractory to both kinds of enzyme. Enzymes that recognize the four-way junction are the most specific probes of cruciform structure, and their use is the method of choice for many experiments such as measurement of extrusion kinetics. The resolving enzymes, such as T4 endonuclease VII,[51] exhibit high specificity for cruciform structures, and very low background levels of cleavage are found elsewhere in the molecule, compared to S1 nuclease, for example. These enzymes were previously available only in very small quantities, restricting their general use, but now the genes for T4 endonuclease VII[74] and T7 endonuclease I[75] have been cloned and overexpressed.

Most chemical probes react with bases that exhibit single-stranded character in the loops, although some such as osmium tetroxide may react with bases at the junction in the absence of sufficient cations to permit folding.[76,77] Many of these probes are discussed in Volume 212 of this series. Once again, some caution is required in the interpretation of results with chemical probes, particularly if only a single probe is employed. The exact nature of the target should be borne in mind, and to describe a chemical probe as single-strand selective may be an oversimplification in some cases. However, if used properly, such probes can be valuable.

[74] J. Tomaschewski, Ph.D. Thesis, Universität Bochum, Germany (1988).
[75] N. Panayotatos and A. Fontaine, *J. Biol. Chem.* **262,** 11364 (1987).
[76] J. A. McClellan and D. M. J. Lilley, *J. Mol. Biol.* **197,** 707 (1987).
[77] D. R. Duckett, A. I. H. Murchie, and D. M. J. Lilley, *EMBO J.* **9,** 583 (1990).

Chemicals such as osmium tetroxide and diethyl pyrocarbonate generate base adducts that are sensitive to alkali and may therefore be analyzed at single-base resolution, generating data of high precision. Others can be converted to the same precision by means of a second chemical modification.[78] Some chemical probes have been used inside cells,[79,80] and cruciform formation inside bacteria has recently been demonstrated, using osmium tetroxide and bipyridine to modify the loops of the extruded cruciform.[81]

Preparation of Supercoiled Plasmid DNA

Supercoiled DNA must be prepared and handled with care for two reasons. First, cleavage anywhere within the entire circular molecule completely changes the topology and releases the superhelical constraint. Second, because of their topology, the molecules are more sensitive to damage than normal DNA, and conditions that lower helix stability should be avoided. When preparing DNA for structural studies we do not use preparative methods based on alkali or thermal denaturation, as such vigorous denaturation of the molecules could well lead to the irreversible formation of alternative structures in the DNA. Instead we adopt procedures based on more gentle lysis of cells, followed by isopycnic gradient centrifugation of the supercoiled plasmid DNA in cesium chloride, in the presence of ethidium bromide. This method of separation is preferred for several reasons. First, the conditions are mild. Second, the separation depends on the topology of the DNA, and the method specifically separates constrained from nicked DNA circles. It therefore gives a higher yield of supercoiled DNA than other methods. Third, the DNA is taken from the gradient as positively supercoiled DNA in the presence of a high concentration of an intercalator, and it is free of cruciform and other structures that require negative supercoiling. Subsequent careful removal of the ethidium ions (we use solvent extraction at low temperature) can leave the DNA free of perturbed structures (such as most cruciforms) so that the kinetics of their formation may be studied.

The following is the procedure used in our laboratory to prepare supercoiled plasmid DNA for structural studies, including the kinetics of cruci-

---

[78] Y. Kohwi and T. Kohwi-Shigematsu, *Proc. Natl. Acad. Sci. U.S.A.* **85,** 3781 (1988).

[79] S. Sasse-Dwight and J. D. Gralla *J. Biol. Chem.* **264,** 8074 (1989).

[80] E. Palecek, E. Rasovska, and P. Boublikova, *Biochem. Biophys. Res. Commun.* **150,** 731 (1988).

[81] J. A. McClellan, P. Boublikova, E. Palecek, and D. M. J. Lilley, *Proc. Natl. Acad. Sci. U.S.A.* **87,** 8373 (1990).

form extrusion. The protocol utilizes pBR322-based plasmids transformed into a common laboratory *Escherichia coli* strain such as HB101.

Growth and Lysis of Bacterial Cells

Exact growth conditions will depend on strain, plasmid, and conditions; the following applies to a pBR322-type plasmid grown in HB101. Prepare 500 ml of M9 medium in a 2-liter culture flask, containing 1.5 g KH_2PO_4, 3.0 g Na_2HPO_4, 0.25 g NaCl, and 0.5 g NH_4Cl, and autoclave. When cool, add the following, using aseptic technique: 2.5 ml of 40% glucose, 25.0 ml of 10% casamino acids, 1.0 ml of 1 mg/ml thiamin, 0.5 ml of 1 M $MgSO_4$, and 2.0 ml of 0.25% thymidine. Any antibiotic required for selection should be added at this stage. Inoculate with 5 ml of a fresh overnight culture of the plasmid-bearing strain, grown in LB medium. Shake the flask with good aeration at 37° until the cell density gives an A_{660} of 0.6 (usually 3–4 hr after inoculation). If chloramphenicol amplification is desired, 75 mg chloramphenicol is added at this stage, and shaking continued for a further 16–18 hr. Longer growth periods can result in lower levels of supercoiling in the extracted plasmid.

The cells are collected by centrifugation at 7500 rpm for 10 min at 4°. The supernatant is decanted into dilute hypochlorite solution. The cellular pellet is resuspended in 7.5 ml of 25% sucrose in 50 mM Tris, pH 8.0, and kept on ice for 10 min. The following additions are made, with constant swirling to ensure good lysis (the vessel is placed on ice for 10 min between each addition). (1) 0.75 ml of 10 mg/ml lysozyme in 25 mM Tris, pH 8.0; (2) 1.5 ml of 250 mM EDTA; (3) 1.5 ml of 5 M NaCl; and (4) 1.5 ml of 10% sodium dodecyl sulfate (SDS). At this point lysis occurs, and the suspension becomes very viscous. The lysate is left for 4–20 hr on ice.

Isopycnic Gradient Centrifugation of Plasmid DNA

The lysate is cleared by centrifugation at 20,000 rpm for 45 min at 4°; the supernatant is recovered and its volume measured. To this is added 0.97 g of CsCl and 25 μl of a 10 mg/ml solution of ethidium bromide, per milliliter of supernatant. The mixture is centrifuged at 10,000 rpm for 10 min, and the clear solution is carefully removed from the dark pellicle and placed in ultracentrifuge tubes. For the fastest density gradient centrifugation we employ a vertical rotor (Beckman VTi 65). The centrifugation is performed at 54,000 rpm at 15° for 7–16 hr. The (lower) plasmid band is recovered by side puncture (it is usually quite visible in daylight; a UV light source can be employed, but there is a danger of UV-induced damage of supercoiled DNA) and combined in a fresh centrifuge tube, which can be filled if necessary with a solution of 0.97 g/ml CsCl, 250 μg/ml ethidium bromide. This is centrifuged at 54,000 rpm at 15° for at least 10 hr. The plasmid DNA is again recovered by side puncture, transferred to a microcentrifuge tube (Eppendorf), and placed on ice.

Preparation of Cruciform-Free Supercoiled DNA

At this stage the DNA is still in the presence of a high concentration of ethidium bromide. As a consequence it is positively supercoiled and free of any structures that are underwound, including cruciforms. To regenerate negatively supercoiled DNA the ethidium bromide must be removed; this must be done using conditions that do not result in the formation of cruciform structures and is therefore done at low temperature throughout, and with avoidance of helix-destabilizing agents. We employ repeated extraction into 1-butanol to remove ethidium bromide, followed by dialysis.

An equal volume of 1-butanol at 0° is added to the plasmid solution and vortexed. The layers are separated, and the organic (upper) phase removed. A further volume of 1-butanol is added, and the entire procedure repeated 7 times. Care should be taken throughout to maintain a low temperature. It should be noted that the extraction also reduces the volume of the aqueous phase to some extent, and if this becomes excessive precipitation of CsCl will occur. This can be prevented by restoration of volume. Following extraction of ethidium bromide, the negatively supercoiled DNA is extensively dialyzed against 10 mM Tris, pH 7.5, 0.1 mM EDTA at 4°, in order to remove CsCl. On removal from the dialysis sac the DNA is not manipulated further, and it can be used directly for kinetic and other measurements.

The yield of DNA depends on the plasmids and strains used, but typical yields range from 200 μg to 2 mg DNA from 1 liter of culture. The concentration after the second gradient is normally in the range 200–1000 μg/ml. The DNA should be over 95% supercoiled at the end of the preparation. We store the supercoiled DNA in aliquots at −70°, and gently thaw the frozen DNA at 4° just before use.

Preparation of Single Topoisomeric Species

For some studies it is useful to work with preparations of a circular plasmid that contain one single topoisomeric species or a narrow Gaussian distribution of topoisomers. For an example, see Lilley and Hallam.[13] Preparation of the latter is described in [5] in Volume 212, this series. The high resolving power of agarose gels is used for the separation of topoisomers, and the individual species are excised and electroeluted.

A topoisomer distribution centered on the linking difference of interest must first be prepared by relaxation of supercoiled plasmid with topoisomerase I in the presence of an appropriate concentration of ethidium bromide (see [5] in Volume 212, this series). Ten to sixty micrograms of DNA is relaxed and loaded on to a 1 cm thick agarose gel in 90 mM Tris–borate, pH 8.3, 1 mM EDTA (TBE buffer). A single lane of 2 cm width can take 12 μg of DNA without overloading. The 25 cm long gel is electrophoresed until the supercoiled DNA has traveled about 80% of the length of the gel (the speed of migration will depend on the size of plasmid, and the time of electrophoresis will have to be determined in a trial experiment). The bands are visualized by staining in 1 μg/ml ethidium bromide and illumination on a transilluminator at 300 nm. We find that a glass plate inserted between the transilluminator and the gel helps to avoid DNA breakage by the UV light, thereby improving the integrity of the topoisomers. The required topoisomers are excised with a scalpel and electroeluted. This DNA is sufficiently pure for most purposes but can be further purified if required, using isopycnic centrifugation. For this we employ a small-angle rotor in a Beckman TL100 centrifuge, taking a tube of 1 ml volume. We find that quantities of DNA as low as 1 μg can be successfully purified using this method.

Rate Measurements of Cruciform Extrusion

In general, cruciform extrusion is highly temperature dependent because of large enthalpies of activation. Hence conditions can be chosen under which the process is relatively slow, that is, on a time scale of minutes to hours. Moreover, the temperature dependence can be exploited as a means of starting and ending the extrusion process at will; thus, despite the use of rather unsophisticated techniques, it is possible to study extrusion during time intervals down to about 30 sec. Two kinds of kinetic experiments may be performed: (1) To examine the time course of cruciform extrusion under a fixed set of conditions, a sample of the plasmid is incubated at a given temperature and buffer conditions, and aliquots are removed at various times for subsequent analysis of cruciform extrusion. (2) For analysis of cruciform extrusion as a function of conditions, a set of samples can be incubated for a fixed period, while varying a particular parameter such as sodium ion concentration. At the end of the period all samples are analyzed for cruciform extrusion. The outline of the procedure for kinetic measurements can be summarized as follows:

Preparation of Cruciform-Free DNA. DNA preparation is performed as above. At the beginning of the kinetic procedure the DNA is diluted into the required buffer at 4°. No extrusion occurs under these conditions.

Extrusion Interval. The supercoiled DNA sample at a concentration of about 30 μg/ml is then placed in a water bath at the required temperature for a certain time, in the buffer conditions (composition, pH, ionic strength) appropriate to the experiment. Aliquots may be removed at various times from a sample incubated at the fixed temperature. Because of the high temperature dependence of these reactions, accurate and precise temperature control is essential. We routinely use water baths that can be set with a precision of 0.01°.

Termination of Extrusion. The sample or aliquot is rapidly transferred to ice to stop any cruciform extrusion. After 5 min, concentrated reaction buffer is added, and the sample is left for a further period.

Assay of Cruciform Formation. The assay of cruciform formation is carried out using a nuclease that selectively cleaves the cruciform structure. In the majority of published studies S1 nuclease has been used to cleave the cruciform loops, but the advent of cloned resolving enzymes makes these the reagents of choice owing to their much higher specificity for the cruciform. Both types of enzymes can be used quite successfully at 15°, under which conditions no cruciform extrusion occurs during the incubation (at least, for the great majority of sequences). At the end of the incubation, the DNA is ethanol precipitated, redissolved, and cleaved to completion with a restriction enzyme. The DNA is then analyzed by gel electrophoresis. It is best to choose a restriction enzyme that cleaves once only in the plasmid, at a site relatively close to the inverted repeat studied.

Typical results of a time course of cruciform extrusion are given in Fig. 4. The upper band (band F) is the full-length linear DNA, which corresponds to restriction cleavage of DNA that had not been cleaved at a cruciform structure. The second band (band C) migrating ahead of the full-length species is that arising from an initial cleavage by the nuclease (S1 nuclease in this example) at the cruciform, followed by restriction cleavage; this band is diagnostic for the presence of the cruciform, and its intensity is related to the amount of cruciform extrusion that occurred during the incubation step. This may be quantified by densitometry. If T4 endonuclease VII is used to assay for cruciform extrusion, the relative intensity of band C can be taken as a good estimate of the amount of cruciform extrusion in absolute terms. However this is not the case for S1 and other less specific nucleases. With these nucleases there is too much nonspecific cleavage elsewhere in the molecule. Under these conditions, the intensity of band C is only *proportional* to the extent of extrusion, and one obtains only a relative measure of the degree of cruciform extrusion. However, this is adequate for the measurement of rate constants.

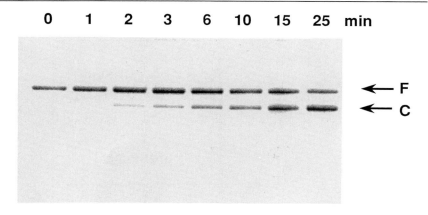

FIG. 4. Example of a time course for the extrusion of a cruciform in a supercoiled plasmid. The plasmid studied is pIRbke8, in which adenine bases at the center of the inverted repeat have been N[6]-methylated using EcoRI methylase and S-adenosylmethionine.[25] Cruciform extrusion proceeds by the S-type mechanism. Cruciform-free supercoiled DNA was incubated at 37° in 10 mM Tris-HCl, pH 7.5, 0.1 mM EDTA, 50 mM NaCl. Aliquots were removed at the times indicated and stored at 0°. They were then digested with 2 units of S1 nuclease in 50 mM sodium acetate, pH 4.6, 50 mM NaCl, 1 mM ZnCl$_2$ at 15° for 30 min, before complete cleavage by HindIII. The DNA was electrophoresed in 1% agarose, stained with 1 μg/ml ethidium bromide, and photographed under UV illumination. Two bands can be distinguished; band F is full-length plasmid, and band C is the shortened species arising from S1 nuclease cleavage at the loop of the extruded cruciform. The relative extent of extrusion may be estimated by densitometry as the ratio of intensities C/(F + C).

Cruciform extrusion is a unimolecular process represented by

$$D_u \xrightarrow{k} D_x$$

where D_u is unextruded plasmid DNA and D_x is cruciform-containing DNA. This is a first-order process that obeys the equation

$$d[D_u] = -k[D_u^t] \, dt \tag{7}$$

where D_u^t is the concentration of unextruded DNA at time t. A time course of the progress of cruciform extrusion is presented in Fig. 5 as a semilogarithmic plot, from the slope of which is obtained the rate constant (k) for the process. The half-time for the reaction may be calculated as

$$t_{1/2} = 0.69/k \tag{8}$$

A

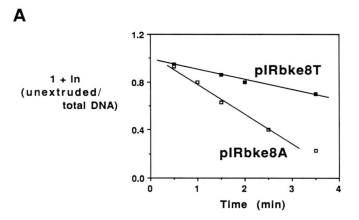

B

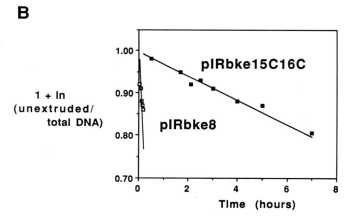

FIG. 5. Rates of cruciform extrusion for inverted repeats of related sequence, as shown by progress plots of ln(unextruded fraction) against time for four closely related plasmids. The inverted repeat sequences are all based on that of pIRbke8, with either one or two mutations in the repeat unit[23]; these plasmids extrude by the S-type mechanism, requiring the presence of salt to allow extrusion to proceed. Unextruded plasmid DNA was incubated at 37° in 10 mM Tris, pH 7.5, 50 mM NaCl, and aliquots were removed at different times and placed on ice. These were later digested with either (A) S1 nuclease or (B) T7 endonuclease I, followed by complete cleavage with HindIII. Note the very different time scales for extrusion for the different inverted repeat sequences. Under these experimental conditions the following reaction half-times were measured: pIRbke8, 70 min; pIRbke8A, 6 min; pIRbke8T, 17 min; pIRbke15C16C, 1649 min. Overall it was observed that the rate of extrusion within this series of sequences varied by a factor of almost 2000.

A

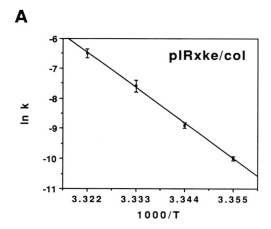

B

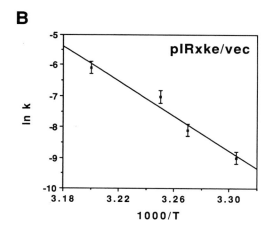

FIG. 6. Examples of Arrhenius plots of the temperature dependence of cruciform extrusion for typical C-type and S-type sequences. (A) Temperature dependence of the extrusion of a C-type plasmid pIRxke/col. Rate constants (k) were measured in 10 mM Tris-HCl, pH 7.5, 0.1 mM EDTA as a function of the absolute temperature (T). From these data an activation energy of 215 kcal mol^{-1} was calculated. (B) Temperature dependence of extrusion of an S-type plasmid pIRxke/vec, in 10 mM Tris-HCl, pH 7.5, 50 mM NaCl. An activation energy of 52 kcal mol^{-1} was calculated. The inverted repeat sequences of pIRxke/col and pIRxke/vec are identical; the plasmids differ in the sequences that flank the inverted repeats. In the case of pIRxke/col the contextual sequences are $(A + T)$-rich sequences derived from the plasmid ColE1, whereas for pIRxke/vec the xke inverted repeat is embedded in pBR322 sequences of normal base composition.[20]

The temperature dependence of cruciform extrusion is well described by the Arrhenius equation,

$$k = A \, e^{-E_a/RT} \tag{9}$$

where E_a is the Arrhenius activation energy, R the gas constant, T the absolute temperature, and A the temperature-independent, preexponential factor. Good straight lines have been obtained when rate constants measured at different temperatures are plotted against reciprocal temperature, as shown in Fig. 6. The slope of the line is $-E_a/R$, from which the Arrhenius activation energy is calculated.

[9] Protonated DNA Structures

By Maxim D. Frank-Kamenetskii

DNA Protonation

Canonical B-DNA has no strong protonation sites, and because of this it does not pick up protons with the exception of totally nonphysiological conditions of extremely low pH. Single-stranded DNA has stronger protonation sites, but it is protonated only below pH 4.3, the pK value for the N-3 position of free cytosines. However, this does not mean that protonation is an insignificant phenomenon for DNA. Some nonorthodox DNA structures, first of all those which include DNA triplexes, provide very strong protonation sites with pK values above 7, which means that these structures are protonated when formed under normal pH conditions.

The best studied examples of protonated DNA structures are H-DNA and intermolecular triplexes. Both types of unusual structures have a common element, the triplex, which carries the protonation sites. In DNA triplexes the protonation site is created because of formation of $C \cdot G*C^+$ base triads. In this triad, the usual Watson–Crick $C \cdot G$ pair is supplemented by protonated cytosine, which is attached to guanine via the Hoogsteen mode of binding. Not all triplexes are necessarily protonated, and protonated triplexes do not exhaust possible protonated structures. For example, the protonated $C \cdot C^+$ noncanonical base pair is another plausible candidate to be an element of an unusual DNA structure. Under neutral and slightly acid conditions (above pH 5), the triplex remains the major element of the protonation structures detected to date.

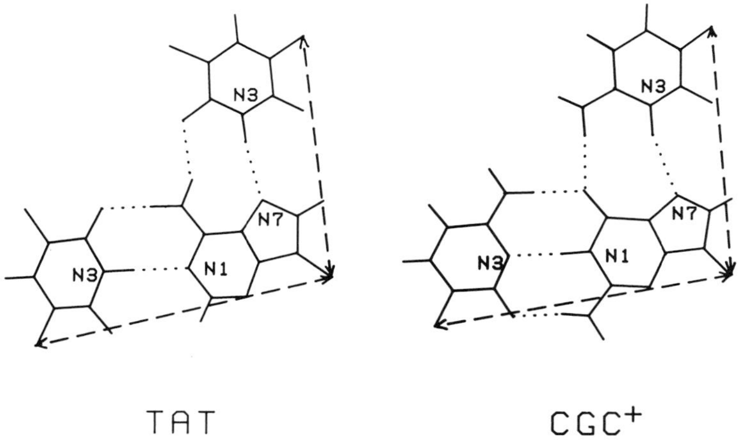

TAT   CGC+

FIG. 1. Isomorphous canonical base triads T·A*T and C·G*C+.

Triplexes and H-DNA

The protonated triplex is formed as a result of association, in the Hoogsteen manner, of a duplex carrying a homopurine–homopyrimidine sequence with the pyrimidine third strand, whose sequence is complementary to the sequence of the purine strand of the duplex. The Hoogsteen complementary principle is identical to the Watson–Crick one, with the exception of polarity of the strands:

$$\begin{array}{ll} \text{"Hoogsteen duplex"} & \begin{cases} 5'\text{-TCCTTCTCCC} \\ 5'\text{-AGGAAGAGGG} \end{cases} \\ \text{Watson–Crick duplex} & \begin{cases} 5'\text{-AGGAAGAGGG} \\ 3'\text{-TCCTTCTCCC} \end{cases} \end{array}$$

Therefore, the sequence of purines in the purine strand can be arbitrary, whereas the sequence in two pyrimidine strands of the triplex are deduced from it in accordance with the Watson–Crick and Hoogsteen complementary principles. In contrast to Watson–Crick duplex, in which the two strands are antiparallel, in the "Hoogsteen duplex" the two strands are parallel.

Such a pyrimidine–purine–pyrimidine (PyPuPy) triplex[1] consists of a regular array of base triads of two types, T·A*T and C·G*C+ (Fig. 1). These two base triads are isomorphous: the respective distances between glycosyl bonds in them are virtually equal (shown by double-headed arrows in Fig. 1).

[1] J. S. Lee, D. A. Johnson, and A. R. Morgan, *Nucleic Acids Res.* **6**, 3073 (1979).

In the Hoogsteen pairs A*T and G*C⁺ the N-3 nitrogens of pyrimidines form hydrogen bonds with the N-7 nitrogens of purines. In case of thymine, the required N-3 hydrogen atom is inherent in the usual tautomer. By contrast, cytosine normally has no hydrogen atom at the N-3 position. However, this position is the most favorable protonation site among all four canonical DNA nucleotides, with the pK value near 4.3. When the Hoogsteen pair between G and C is formed in a triplex, a much stronger attachment site for a proton at the N-3 position of C is created, and this significantly shifts the pK value as compared to the figure for free cytosine, pK_f(4.3), up to the pK_t value, which lies somewhere between 7 and 8.[2] As a result, the reaction of triplex formation appears to be strongly pH dependent within the whole region below neutral pH down to pH 4.3 for the case of an intermolecular triplex and down to even lower pH for the case of an intramolecular triplex (H-DNA).

H-DNA is formed in homopurine–homopyrimidine mirror repeats as a result of internal disproportionation to form a triplex of the original duplex and a single-stranded region.[3-5] Figure 2 shows H-DNA for a mirror repeat:

$$5' \cdots \text{TCCTTCTCCC} \cdots\cdots\cdots \text{CCCTCTTCCT} \cdots 3'$$

The mirror-repeat requirement for H-DNA-forming sequences immediately follows from the Hoogsteen complementarity principle. Two isomeric forms of H-DNA are possible,[3-6] with the triplex on the 3' end of the purine strand or on the 5' end. They are designated as H-y3 and H-y5, respectively,[6] where "y" is incorporated to indicate that the third strand in the triplex is the pyrimidine one.

Theoretical Background

We first consider an intramolecular protonated structure. Let m be the number of base pairs within DNA undergoing the transition from canoni-

[2] F. B. Howard, J. Frazier, M. N. Lippsett, and H. T. Miles, *Biochem. Biophys. Res. Commun.* **17**, 93 (1964).
[3] V. I. Lyamichev, S. M. Mirkin, and M. D. Frank-Kamenetskii, *J. Biomol. Struct. Dyn.* **3**, 667 (1986).
[4] S. M. Mirkin, V. I. Lyamichev, K. N. Drushlyak, V. N. Dobrynin, S. A. Filippov, and M. D. Frank-Kamenetskii, *Nature (London)* **330**, 495 (1987).
[5] M. D. Frank-Kamenetskii, *in* "DNA Topology and Its Biological Effects" (N. R. Cozzarelli and J. C. Wang, eds.), p. 185. Cold Spring Harbor Laboratory, Cold Spring Harbor, New York, 1990.
[6] H. Htun and J. E. Dahlberg, *Science* **243**, 1571 (1989).

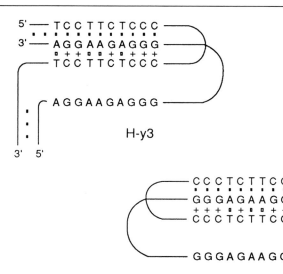

FIG. 2. Two isoforms of H-DNA. Watson–Crick pairing is labeled by the filled symbols, nonprotonated Hoogsteen pairing is shown by open symbols, and protonated Hoogsteen pairing is shown by plus symbols.

cal B-DNA into a protonated structure. The free energy (ΔF) of transition per base pair will be[7,8]

$$\Delta F = \Delta F_o - (RT/r) \ln(1 + 10^{pK_t - pH}) \tag{1}$$

where r is the number of base pairs per one protonated site in our protonated structure; pK_t is the pK value of cytosine within the structure. ΔF_o is the free energy of formation of the structure under consideration with all protonation sites unoccupied (i.e., at $pH > pK_t$). Quite naturally, at this region the stability of the "protonated structure" does not depend on pH [see Eq. (1)]. By contrast, the stability of the protonated structure is strongly pH dependent at pH values below pK_t:

$$\Delta F = \Delta F_o^* + pH(RT/r) \, 2.3 \tag{2}$$

[7] V. I. Lyamichev, S. M. Mirkin, and M. D. Frank-Kamenetskii, *J. Biomol. Struct. Dyn.* **3**, 327 (1985).

[8] V. I. Lyamichev, S. M. Mirkin, and M. D. Frank-Kamenetskii, *J. Biomol. Struct. Dyn.* **5**, 275 (1987).

Theoretically, Eq. (2) is valid down to pH values that correspond to protonation of the DNA double helix, in other words, within a very wide range of pH.

Equation (2) predicts a linear dependence of the free energy of stabilization of the intramolecular protonated structure on pH. This manifests itself experimentally, most significantly in the pH dependence of supercoiling that induces a transition into the protonated structure under superhelical stress. Equation (2) leads to a simple equation for superhelical density (σ) of the transition[7,8]:

$$\sigma_{tr} = (0.1/r)(pH_o - pH) \tag{3}$$

The pH_o value is the value of pH at which the H form is extruded in linear, topologically unconstrained DNA. The range of validity of Eq. (3) is the same as that of Eq. (2). Moreover, because the superhelical free energy as a function of superhelical density is well known (see [5] in Volume 212 of this series), one can determine the energetic characteristics of a protonated structure by inducing the transition to it from B-DNA under superhelical stress.[5]

In the case of an intermolecular protonated structure (we mean the intermolecular triplex in particular), the reference state corresponds to the double helix plus the single-stranded pyrimidine oligonucleotide. Therefore, instead of Eq. (1) one has

$$\Delta F = \Delta F_o - (RT/r) \ln[(1 + 10^{pK_f - pH})/(1 + 10^{pK_t - pH})] \tag{4}$$

where $pK_f < pK_t$ is the pK value for the single-stranded oligonucleotide (pK_f 4.3). Again, at $pH > pK_t$ we have a structure with unoccupied protonation sites. Its stability does not depend on pH. Equation (2) is also valid but within a narrower range: $pK_t > pH > pK_f$. Below pK_f both states, the intermolecular triplex and the duplex plus oligonucleotide, are equally protonated, and equilibrium between them ceases to depend on pH [see Eq. (4)].

The lifetime (τ) of an intramolecular triplex should be an extremely sharp function of pH, in the range $pK_f < pH < pK_t$, as follows from Eq. (2)[9]:

$$\tau_t = \tau_o \, 10^{-NpH/p} \tag{5}$$

where N is the number of residues in the oligonucleotide. It follows from Eq. (5) that small changes of pH may entail a dramatic conversion of a weak and reversible complex to a virtually irreversible one.

[9] V. I. Lyamichev, S. M. Mirkin, M. D. Frank-Kamenetskii, and C. R. Cantor, *Nucleic Acids Res.* **16**, 2165 (1988).

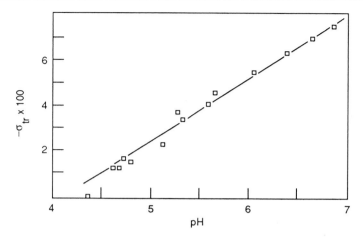

FIG. 3. Dependence of superhelical density of the B-to-H transition on pH for the $(TC)_n \cdot (GA)_n$ insert. (Data are from Lyamichev *et al.*[7])

Detection of Intramolecular Protonated Structures

Two-Dimensional Gel Electrophoresis

To be detected by two-dimensional (2-D) electrophoresis, the transition of a DNA region from canonical B-DNA to a protonated structure should be associated with a significant unwinding of the double helix. For example, if a homopurine–homopyrimidine insert adopts the H form, topologically this is the same as unwinding of the whole insert. Therefore H-DNA can be easily detected by 2-D gel electrophoresis. The method is described at length in [5] in Volume 212 of this series. The formation of a protonated structure is easily recognizable by this method because of a strong dependence on pH of the discontinuity position on the 2-D patterns. Figure 3 shows such data for H-DNA first reported by Lyamichev *et al.*[7] It is very convenient, when possible, to present these data in the form of a dependence of the superhelical density of transition on pH (Fig. 3). A strong dependence of σ_{tr} on pH unambiguously testifies to the effect that the structure under study is protonated. Moreover, the slope of the linear part of the dependence curve yields the number r of base pairs in the insert per one protonation site [see Eq. (3)]. Available experimental data agree with Eq. (3).[4,7,8]

Two-dimensional gel electrophoresis proved to be a very efficient

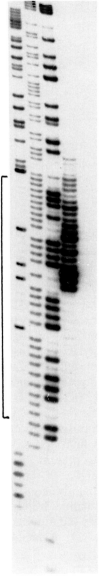

method to study protonated structures. The initial proposal of H-DNA stemmed from data from 2-D gels.[3,7] This method has also detected a protonated structure in telomeric sequences from *Tetrahymena*[11] and human.[12] However, 2-D gels are not enough to arrive at definite conclusions about the nature of a transition they detect.

Enzymatic and Chemical Probing

Historically, intramolecular protonated structures were first detected owing to their preferential cleavage by the single-strand-specific nuclease S1. This enzyme is known to prefer acid conditions, and because of this it detected protonated structures. Ironically, in spite of the popularity of S1 probing of unusual DNA structures in general and protonated ones in particular, until recently the method did not permit one to obtain reliable results at sequence resolution. As was shown[10,12] the major cause of this stemmed from the fact that in a standard protocol for S1 probing 1 mM Zn^{2+} is used (zinc ions are essential for functioning of the enzyme). Such high zinc ion concentration can affect the formation of unusual structure. To avoid this, it is recommended to decrease sharply the amount of the Zn^{2+} added.[10,12] When 30 μM of $ZnCl_2$ or $ZnSO_4$ is used in the standard S1 digestion protocol (30 mM sodium acetate, pH 4.0–5.0, 50 mM NaCl), very good results are obtained[10,12] (see Fig. 4).

All usual chemical probes, which are used to study unusual DNA structures, are applicable to the protonated structures as well because all of them work quite well at acid pH. Most popular are diethyl pyrocarbonate (DEP), which detects unpaired purines, preferentially adenines; dimethyl sulfate (DMS), which detects guanines (both in single strands and in

---

[10] B. P. Belotserkovskii, A. G. Veselkov, S. A. Filippov, V. N. Dobrynin, S. M. Mirkin, and M. D. Frank-Kamenetskii, *Nucleic Acids Res.* **18,** 6621 (1990).

[11] V. I. Lyamichev, S. M. Mirkin, O. N. Danilevskaya, O. N. Voloshin, S. V. Balatskaya, V. N. Dobrynin, S. A. Filippov, and M. D. Frank-Kamenetskii, *Nature (London)* **339,** 634 (1989).

[12] O. N. Voloshin, A. G. Veselkov, B. P. Belotserkovskii, O. N. Danilevskaya, M. N. Pavlova, V. N. Dobrynin, and M. D. Frank-Kamenetskii, *J. Biomol. Struct. Dyn.* **9,** 643 (1992).

FIG. 4. S1 probing of H-DNA within a supercoiled plasmid (lane 4). The data for the purine strand are shown. The bracketed sequence, 5'-AAGGGAGAATGGGGTA-TAGGGGTAAGAGGGAA, forms the H-y3 structure (see Fig. 2). The 3' end is at the bottom. Lanes 1–3 are the Maxam–Gilbert ladders for pyrimidines, purines, and guanines. Lane 5 corresponds to a control experiment without S1 nuclease added. (Data are from Ref. 10.)

duplexes) with unprotected N-7 positions; osmium tetroxide (OT), which detects unpaired thymines (see [8] and [17] in Volume 212 of this series); and potassium permanganate (PP), which detects unpaired pyrimidines, preferentially thymines. Osmium tetroxide is more effective in alkaline media (O. Voloshin, unpublished), whereas PP works better in acidic media. Examples of the application of these and other probes to study intramolecular protonated structures can be found in Refs. 5, 6, 11, and 12, and in corresponding chapters in this volume.

Detection of Intermolecular Protonated Structures

Comigration Assay

The comigration assay was developed by Lyamichev *et al.*[9] It is based on the fact that with decreasing pH the lifetime of the complex between a single-stranded pyrimidine oligonucleotide and the corresponding duplex dramatically increases [see Eq. (5)]. As a result conditions could be easily reached at which even short pyrimidine oligonucleotides (dodecamers and probably even shorter) form virtually irreversible complexes, which would remain intact during the gel electrophoresis experiment.

Several micrograms of cold plasmid DNA are mixed with a picomolar quantity of end-labeled oligonucleotide in the electrophoretic sodium citrate buffer (100 mM sodium, pH 4.8), and the mixture is incubated for 10–15 min at 40–45°. Then the sample is cooled and layered on a 1.5% agarose gel. Electrophoresis is performed with an electric field strength of 1.3 V/cm at room temperature for 15 hr. To maintain constant pH, the buffer is constantly pumped between the two chambers of the apparatus during electrophoresis. After electrophoresis the gel is radioautographed, stained with ethidium bromide, and photographed.

As a result one obtains two images of the same gel (Fig. 5). The radioautograph indicates where the oligonucleotide moved, whereas the photograph of the dye-stained gel shows the position of unlabeled DNA. If the two images coincide, one concludes that the oligonucleotide was bound to DNA. In the opposite case one sees no radiolabel at the place of DNA but rather a smear at the lower part of the gel.

The comigration method proved to be efficient in detecting intermolecular triplexes between DNA carrying homopurine–homopyrimidine tracts and corresponding pyrimidine oligonucleotides. On the basis of the comigration assay Moores has developed a vector for gene mapping.[13] In

[13] J. C. Moores, *Strategies Mol. Biol.* **3**, 23 (1990).

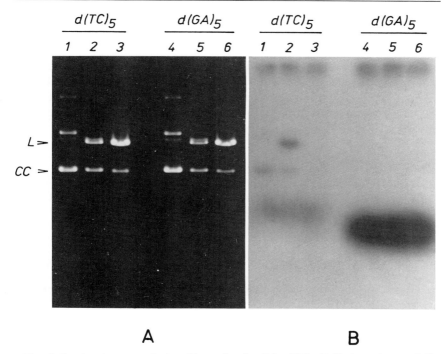

A   **B**

FIG. 5. Comigration assay. A plasmid carrying the 45-bp $(CT)_n \cdot (AG)_n$ insert in superhelical (lanes 1, 4) and linear (lanes 2, 5) forms as well as the vector plasmid (without the insert)
(lanes 3, 6) were incubated either with $d(TC)_5$ (lanes 1–3) or $d(GA)_5$ (lanes 4–6) under
conditions indicated in the text. (A) Ethidium-stained gel (L indicates linear DNA, CC
indicates closed circular DNA). (B) Radioautograph of the same gel as in (A). (Data are from
Ref. 9.)

this triple helix vector a polylinker, into which genomic DNAs are cloned,
is flanked by two different homopurine–homopyrimidine tracts. Applying
the above technique using either one oligonucleotide or the other, it is
possible to identify lengths of fragments either from one tract or from the
other. Such an approach permits one to avoid blotting hybridization.

Footprinting

The general idea of footprinting methods consists of modification of
the chemical groups on DNA, which are protected in case of binding. After
such modification the binding site manifests itself as an unmodified region.

In case of protonated triplexes, an appropriate agent is DMS. Other agents may be also used, such as copper phenanthroline.[14]

Lyamichev et al.[15,16] applied a photofootprinting assay, which is based on the fact that triplex formation strongly protects the duplex against photodamage, particularly the formation of [6 − 4]-photoproducts. Experimentally, these photoproducts are converted to chain breaks in the same way as sites of chemical modification, that is, by the hot piperidine treatment (see [14] in Volume 212 of this series).

Affinity Cleavage

In the affinity cleavage method, oligonucleotides are tagged with chemically (or photochemically) active groups. Complex formation manifests itself in cleavage of the DNA target in the presence of the tagged oligonucleotide.

Triplex Mode of Recognition

Numerous techniques in modern molecular biology and biotechnology are grounded on the Watson–Crick complementary recognition. The sequence specificity of the recognition stems from the fact that even solitary mismatches significantly weaken the binding of an oligonucleotide as compared with the complementary one.

Triplex formation provides the unique opportunity to recognize a specific DNA sequence without separation of the complementary strands in the duplex, owing to the Hoogsteen complementary recognition. Solitary mismatches produce in this case virtually the same weakening effect as in the case of Watson–Crick recognition.[10]

The triplex mode of recognition has its advantages and disadvantages as compared to the Watson–Crick one. Obviously, whereas the triplex mode of binding permits one to recognize duplex DNA, its application is restricted to only homopurine–homopyrimidine sequences. The major advantage of the protonated triplexes in practical applications stems from the opportunity to monitor their stability and specificity by changing pH, which is impossible in the case of Watson–Crick recognition. In the triplex mode of recognition, pH manifests itself as an extremely powerful and sensitive lever of regulation. At low pH one reaches virtually irreversible

[14] J.-C. Francois, T. Saison-Behmoaras, and C. Helene, Nucleic Acids Res. 16, 11431 (1988).
[15] V. I. Lyamichev, M. D. Frank-Kamenetskii, and V. N. Soyfer, Nature (London) 344, 568 (1990).
[16] V. I. Lyamichev, O. N. Voloshin, M. D. Frank-Kamenetskii, and O. N. Soyfer, Nucleic Acids Res. 19, 1633 (1991).

attachment of an oligonucleotide to its Hoogsteen-complementary sequence on DNA [see Eq. (5)]. This opens the opportunity of site-specific radiolabeling of duplex DNA[9,13] (see also Fig. 5). On the other hand, at neutral pH, where triplex formation process is highly reversible, one can achieve a fantastic specificity of recognition of DNA sequences, as demonstrated by Strobel and Dervan.[17]

[17] S. A. Strobel and P. B. Dervan, *Nature (London)* **350,** 172 (1991).

[10] Guanine Quartet Structures

By Dipankar Sen and Walter Gilbert

Introduction

Of the five nucleosides commonly found in DNA and RNA, only guanosine is capable of forming extensive "self"-structures in solution, hydrogen bonding with itself to give G-G base pairs and guanine quartets (reviewed in Ref. 1). This phenomenon was initially studied with guanosine itself, but also with 5′- and 3′-GMP and poly(I) and poly(G).[2-8] The formation of these structures was found to be uniquely dependent on monovalent cations and on the identity of the cation.[9-13] Recently, natural sequences from chromosomal telomeres and immunoglobulin switch regions were shown to form two different kinds of noncanonical structures: parallel quadruplexes containing G quartets[14] and also fold-back duplexes

[1] W. Saenger, "Principles of Nucleic Acid Structure," p. 315. Springer-Verlag, New York, Berlin, Heidelberg, and Tokyo, 1984.
[2] M. Gellert, M. N. Lipsett, and D. R. Davies, *Proc. Natl. Acad. Sci. U.S.A.* **48,** 2013 (1962).
[3] P. Tougard, J.-F. Chantot, and W. Guschlbauer, *Biochim. Biophys. Acta* **308,** 9 (1973).
[4] S. Arnott, R. Chandrasekaran, and C. M. Martilla, *Biochem. J.* **141,** 537 (1974).
[5] S. B. Zimmerman, G. H. Cohen, and D. R. Davies, *J. Mol. Biol.* **92,** 181 (1975).
[6] S. B. Zimmerman, *J. Mol. Biol.* **92,** 181 (1975).
[7] V. Sasisekharan, S. B. Zimmerman, and D. R. Davies, *J. Mol. Biol.* **92,** 171 (1975).
[8] P. Mariani, *et al. J. Am. Chem. Soc.* **111,** 6369 (1989).
[9] H. T. Miles and J. Frazier, *Biochim. Biophys. Res. Commun.* **49,** 199 (1972).
[10] T. J. Pinnavaia, *et al. J. Am. Chem. Soc.* **100,** 3625 (1978).
[11] H. T. Miles and J. Frazier, *J. Am. Chem. Soc.* **100,** 8037 (1978).
[12] F. B. Howard and H. T. Miles, *Biochemistry* **21,** 6736 (1982).
[13] F. B. Howard and H. T. Miles, *Biopolymers* **21,** 147 (1982).
[14] D. Sen and W. Gilbert, *Nature (London)* **334,** 364 (1988).

containing G-G base pairs.[15] Further investigation has shown that quartet structures constitute a closely related family,[16-20] whose individual members differ in their overall molecularity and in the relative orientations of their strands (either parallel or antiparallel). The formation of these alternative structures by specific sequences depends on such factors as temperature, DNA concentration, and, most importantly, on the molar ratios of sodium and potassium ions in the reaction buffers.

The major G-quartet structures (illustrated in Fig. 1) can be classified as follows: (1) tetramolecular parallel quadruplexes (G4-DNA)[14,18]; (2) unimolecular antiparallel quadruplexes (which we shall term G4'-DNA)[16,19,20]; and (3) bimolecular antiparallel quadruplexes (G'2-DNA).[16-18] In addition, there are unimolecular hairpin structures containing G-G base pairs.[15] Antiparallel tetramolecular quadruplexes have not been described. Table I lists a number of G-rich sequences and the various quartet structures formed by each.

G4-DNA is formed by single-stranded DNA molecules that contain either simple or complex guanine motifs, meaning, respectively, short single stretches of 3–10 contiguous guanines in a DNA molecule or multiple stretches of contiguous guanines separated from each other by other bases. G4-DNA complexes, even those formed by simple guanine motifs, are exceedingly stable. Thus, the complex formed by a 12-base oligomer that contained four contiguous guanines (and thus four quartets) had a melting point of above 90° in 50 mM KCl.[18] In general, the fraction of "useful" guanines (i.e., those that participate in quartet formation) out of the total number of bases in an oligomer correlates positively with the rate of G4-DNA formation by that oligomer and also with the thermal stability of the complex. A rigorous study on this dependence is not yet available; however, the presence of long stretches of bases not involved in base pairing should contribute unfavorably to the stability of quartet structures formed by an oligomer. Thus, the likelihood of formation of a stable G4-DNA complex by a given oligomer will depend on the balance between the favorable free energy of quartet formation and the unfavorable contributions of bases not involved in pairing or quartet formation. So far, the lowest useful guanine content in an oligomer shown to form a G4-DNA

[15] E. R. Henderson, C. C. Hardin, S. K. Wolk, I. Tinoco, Jr., and E. H. Blackburn, *Cell (Cambridge, Mass.)* **51**, 899 (1987).

[16] J. R. Williamson, M. K. Raghuraman, and T. R. Cech, *Cell (Cambridge, Mass.)* **59**, 899 (1989).

[17] W. I. Sundquist and A. Klug, *Nature (London)* **342**, 825 (1989).

[18] D. Sen and W. Gilbert, *Nature (London)* **344**, 410 (1990).

[19] I. G. Panyutin, O. I. Kovalsky, and E. I. Budowsky, *Nucleic Acids Res.* **17**, 8257 (1989).

[20] I. G. Panyutin, O. I. Kovalsky, E. I. Budowsky, R. E. Dickerson, M. E. Rikhirev, and A. A. Lipanov, *Proc. Natl. Acad. Sci.* **78**, 3015 (1990).

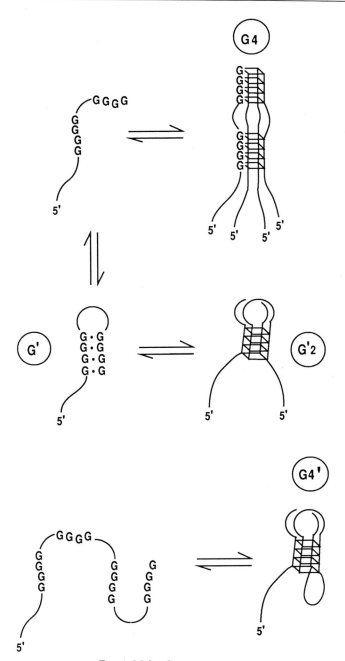

FIG. 1. Major G-quartet structures.

TABLE I
G-Rich Sequences and Quartet Structures

Sequence[a]	Structure	Ref.
ACTGTCGTACTTGATATT**GGGGC**	G4	18
GGGACCAGACCTAGCAGCTAT**GGGGG**AGCT**GGGG**AAG**GTGGG**AATGTGA	G4, G'2	14,18
ACTGTCGTACTTGATATTTT**GGGG**TTTT**GGGG**	G4, G'2	17,18
TT**GGGG**TT**GGGG**TT**GGGG**TT**GGGG**	G4'	16
TTTT**GGGG**TTTT**GGGG**TTTT**GGGG**TTTT**GGGG**	G4'	16
. . . G**GGGGGGGGGGGGGGGGGGGGGGGGGGGGG** . . .	G4, G4'	5,20

[a] Guanines that participate in quartet formation in G4-DNA are shown in bold type. Guanines that participate in quartet formation in other classes of quadruplex structures are underlined.

complex stable in 50 mM Tris–borate, 1 mM EDTA (TBE) at 22° are nine guanines within the complex motif G_5AGCTG_4 of a 71-base oligomer.[18]

Unimolecular antiparallel quadruplexes (G4'-DNA) can only be formed by oligomers that contain a complex guanine motif made up of at least four separated stretches of three or four contiguous guanines. Existing studies have focused primarily on telomeric sequences in which the intervening bases have been thymine or thymine and adenine.[16,17] DNA molecules with very long stretches (i.e., >27 bases) of contiguous guanines will also form these structures.[19,20]

Bimolecular antiparallel quadruplexes (G'2-DNA) are structurally analogous to G4'-DNA, the difference between them being that in G'2-DNA two distinct DNA molecules, each containing two separated stretches of guanines, have bonded together to form an antiparallel quadruplex.[16,17]

Specific Methods

In general, the optimal conditions for forming G-quartet structures are at neutral pH in buffers that contain monovalent ions. In general, sodium and potassium ions favor the formation of quartet structures, whereas lithium does not. Thus, where the formation of guanine-quartet structures by G-rich single-stranded DNAs is *not desired,* one should exclude sodium and potassium and substitute lithium, or even tetramethylammonium, in buffers.

In this chapter we do not deal with the details of DNA oligomer synthesis and purification, nor with oligomer end-labeling or the methods of denaturing and nondenaturing gel electrophoresis. Detailed protocols for these may be found elsewhere in this series.

G4-DNA: Simple Motifs (Oligomers Containing Three to Six Contiguous Guanines)

1. Dissolve the DNA oligomer (end-labeled as desired) in 10 μl of 10 mM Tris, pH 7.5, 1 mM EDTA (TE buffer) at DNA concentrations in the range of 100 μM – 1 mM oligomer.

2. Heat at 95° for 90 sec in a closed Eppendorf tube and chill on ice. Spin briefly.

3. Add 10 μl of TE plus 100–500 mM NaCl or KCl, and mix by vortexing. The use of sodium or potassium is a matter of choice, but it should be noted that the potassium complexes are significantly more stable ($\Delta T_m \sim 20°$) than the sodium complexes.[18]

4. Leave at room temperature for 30 min for G4-DNA to form.

The optimal electrophoresis conditions for resolving G4-DNA of simple motif oligomers from their single-stranded forms (usually the two are at equilibrium under the above buffer conditions at 20° or higher) are to chill samples on ice after incubation and to load in 6–12% nondenaturing gels run in 50 mM TBE plus 5 mM KCl, at 4° (if there are five or more contiguous guanines in the oligomer, the gel can be run at room temperature). To obtain clean product bands, it is recommended that even nondenaturing gels be prerun for a period prior to loading samples. G4-DNA generally runs significantly behind the single-stranded form of the oligomer in nondenaturing gels.

G4-DNA: Complex Motifs (Oligomers Containing Multiple Sets of Contiguous Guanines). An important feature of the formation of complex G4-DNA motifs is that the formation is promoted by sodium and inhibited by potassium. However, the optimal conditions for formation is a mixture of the two, usually Na:K in a 10:1 to 20:1 ratio.[18]

1. Dissolve the DNA oligomer (end-labeled as desired) in 10 μl of TE buffer (see above) at DNA concentrations in the range of 50 μM – 1 mM oligomer.

2. Heat at 95° for 120 sec in a closed Eppendorf tube and chill on ice. Spin briefly.

3. Add 10 μl of TE plus 1.9 M NaCl and 0.1 M KCl. Mix by vortexing and spin to collect 20 μl total solution at the bottom of the tube.

4. Draw up the solution into the central portion of a thin-walled 50 μl capillary or micropipette, using a hand-held micropipette filler.

5. Keeping one end of the micropipette attached to the filler, seal its free end by heating it to red-heat in the flame of a Bunsen burner.

6. After the sealed end has cooled, hold the micropipette by its sealed end and seal the other end.

7. Incubate the sealed capillary at 60° for 12–48 hr (for oligomers

consisting of stretches of only ~ 3 guanines or containing large separations between stretches of guanines, it may be advisable to incubate at 37°, through, obviously, the rate of G4-DNA formation will be significantly slower).

8. After incubation, allow the capillaries to cool to room temperature. Snap off both ends of the capillaries after scoring with a glass cutter, then release the DNA solution into a clean Eppendorf tube. Spin briefly to collect liquid at the bottom of the tube.

In general, G4-DNAs containing upward of eight or nine guanine quartets are exceedingly stable at room temperature and are not in rapid equilibrium with their single-stranded forms. Consequently, these complexes can usually be diluted up to 10-fold without a significant change in the relative ratio of G4 to single-stranded DNA; the products can be analyzed in prerun, nondenaturing gels in 50 mM TBE and at room temperature. The very slow interconversion of the G4-DNA and single-stranded forms also permits their modification in a crude mixture by such reagents as dimethyl sulfate.[18] The partially methylated mixtures can then be separated in prerun, nondenaturing gels, eluted separately, processed with piperidine[21] or pyrrolidone,[22] and run on denaturing gels to study methylation protection.

The stability of G4-DNA complexes formed from sodium-only or sodium-dominant solutions can be enhanced by a subsequent addition of potassium chloride. Thus, G4-DNA formed from 1 M NaCl or 0.95 M NaCl plus 50 mM KCl can be diluted into TE plus 50 mM NaCl and 25 mM KCl, or even TE plus 10 mM KCl, for a satisfactory stability of the G4-DNA complex. It is usually also safe to ethanol precipitate G4-DNA samples from solutions containing 0.3 M sodium or potassium acetate.

G′2-DNA. The formation of G′2-DNA structures requires only moderate concentrations of DNA. Usually both sodium- and potassium-containing buffers are useful, except that, with sodium, one may also form a certain proportion of G4-DNA (which will not form in purely potassium buffers). Thus, one may use either sodium or potassium to form G′2-DNA when the DNA concentration is moderate (up to 1 μM); at higher concentrations of DNA, potassium alone is recommended. As with G4-DNA from simple-motif oligomers, the potassium G′2 complexes are significantly more stable than their sodium counterparts.

1. Dissolve the DNA (end-labeled as desired) in 10 μl TE buffer (see above) to 0.5 – 100 μM oligomer.

[21] A. M. Maxam and W. Gilbert, this series, Vol. 65, p. 499.
[22] Y. Shi and B. M. Tyler, *Nucleic Acids. Res.* **17**, 3317 (1989).

2. Heat in a sealed Eppendorf tube at 90° for 60 sec. Chill on ice. Spin briefly.

3. Add 10 μl TE plus 2 M KCl. Mix by vortexing. Spin.

4. Incubate at 37° for 12–48 hr in the sealed Eppendorf tube; for relatively G-rich oligomers, it may be profitable to incubate at 60° for 12–24 hr in a sealed capillary, as described in the protocol for G4-DNA.

G'2-DNA complexes made in 1 M KCl solutions may be diluted with TE buffer to a final KCl concentration of 50 mM. Electrophoretic analysis of G'2 complexes is best achieved in prerun, nondenaturing gels in 50 mM TBE plus 5–10 mM KCl at 4°; G'2 structures are generally not as stable as G4-DNA formed from the same sequence, containing fewer guanine quartets. It is useful to run the untreated single-stranded oligomer as a size control in the gel to identify the G'2 complexes, which usually run slower than the single-stranded DNA in nondenaturing gels, though not as slow as G4-DNA complexes.

G4'-DNA. To form the unimolecular folded-back structure G4'-DNA, one needs either a stretch of at least 28 contiguous guanines or else a minimum of four separated guanine motifs [the telomere-derived oligomers $(T_4G_4)_4$ and $(T_2G_4)_4$ are good examples of oligomers that form this structure]. Being unimolecular, the formation of G4'-DNA is independent of DNA concentration.[16] Like all other G-quartet-containing structures, it will not form in lithium solutions but will form in sodium, potassium, rubidium, and cesium solutions (although it should be noted that there are subtle structural differences between, e.g., the Na^+ and K^+ complexes, as determined by methylation probing[16]).

To avoid generating side products of the G'2 and G4 classes, it is best to work with low concentrations (1–20 nM) of end-labeled DNA.

1. Dissolve the DNA in 5 μl TE buffer (see above) in a sealable Eppendorf tube.

2. Heat to 94° for 60 sec. Chill on ice. Spin briefly.

3. Add 5 μl of TE plus 100 mM NaCl or KCl. Vortex to mix.

Analyze G4'-DNA on nondenaturing gels run at 4° in 50 mM TBE plus 50 mM NaCl or KCl, depending on which cation has been added to sample. Owing to the compactness of G4' structures, they often move significantly ahead of the input "unfolded" oligomer during nondenaturing gel electrophoresis. As with the other quartet-containing structures, the potassium-assisted G4' complexes are significantly more stable than their sodium counterparts.[16]

G-G Base-Paired Structures. Although G-G base-paired structures do not contain G-quartets, they nevertheless are putative pathway intermediates toward the formation of G'2 and G4' structures. The stability of G-G foldback structures (G'-DNA) of this type is not as great as that of quartet-containing structures. G'-DNA structures are best defined by their greater mobility (relative to input oligomer) on nondenaturing gels.[15] To generate G' structures, it is important to work with low DNA concentrations and to exclude sodium and potassium salts from buffers wherever possible, substituting lithium where appropriate.

1. Dissolve the DNA (end-labeled as appropriate) at 10 nM – 1 μM in 10 μl TE buffer.
2. Heat at 94° for 60 sec. Chill on ice. Spin down briefly.

Load on a nondenaturing gel run at 4° in 50 mM TBE (plus 50 mM LiCl, if desired).

Discussion

Identifying Quartet-Containing Structures

Certain properties can help identify an otherwise undefined complex of DNA as a species putatively containing guanine quartets. The first of these is methylation protection of the N-7 positions of some or all of its constituent guanines. Methylation protection is not unique to quartet-containing structures per se; guanines involved in Hoogsteen-type base pairing in other structures will also be protected from methylation. Conversely, however, a total absence of G methylation protection in a complex does indicate the absence of stable G quartets within its structure.

A second correlate of quartet-containing complexes is that sodium and potassium (and not lithium) are likely to have been present in the buffer during their formation. Lithium ions *do not* favor the formation of quartet structures. Thus, if an unidentified complex can be shown to form in sodium or potassium solutions but *not* in lithium solutions, and if some of its constituent guanines are found to be methylation protected, the complex may reasonably be assumed to contain G quartets.

Determination of Molecularity

Once a complex is determined to contain G quartets, its molecularity may still be unknown. Thus, it may not be possible to decide whether it belongs to the G'2 or the G4 class (it is easier to classify G4', for it forms in a concentration-independent manner).

The simplest way to determine the molecularity of a G-quartet-containing complex is to generate and quantify heterogeneous complexes from two oligomers of different size that nevertheless contain the same quartet-forming guanine motifs. The number of independent products reveals the molecularity of the complex. Thus, if an equimolar mixture of the oligomers A and B (which contain the same G motifs, but are of different size) forms G4-DNA on appropriate treatment, then five different G4 products, corresponding to A_4, A_3B, A_2B_2, AB_3, and B_4, should form and be resolved in a nondenaturing gel.[18] Similarly a G'2-type complex should yield three product bands, corresponding to A_2, AB, and B_2.[17]

Influence of pH and Melting of Refractory Quartet Structures

G-quartet structures form within a fairly wide range of pH, centered around neutral. In general, pH 6.0–8.0 is found to be optimal. G4-DNA, once formed, is stable up to an alkaline pH of 12.0; however, at the acidic extreme, there exists a danger of depurination and precipitation. The alkali lability of G4-DNA above pH 12.0 may be exploited to denature complexes that are stable even to boiling in 90% formamide. To denature such a complex, (1) ethanol precipitate it and dry the pellet; (2) add 10 μl of ice-cold 25 mM NaOH, dissolve the DNA by vortexing, and store it on ice for 30 sec; and (3) neutralize the solution by promptly adding, in succession, 10 μl of 100 mM Tris, pH 7.0, and 10 μl of 25 mM HCl, vortexing briefly after each step to mix.

[11] Parallel-Stranded Duplex DNA

By KARSTEN RIPPE and THOMAS M. JOVIN

Introduction

A number of alternative DNA structures have been described in the literature in which some or all of the strands (2, 3, or 4 in number) have the same polarity, that is, a $5'-3'$ orientation.[1] Most of these require modifications of the chemical structure of the DNA or certain solution conditions such as low pH. The parallel-stranded (ps) double helical structure denoted here as parallel-stranded DNA (ps-DNA) forms readily under physiological conditions and in the absence of chemical alterations of the bases or backbone, given the appropriate sequences.[1,2] The structural features of

[1] T. M. Jovin, *in* "Nucleic Acids and Molecular Biology" (F. Eckstein and D. M. Lilley, eds.), Vol. 5, p. 25. Springer-Verlag, Berlin and Heidelberg, 1991.
[2] T. M. Jovin, K. Rippe, N. B. Ramsing, R. Klement, W. Elhorst, and M. Vojtíškovȧ, *in* "Structure and Methods, Volume 3: DNA and RNA" (R. H. Sarma and M. H. Sarma, eds.), p. 155. Adenine Press, Schenectady, New York, 1990.

ps-DNA, originally proposed in a theoretical study by Pattabiraman,[3] are the trans-Crick–Watson (reverse Watson–Crick) base pairing with the glycosidic bonds in a trans orientation[4-6] and the lack of distinct major and minor grooves.[2,4,7] The ps helix is right-handed with a helical twist comparable to that of B-DNA, as determined from studies of a ps insert incorporated into a plasmid vector.[8] Parallel-stranded duplexes exhibit spectroscopic, thermodynamic, and biochemical properties different from those of conventional antiparallel B-DNA, and they are remarkably stable.[1,2,9] In the following we present the methods currently available for the synthesis and characterization of ps-DNA.

Design of Oligonucleotides

Strategies for Constructing Parallel-Stranded Duplexes

The stability of the ps duplexes examined to date decreases in the order homopolymeric $d(A) \cdot d(T)$ sequences,[4,5,9,10] mixed homopolymeric and alternating $d(A \cdot T)$ sequences,[7,9,11-13] and strictly alternating $d(A-T)$ sequences.[10,14] Interspersed $d(G \cdot C)$ base pairs can be incorporated in mixed $d(A \cdot T)$ sequences but destabilize the ps duplex.[13,15] Other base pairing schemes in a trans orientation with at least two hydrogen bonds have been analyzed theoretically[16] and are under current investigation; they are not treated here.

[3] N. Pattabiraman, *Biopolymers* **25**, 1603 (1986).

[4] J. H. van de Sande, N. B. Ramsing, M. W. Germann, W. Elhorst, B. W. Kalisch, E. von Kitzing, R. T. Pon, R. M. Clegg, and T. M. Jovin, *Science* **241**, 551 (1988).

[5] M. W. Germann, H. J. Vogel, R. T. Pon, and J. M. van de Sande, *Biochemistry* **28**, 6220 (1989).

[6] C. Otto, G. A. Thomas, K. Rippe, T. M. Jovin, and W. L. Peticolas, *Biochemistry* **30**, 3062 (1991).

[7] N. B. Ramsing and T. M. Jovin, *Nucleic Acids Res.* **16**, 6659 (1988).

[8] J. Klysik, K. Rippe, and T. M. Jovin, *Nucleic Acids Res.* **19**, 7145 (1991).

[9] N. B. Ramsing, K. Rippe, and T. M. Jovin, *Biochemistry* **28**, 9528 (1989).

[10] O. F. Borisova, Y. B. Golova, B. P. Gottikh, A. S. Zibrov, I. A. Il'ychova, Y. P. Lysov, O. K. Mamayeva, B. K. Chernov, A. A. Chernyi, A. K. Shchyolkina, and V. L. Florentiev, *Mol. Biol. (U.S.S.R.)* **23**, 1535 (1989).

[11] M. W. Germann, B. W. Kalish, and J. M. van de Sande, *Biochemistry* **27**, 8302 (1988).

[12] K. Rippe, N. B. Ramsing, and T. M. Jovin, *Biochemistry* **28**, 9536 (1989).

[13] D. Rentzeperis, K. Rippe, T. M. Jovin, and L. A. Marky, *J. Am. Chem. Soc.,* in press.

[14] M. W. Germann, B. W. Kalish, R. T. Pon, and J. H. van de Sande, *Biochemistry* **29**, 9426 (1990).

[15] K. Rippe, N. B. Ramsing, R. Klement, and T. M. Jovin, *J. Biomol. Struct. Dyn.* **7**, 1199 (1990).

[16] I. A. Il'Ychova, Y. P. Lysov, A. A. Chernyi, A. K. Shchyolkina, B. P. Gottikh, and V. L. Florentiev, *J. Biomol. Struct. Dyn.* **7**, 879 (1990).

With mixed $d(A \cdot T)$ sequences, the parallel-stranded helix conformation is 20–30% less stable (in terms of the transition enthalpy) than an equivalent antiparallel-stranded (aps) molecule (Table I). Thus, it is necessary to restrict the degree to which alternative double helical configurations can arise within a designed sequence. Examples of different strategies for achieving this are shown in Fig. 1. The first experimental demonstration of ps-DNA[4] was with a set of hairpins having a $d(A)_{10} \cdot d(T)_{10}$ stem and $d(G)_4$ or $d(C)_4$ loops, as shown in Fig. 1A (ps-C10). These hairpins had a $5'-p-5'$ or $3'-p-3'$ linkage in the loop such that the strands in the stem were forced into a parallel orientation. Other homopolymeric $d(A) \cdot d(T)$ ps hairpins with a stem length of 8, 10,[5,9] or 15 base pairs (bp) (Table II,

TABLE I

THERMODYNAMIC PARAMETERS FOR PS AND APS HELIX INITIATION AND HELIX GROWTH[a]

Parameter	ps-DNA (kJ mol⁻¹)	aps-DNA (kJ mol⁻¹)
ΔH_β (hairpins)	25	25
ΔH_β (linears)	44	44
ΔH_s (AA/TT)	−24	−29
ΔH_s (AT/TA)	−18[b]	−36[c]
ΔH_s (TA/AT)	—	−23[c]
	(kJ mol⁻¹ K⁻¹)	(kJ mol⁻¹ K⁻¹)
ΔS_β (hairpins)	0.053	0.053
ΔS_β (linears)	0.152	0.152
ΔS_s (AA/TT)	−0.072	−0.084
ΔS_s (AT/TA)	−0.058	−0.096[c]
ΔS_s (TA/AT)	—	−0.084[c]

[a] The values for helix initiation are indexed with a β and those for helix elongation with an S. The nearest-neighbor interactions (in the $5' \rightarrow 3'$ direction) as well as the initiation contribution were derived from a least-squares fit of the experimentally determined ΔH_{vH} and ΔS values given in Table II. It was assumed that the enthalpy and entropy transition terms are a linear combination of the nearest-neighbor values corresponding to the oligonucleotide sequence. The initiation contribution for hairpins and linear duplexes was set to be the same for the parallel and antiparallel stranded molecules in the fit. The data refer to helix growth rather than dissociation.

[b] Note that in ps-DNA AT/TA and TA/AT are equivalent because of symmetry.

[c] The aps AT/TA and TA/AT enthalpy and entropy interactions could not be determined accurately because all sequences had only a difference of one in the number of AT/TA and AT/TA nearest-neighbor interactions and the equation system for the aps molecules has an extra variable. See K. J. Breslauer, R. Frank, H. Blöcker, and L. A. Marky, *Proc. Natl. Acad. Sci. U.S.A.* **83**, 3746 (1986), for other values.

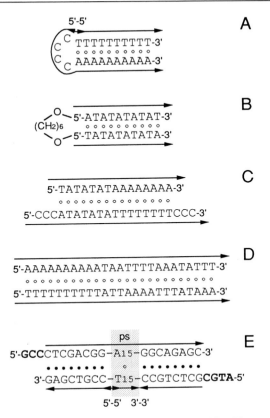

FIG. 1. Different strategies for obtaining parallel-stranded DNA. The trans-Crick–Watson base pairs (reverse Watson–Crick) of the parallel-stranded duplex are shown by the symbol (∘). (A) Hairpin structure designated as ps-C10 with a 5′-p-5′ linkage in the loop and 10 d(A · T) base pairs in the stem.[4,5,9] (B) Hairpins formed by chemical cross-linking of two DNA strands with a hexanediol bridge.[10,17] (C) Duplexes formed with strands of unequal length so as to constrain hybridization in the parallel orientation.[11] (D) Duplexes formed by hybridization of sequences with perfect complementarity only in the parallel orientation. The 25-nt duplex ps-D1 · D2 is shown.[12] D1 is the upper strand, and D2 is the bottom strand. (E) Duplexes consisting of aps–ps–aps segments and terminal overlaps for insertion into a vector and circularization.[8] The bottom strand, designated as P5, contains 3′-p-3′ and 5′-p-5′ linkages that lead to inversion of polarity of the central 15 dT residues. The conventional cis-Watson–Crick base pairs of the antiparallel flanking regions are shown by the symbol (•).

ps-C15) and alternating d(A-T) hairpins with a stem of 8 or 10 bp[14] have been characterized. The introduction of 5'-p-5' and 3'-p-3' phosphodiester bonds can also be used to construct duplexes that have a ps insert with aps flanking regions (Fig. 1E, ps-P3·P5).[8] With appropriate overhangs such a duplex can be ligated into suitable plasmid vectors according to protocols described by Klysik et al.[8] in order to study the influence of topological stress on the insert.

An alternative for synthesizing ps-DNA hairpins is to chemically cross-link the ends of two DNA strands. This can be accomplished by introducing a 1,6-hexanediol linker[10,17] (Fig. 1B) or with bifunctional cross-linking reagents (M. Vojtísková and T. Jovin, unpublished) acting on 5'-terminal amino or sulfhydryl groups. Cross-linking via the 1,6-hexanediol linker has been reported for alternating d(A-T) sequences,[10,17] homopolymeric d(A)·d(T) sequences,[10,17] and mixed sequences also containing d(G·C) base pairs.[18]

It is not necessary to use the chemical modifications of the primary DNA structure mentioned above if the sequence is chosen such that complementarity in the parallel direction is more favorable than that in the case of alternative antiparallel secondary structures. Examples of appropriate sequences are given in Fig. 1C,D. With the ps construct in Fig. 1C the overhanging dC residues in the bottom strand are necessary to prevent the formation of antiparallel concatamers.[11] In the ps linear duplex ps-D1·D2 (Fig. 1D) the strands are fully base paired, and potential aps secondary structures are suppressed by the selected sequence.[12] A strategy of computer-aided search for such a sequence through all possible combinations of dA and dT in a segment of a given length is outlined in the next section.

Evaluation of Sequences for Parallel-Stranded Linear Duplexes

The formation of alternative aps double helical configurations in selected sequences can be minimized with a computer program. The following input parameters are required: (i) length n of the desired sequence: (ii) maximal number of base pair interactions m_{hom} tolerated in aps homodimeric complexes, that is, through self-pairing of two copies of the given sequence such that the strands can adopt any staggered relative position; (iii) maximal number of base pair interactions m_{het} tolerated in aps heterodimer complexes between the given sequence and its inverted complement

[17] A. K. Shchyolkina, Y. P. Lysov, I. A. Il'Ychova, A. A. Chernyi, Y. B. Golova, and B. K. Chernov, B. P. Gottikh, and V. L. Florentiev, *FEBS Lett.* **244,** 39 (1989).
[18] N. A. Tchurikov, B. K. Chernov, Y. B. Golova, and Y. D. Nechipurenko, *FEBS Lett.* **257,** 415 (1989).

(i.e., the complementary strand constituting the idealized fully paired ps duplex).

We evaluate sequences consisting only of A and T and represent them as a binary number, assigning the values 0 and 1 to T and A, respectively. Consider, for example, the 25-nucleotide (nt) sequence D1 (Fig. 1D). We test for condition (ii) above by using two D1 strands aligned in all possible aps orientations (without bulges) and score for the maximal number of aps A · T base pairs (14) and the longest run without mismatches (8). If this number is less than or equal to m_{hom}, the program performs the same test for condition (iii) using the D1 + D2 combination and the parameter m_{het}. In our example, a maximum of 12 bp can form, and the longest run of base pairs is 10 (Fig. 2). In general, upon specifying the length n, all possible sequences (of which 2^{n-1} are nonredundant) are examined and ranked in a hierarchy of desirability based on the lowest number of the possible aps homodimeric and heterodimeric interactions, including those that can

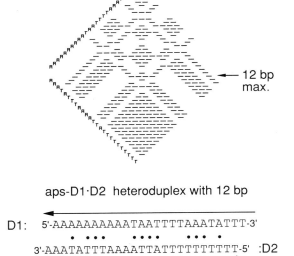

aps-D1·D2 heteroduplex with 12 bp

D1:   5'-AAAAAAAAAAATAATTTTAAATATTT-3'
        • • • •   • • • •   • • • •
       3'-AAATATTTAAAATTATTTTTTTTTT-5' :D2

FIG. 2. Matrix plot for graphic display of competitive aps secondary structures. The antiparallel secondary structures (aps heteroduplexes) that can be formed from strands D1 and D2 (Fig. 1D are shown). The positions of potential base pairing are given by lines. All possibilities arising from pairing of the two strands in an antiparallel orientation are displayed. The combination with the maximum number of base pairs (12) is indicated by an arrow and is reproduced at the bottom. In the same way D1 can be plotted against D1 to evaluate potential aps homoduplexes.

arise in intermolecular interactions, also considering overhangs. Such an approach yields a family of sequences which then can be displayed graphically (Fig. 2) for studying the localization of the base-paired parts within the sequence. In our experience, the best ps versus aps discrimination is provided by sequences consisting of a homopolymeric segment ($\sim 0.4n$ in length) juxtaposed to a mixed A,T stretch, such that the maximal number of potential aps interactions is limited to about $n/2$.

The general strategy has been to generate an oligonucleotide quartet consisting of the selected sequence [e.g., D1], its ps complement [D2], and the two respective inverses [D4 and D3], such that two pairs of ps and aps duplexes can be formed, [ps-D1·D2 and ps-D3·D4] and [aps-D1·D3 and aps-D2·D4].[12] These can be studied and compared by numerous analytical techniques in order to elucidate the significant features common to or distinctive between the two conformations. The sequence repertoire can be extended by introduction of interesting substitutions, for example, the replacement of d(A·T) by d(G·C) base pairs.[15]

Prediction of Thermodynamic Parameters from Sequence

Concept of Nearest-Neighbor Approximation

Given the sequence of a potential ps duplex structure one can estimate its stability. The standard enthalpy $\Delta H°$ and the entropy $\Delta S°$ differences for the coil–helix transition are determined mainly by the stacking interactions of two adjacent bases, that is, the nearest-neighbors. Thus it is possible to decompose a duplex into its different nearest-neighbor quartets and derive $\Delta H°$ and $\Delta S°$ by adding the contributions of nearest-neighbors and an additional nucleation term[19-21] according to Eqs. (1) and (2).[22] Next-nearest-neighbor interactions are not considered. The indices β and s refer to helix initiation and elongation, respectively. The total number of base pairs formed is n.

$$\Delta H° = \Delta H_\beta° + \sum_{i=1}^{n-1} \Delta H_{s_i}° \qquad (1)$$

[19] K. J. Breslauer, R. Frank, H. Blöcker, and L. A. Marky, *Proc. Natl. Acad. Sci. U.S.A.* **83,** 3746 (1986).
[20] S. Freier, R. Kierzek, J. A. Jaeger, N. Sugimoto, M. H. Caruthers, T. Neilson, and D. H. Turner, *Proc. Natl. Acad. Sci. U.S.A.* **83,** 9373 (1986).
[21] O. C. Uhlenbeck, P. N. Borer, B. Dengler, and I. Tinoco, Jr., *J. Mol. Biol.* **73,** 483 (1973).
[22] In the expression for the entropy change an additional term ΔS_{sym} has to be included for self-complementary DNA according to Breslauer *et al.*[19]

$$\Delta S° = \Delta S_\beta° + \sum_{i=1}^{n-1} \Delta S_{s_i}° \tag{2}$$

Based on Eqs. (1) and (2), a least-squares algorithm was applied to the measured data presented in Table II in order to determine the best global estimate for the different nearest-neighbor interactions and initiation parameters for the coil–helix transition of ps hairpins and ps linear duplexes. The resultant values are presented in Table I and were used to calculate the thermodynamic quantities given in Table II in the form of a comparison with the measured values. The agreement is generally satisfactory.

In the following we use the designation ΔH_{vH} (van't Hoff enthalpy of melting) instead of $\Delta H°$ to indicate that the analyzed values were primarily determined from thermal transitions using optical measurements and not calorimetrically. The parameters in Eqs. (3)–(6) and in Table II are given for the helix–coil transition rather than for the inverse process referred to in Eqs. (1) and (2) and in Table I. The entropy change ΔS was referenced to 0.1 M NaCl by using the melting point T_m in 0.1 M NaCl and a total strand concentration C_t of 2 μM for the linear duplexes. The melting enthalpy ΔH_{vH} was assumed to be salt independent in the range from 0.1 to 1.0 M NaCl.[9,12]

Procedure: Estimation of ΔH_{vH}, ΔS, and T_m for Helix–Coil Transition of a Given Parallel-Stranded Hairpin or Linear Duplex and Its Antiparallel-Stranded Reference

Estimation of ΔH_{vH} and ΔS. The number of AA/TT (n_{AA}) and AT/TA (n_{AT}) nearest-neighbors [and in the case of the aps reference molecules the number of TA/AT (n_{TA}) as well] are counted for the given sequence. In the ps-D1·D2 duplex, for example, $n_{AA} = 17$ and $n_{AT} = 7$. Note that in ps-DNA there is no distinction between AT/TA and TA/AT nearest neighbors, since the strand backbone orientation is identical. Then, ΔH_{vH} and ΔS are calculated according to the following equations:

Parallel-stranded hairpin (linear) transition

$$\Delta H_{vH} = [-25(-44) + 23.6 n_{AA} + 18.3 n_{AT}] \quad \text{kJ mol}^{-1} \tag{3}$$

$$\Delta S = [-0.05(-0.15) + 0.072 n_{AA} + 0.058 n_{AT}] \quad \text{kJ mol}^{-1} \tag{4}$$

Antiparallel-stranded hairpin (linear) transition

$$\Delta H_{vH} = [-25(-44) + 28.7 n_{AA} + 35.5 n_{AT} + 22.6 n_{TA}] \quad \text{kJ mol}^{-1} \tag{5}$$

$$\Delta S = [-0.05(-0.15) + 0.084 n_{AA} + 0.096 n_{AT} + 0.084 n_{TA}] \\ \text{kJ mol}^{-1} \text{ K}^{-1} \tag{6}$$

TABLE II

THERMODYNAMIC DATA FOR HELIX–COIL TRANSITIONS OF PS-DNA AND APS-DNA[a]

DNA duplex	$\Delta H_{vH}{}^b$ (kJ mol^{-1})		ΔS^c (kJ mol^{-1} K^{-1})		$T_m{}^d$ (°C)	
	Exp.	Calc.	Exp.	Calc.	Exp.	Calc.
ps-C8[e]	153	140	0.50	0.45	35	38
ps-C10[e]	202	188	0.64	0.60	40	42
ps-C15[f]	250	306	0.79	0.96	43	47
ps-C8_alt[g]	106	103	0.36	0.35	19	19
ps-C10_alt[g]	137	140	0.47	0.47	21	25
ps-F5·F6[h]	235	249	0.72	0.76	8	10
ps-15-mer[i]	309	246	0.97	0.76	11	10
ps-C2·C3[j]	353	401	1.06	1.22	25	26
ps-D1·D2[k]	500	485	1.40	1.48	31	30
ps-D3·D4[j]	507	485	1.56	1.48	29	30
aps-C8[e]	184	176	0.58	0.56	45	38
aps-C10[e]	230	234	0.71	0.73	51	45
aps-C15[f]	362	377	1.10	1.15	57	52
aps-C10_alt[g]	244	243	0.78	0.76	39	47
aps-F5·F7[h]	352	354	1.11	1.06	27	26
aps-15-mer[i]	487	366	1.47	1.07	33	34
aps-C3·C7[j]	546	523	1.62	1.55	40	39
aps-D1·D3[k]	654	653	1.92	1.92	49	48
aps-D2·D4[j]	628	641	1.86	1.91	44	43

[a] The listed parameters were obtained from experiments with 0.1–1.0 M NaCl and 10 mM sodium cacodylate, pH 7.2. The calculated values (Calc.) were determined according to the Eqs. (3)–(6) given in the text.

[b] ΔH_{vH} is the average van't Hoff enthalpy of melting. No correlation with salt concentration in the range 0.1–1.0 M NaCl was found.[9,12] The ΔH_{vH} values are within ±10% experimental error as estimated from completely independent experiments.

[c] The entropy was calculated by using the average ΔH_{vH} and the T_m in 0.1 M NaCl calculated from linear regressions in 0.05–0.8 M NaCl.

[d] T_m in 0.1 M NaCl (including the buffer contribution of 0.11 M NaCl) and scaled for the linear duplexes to a strand concentration of 2.0 μM according to Eq. (8). The T_m values are within ±4° experimental error as estimated from completely independent experiments.

[e] Average values from Ramsing et al.[9] and Germann et al.[5]

[f] K. Rippe, Ph.D. Thesis, University of Göttingen, 1991. The molecule ps-C15 is a hairpin with 15 homopolymeric d(A·T) base pairs in the stem and a C$_4$ loop.

[g] Germann et al.[14] The T_m in 0.1 M NaCl was estimated from given values in 0.8 M NaCl using the salt dependence of T_m determined in Ramsing et al.[9]

[h] K. Rippe, Ph.D. Thesis, University of Göttingen, 1991. The duplexes ps-F5·F6 and aps-F5·F7 have the same sequence as ps-F1·F2 and aps-F1·F4, respectively[2] but no 5'-terminal amino group.

[i] Germann et al.,[11] not included in the least squares fit because of single-stranded regions present at the ends of the duplex (Fig. 1C).

[j] Ramsing et al.[9]

[k] Average values from Rippe et al.[12] and Rentzeperis et al.[13]

The initiation contribution is different for hairpins and linear duplexes. In Eqs. (3)–(6) the values for the initiation of the linear duplexes are given in parentheses. It is possible to incorporate $d(G \cdot C)$ base pairs into a parallel-stranded duplex, but they destabilize the structure.[15] The database is yet too small to derive ΔH_s and ΔS_s values of the different AG/TC, GA/CT, and GT/CA nearest-neighbor interactions in ps-DNA. However, based on available data[13,15] we have estimated an average contribution of -14 kJ mol^{-1} for ΔH_s and -0.05 kJ mol^{-1} K^{-1} for ΔS_s.

Estimation of melting point T_m from ΔH_{vH} and ΔS. The melting point T_m, the midpoint of the helix–coil transition, can be calculated according to a two-state model from Eqs. (7) and (8)[9]:

Hairpin transition:

$$T_m = \frac{\Delta H_{vH}}{\Delta S} \tag{7}$$

Linear duplex transition:

$$T_m = \frac{\Delta H_{vH}}{\Delta S - R \ln(C_t/4)} \tag{8}$$

where R is the gas constant (8.314 J mol^{-1} K^{-1}) and C_t the total strand concentration in M units. As evident from Eqs. (7) and (8), T_m of the hairpins is independent of concentration, whereas T_m of the linear duplexes depends on C_t. The derived values of T_m in Table II correspond to $C_t = 2 \times 10^{-6}$ M for the linear duplexes. At this concentration $R \ln(C_t/4) = -0.121$ kJ mol^{-1} K^{-1}.

If the concentration of the two strands is not equimolar, as, for example, in experiments in which only on strand is ^{32}P-labeled and the complementary unlabeled strand with the concentration C_i is in great excess, T_m can be estimated according to the expression

$$\frac{1}{T_m} = \frac{1}{T_{mref}} - \frac{R}{\Delta H_{vH}} \ln\left(\frac{4C_i}{C_t}\right) \tag{9}$$

In Eq. (9) T_{mref} is the reference T_m of an equimolar mixture of strands with a total strand concentration C_t under the same salt conditions.[23]

Effects of Salt on Thermodynamic Stability. The T_m values calculated according to Eqs. (3)–(8) correspond to the reference condition of 0.1 M NaCl. At other salt concentrations, T_m can be derived from the slope of the T_m versus log[Na] curve. The salt dependence, namely, the change of T_m per decade of salt concentration $\Delta T_m / \Delta \log[\text{Na}]$, is on average $17 \pm 2°$ for the ps and $16 \pm 1°$ for the aps references.[9] Thus a change in salt concentration from 0.05 to 0.5 M increases the T_m of a ps duplex by approximately $17°$. Mg^{2+} also exerts a stabilizing effect on both the ps and aps

[23] J. Klysik, K. Rippe, and T. M. Jovin, *Biochemistry* **29**, 9831 (1990).

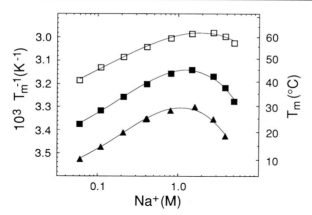

FIG. 3. Melting temperature of ps and aps linear duplexes as a function of salt concentration. Displayed are the duplexes ps-D1·D2 (■) (Fig. 1D); ps-D5·D6 (▲), a variant of ps-D1·D2 in which four d(A·T) base pairs are replaced by d(G·C) base pairs[15] and aps-D1·D3 (□), the aps reference for ps-D1·D2. The solid lines are a third order regression. The T_m values were corrected to a total concentration of 1.6 μM strands according to Eq. (8). The Na+ concentration includes the contribution of the buffer.

duplexes, with the T_m in 2 mM MgCl$_2$ being equal to that in 0.2–0.3 M NaCl.

The linear dependence of T_m (or $1/T_m$) on log[Na+] is only valid for Na+ concentrations below 0.8 M. Above this value, T_m passes through a maximum and then decreases with a further increase of [Na+] (Fig. 3). This effect is known for B-DNA[24] but is more pronounced with the ps linear duplexes (Fig. 3). Measurements of absorption hyperchromicity and chemical reactivity with osmium tetroxide–pyridine (Os,py) show that high salt concentrations reduce the thermodynamic stability of the ps helix.[23] For example, the T_m of ps-D1·D2 in 1.6 M NaCl is 46° but decreases to 32° in 4.9 M NaCl (Fig. 3). In addition, the helix–coil transition is less cooperative; ΔH_{vH} decreases by about 30%. Thus the fractional transition to the single-stranded state (θ) calculated for a two-state model according to Eqs. (10) and (11)[9] can increase considerably at salt concentrations in the molar range:

Hairpin transition:

$$\theta = \frac{1}{1 + E} \tag{10}$$

Linear duplex transition:

$$\theta = \frac{(1 + 8E)^{1/2} - 1}{4E} \tag{11}$$

[24] C. F. Anderson and M. T. Record, Jr., *Annu. Rev. Phys. Chem.* **33**, 191 (1982).

with

$$E \equiv \exp\left[\frac{\Delta H_{vH}}{R}\left(\frac{1}{T}-\frac{1}{T_m}\right)\right]$$

Synthesis and Purification of Oligonucleotides

The synthesis of the dT_{10} tract in the ps-C10 hairpin molecule with $5'-p-5'$ linkage (Fig. 1A) and the synthesis of the dT_{15} tract within the $3'-p-3'$ and $5'-p-5'$ phosphodiester bonds (Fig. 1E) were carried out with 5'-(β-cyanoethyl-N,N-diisopropylamino)phosphinyl-3'-(4,4'-dimethoxytrityl)thymidine [5'-(β-cyanoethyl)-3'-dimethoxytritylthymidine phosphoramidite, 3'-DMTr-βdT]. This compound is synthesized as described below. In contrast to the procedure outlined by van de Sande *et al.*,[4] it is advantageous to use *tert*-butyldiphenylchlorosilane instead of *tert*-butyldimethylchlorosilane for the protection of the 5'-OH group of thymidine because of its higher specificity for the 5'-OH group. N^6-Benzoyl-5'-(β-cyanoethyl)-3'-dimethoxytrityladenosine phosphoramidite (3'-DMTr-βdA^{bz}) can be synthesized in a similar way by using commercially available N^6-benzoyl-2'-deoxyadenosine instead of thymidine as the starting material[12] or by introducing an extra step for the protection of the N-6 exocyclic amino group of adenine. Starting the synthesis in the $5' \rightarrow 3'$ direction requires loading of the solid controlled-pore glass (CPG) support with the 3'-DMTr protected nucleoside. This can be done according to procedures used for the 5'-DMTr protected nucleoside.[25]

Procedure: Synthesis of 5'-(β-Cyanoethyl-N,N-diisopropylamino)-phosphinyl-3'-(4,4'-dimethoxytrityl)thymidine

Protection of 5'-Hydroxyl Group of Thymidine with tert-Butyldiphenylchlorosilane. Dissolve 5.0 g (20.7 mmol) thymidine in 100 ml dry pyridine and evaporate to remove traces of water. Then redissolve in 100 ml fresh pyridine. Add 2.9 ml (3.1 g, 11.4 mmol, 0.55 Eq) *tert*-butyldiphenylchlorosilane and after 30 min add another 2.9 ml *tert*-butyldiphenylchlorosilane. Monitor the reaction by thin-layer chromatography (TLC) with 9:1 (v/v) dichloromethane–methanol on a silica gel 60 plate with a fluorescent UV indicator 254 nm. The product 5'-(*tert*-butyldiphenyl)thymidine is depicted in Fig. 4A.

[25] M. J. Damha, P. A. Giannaris, and S. V. Zabarylo, *Nucleic Acids Res.* **18**, 3813 (1990). A detailed description of the synthesis of 3'-DMTr-βdA^{bz} and the loading of the CPG support with 3'-DMTr thymidine and N^6-benzoyl-3'-DMTr adenosine will be published elsewhere (Wodarski *et al.*, in preparation).

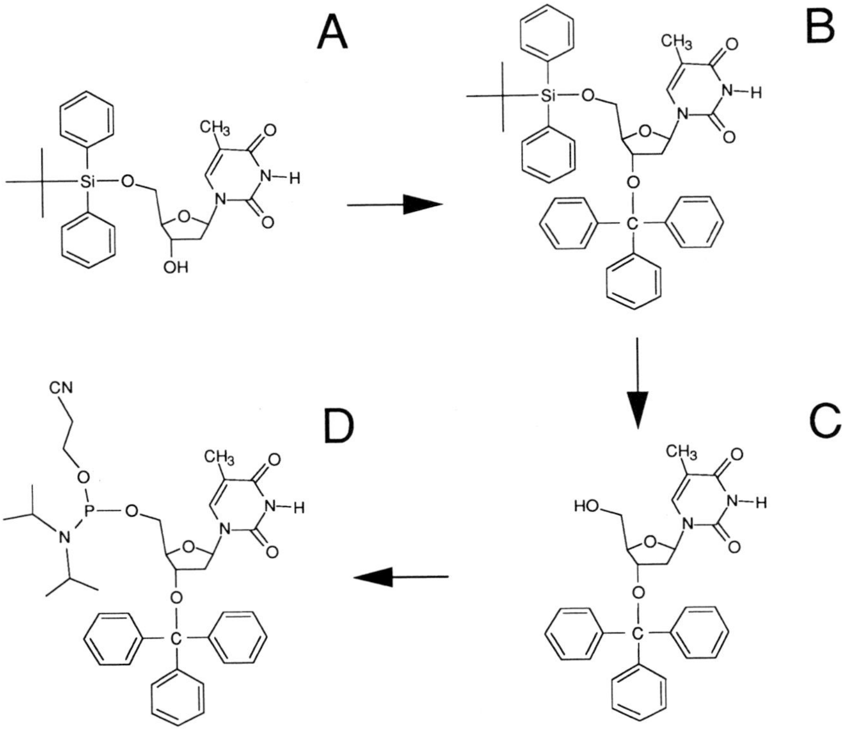

FIG. 4. Synthesis of 5'-(β-cyanoethyl-N,N-diisopropylamino)phosphinyl-3'-dimethoxytri-tylthymidine (3'-DMTr-βdT). (A) 5'-(tert-Butyldiphenyl)thymidine. (B) 5'-(tert-Butyldiphen-ylsilyl)-3'-(4,4'-dimethoxytrityl)thymidine. (C) 3'-(4,4'-Dimethoxytrityl)thymidine. (D) 5'-(β-Cyanoethyl-N,N-diisopropylamino)phosphinyl-3'-(4,4'-dimethoxytrityl)thymidine.

Protection of 3'-Hydroxyl Group with 4,4'-Dimethoxytrityl Chloride.
Add 8.4 g (24.8 mmol, 1.2 Eq) 4,4'-dimethoxytrityl chloride (DMTr-Cl) to convert the 5'-(*tert*-butyldiphenylsilyl)thymidine (Fig. 4A) to 5'-(*tert*-bu-tyldiphenylsilyl)-3'-(4,4'-dimethoxytrityl)thymidine (Fig. 4B). Monitor the reaction by TLC with 9 : 1 (v/v) dichloromethane–methanol and add more DMTr-Cl after 2 hr if necessary. If the reaction is complete, add 50 ml of distilled water, incubate for 10 min, evaporate to dryness, and dissolve in 250 ml ethyl acetate with 0.4% (v/v) triethylamine (TEA). The TEA is included here and in the following steps to prevent acid-catalyzed cleavage of the DMTr group. Extract the ethyl acetate phase first with 10% (w/v) aqueous $NaHCO_3$ and then with saturated aqueous NaCl. Dry the organic phase over Na_2SO_4, filter, evaporate, and reevaporate after addition of toluene.

Removal of 5'-tert-Butyldiphenylsilyl Group with 1 M Tetrabutylam-monium Fluoride in Tetrahydrofuran. The product from the previous step, 5'-(*tert*-butyldiphenylsilyl)-3'-(4,4'-dimethoxytrityl)thymidine (Fig. 4B) is dried and then dissolved in 100 ml tetrahydrofuran (THF) containing 0.4% (v/v) TEA. Add 22.7 ml (20.7 mmol, 1.0 Eq) of 1.1 *M* tetrabutylam-monium fluoride in THF and monitor the reaction by TLC with 9 : 1 (v/v) dichloromethane–methanol and with commercially available 5'-(4,4'-di-methoxytrityl)thymidine as reference. If the reaction is complete, add 50 ml of distilled water and proceed with the purification as described above. Dissolve the dried product in a small volume (~ 10 ml) of dichloro-methane containing 1% (v/v) TEA and purify by chromatography with Kieselgel 60 (Merck, Darmstadt, Germany) as described by Sproat and Gait.[26] Use 20 g Kieselgel 60 per gram of crude material. Pack the column and elute the product 3'-(4,4'-dimethoxytrityl)thymidine (Fig. 4C) with dichloromethane containing 1% (v/v) TEA. If the product stays on the column, supplement the solvent with methanol in 1% increments until the product elutes. Evaporate to dryness, dissolve in dichloromethane, and precipitate by addition to rapidly stirred *n*-hexane (20–25 volumes); filter the solid.

Phosphitylation of 5'-Hydroxyl Group with Chloro(β-cyanoethoxy)di-isopropylaminophosphine.[27] Dry diisopropylethylamine by distillation from calcium hydride. Distill dichloromethane from P_2O_5 and incubate it overnight with Al_2O_3 to remove traces of acid. Then weigh 3.0 g (12.4 mmol) of the silica gel-purified 3'-(4,4'-dimethoxytrityl)thymidine (Fig. 4C) in a dry 250-ml round-bottomed flask and coevaporate twice with dry dichloromethane containing 10% (v/v) dry pyridine. Add a magnetic stir-rer, cap the flask with a rubber septum, insert an 18-gauge needle through the rubber septum, place the flask in a desiccator, and dry under reduced pressure overnight. Then break the vacuum in the desiccator by slowly introducing argon from a balloon into the desiccator. Open the desiccator and withdraw the venting needle. Inject a needle attached to an argon-filled balloon through the rubber septum of the round-bottomed flask as well as through the cap of the phosphitylating reagent chloro(β-cyanoethoxy)di-isopropylaminophosphine.

In the following steps, all reagents are injected through the rubber septum with an oven-dried syringe. Start by injecting 8.6 ml diisopropyl-

---

[26] T. Sproat and M. J. Gait, *in* "Oligonucleotide Synthesis: A Practical Approach" (M. J. Gait, ed.), p. 199. IRL Press, Oxford, 1984.

[27] The procedure is similiar to the one described by T. Atkinson and M. Smith [*in* "Oligonu-cleotide Synthesis: A Practical Approach" (M. J. Gait, ed.), p. 35. IRL Press, Oxford, 1984], but a different phosphitylating reagent is used.

ethylamine (6.4 g, 49.6 mmol, 4 Eq) and then 30 ml dichloromethane, and dissolve the 3′-(4,4′-dimethoxytrityl)thymidine. Then add 4.2 ml (4.4 g, 18.6 mmol, 1.5 Eq) of the phosphitylating reagent chloro(β-cyanoethoxy) diisopropylaminophosphine. Stir the mixture and monitor the reaction by TLC in 45:45:10 (v/v) ethyl acetate–dichloromethane–triethylamine by withdrawing aliquots with a dry syringe. If the reaction is complete quench the excess phosphitylating reagent with 0.2 ml anhydrous methanol. Evaporate to dryness, dissolve in dichloromethane, and precipitate by addition to rapidly stirred n-hexane. Filter the solid, dissolve in a small volume (~10 ml) of 45:45:10 (v/v) ethyl acetate–dichloromethane–triethylamine, and purify by chromatography with Kieselgel 60 (Merck) as described by Sproat and Gait.[26] Use 20 g silica gel 60 per gram of crude material. Pack the column and elute the product with 45:45:10 (v/v) ethyl acetate–dichloromethane–triethylamine. Alternatively the column can be packed with n-hexane containing 0.4% (v/v) triethylamine and then equilibrated and eluted with 49:49:2 (v/v) n-hexane–dichloromethane–triethylamine, which is a less polar solvent that gives better resolution.

Evaporate the product-containing fractions to dryness, dissolve in dichloromethane, precipitate by addition to rapidly stirred n-hexane, and filter the solid. Then dry under reduced pressure in a desiccator over P_2O_5. Analyze the product by ^{31}P NMR in CD_3CN from -10 to 190 ppm with 85% phosphoric acid as external standard. The two stereoisomers of the product 5′-(β-cyanoethyl-N,N-diisopropylamino)phosphinyl-3′-(4,4′-dimethoxytrityl)thymidine (Fig. 4D) appear around 150 ppm.

Procedure: Automated Synthesis of 5′-p-5′ and 3′-p-3′ Phosphodiester Bonds with 3′-Dimethoxytrityl-Protected β-Cyanoethylthymidine

Weigh the amount of 5′-(β-cyanoethyl)-3′-dimethoxytritylthymidine phosphoramidite (3′-DMTr-βdT) needed for the synthesis (usually 0.1–1.0 g) into a dark glass vial that can be attached to the synthesizer and cap it with a rubber septum. Insert an 18-gauge needle through the rubber septum, place the vial in a desiccator, and purge several times with argon. Then dry overnight under reduced pressure with the desiccator, connected to an oil pump. The next day slowly introduce argon from a balloon into the desiccator, open the desiccator, and withdraw the venting needle immediately. The 3′-DMTr-βdT is dissolved to give a concentration of 0.1 M in dry acetonitrile as for normal 5′-DMTr β-cyanoethyl-protected phosphoramidites. It is important to check the solubility of the 3′-DMTr-βdT phosphoramidite. If insoluble particles remain in the solution severe problems may arise during synthesis because of blocking of delivery lines.

Coupling in the 5′ → 3′ direction is about 4 times slower because of the lower reactivity of the 3'-OH group.[28] To account for the difference, two modifications are introduced into the synthesis cycle. First, the 3'-DMTr-βdT + tetrazole delivery is increased by about 25% and the tetrazole delivery by about 20%. For the 1 μmol scale synthesis on an Applied Biosystems (Foster City, CA) synthesizer this is done by introducing an extra 3'-DMTr-βdT + tetrazole delivery step for 5 sec followed by an extra tetrazole delivery step for 3 sec. Then, an additional wait step of 2 min is introduced for the 3'-DMTr-βdT port. The efficiency of coupling can be monitored by spectrophotometrically measuring the liberated DMTr cation and is as high as with the 5'-DMTr β-cyanoethyl-protected phosphoramidites. Incorporation of the 5'-p-5' and 3'-p-3' phosphodiester bonds at the desired positions in the sequence is accomplished by switching from the conventional 5'-DMTr β-cyanoethyl-protected phosphoramidite to 3'-DMTr-βdT and vice versa during synthesis.

Oligonucleotide Purification

Purification is according to standard protocols. We have no indication that the 5'-p-5' or 3'-p-3' phosphodiester bonds lead to unusual chromatographic behavior. We use the following procedure. After synthesis with a terminal DMTr group, the protecting groups liberated by treatment with concentrated ammonia are removed by a Sephadex G-25 column. After purification by reversed-phase chromatography and detritylation, anion-exchange chromatography on a fast-protein–high-performance liquid chromatography (FPLC/HPLC) system is carried out.[29] The last step consists of desalting on a Sephadex G-25 column.

Characterization of Oligonucleotides and Parallel-Stranded Duplexes

Ultraviolet Absorbance Spectroscopy

Several spectroscopic methods have been used for characterization of the ps conformation.[1,2] They include ultraviolet absorption,[2,4,7,11,12,14,15] circular dichroism,[4,5,7,11,12,15] Raman,[6] ^{31}P NMR,[2,5] and ^{1}H NMR[5] spectroscopy. In this chapter we restrict ourselves to measurements of UV absorption because it requires only simple instrumentation and gives characteristic differences for the ps and aps duplexes.

[28] B. Kaplan and K. Itakura, in "DNA Synthesis on Solid Supports and Automation" (S. A. Narang, ed.), p. 9. Academic Press, Orlando, Florida, 1987.

[29] For details on HPLC purification of synthetic DNA, see [2] in this volume.

Extinction Coefficients. The UV absorbance extinction coefficient for dA · dT oligonucleotides was determined by phosphate analysis of a 21-nucleotide sequence: 8.6 mM (base)$^{-1}$ cm^{-1} at 264 nm in the denatured state at 70°; buffer 0.1 M NaCl, 10 mM sodium cacodylate, pH 7.2.[7] The wavelength of 264 nm was chosen to compensate for the different ratios of dA and dT with other sequences, because it corresponds to an isosbestic point for poly[d(A)] and poly[d(T)] under approximately the same conditions.[30] We have also calculated an idealized value of 8.6 mM (base)$^{-1}$ cm^{-1} at 264 nm and 70° for a dG residue in a mixed sequence from the expression $2\varepsilon\{poly[d(G-C)]\}_{70°,264nm} - \varepsilon\{poly[d(C)]\}_{70°,264nm}$.[15] With the extinction coefficients in the denatured state at 70° of 8.6 mM (base)$^{-1}$ cm^{-1} for dA, dT, and dG residues and 7.3 mM (base)$^{-1}$ cm^{-1} for dC,[31] one can calculate the extinction coefficient of a given sequence, assuming additivity of unit contributions.

The linear ps-DNA duplexes are prepared by mixing equimolar amounts of the desired oligonucleotides in the specified salt solutions, heating to 70° for 5 min, and cooling slowly to room temperature. The heating step might not be necessary if the single strands cannot form stable secondary structures. Solutions of sodium cacodylate are used as buffers because they are antibacterial, as opposed to phosphate buffers, and their pH does not vary appreciably with temperature, as with Tris-HCl.

Ultraviolet Spectra of Parallel-Stranded Duplexes and Their Antiparallel-Stranded References. The UV spectra of the ps-D1 · D2 (Fig. 1D) and aps-D1 · D3 molecules are shown in Fig. 5 in different representations. The native ps-D1 · D2 duplex shows a blue shift of the absorbance maximum relative to the aps-D1 · D3 reference, whereas the spectra in the denatured state are nearly identical (Fig. 5A and 5B). The difference presumably reflects particular features of the electronic structure and stacking geometry of ps-DNA. A plot of the ratio of the absorbances in the denatured and native states versus wavelength emphasizes the characteristic differences between ps-DNA and aps-DNA in a manner independent of concentration (Fig. 5C). By choice of appropriate wavelength, absorbance ratios can be derived that demonstrate the qualitative differences in the spectra and melting temperature of ps- and aps-DNAs (Fig. 5D). Thus the A_{250}/A_{284} ratio decreases in the case of ps-DNA but increases on denaturation of aps-DNA. Other spectral data, including that of DNAs with trans-d(G · C) base pairs, are given elsewhere.[2,15]

The thermally induced helix–coil transitions measured by UV absorption spectroscopy can be visualized by normalizing the absorbances at each

[30] M. Riley, B. Maling, and M. J. Chamberlin, *J. Mol. Biol.* **20,** 359 (1966).
[31] R. Inman, *J. Mol. Biol.* **9,** 624 (1964).

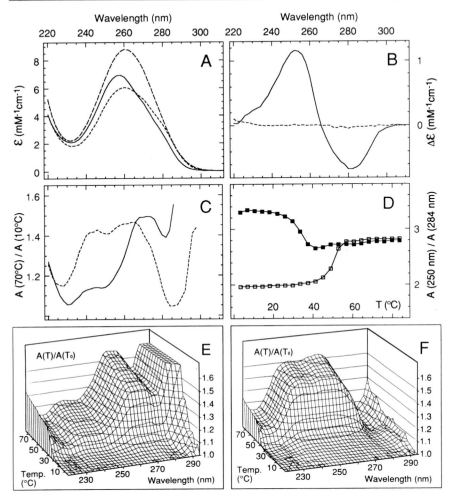

FIG. 5. Various ways to visualize the difference in UV absorption between ps- and aps-DNA. (A) UV spectra at 10° of ps-D1·D2 (—) and aps-D1·D3 (---); at 70° melted ps-D1·D2 duplex (– – –) and melted aps-D1·D3 duplex (····). (B) UV difference spectra (ps-D1·D2 − aps-D1·D3) at 10° (—) and at 70°(---). (C) Hyperchromicity spectra expressed as the ratio of absorbances at 70° and 10° of ps-D1·D2 (—) and aps-D1·D3 (---). (D) Wavelength-dependent melting curve of ps-D1·D2 (■) and aps-D1·D3 (□). The ratio of UV absorbances at 250 and 284 nm is plotted versus temperature. (E, F) Thermally resolved ultraviolet absorption spectra of ps and aps helices. The ratio $A_\lambda(T)/A_\lambda(T_0)$, where T_0 is the initial temperature, is plotted versus temperature for (E) ps-D1·D2 and (F) aps-D1·D3. Spectra (A)–(C) were recorded in 0.1 M NaCl, 10 mM sodium cacodylate, pH 7.2, and spectra (D)–(F) were measured in 2 mM MgCl$_2$ and 10 mM sodium cacodylate, pH 7.2. (Data in part are from Rippe et al.[12])

temperature T and wavelength λ to the values at the initial (low) temperature T_0. The resultant three-dimensional plots (Fig. 5E,F) reveal a characteristic pattern for both the ps and the aps conformation.[2,7,12,15] Whereas the ps duplexes show a peak around 270 nm and a strong increase of the absorbance ratio beginning at 280 nm, the aps reference molecules have a plateau from 240 to 270 nm and show little hyperchromicity from 280 to 290 nm.

Native Gel Electrophoresis

Gel electrophoresis under native conditions for ps and aps duplexes in the range of 15–40 base pairs can be carried out on a 14% polyacrylamide (acrylamide–bisacrylamide, 19:1) gel in 89 mM Tris–borate (pH ~ 8) and 2 mM MgCl$_2$ (TBM), that is, in the normally used TBE buffer but with EDTA replaced by 2 mM MgCl$_2$. As loading solutions 5× Ficoll loading buffer [containing 15% Ficoll (w/v), 0.3% bromphenol blue, and 0.3% xylene cyanole in TBM buffer] or 5× sucrose loading buffer [containing 40% sucrose (w/v), 0.3% bromphenol blue, and 0.3% xylene cyanole in TBM buffer] are appropriate. It is important to achieve an effective temperature control of the gel during electrophoresis in order to avoid denaturation of the duplexes being resolved. We use a system in which the gel is completely immersed in the lower buffer, which in turn is cooled by a water bath. The voltage applied should not exceed 10 V/cm. The gel shown in Fig. 6 was run under these conditions at 20°.

An important finding from the electrophoretic analysis is that the ps duplexes have slightly higher electrophoretic mobilities on gels than their aps references.[4,7,11,12] With the duplexes analyzed in Fig. 6 this difference is less evident because the ps-P3·P5 duplex (Fig. 1E) has aps flanking sequences.

Drug Binding

The intercalating agents ethidium bromide, propidium iodide, and actinomycin D bind readily to ps-DNA, in some cases with greater affinity than to aps-DNA, whereas the nonintercalators bisbenzimidazole Hoechst 33258 (BBI-258) and DAPI form complexes with much reduced stability compared to aps-DNA[1,2,4,5,7,10,11,32] This difference can be exploited to distinguish between the two DNA conformations, either by direct measurements of the fluorescence or by analysis on native polyacrylamide gels followed by staining with BBI-258 and ethidium bromide. An example is

[32] J. B. Chaires, F. G. Loontiens, M. M. Garner, J. Klysik, R. M. Wadkins, and T. M. Jovin, in preparation.

ps-P3·P5 aps-P3·P4   P3   P4   P5
Et   BBI   Et   BBI   Et   Et   Et

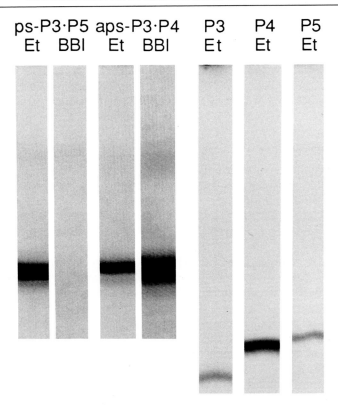

FIG. 6. Native gel electrophoresis and drug binding of oligonucleotides and their ps and aps duplexes. Gel electrophoresis under native conditions was carried out as described in the text. The amount of DNA applied to each lane was 0.2 nmol of strands. The gel was first stained with Hoechst 33258 (BBI-258, lanes denoted by BBI) followed by staining with ethidium bromide (lanes denoted by Et). The lanes with the single strands showed no fluorescence with BBI and are not included here. The digitized pictures of the fluorescent bands were obtained by illuminating the gel with a UV transilluminator and recording the images with a Series 200 slow-scan digital CCD camera (Photometrics, Tucson, AZ) and appropriate band-pass filters. (Data are from Klysik *et al.*[8])

shown in Fig. 6 for the ps-P3·P5 duplex (Fig. 1E) and its aps reference aps-P3·P4. It is apparent from the ethidium bromide-stained gel (Fig. 6, Et designated lanes) that the ps-P3·P5 duplex as well as the aps-P3·P4 reference molecules migrate as duplexes compared with the migration of the single-stranded P3, P4, and P5. No fluorescence is perceived on staining ps-P3·P5 with BBI-258, whereas the aps-P3·P4 reference shows the normal enhancement of fluorescence on drug binding (Fig. 6, BBI designated lanes). BBI-258 has a strong preference for (dA · dT)-rich regions, and thus it is also excluded from the aps flanking regions of ps-P3 · P5 and ps-P5 · P6 in which d(G · C) base pairs predominate.

Procedure: Double Staining with Hoechst 33258 (BBI-258) and Ethidium Bromide

To reduce background staining it is important to use pure acrylamide and to electrophorese the gel for 60 min at 10 V/cm to remove traces of residual ammonium persulfate and acrylic acid prior to loading the samples. About 0.2–5 μg DNA per band can be loaded for the BBI-258 and ethidium bromide staining. The times for staining and destaining depend on the thickness of the gel and are given for a gel 1 mm thick. If the desired resolution is obtained, remove the gel carefully from the plate, immerse it in 0.15 μg/ml Hoechst 33258 (BBI-258) in TBM, and gently agitate for 5–20 min. Bands can be visualized by placing the gel on a standard transilluminator with UV light (~302 nm) or by illumination with a hand-held Hg-lamp (365 nm). The excitation maximum of BBI-258 is at 350 nm, and the emission maximum is at 465 nm. To photograph the gel use band-pass filters centered around the latter wavelength.

After destaining the gel in TBM without dye for about 30–60 min the gel is incubated in 1 μg/ml ethidium bromide in TBM for 10–20 min with gentle shaking. If the background is too high, destain for 20–30 min. Excitation is with a transilluminator at 302 nm. The emission maximum is at 610 nm. A suitable red cutoff filter can be used for photography.

DNA Processing Enzymes and Chemical Probing

Parallel-stranded DNA shows different substrate properties than B-DNA with several DNA processing enzymes (Table III). It is insensitive to the restriction enzymes DraI, MseI, and SspI and to degradation with pancreatic DNase I and Escherichia coli exonuclease III under standard conditions.[33] Chemical probing with a variety of reagents [osmium–pyridine, diethyl pyrocarbonate (DEPC), $KMnO_4$, and the phenanthroline–copper complex $(OP)_2Cu^+$] confirms that ps-DNA exists in an ordered duplex structure, but one that differs from aps-DNA[23] (Table III). For example, $(OP)_2Cu^+$ cleaves the aps but not the ps duplex, attesting to the specificity of this reagent (as well as of DNase I) for the minor groove geometry of B-DNA, a feature absent in the model of ps-DNA.[2,4,7]

---

[33] K. Rippe and T. M. Jovin, Biochemistry 28, 9542 (1989).

TABLE III
SUBSTRATE PROPERTIES OF LINEAR PARALLEL-STRANDED
DUPLEXES[a]

Agent	NaCl[b]	Activity		
		ps	aps	ss
Restriction enzymes				
DraI	Low	−	+++	++
MseI	Low	−	+++	−
SspI	Low	−	+	−
DNA nucleases				
DNase I	Low	−	+++	+++
Exonuclease III	Low	−	++	+++
Chemical reagents				
Os,py	Low	−	−	++
	High	+++[c]	+	+++
DEPC	Low	+[d]	+[d]	++
	High	+	+	+
KMnO₄	Low	+	+	+++
	High	+++	++	+++
(OP)₂Cu⁺	Low	−	++[e]	+

[a] Relative activities: +++, very high; ++, high; +, moderate; −, none or very low. Data for enzymatic activities were adopted from Rippe and Jovin[33] and those for chemical reagents are from Klysik et al.[23] obtained with the 25-bp linear ps and aps duplexes.[12] ss, Single strands or nonpairing combinations of oligonucleotides.

[b] In all cases 10 mM MgCl₂ was present. Low; <2 M NaCl; high, >4 M NaCl.

[c] High activity is attributable to reaction with single strands formed on dissociation of the duplex.

[d] Low reactivity was found, but the modification patterns of the same strand involved in ps or aps duplex formation are different.

[e] Preferred cutting sites are the TAT steps.

Acknowledgments

We are indebted to N. B. Ramsing for long-standing participation and collaboration in many phases of this work. He is responsible for setting up the equation system for deriving the thermodynamic parameters in MATHEMATICA and for the graphic display program of secondary structures. We also thank our colleagues for critical reading of the manuscript.

[12] Crystallographic Studies of DNA Containing Mismatches, Modified and Unpaired Bases

By WILLIAM N. HUNTER

Introduction

The replication of DNA must occur with a high degree of precision in order for genetic information to be faithfully transmitted from one generation to the next. It was recognized by Watson and Crick[1,2] that a complementary base-pairing scheme (A with T, G with C) provided a mechanism whereby this may be accomplished. The Watson–Crick base pairs are shown in Fig. 1. The incorporation of non-Watson–Crick base pairs, such as G pairing with T, into duplex DNA does occur during replication, and this may induce errors in the processing of the genetic code. Such pairings are referred to as mismatches or mispairs. In addition, the integrity of genomic DNA is constantly pressured by chemical and physical forces present in our environment. Cancerous chemicals, ultraviolet light, and ionizing radiation are but some of the environmental factors which directly damage DNA and which may account for biosynthetic errors during life processes. These errors may be strand breaks, missing or extra unpaired bases, chemically modified bases, or mismatches. Single-crystal X-ray diffraction methods have been applied to a number of synthetic oligonucleotide fragments which with duplex formation contain examples of these structural "errors." Table I[3-23] lists the sequences so far studied and the

[1] J. D. Watson and F. H. C. Crick, *Nature (London)* **171**, 737 (1953).

[2] J. D. Watson and F. H. C. Crick, *Nature (London)* **171**, 964 (1953).

[3] O. Kennard, *J. Biomol. Struct. Dyn.* **3**, 205 (1985).

[4] T. Brown, G. Kneale, W. N. Hunter, and O. Kennard, *Nucleic Acids Res.* **14**, 1801 (1986).

[5] M. Coll, D. Saal, C. A. Frederick, J. Aymami, A. Rich, and A. H.-J. Wang, *Nucleic Acids Res.* **17**, 911 (1989).

[6] M. Coll, A. H.-J. Wang, G. A. van der Marel, J. H. van Boom, and A. Rich, *J. Biomol. Struct. Dyn.* **4**, 157 (1986).

[7] L. Van Meervelt, M. H. Moore, L. P. Kong Thoo, D. M. Brown, and O. Kennard, *J. Mol. Biol.* **216**, 773 (1990).

[8] P. S. Ho, C. A. Frederick, G. Quigley, G. A. van der Marel, J. H. van Boom, A. H.-J. Wang, and A. Rich, *EMBO J.* **4**, 3617 (1985).

[9] S. C. Ginell, S. Kuzmich, R. A. Jones, and H. M. Berman, *Biochemistry* **46**, 10461 (1990).

[10] G. Kneale, T. Brown, O. Kennard, and D. Rabinovich, *J. Mol. Biol.* **186**, 805 (1985).

[11] T. Brown, O. Kennard, G. Kneale, and D. Rabinovich, *Nature (London)* **315**, 604 (1985).

[12] Z. Shakked, T. E. Haran, M. Eisenstein, F. Frolow, and D. Rabinovich, *J. Mol. Biol.* **200**, 151 (1988).

[13] W. B. T. Cruse, J. Aymami, O. Kennard, T. Brown, A. G. C. Jack, and G. A. Leonard, *Nucleic Acids Res.* **17**, 55 (1989).

METHODS IN ENZYMOLOGY, VOL. 211

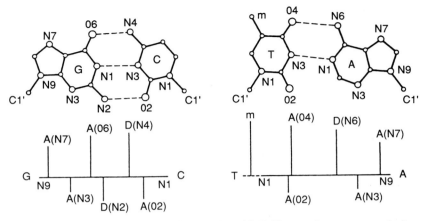

FIG. 1. Watson–Crick pairing, G with C and A with T. The atomic arrangement is shown at top, and below is a stick representative to point out the disposition of the functional groups. A stands for a hydrogen bond acceptor, D for a donor. Note the approximately symmetrical relationship of the bases, be they purine or pyrimidine, with respect to a vector linking the C-1' atoms. The stick drawing has the major groove above, the minor groove below the horizontal line.

points of structural interest. These studies have given information which aids our understanding about DNA structure and have led to proposals about the nucleic acid–protein interactions which are important for maintaining the fidelity of replication.[24]

[14] G. G. Privé, U. Heinemann, S. Chandrasegaran, L. S. Kan, M. L. Kopka, and R. E. Dickerson, *Science* **238**, 498 (1987).

[15] W. N. Hunter, T. Brown, G. Kneale, N. N. Anand, D. Rabinovich, and O. Kennard, *J. Biol. Chem.* **262**, 9962 (1987).

[16] T. Brown, W. N. Hunter, G. Kneale, and O. Kennard, *Proc. Natl. Acad. Sci. U.S.A.* **83**, 2402 (1986).

[17] W. N. Hunter, T. Brown, N. N. Anand, and O. Kennard, *Nature (London)* **320**, 552 (1986).

[18] P. W. R. Corfield, W. N. Hunter, T. Brown, P. Robinson, and O. Kennard, *Nucleic Acids Res.* **15**, 7935 (1987).

[19] T. Brown, G. A. Leonard, E. D. Booth, and J. Chambers, *J. Mol. Biol.* **207**, 455 (1989).

[20] G. A. Leonard, J. Thompson, W. P. Watson, and T. Brown, *Proc. Natl. Acad. Sci. U.S.A.* **87**, 9573 (1990).

[21] L. Joshua-Tor, D. Rabinovich, H. Hoppe, F. Frolow, E. Appella, and J. L. Sussman, *Nature (London)* **334**, 82 (1988).

[22] M. Miller, R. W. Harrison, A. Wlodawer, E. Appella, and J. L. Sussman, *Nature (London)* **334**, 85 (1988).

[23] R. Chattopadhyaya, S. Ikuta, K. Grzeskowiak, and R. E. Dickerson, *Nature (London)* **334**, 175 (1988).

[24] P. Modrich, *Annu. Rev. Biochem.* **56**, 435 (1987).

TABLE I
STRUCTURES WITH NON-WATSON–CRICK BASE PAIRS, MODIFIED BASES, OR EXTRA
HELICAL BASES[a]

Sequence	Comment	Global form	Ref.
d(TGCGCG)	G·T	Z	3
d(BrUGCGCG)	BrU·G	Z	4
d(CGCGFUG)	FU·G	Z	5
d(CGTDCG)	T·D	Z	6
d(CDCGTG)	T·D	Z	6
d(CGCGXG)	G·X	Z	7
d(CGCGTG)	G·T	Z	8
d(CGCOCG)	G·O	Z	9
d(GGGGTCCC)	G·T	A	10
d(GGGGCTCC)	G·T	A	11
d(GGGTGCCC)	G·T	A	12
d(GGIGCTCC)	I·T	A	13
d(CCAAGATTGG)	G·A	B	14
d(CGCGAATTTGCG)	G·T	B	15
d(CGCGAATTAGCG)	G·A	B	16
d(CGCAAATTCGCG)	C·A	B	17
d(CGCIAATTAGCG)	I·A	B	18
d(CGCAAATTGGCG)	G·A	B	19
d(CGCOAATTTGCG)	O·T	B	20
d(CGCAGAATTCGCG)	Extra A	B	21
d(CGCGAAATTTACGCG)	Extra A	B	22
d(CGCGCGTTTTCGCGCG)	Hairpin	Z	23

[a] BrU, 5-Bromouridine; FU, 5-fluorouridine; D, 2-aminoadenine; X, N^4-methoxycytosine; and O, O^6-methylguanine.

Crystallographic Analyses

The single-crystal diffraction methods applied to structure analyses of mismatches and DNA fragments with extra bases are essentially the same as applied to any biological macromolecule and will only be summarized. Comprehensive details can be found in Refs. 25–28. The single-crystal X-ray analysis can be divided into four stages.

[25] T. L. Blundell and L. N. Johnson, "Protein Crystallography." Academic Press, New York 1976.

[26] A. McPherson, "Preparation and Analysis of Protein Crystals." Wiley, San Francisco, California, 1982.

[27] H. Wyckoff, C. H. W. Hirs, and S. N. Timasheff, eds., this series, Vol. 114.

[28] H. Wyckoff, C. H. W. Hirs, and S. N. Timasheff, eds., this series, Vol. 115.

Crystallization

The oligonucleotide fragment is synthesized and purified. It must then be coerced into an ordered precipitation. This is achieved by driving a solution into a supersaturated state where crystal nucleation may occur and then maintaining the conditions around the nucleus so that crystal growth can take place. The most common methods for crystal growth are to diffuse a precipitating agent into an oligonucleotide solution or to diffuse out water. There are a number of considerations. To bring together negatively charged nucleic acids into an ordered crystal lattice requires the presence of cations. The nature and concentration of the cation can have a profound effect on crystallization. Other variables are the concentration and type of precipitant, temperature, and pH. In most of the cases listed in Table I the crystallization conditions for one of the Watson–Crick parent sequences had already been determined and they were crystallized in a similar manner. Several samples did, however, give completely new crystal forms for which the crystallization conditions had to be determined.

Data Collection

Oligonucleotide crystals, like those of proteins, are extensively hydrated. To preserve the crystalline order they must be kept in a humid environment. For the purpose of data collection this is achieved by putting the sample in a capillary which is then sealed with wax. A drop of crystallization solution is often placed in the capillary as well. Most samples listed in Table I had the diffraction data measured on automated four-circle diffractometers, but the availability of area detectors is speeding up this process. Selected crystallographic details are given in Table II. The resolution quoted in Table II is an indication of how ordered the crystal lattice is: the smaller the resolution given, the more accurate the final structure.

Structure Solution

The structure solution part of the analysis aims to place a model structure at the correct position in the unit cell and with the correct orientation. This may be achieved by using multiple isomorphous replacement methods to generate the electron density map and then constructing a molecular model to fit this density. Alternatively an oligonucleotide model structure can be constructed from fiber diffraction results, or a previously determined crystal structure can be used in molecular replacement methods to position and orient this known structure in the unknown unit cell.

With few exceptions most of the "error" structures crystallize isomorphously with one of the parent compounds for which the structure had

TABLE II

CRYSTALLOGRAPHIC DETAILS OF SELECTED STRUCTURES CONTAINING NON-WATSON–CRICK BASE PAIRS AND EXTRA BASES[a]

Sequence	Mispair	DNA form	Space group	Unit cell dimensions (Å)			Resolution (Å)	R-factor (%)
				a	b	c		
(GGGGCTCC)	G·T	A	$P6_1$	45.2	45.2	42.9	2.2	14
(GGGGTCCC)	G·T	A	$P6_1$	44.7	44.7	42.4	2.1	14
(GGGTGCCC)	G·T	A	$P6_1$	45.6	45.6	41.0	2.5	15
(GGIGCTCC)	I·T	A	$P6_1$	45.1	45.1	45.5	1.7	14
(CCAAGATTGG)	G·A	B	$C2, \beta = 119°$	32.5	26.2	34.3	1.3	17
(CGCGAATTAGCG)	G·A	B	$P2_12_12_1$	25.7	41.9	65.2	2.5	17
(CGCAAATTGGCG)	G·A	B	$P2_12_12_1$	25.2	41.2	65.0	2.3	16
(CGCIAATTAGCG)	I·A	B	$P2_12_12_1$	25.8	41.9	65.1	2.5	19
(CGCAAATTCGCG)	C·A	B	$P2_12_12_1$	25.4	41.4	65.2	2.5	19
(CGCGAATTTGCG)	G·T	B	$P2_12_12_1$	25.5	41.2	65.6	2.5	18
(CGCOAATTTGCG)	O·T	B	$P2_12_12_1$	25.4	40.7	65.9	2.0	18
(TGCGCG)	G·T	Z	$P2_12_12_1$	17.9	30.7	45.1	1.5	18
(BrUGCGCG)	BrU·G	Z	$P2_12_12_1$	17.9	30.9	49.9	2.2	16
(CGCGTG)	G·T	Z	$P2_12_12_1$	17.5	31.6	45.6	1.0	20
(CGCGFUG)	FU·G	Z	$P2_12_12_1$	17.8	31.3	45.4	1.5	20
(CGCAGAATTCGCG)	Extra A	B	$C2, \beta = 99°$	78.5	42.8	25.2	2.8	15
(CGCGAAATTTACGCG)	Extra A	B	$I222$	37.0	53.7	101.6	3.0	24
(CGCGCGTTTTCGCGCG)	Hairpin	Z	$C2, \beta = 95°$	57.2	21.6	36.4	2.1	23

[a] O, O^6-Methylguanosine; BrU, 5-bromouridine; FU, 5-fluorouridine; and X, N^4-methoxycytosine.

previously been determined. There are notable exceptions; the decamer d(CCAAGATTGG), the looped out A structure d(CGCAGAATTCGCG) and the hairpin d(CGCGCGTTTTCGCGCG) were solved with molecular replacement. In the case of d(CGCGAAATTTACGCG) novel maximum entropy methods were applied. Once a starting model is positioned correctly in the unit cell the next stage, refinement, proceeds.

Refinement

The aim in refinement is to reduce differences between calculated structure factors (based on an atomic model) and the experimental observations, that is, the diffraction data. This is achieved by a combination of large- and small-scale alterations of atomic positions. Gross alterations of the model to optimize the fit to generated electron density maps are often required. Access to computer graphics greatly speeds up this process. The smaller manipulations of the model are generally carried out by a series of least-squares optimization calculations. Because the nucleic acid crystals do not often diffract to very high resolution, it is necessary to utilize the accurate structural information generated from small molecule structure determinations of the nucleic acid components. In this way the least-squares calculations are closely linked to known geometrical parameters. This compensates for the limited diffraction data that are available. The two most commonly used programs in the least-squares calculations are CORELS[29] and NUCLSQ.[30]

Oligonucleotide crystals are highly hydrated, and water or cations must be considered to be part of the molecular model. A proper refinement using diffraction data beyond about 2.8 Å will include the location of solvent positions. Although cations are present, it is often not possible to positively identify them. A safe way to treat the cations is to include them as oxygen atoms of water molecules. The careful identification of solvent positions and treatment of the model is necessary to avoid errors of interpretation and to pinpoint solvent positions correctly. As discussed later, the identification of solvent positions in the $G \cdot T$ and $C \cdot A$ mispairs was important because solvent molecules are closely associated with the actual mispairs.

Refinement ceases when no further improvement of the model fit to the electron density maps can be accomplished, when all well-ordered solvent positions are located, and when no further improvement in the crystallographic residual can be obtained. This residual or reliability index, $R = \Sigma F_o - F_c/F_o$, where F_o is the observed and F_c the calculated structure factors. Most of the structures of relevance to this chapter refine to R values of less than 20%, which is perfectly acceptable for molecules of this size.

During the refinement of macromolecules it is common practice to omit from the electron density map calculations parts of the structure either for which no clear structural interpretation of the electron density can be made or which the researcher is prepared to leave out until the rest of the model is accurately refined. An interpretation of the density can then be made toward the end of the refinement. This method was applied by the author in a number of cases and has subsequently been used for most of the structures listed in Table I. The atoms of the mismatched bases are omitted from all calculations while other parts of the molecule are dealt with and solvent positions located. As the refinement proceeds the quality of the electron density maps improves, and an unambiguous interpretation of the electron density can be made. Figure 2 provides an example of an omit electron density map for a $G \cdot T$ mispair. Care is then taken to check that after the structural interpretation the model behaves normally in further refinement.

[29] J. L. Sussman, S. R. Holbrook, G. M. Church, and S. H. Kim, *Acta Crystallogr. Sect. A: Cryst. Phys. Diffr. Theor. Gen. Crystallogr.* **A33**, 800 (1977).

[30] E. Westhof, P. Dumas, and D. Moras, *J. Mol. Biol.* **184**, 119 (1985).

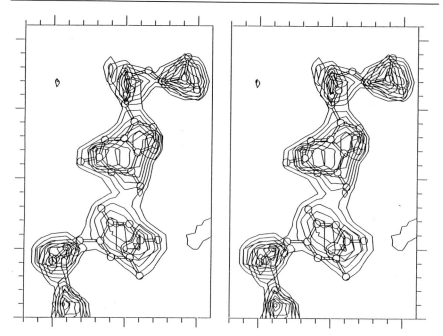

FIG. 2. Stereoview of an omit electron density map at 2.5 Å resolution for the G·T mispair in the structure d(CGCGAATTTGCG). The continuous lines represent increases in the electron density level. Atomic positions are depicted as spheres. The location of the bases is unambiguous.

The accuracy of the final structures produced depends to a large extent on how ordered the crystal is and to what resolution it diffracts (see above). In most cases cited in Table I, the errors in atomic position will be between 0.1 and 0.5 Å.

Results and Some Implications for Recognition and Repair

The X-ray analyses of DNA fragments with potential mispairs does not provide information about the occurrence of rare tautomeric forms at the time of replication, but they do define the structure of the duplex which incorporates a mispair or modified base. They show how functional groups interact to form hydrogen bonds and give some idea about the van der Waals and hydrophic interactions that stabilize a variety of molecular conformations. The purine–pyrimidine pairings G·T and A·C form what

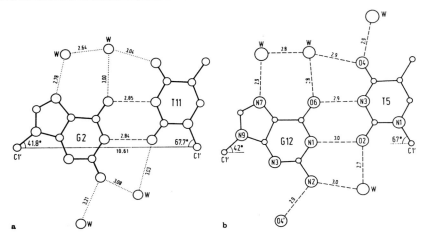

FIG. 3. G·T mismatches with the distinctive hydration pattern observed in two crystal structures, (a) d(CGCGTG) and (b) d(GGGGTCCC). Distances are given in angstroms.

are termed "wobble" pairs. The pyrimidine moves out into the major groove and the purine to the minor groove. Solvent positions form hydration bridges between the bases in the major groove for both of the pairings. The G·T mispair has an additional solvent bridge in the minor groove. Figure 3 illustrates two G·T pairings observed in different sequences showing the well-defined hydration network. Figure 4 shows structural formulas for G·T, C·A, and one type of G·A pairing. The purine–purine pairing of G with A has been observed in three different conformations (see Fig. 5). This structural variability may explain why this mismatch is the most likely to escape detection by polymerase III.[31]

Efforts to crystallize the A·A, G·G, C·C, C·T, C·T, and T·T mispairs have so far not met with success. The T·T pairing has been observed not in a duplex but formed by intermolecular interactions in the sequence d(CGCGCGTTTTCGCGCG). A symmetry related molecule is positioned so that two thymines are close enough to pair. The enol tautomer form may be present for one of the bases. Two structures which have looped out adenines have been described in the literature. Crystal packing forces have been implicated in determining the positions of these extrahelical bases. Nevertheless these structures show how unpaired bases can be associated with a stable duplex.

[31] A. R. Fersht, J. W. Knill-Jones, and W. C. Tsui, *J. Mol. Biol.* **156**, 37 (1982).

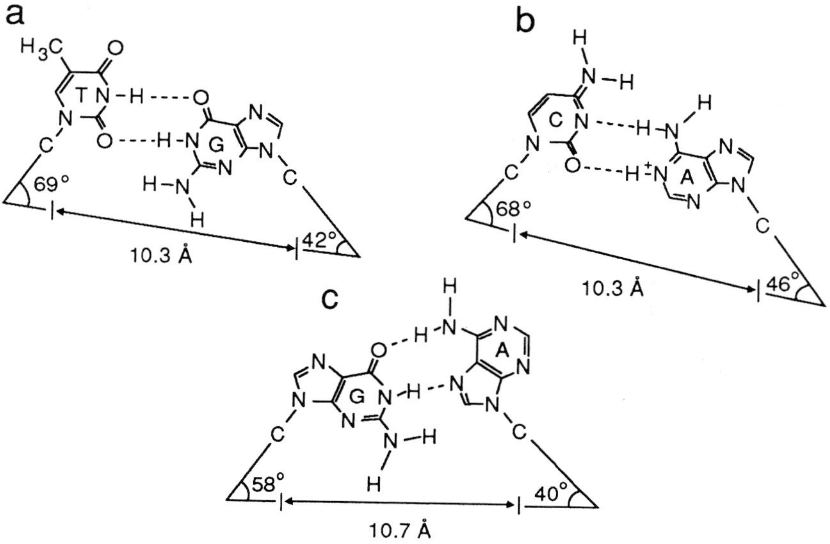

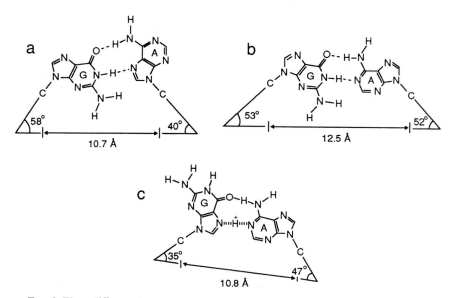

FIG. 4. Structural formulas and selected geometric details for three mispairs analyzed in a B-form dodecamer structure. (a) G·T, (b) C·A, and (c) G(anti)·A(syn). The C-1' separation is given, as are the angles representing the N(base)–C-1'–C-1' vector. The latter parameter highlights the asymmetry of the mispair bases with respect to the Watson–Crick pairs.

FIG. 5. Three different G·A pairings depicted in a similar manner to the mispairs in Fig. 3. Note the quite different atomic arrangements.

There are now several studies concentrating on the pairing properties of chemically modified bases. Some of those studied are 5-bromouridine, 5-fluorouridine, N^4-methoxycytosine, and O^6-methylguanine. These structures provide conformational detail that may assist in the design of hybridization probes and show potential mutagenic lesions at the molecular level. There is a paucity of structural information on the enzymes responsible

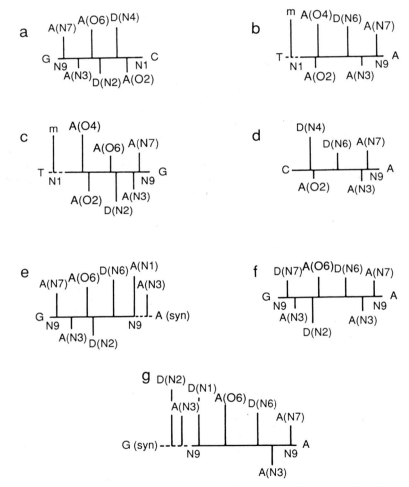

FIG. 6 Stick representations for (a) G·C, (b) A·T, (c) G·T, (d) C·A, (e) G(anti)·A(syn), (f) G(syn)·A(anti), and (g) G(anti)·A(anti). The variety of functional group arrangements of mispaired bases are highlighted.

for maintaining the fidelity of replication. The structures discussed in this chapter are able to provide clues to what structural features may be important in the nucleic acid–protein interactions involved in mismatch recognition and repair. A number of conclusions arise from the crystallographic results and from consideration of the relevant biochemical studies. These are now summarized. Readers are referred to the original publications for full structural details.

First, non-Watson–Crick base pairs have been observed to adopt a variety of hydrogen-bonded schemes. Some of these pairings are quite stable. Second, the overall conformation of the duplex, A, B, or Z form is unaltered by mispair formation or the presence of extra bases. Indeed, changes are highly localized in the structures analyzed to date. The G · T mispair has been studied in each duplex type and assumes the same conformation even to the extent that the same localized hydration pattern is observed. Third, the crystallographic results indicate the variety of functional group arrangements on the DNA (see Fig. 6). This, in conjunction with points raised above, suggests that the high degree of enzyme specificity shown in mispair recognition and correction procedures must in large part be due to the protein structure. This clearly points the way forward to structural research on these enzymes and complexes of these enzymes with oligonucleotides.

Acknowledgments

The author is supported by the University of Manchester, the Wellcome Trust, the Royal Society, and the Science and Engineering Research Council (U.K.). We thank Springer-Verlag for permission to use some figures from the Landolt–Bornstein tables giving crystallographic and structural data for nucleic acids.

Section III

Spectroscopic Methods for Analysis of DNA

[13] ¹H NMR Spectroscopy of DNA

By Juli Feigon, Vladimír Sklenář, Edmond Wang,
Dara E. Gilbert, Román F. Macaya, and Peter Schultze

Introduction

Advances in NMR technology and instrumentation since the mid-1980s have led to a revolution in the use of NMR spectroscopy for the determination of macromolecular structures.[1] Protein structures determined from data obtained by NMR methods are now accepted by both NMR spectroscopists and crystallographers. Application of these methods to the determination of nucleic acid structures has lagged somewhat behind the progress made for proteins. This is partially because nucleic acids, with only four different bases (usually) and generally linear structures for DNA, require somewhat different approaches from proteins in obtaining appropriate spectra, in assignment methodology, and in using the information derived from the NMR data to generate structures. The accuracy and utility of structures of DNA that can be obtained using NMR data are still a matter of investigation and debate, primarily owing to the lower proton density of nucleic acids relative to proteins and the difficulty in constraining the phosphodiester backbone from ¹H NMR data alone. This structure determination is the ultimate goal of NMR studies of nucleic acids. To achieve this goal, it is necessary first to obtain high quality NMR data to use for input into whatever approach is being used to generate the final structure. This chapter focuses primarily on the methods currently being used in our laboratory to obtain, process, assign, and analyze NMR spectra of DNA oligonucleotides. The utility of these studies for qualitative analysis of DNA structures will be discussed. A brief overview of current approaches to using the NMR data for three-dimensional (3-D) structure determination will be presented at the end.

NMR spectroscopy of DNA oligonucleotides was largely made possible by the advent of convenient DNA synthesis methods at about the same time that two-dimensional (2-D) NMR was beginning to be applied to the study of proteins. Prior to that, most ¹H NMR spectroscopy of nucleic acids was done on tRNA and synthetic RNA polymers.[2-4] Much of the

---

[1] K. Wüthrich, "NMR of Proteins and Nucleic Acids." Wiley, New York, 1986.
[2] D. R. Kearns and P. H. Bolton, *in* "Biomolecular Structure and Function," (P. F. Agris, R. N. Loeppky, and B. D. Sykes, eds.), p. 493. Academic Press, New York, 1978.
[3] D. R. Kearns, *Annu. Rev. Biophys. Bioeng.* **6**, 477 (1977).
[4] D. R. Kearns, *Prog. Nucleic Acid Res. Mol. Biol.* **18**, 91 (1976).

METHODS IN ENZYMOLOGY, VOL. 211

early work focused on observation of exchangeable imino resonances, and assignments were based largely on chemical shift and ring current shift arguments. The application of one-dimensional nuclear Overhauser effects (NOEs) was the first reliable assignment method for imino AH2 resonances.[5] Assignments of nonexchangeable resonances were still largely based on chemical shifts. The first two-dimensional spectra of double-stranded DNA were published in 1982.[6] It was apparent from these early spectra that they contained the information for making assignments of the nonexchangeable protons in right-handed B-DNA structures,[6] and sequential assignment strategies for B-DNA oligonucleotides were soon published.[7-9] Two-dimensional NMR techniques have since been applied to study a wide range of oligonucleotide structures (reviewed in Refs. 1 and 10-15), including B-DNA duplexes containing mismatches, bulges, and modified bases, drug–DNA complexes, and protein–DNA complexes. In addition, non-B-DNA structures, such as Z-DNA, triplexes, quadruplexes, dumbbells, hairpin loops, and Holliday junctions have also been characterized by NMR.

Sample Preparation

Synthesis and Purification

DNA in the milligram quantities needed for ^{1}H NMR spectroscopy can be conveniently synthesized on commercial DNA synthesizers. Because large quantities are needed and purity is a more important issue for NMR studies than for uses like cloning, it is necessary to ensure that the machine is working optimally and chemicals are fresh. We have found that one 10 μmol synthesis is adequate for about two NMR samples of DNA of

[5] V. Sanchez, A. G. Redfield, P. D. Johnston, and J. Tropp, *Proc. Natl. Acad. Sci. U.S.A.* **77**, 5659 (1980).

[6] J. Feigon, J. M. Wright, W. A. Denny, W. Leupin, and D. R. Kearns, *J. Am. Chem. Soc.* **104**, 5540 (1982).

[7] J. Feigon, W. Leupin, W. A. Denny, and D. R. Kearns, *Biochemistry* **22**, 5943 (1983).

[8] R. M. Scheek, N. Russo, R. Boelens, and R. Kaptein, *J. Am. Chem. Soc.* **105**, 2914 (1983).

[9] D. R. Hare, D. E. Wemmer, S. H. Chou, G. Drobny, and B. R. Reid, *J. Mol. Biol.* **171**, 319 (1983).

[10] D. E. Wemmer and B. R. Reid, *Annu. Rev. Phys. Chem.* **36**, 105 (1985).

[11] F. J. M. van de Ven and C. W. Hilbers, *Eur. J. Biochem.* **178**, 1 (1988).

[12] B. R. Reid, *Q. Rev. Biophys.* **20**, 1 (1987).

[13] D. J. Patel and L. Shapiro, *Q. Rev. Biophys.* **20**, 35 (1987).

[14] D. J. Patel, L. Shapiro, and D. Hare, *Annu. Rev. Biophys. Biophys. Chem.* **16**, 423 (1987).

[15] D. E. Gilbert and J. F. Feigon, *Curr. Opin. Struct. Biol.* **1**, 439 (1991); D. E. Wemmer, *Curr. Opin. Struct. Biol.* **1**, 452 (1991).

length 16 bases or less. For longer DNA, we get better results by combining several 1 μmol syntheses.

The most difficult part of obtaining DNA for NMR samples is purification. Many laboratories use high-performance liquid chromatography (HPLC) with adequate results. The method we have found to be overall most satisfactory, in terms of final purity, yield, and time, is to separate full-length sequences from failure sequences and other by-products of the synthesis on Sephadex gel filtration columns using only water as the eluant, following the method of Kintanar et al.[16] Typical column sizes are 2.4 cm wide and 110 cm long. Sephadex G-25 superfine is used for DNA oligonucleotides 12 bases or less, and Sephadex G-50 is used for more than 12 bases. The peak fractions are then assayed on DNA sequencing gels after kinasing with [γ-^{32}P]ATP in order to see which fractions contain only the desired DNA. Because the DNA comes off the column as a very broad peak, many fractions will contain some amount of full-length as well as failure sequences; because these are discarded (or, in some cases, repooled and rerun on the column) a fair amount of the total yield is wasted by this method. Nevertheless, we find that our overall yields of product of NMR quality are still higher than those we have obtained by HPLC, since we do not need to do any additional purification to remove protonated solvents, salts, etc. Once the fractions containing only full-length oligonucleotide are determined, they are pooled and lyophilized. The lyophilized DNA is dissolved in a small amount of water and pooled into a single Eppendorf tube, relyophilized, and stored frozen in the presence of a desiccant until ready for use.

NMR Sample

Our typical NMR sample contains 2–4 mM in strand in 400 μl buffer. Samples are generally prepared by dissolving the lyophilized powder in a small amount of H_2O or D_2O and removing the appropriate amount for the desired final DNA concentration to an Eppendorf tube. Alternatively, the desired amount of the lyophilized DNA may be weighed out. Salts and buffer solutions are added to the sample to the desired final concentration. Because a nonprotonated buffer is needed, phosphate has generally been the buffer of choice. However, this buffer can be a problem if divalent metal ions are also used or if one also wishes to acquire ^{31}P NMR spectra on the same sample. The best solution is frequently just to add the desired salts and adjust the pH of the sample to the desired pH without adding any buffer.

---

[16] A. Kintanar, R. E. Klevit, and B. R. Reid, *Nucleic Acids Res.* **15**, 5845 (1987).

In any case, the pH of the DNA sample should always be checked. We have found that dissolved DNA often has a very low pH (<5). This may also lead to problems in dissolving the sample (owing to aggregation), which are relieved when the pH is adjusted toward neutrality. Even with added phosphate buffer the pH may remain low (or high); this has undoubtedly led to errors in the literature on sample pH and interpretation of the resulting spectra. Because the sample can be dried again, the change in volume from adjusting the pH does not matter.

For spectra of the nonexchangeable protons, the sample is then lyophilized and redissolved in D_2O. This can be done once or twice, after which the sample is redissolved in D_2O (99.98% or better) and transferred to an NMR tube which has been previously soaked in D_2O and dried under a stream of $N_{2(g)}$ just before use. It is advisable to recheck the pH of the sample in the NMR tube with an NMR pH electrode. The sample can then be dried one more time in the NMR tube with a stream of filtered $N_{2(g)}$ by inserting a long syringe needle into the NMR tube just above the level of the solution. The sample is then redissolved in 99.996% D_2O. For spectra including the exchangeable protons, the DNA is dissolved in the appropriate solution with 90% H_2O/10% D_2O.

If the DNA sample is not self-complementary, the two strands need to be mixed in the proper ratios. Approximate sample concentrations can be determined by absorbance at 260 nm or by weight. The duplex can be separated from any single strand by chromatography on a hydroxyapatite column, provided the DNA is long enough to be stable as a duplex at room temperature on the column. The fractions containing the duplex are pooled and concentrated by lyophilization. The excess salt can then be removed by dialysis against water or by chromatography on a Sephadex G-10 column. An alternative method to forming duplexes (or triplexes) from non-self complementary single strands is to titrate one strand into an NMR tube containing the other and monitoring the nonexchangeable resonances for duplex formation. The easiest resonances to monitor are usually the methyl resonances.

Many laboratories use much higher concentrations of DNA than that suggested above. This may result in sample aggregation and increased solution viscosity, both of which will lead to wider linewidths. We have found that above about 3 mM strand for many of our samples the signal-to-noise ratio (S/N) actually decreases due to line broadening. In addition, spin diffusion will become a problem in NOESY (nuclear Overhauser effect spectroscopy) experiments at shorter mixing times as the sample concentration is increased. Although with high sample concentrations it has been reported that mixing times of 40 msec and less must be used to stay in the linear range of NOE buildups,[17] we have found that for a 12-base pair (bp) DNA oligonucleotide at 2 mM duplex sample concentration and 35° NOE buildup rates are linear out to at least 75 msec.[18]

Obtaining High Quality ¹H NMR Spectra of DNA

One-Dimensional ¹H NMR Spectra

Before beginning a lengthy two-dimensional NMR study of a DNA oligonucleotide, it is useful to obtain one-dimensional spectra in D_2O and H_2O at several temperatures in order to assess the thermal stability of the molecule under NMR conditions. Resonances in D_2O cover the range from about 9.5 to 0.5 ppm. Much of the early NMR work on nucleic acids monitored the melting of DNA by chemical shifts as a function of temperature.[1] The one-dimensional spectra also allow a quick assessment of sample purity as well as the possible presence of more than one conformation.

In general, both the imino and amino resonances of DNA exchange too fast to be observed in D_2O, and they must be observed in H_2O solution.[19] For proteins, spectra in water are generally acquired with a presaturation pulse on the H_2O. This is not usually the method of choice for nucleic acids, since the amino and imino resonances will exchange with water during the presaturation pulse and therefore their intensity will usually be greatly diminished. For some molecules, enough resonance intensity remains that this may still be a useful technique for qualitative results in a 2-D experiment[20], or in a protein–DNA complex. In general, however, some sort of selective excitation which leaves the water unperturbed in the equilibrium position is used. These water suppression techniques have been recently reviewed.[21-23] The best method for water suppression appears to be instrument dependent, and new methods are still being proposed.

In our laboratory, on our GN 500 MHz (GE NMR; Freemont, CA)

[17] B. R. Reid, K. Banks, P. Flynn, and W. Nerdal, *Biochemistry* **28**, 10001 (1989).
[18] E. Wang and J. Feigon, unpublished results (1989).
[19] D. R. Kearns, D. J. Patel, and R. G. Shulman, *Nature (London)* **229**, 338 (1971).
[20] P. Rajagopal, D. E. Gilbert, G. A. van der Marel, J. H. van Boom, and J. Feigon, *J. Magn. Reson.* **78**, 526 (1988).
[21] P. J. Hore, this series, Vol. 176, p. 64.
[22] M. Guéron, P. Plateau, and M. Decorps, *Prog. NMR Spectrosc.* **23**, 135 (1991).
[23] V. Sklenář, *in* "NMR Applications in Biopolymers" (J. W. Finley, S. J. Schmidt, and A. S. Serianni, eds.), Vol. 56, p. 63. Plenum, New York, 1990.

spectrometer, we have found that consistently satisfactory results (even for nonexperienced operators) are obtained using the $1\bar{1}$ spin echo pulse sequence proposed by Sklenář and Bax.[24] This technique can tolerate relatively large inhomogeneities of both static magnetic and radiofrequency field and can easily be incorporated into various phase-sensitive 2-D NMR experiments. The pulse sequence $90_x\text{-}t\text{-}90_{-x}\text{-}d_1\text{-}90_f\text{-}2t\text{-}90_f\text{-}d_2\text{-}\mathrm{Acq}_p[\,f = x,y,-x,-y;\ p = x,-x,x,-x;\ d_1 \sim d_2 \sim 30\text{-}50\ \mu\mathrm{sec}]$ gives an excitation profile proportional to $\sin^3$ with the maxima at offsets $+1/[4(t + 2/3t_{90})]$, where t_{90} is the length of the $90°$ pulse. There is no excitation at the carrier frequency, and spectra with very flat baselines can be obtained. The refocusing delay d_2 in the $1\bar{1}$ spin echo pulse sequence should be adjusted so that there is a near zero linear phase correction of the Fourier-transformed spectrum; this compensates for the signal delay on the filters which distorts the intensities of the first few points in the free induction decay (FID) resulting in baseline distortions.

It is possible to obtain one-dimensional spectra in H_2O with accurate intensities over almost the entire spectral range using the NEWS pulse sequence.[25] However, this pulse sequence will not work for two-dimensional NMR experiments, so its utility is limited.

Obtaining Two-Dimensional Spectra in D_2O

Today, setting up various phase-sensitive two-dimensional NMR experiments in D_2O is fairly routine. The most valuable data for DNA samples are obtained from NOESY,[26] ROESY,[27] TOCSY,[28,29] P.(E).COSY,[30,31] 2Q,[32] and HOENOE[33] experiments. A flat baseline in both frequency dimensions is a basic precondition for reliable analysis of phase-sensitive two-dimensional NMR spectra. This requires both optimization of the acquisition parameters, discussed here, and the processing parameters, discussed below. In the f_2 domain the appropriate choice of audiofrequency filters and of the preacquisition delay are important. For instruments where there is a choice between Bessel and Butterworth filters,

[24] V. Sklenář and A. Bax, *J. Magn. Reson.* **74**, 469 (1987).

[25] L. Pùček, E. Wang, J. Feigon, Z. Starčuk, and V. Sklenář, *J. Magn. Reson.* **91**, 120 (1991).

[26] A. Kumar, R. R. Ernst, and K. Wüthrich, *Biochem. Biophys. Res. Commun.* **95**, 1 (1980).

[27] A. A. Bothner-By, R. L. Stephens, J. Lee, C. D. Warren, and R. W. Jeanloz, *J. Am. Chem. Soc.* **106**, 811 (1984).

[28] A. Bax and D. G. Davis, *J. Magn. Res.* **65**, 355 (1985).

[29] L. Braunschweiler and R. R. Ernst, *J. Magn. Reson.* **53**, 521 (1983).

[30] D. Marion and A. Bax, *J. Magn. Reson.* **80**, 528 (1988).

[31] L. Mueller, *J. Magn. Reson.* **72**, 191 (1987).

[32] L. Braunschweiler, G. Bodenhausen, and R. R. Ernst, *Mol. Phys.* **48**, 535 (1983).

[33] V. Sklenář and J. Feigon, *J. Am. Chem. Soc.* **112**, 5644 (1990).

the Bessel filter usually gives a flatter baseline with only a small decrease in signal-to-noise ratio. For magnitude experiments or spectra processed with an unshifted sine-bell or similar function, the Butterworth filter is acceptable. For spectrometers where a flat baseline cannot be obtained with these methods, application of a refocusing pulse at the end of the pulse sequence is useful.[34] The preacquisition delay which separates the last pulse in the pulse sequence and the start of the data acquisition should be carefully adjusted until the best baseline is obtained; the optimum situation gives a one-dimensional spectrum for which the first-order phase correction is near zero.

In the f_1 dimension, different procedures must be applied for spectra obtained with TPPI or hypercomplex Fourier transform acquisition schemes. As has been recently shown by Bax and co-workers,[35] the following approach gives a flat baseline in f_1 for data acquired with the hypercomplex method. The first t_1 increment is set to $[DW_1/2 - k(2/\pi)t_{90}]$, where DW_1 is the dwell time in the t_1 dimension, t_{90} is the length of the 90° pulse, and $k = 2$ for NOESY and $k = 1$ for TOCSY experiments. No scaling factor is applied to the first points of the t_1 FIDs, and the spectra in the f_1 domain are phase corrected with a linear phase correction of 180°.

The most effective method for tracing out single-step J connectivities and for measuring J coupling constants in the DNA sugar residues is the P.COSY experiment.[30] If a mixing pulse of 90° is applied, the cross-peak patterns are essentially the same as those obtained with DQF-COSY[31] but can be obtained in one-fourth the time. A mixing pulse of 45° or less results in E.COSY[36] type spectra from which the J coupling constants for H1′–H2′, H1′–H2″, and H2′–H2″ can be directly measured.[37] The artificial generation of the dispersive diagonal which is subsequently subtracted from the phase-sensitive COSY spectrum reduces the experimental time to one-quarter for the P.COSY alternative to DQF-COSY and to one-sixth for the P.E.COSY alternative to E.COSY experiments. Sixteen scans are required to fulfill the coherence transfer selection and CYCLOPS[38] phase cycles per t_1 complex point.

If measuring time is a limiting factor, the experimental protocols for many of these experiments can be optimized to obtain the highest quality spectra in the shortest possible time. If the rf channels are well balanced, the CYCLOPS phase cycling can be omitted in TOCSY, HOENOE, and

[34] D. G. Davis, *J. Magn. Reson.* **81**, 603 (1989).
[35] A. Bax, M. Ikura, L. E. Kay, and G. Zhu, *J. Magn. Reson.* **91**, 174 (1991).
[36] C. Griesinger, O. W. Sørensen, and R. R. Ernst, *J. Am. Chem. Soc.* **107**, 6394 (1985).
[37] A. Bax and L. Lerner, *J. Magn. Reson.* **79**, 429 (1988).
[38] D. I. Hoult and R. E. Richards, *Proc. R. Soc. London, A* **344**, 311 (1975).

NOESY experiments. Axial peaks and single-quantum coherence can be suppressed in NOESY spectra by applying a short homospoil pulse during the mixing period. Two to four dummy scans are usually used at the beginning of each t_1 increment to establish a dynamic equilibrium in the spin system; these can usually be eliminated if a "pulse equilibrium cascade" is applied at the beginning of the pulse sequence. This is done by putting two short spin-lock pulses (2–3 msec) applied in perpendicular directions followed by a short homospoil pulse (3–5 msec) at the beginning of the relaxation delay. As a result, the spin system is completely incoherent at the beginning of the relaxation interval, and the artifacts resulting from different magnetization magnitudes at the beginning of the t_1- preparation period are effectively suppressed. With these schemes, good TOCSY and NOESY spectra on 2–3 mM DNA samples can be obtained with only four and eight scans per t_1 complex point, respectively. A 256 × 1024 complex data matrix can be obtained in less than one hour. This approach is especially useful for quickly checking a new sample and is essential if three-dimensional spectra are to be obtained. For spectra with a high signal-to-noise ratio, the full phase cycle and acquisition times up to 12–24 hr are recommended.

Obtaining Two-Dimensional Spectra in H₂O

NMR studies of the exchangeable resonances of DNA have been largely limited until recently to one-dimensional spectra and assignments of the imino resonances only via one-dimensional NOEs.[5] The imino and amino resonances are very important indicators of the hydrogen bonding of the bases and provide some of the few cross-strand connectivities, and they are therefore very important for structure determination. Most of these resonances can be readily assigned from two-dimensional NOESY spectra taken in water,[20,39,40] and the NOESY experiment is the most informative two-dimensional experiment for DNA samples in H₂O. Most studies on DNA conformation have been done at neutral pH. However, the exchange rate of the imino resonances decreases with lower pH, so it is useful to obtain spectra in H₂O at somewhat lower pH (~6) in order to see cross-peaks to something other than H₂O. Some DNA sequences have pH-dependent structural transitions, however, so it is important to determine if this is the case before changing the pH. Running at low temperature will also help, although in general the peaks are broader and poorer water suppression is obtained at low temperature.

[39] R. Boelens, R. M. Scheek, K. Dijkstra, and R. Kaptein, *J. Magn. Reson.* **63**, 378 (1985).
[40] V. Sklenář and J. Feigon, *Nature (London)* **345**, 836 (1990).

Obtaining good spectra of DNA in water depends both on using optimal acquisition conditions and on processing data appropriately. Data processing will be discussed in the next subsection. The most important factors in terms of acquisition are proper shimming to decrease the water "hump" and appropriate adjustment of parameters so that a flat baseline is obtained with minimum first-order baseline correction. The standard pulse sequences are used except that the nonselective read pulse is replaced by a selective pulse sequence. The selective read pulse can be the 1T̄ spin echo,[24] jump and return,[41] or other pure selective excitation sequence. This gives a spectrum with accurate intensities along f_1, so that with proper processing most of the cross-peaks can be seen at least on one side of the diagonal. The water magnetization which survives in the transverse plane during the mixing period must be destroyed by application of a homospoil pulse. Problems with radiation damping which may occur in scans when the water magnetization is aligned along the $-z$ axis at the beginning of an acquisition can be solved by modifying the phase cycle[42] or by detuning of the probe. In our experience, slight detuning which lowers the sensitivity by only about $5-10\%$ is usually sufficient to prevent receiver overload during the acquisition.

In certain cases, NOESY spectra of DNA in H_2O can be obtained using a selective saturation pulse on water during the recycle delay and mixing time,[20] as is done for proteins, which can be useful for assignment purposes but not for quantitation. The exchangeable resonance intensity will be decreased, and the intensity of observable exchangeable resonances and their cross-peaks will depend on their exchange rate with water. In general, this method results in loss of all of the resonance intensity of the G amino and terminal exchangeable resonances.

Processing Two-Dimensional NMR Spectra

Processing two-dimensional spectra of DNA in D_2O is essentially like processing spectra of proteins. Sources of and approaches to eliminating various artifacts in t_1 and t_2 have been discussed elsewhere.[1,43] Optimal apodization functions to apply will depend on the type of experiment, the application (i.e., assignments versus quantitation), and the sample. In general, spectra are processed several different ways until the best results are obtained. In some cases, it is beneficial to baseline correct the spectrum in f_2 and/or f_1, usually with a first-order or first- or second-order polynomial fit, respectively.

[41] P. Plateau and M. Guéron, *J. Am. Chem. Soc.* **104,** 7310 (1982).
[42] P. R. Blake and M. F. Summers, *J. Magn. Reson.* **86,** 622 (1990).
[43] G. Otting, H. Widmer, G. Wagner, and K. Wüthrich, *J. Magn. Reson.* **66,** 187 (1986).

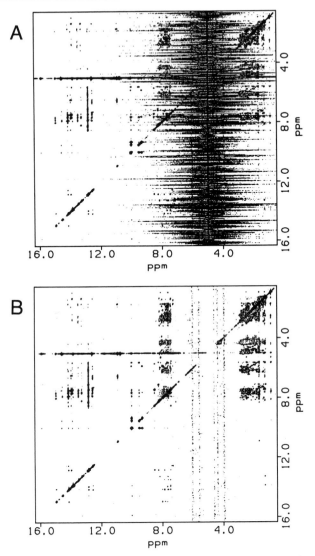

FIG. 1. Comparison of 500 MHz ^{1}H NMR NOESY spectrum in H_2O of the DNA oligonucleotide d(AGATAGAACCCCTTCTATCTTATATCTGTCTT) processed (A) without time domain convolution difference (smooth) and without polynomial baseline correction and (B) smoothed with a Gaussian window (± 32 points) in t_2 and baseline corrected with a sixth-order polynomial in f_2 and a second-order polynomial in f_1. The NMR sample was 400 μl of 2.3 mM DNA oligonucleotide in 100 mM NaCl, 5 mM MgCl$_2$, pH 5.5. The standard NOESY pulse sequence[26] was used, except the read pulse was replaced with a 1T spin echo observe pulse.[24] Acquisition parameters were τ_m = 150 msec, with a 5 msec homospoil

The visual presentation of NOESY spectra in H_2O can be improved by various postacquisition data processing methods which further reduce the residual water signal.[44-46] The effect of the time domain convolution difference routine[44] and baseline correction on a NOESY spectrum of a 32-base DNA oligonucleotide which folds to form an intramolecular triplex is shown in Fig. 1. As can be seen, this processing not only substantially improves the appearance of the spectrum, but also enhances the information content of the spectrum. This will be the case whenever the dispersive residual water impairs the baseline along the f_2 dimension.

Assignment Strategies

B-DNA Assignments

Assignment strategies for the nonexchangeable proton resonances of B (or A) DNA are straightforward and have been well documented in the literature.[1,7-9,12,47-52] The base pairs and the deoxyribose sugar are shown in Fig. 2 with the numbering system used. Examination of the B-DNA helix reveals that there should be a series of sequential connectivities (i.e., short interproton distances which will give rise to NOEs) from base H8 or H6 to sugar H1', H2', H2" to base and so on in the 5'−3' direction along each strand of the helix. These sequential connectivities are illustrated schematically in Fig. 3A. Connectivities between protons in a given sugar can be obtained from correlated spectra (COSY, TOCSY, 2Q, etc). Be-

[44] D. Marion, M. Ikura, and A. Bax, *J. Mag. Res.* **84**, 425 (1989).

[45] Y. Kuroda, A. Wada, T. Yamazaki, and K. Nagayama, *J. Magn. Reson.* **84**, 604 (1989).

[46] P. Tsang, P. Wright, and M. Rance, *J. Magn. Reson.* **88**, 210 (1990).

[47] M. S. Broido, *Biochem. Biophys. Res. Commun.* **119**, 663 (1984).

[48] M. A. Weiss, D. J. Patel, R. T. Sauer, and M. Karplus, *Proc. Natl. Acad. Sci. U.S.A.* **81**, 130 (1984).

[49] R. M. Scheek, R. Boelens, N. Russo, J. H. van Boom, and R. Kaptein, *Biochemistry* **23**, 1371 (1984).

[50] C. A. G. Haasnoot, H. P. Westerink, G. A. van der Marel, and J. H. van Boom, *J. Biomol. Stereodyn.* **1**, 131 (1983).

[51] W. J. Chazin, K. Wüthrich, M. Rance, S. Hyberts, W. A. Denny, and W. Leupin, *J. Mol. Biol.* **109**, 439 (1986).

[52] R. Grütter, G. Otting, K. Wüthrich, and W. Leupin, *Eur. Biophys. J.* **16**, 279 (1988).

at the beginning; 2 sec recycle delay, 11364 Hz spectral width in both dimensions (88 μsec dwell); acquisition time 180.224 msec; 2048 complex points in t_2; 445 t_1 increments collected; τ for the 1T echo pulse = 56 μsec; 32 scans per t_1 with two dummy scans. The spectra were processed with 2048 complex points in both dimensions and apodized with a shifted sine-bell squared function of 65° and 600 points in t_2 and 75° and 445 points in t_1.

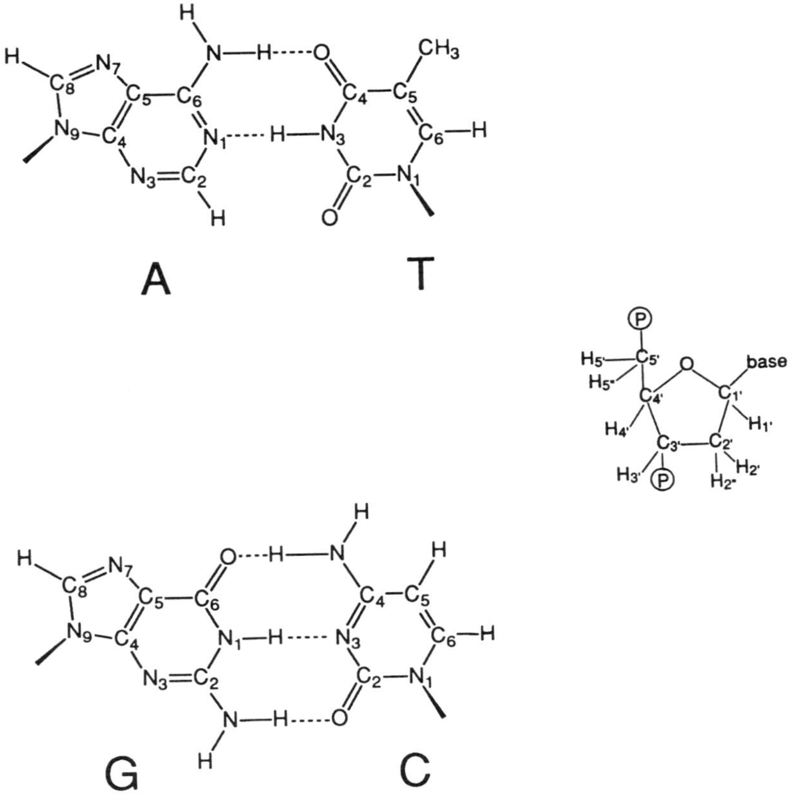

FIG. 2. Watson–Crick A·T and G·C base pairs and the deoxyribose sugar with numbering systems.

cause the assignment procedures have already been well described elsewhere, in this section we simply point out some details which may be useful.

B-DNA, with all anti glycosidic torsion angles, gives a pattern of strong H6,H8–H2′,H2″ cross-peaks and weaker H6,H8–H1′ cross-peaks which are used to make sequential assignments. Although the shortest intra- and internucleotide base–sugar connectivities are H6,H8–H2′ and H6,H8–H2″, respectively, it is usually easiest to make sequential assignments via the H6,H8–H1′ cross-peaks. The more complicated H6,H8–H2′,H2″ cross-peak region can be used to confirm these assignments and to resolve ambiguities. Although some of the intensity in the base–H1′ cross-peak region arises from spin diffusion via the H2′,H2″ protons, this is not a

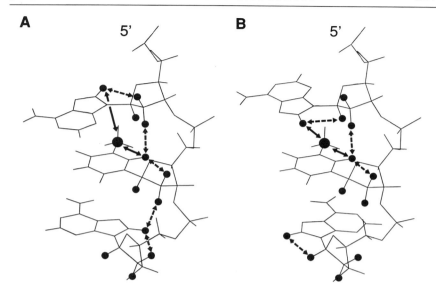

FIG. 3. (A) Schematic of one strand of a B-DNA duplex with the sequence d(ATA). Some of the sequential connectivities which should give rise to NOEs between purine H8 or pyrimidine H6 and the sugar protons are indicated. Although only the shortest intra- and internucleotide base–sugar connectivities are indicated, each base H8 or H6 proton is within NOE distance (<5 Å) of both its own H1′, H2′, H2″ protons and those on the 5′ neighboring base. Note the directionality of the connectivities. Base–base connectivities from thymine methyl to the 5′ neighboring base (H6 or H8) are also indicated. (B) Schematic of one strand of a right-handed duplex with sequence d(ATA) with A nucleotides in the syn orientation. Sequential connectivities are observed 5′ syn–anti 3′ but not 5′ anti–syn 3′. This is generally true for any structure containing syn bases. [J. Feigon, A. H.-J. Wang, G. A. van der Marel, J. H. van Boom, and A. Rich, *Nucleic Acids Res.* **12**, 1243 (1984); D. E. Gilbert and J. Feigon, *Biochemistry* **30**, 2483 (1991)].

problem in terms of assignments since the indirect pathway still always gives the correct connectivities.

It is not always possible to identify the 5′ end of the molecule readily. This is not a problem, since assignment can start at any point in the molecule and proceed in either direction. It is useful first to identify as many of the aromatic resonances as possible as to base type, to use as checkpoints in the sequential assignment. The CH5,CH6 and TMe,TH6 resonances can be readily identified by their COSY cross-peaks. In general, AH8 resonances occur downfield of GH8, but this is variable. It is useful to identify the AH2 resonances also, so that the occasional AH2–H1′ cross-peaks are not confused with sequential connectivities. The AH2 resonances can be identified in standard Watson–Crick base pairs by their strong imino–AH2 cross-peak (observed in NOESY spectra in H_2O) and also by the general lack of cross-peaks to sugar protons.

Assignments of the exchangeable imino, amino, as well as adenine AH2 resonances, can be obtained from NOESY spectra in H_2O.[20,39,40] As already discussed, the most important factor here is to obtain good spectra so that all the cross-peaks can be seen. Iminos are assigned by sequential imino–imino connectivities between adjacent base pairs. Amino resonances are assigned via cross-peaks from the iminos. In most cases the cytosine aminos are the easiest to identify, since they exchange slowly on the NMR time scale and give rise to sharp cross-peaks. The adenine aminos can be intermediate to slow, and they can usually be assigned. The guanine aminos are generally exchanged more rapidly and give rise to very broad resonances which may be difficult to identify and assign.

Assignments in Non-B-DNA and Larger Oligonucleotide Structures

For non-B-DNA structures and drug–DNA complexes the standard sequential assignments procedures will often fail. This is particularly true if any of the bases are in the syn conformation, for example, in alternate bases of Z-DNA, because the sequential connectivities from base to H1′, H2′, H2″ between the anti base and the sugar protons on the 3′ neighboring syn nucleotide are disrupted[53] (see Fig. 3B). Spectral overlap in larger oligonucleotides may also obscure sequential connectivities. In these cases a variety of approaches may be needed first to identify the base protons as to type and then to connect them in sequence to their sugar proton resonances.

Varying Conditions. If there are ambiguities owing to spectral overlap, they can frequently be resolved by acquiring two-dimensional spectra at several different temperatures. This should be obvious, but it is sometimes overlooked. It is also useful to acquire NOESY spectra both with a long mixing time, so that small cross-peaks in the base–H1′ region can be more readily identified, and with shorter mixing times, to help resolve overlap and get more accurate intensities in the base–H2′,H2″ region. Overlapping cross-peaks can be more readily discerned in phase-sensitive COSY experiments than magnitude COSY because of the coupling patterns. The P.COSY experiment gives phase-sensitive spectra without the loss of sensitivity which one gets with DQF-COSY. Finally, it is worthwhile to try various different salt conditions for a sample before settling on final condi-

---

[53] J. Feigon, A. H.-J. Wang, G. A. van der Marel, J. H. van Boom, and A. Rich, *Nucleic Acids Res.* **12**, 1243 (1984).

tions. The stability of DNA is highly dependent on cation concentration. Addition of a divalent cation, such as Mg^{2+}, may make it possible to run experiments at a higher temperature and result in increased resolution.

Comparison of Spectra of Similar Molecules. Ambiguities in assignments can sometimes be resolved by comparing the cross-peak patterns obtained from spectra of two similar molecules. These can be oligonucleotides which differ by a base substitution or by a base deletion at the end.

Chemical Exchange. Frequently, in studying non-B-DNA structures, sample conditions can be adjusted so that there is a mixture of B-DNA and the nonstandard structures. If these are in slow exchange on the NMR time scale, then it may be possible to assign some or all of the resonances in the non-B-DNA structure via their exchange cross-peaks[54] with the previously assigned B-DNA structure. The first application of this was to assign the Z-DNA form of $d(m^5CG)_3$.[53] In favorable cases the method can be used for drug–DNA[55] or drug–protein complexes. Depending on the chemical exchange rate and resolution, it may be possible to distinguish clearly between NOE and exchange cross-peaks in a NOESY spectrum. When this is not the case, a ROESY[27] experiment can be used to separate NOE and exchange cross-peaks by sign.[56]

Identification of Base-Pairing Scheme. Spectra of exchangeable proton resonances of DNA are important indicators of base pair formation and stability. For any DNA structures containing non-Watson–Crick base pairs, NOESY spectra in water are essential for a definitive identification of the base pairing scheme. Such spectra may also be invaluable in assigning sequence-specific base protons. Structural variations in the DNA are usually manifested in the exchangeable resonances. For example, intercalation will result in large upfield shifts of imino resonances adjacent to the intercalation site,[57] protonation of bases will usually result in downfield shifts of the amino resonances and the appearance of new imino resonances,[40,58] and additional base pairs as in triplex formation will result in additional imino resonances being observed.[40,58] For Watson–Crick base-paired DNA, the imino proton resonances are assigned sequentially along the helix by imino–imino NOEs. Cross-peaks from the imino to the aromatic, amino region can then be used to identify the AH2, C amino, A amino, and occasionally the G amino resonances as discussed above. The strong cross-peak between the T imino and AH2 is used to identify the

[54] J. Jeener, B. H. Meier, P. Bachmann, and R. R. Ernst, *J. Chem. Phys.,* 4546 (1979).
[55] R. E. Klevit, D. E. Wemmer, and B. R. Reid, *Biochemistry* 25, 3296 (1986).
[56] P. Rajagopal and J. Feigon, *Biochemistry* 28, 7859 (1989).
[57] J. Feigon, W. A. Denny, W. Leupin, and D. R. Kearns, *J. Med. Chem.* 27, 450 (1984).
[58] P. Rajagopal and J. Feigon, *Nature (London)* 339, 637 (1989).

AH2 resonances in B-DNA. Therefore, it is important to recognize that this may not be appropriate if there are non-Watson–Crick base pairs. For example, an A·T Hoogsteen base pair is identified by a strong NOE cross-peak from the T imino to AH8.[59] Obviously, one must unambiguously distinguish between AH8 and AH2 resonances to identify an A·T Watson–Crick versus a Hoogsteen base pair. This can be reliably done by deuteration of the purine H8 resonances as discussed below.

Deuteration of Purines at C-8. For non-B-DNA structures and for larger oligonucleotides it is often helpful or necessary to identify unambiguously the purine H8 resonances. This can be done by comparing cross-peak patterns in NOESY spectra taken in D_2O and H_2O both before and after deuterating the purines at C-8.[60] Deuteration is also useful to resolve overlapping cross-peaks, since it simplifies the spectrum. Deuteration of DNA samples is straightforward and simply involves heating the sample for a period of time in D_2O to exchange the H8 atoms.[59] We generally adjust the sample pH to 8 and then heat the sample at 60° for 2–3 days. We have had some problems with sample degradation if we heat the sample for this period of time at a pH below 6. The GH8 exchange faster than AH8, so usually some resonance intensity will remain for the AH8 after the GH8 are completely deuterated. It is also possible to deuterate the cytosine CH5 position following the procedure of Brush *et al.*[61]

Selective Excitation. The CH6 and CH5 resonances can usually be readily identified by their strong COSY cross-peaks. However, cross-peaks between CH6 and the sugar protons may be obscured by spectral overlap with other base–sugar cross-peaks. Cross-peaks between the CH6 and other resonances can be selectively observed using the HOENOE experiment.[33] In this experiment, magnetization is transferred from the CH5 to the CH6 resonances by selective excitation of the region of the spectrum containing the CH5 resonances prior to the t_1 evolution period of a regular 2-D NOE experiment. This experiment allows straightforward identification of all of the cross-peaks arising from interactions of the CH6 resonances.

Three-Dimensional NMR. We have recently begun using three-dimensional NOESY–NOESY[62] experiments to assign resonances in

[59] D. E. Gilbert, G. A. van der Marel, J. H. van Boom, and J. Feigon, *Proc. Natl. Acad. Sci. U.S.A.* **86**, 3006 (1989).
[60] M. P. Schwiezer, S. I. Chan, G. K. Helmkamp, and P. O. P. Ts'o, *J. Am. Chem. Soc.* **86**, 696 (1964).
[61] C. K. Brush, M. P. Stone, and T. M. Harris, *Biochemistry* **27**, 115 (1988).
[62] R. Boelens, G. W. Vuister, T. M. G. Koning, and R. Kaptein, *J. Am. Chem. Soc.* **111**, 8525 (1989).

larger oligonucleotide structures. The advantage of this method is that all of the sugar proton resonances including H4', H5', and H5" can be assigned.[63] The three-dimensional NOESY–TOCSY is also useful in this regard.[64,65]

Structure Determination

Qualitative Analysis of DNA Structure

As is the case with proteins, the assignment procedure for DNA oligonucleotides also gives qualitative information on the DNA secondary structure. Base-pairing schemes and information on base-pair stability can be readily deduced from one- and two-dimensional spectra in H_2O. Basic helix types such as A-, B-, and Z-DNA can be identified from analysis of two-dimensional NOESY and COSY spectra.[1] This analysis also gives information about glycosidic torsion angles and sugar puckers. The information on sugar pucker can be indirectly obtained from NOE spectra and more directly obtained from analysis of coupling constants extracted from phase-sensitive correlated spectroscopy. The analysis of sugar puckers is addressed in [15] in this volume.[66] NMR experiments can provide many structural details on the formation of DNA tertiary structures such as hairpins, bulges, triplexes, and quadruplexes. Binding modes and sequence specificity of drugs with DNA can usually be readily identified. Thus, a great deal of valuable structural information can be obtained from a qualitative analysis of relative NOE cross-peak intensities and coupling constants. Some information about local helical twist and minor groove widths can also be deduced. However, more precise details about sequence-specific variation in base tilt, base propellering, helical twist, groove widths, etc., cannot be obtained without combining the NOE and dihedral bond angle information with some sort of distance geometry, relaxation matrix, and/or molecular dynamics approach.

Approaches to Determining Complete Three-Dimensional Structures

The final goal of the NMR work with DNA, namely, the establishment of accurate and precise molecular structures at atomic resolution, encounters some fundamentally different problems than in the case for pro-

[63] E. Wang, V. Sklenář, R. F. Macaya, and J. Feigon, unpublished results (1991).

[64] M. E. Piotto and D. G. Gorenstein, *J. Am. Chem. Soc.* **113**, 1438 (1991).

[65] M. M. W. Mooren, C. W. Hilbers, G. A. van der Marel, J. H. van Boom, and S. S. Wijmanga, *J. Magn. Reson.* **94**, 101 (1991).

[66] J. van Wijk, B. D. Huckriede, J. H. Ippel, and C. Altona, this volume [15].

teins. A network of long-range NOEs, that is, NOEs connecting protons of residues further apart than their sequential neighbors, is suitable to elucidate the tertiary structure of a protein through distance geometry calculations. To obtain good precision it is important to have a large density of this network; in other words, a large number of NOEs is more crucial than the precise quantification of each individual distance constraint.[1]

In double helical DNA, long-range NOEs are generally not available. It has been argued, by Pardi and co-workers,[67] that in this case the precision of structural data is not sufficient to determine the sequence-dependent local variations of DNA. Appropriate reservations about some of the early structure determinations of DNA oligonucleotides by NMR have also been expressed by van de Ven and Hilbers.[11] However, new developments in computational software as well as in spectroscopic approaches can be expected to yield higher precision in the near future. The abundant pathways for spin diffusion in the deoxyribose spin systems of nucleotides make the conventional two-spin approximation used for distance geometry calculations[68,69] especially error prone when applied to DNA. Approaches that take into account the full relaxation matrix have been developed.[70-72] In these methods, consistency of the relaxation matrix with the structural model is achieved by a stepwise procedure of structural refinement and back calculations to improve the distance constraints used in turn in the refinement. Yip and Case[73] have proposed a method for the direct incorporation of the dependence of NOE intensities into a target function to be minimized. This has been added into existing refinement programs.[74,75]

A further increase of precision especially for the determination of the deoxyribose ring conformations should be achievable by including dihedral bond angle constraints as well as NOE constraints in the structure calculations. The latter can be obtained by a detailed analysis of coupling constants. Coupling constants for individual spin systems can be simulated using the programs SPHINX and LINSHA.[76] We have recently developed

[67] W. J. Metzler, C. Wang, D. B. Kitchen, R. M. Levy, and A. Pardi, *J. Mol. Biol.* **214**, 711 (1990).
[68] T. F. Havel and K. Wüthrich, *J. Mol. Biol.* **182**, 281 (1985).
[69] W. Braun, *Q. Rev. Biophys.* **19**, 115 (1987).
[70] R. Boelens, T. M. G. Koning, G. A. van der Marel, J. H. van Boom, and R. Kaptein, *J. Magn. Reson.* **82**, 290 (1989).
[71] B. A. Borgias and T. L. James, *J. Magn. Reson.* **87**, 475 (1990).
[72] C. B. Post, R. P. Meadows, and D. G. Gorenstein, *J. Am. Chem. Soc.* **112**, 6796 (1990).
[73] P. Yip and D. A. Case, *J. Magn. Reson.* **83**, 643 (1989).
[74] J. E. Mertz, P. Güntert, K. Wüthrich, and W. Braun, *J. Biomol. NMR* **1**, 257 (1991).
[75] M. Nilges, J. Habazettl, A. T. Brünger, and T. A. Holak, *J. Mol. Biol.* **219**, 499 (1991).
[76] H. Widmer and K. Wüthrich, *J. Magn. Reson.* **74**, 316 (1987).

a program which can also automatically simulate and iteratively refine larger regions on a COSY spectrum which may contain overlapping cross-peaks.[77] This program should give a more accurate determination of the values of these coupling constants, since a correlation coefficient is computed between experimental and simulated spectra and maximized using Powell's algorithm.[78]

Finally, it is worth reiterating that the quality of the structures obtained will depend on the quality of the input data. Accurate integration of NOESY cross-peak intensities requires that properly processed spectra with good S/N and flat baselines are used. With a relaxation matrix refinement approach, a measure for the agreement between observed and calculated data analogous to the crystallographic R-factor can be defined.[70,75] However, a simple residual involving NOE peak integral has to be interpreted with caution. First, the "observed" data set is not an unbiased experimental result, but is subject to considerable systematic error through peak overlap and baseline distortion. Second, whereas in principle each atom contributes to each reflection in an X-ray diffraction data set, the NOE data are not necessarily evenly distributed throughout the molecule and may leave parts of the structure ill-defined despite a low R-factor. We are optimistic that with state-of-the-art spectroscopic methods and high quality data together with full usage of relaxation matrix refinement software and precise analysis of coupling constants, including possibly ones derived from heteronuclear experiments,[79] reliable high resolution structures of DNA may be obtained.

Acknowledgments

This work was supported by grants from the National Institutes of Health (R01 GM 37254-01) and a National Science Foundation Presidential Young Investigator Award with matching funds from AmGen Inc., E. I. Du Pont de Nemours & Company, Monsanto Co., and Sterling Drug Inc. to J.F. D.E.G. and E.W. were supported in part by an NIH Predoctoral Cell and Molecular Biology training grant (GM 07185), and R.F.M. was supported in part by an NIH Predoctoral Biotechnology training grant (GM 08375).

[77] R. F. Macaya, P. Schultze and J. Feigon, *J. Am. Chem. Soc.,* **114,** 781 (1992).
[78] W. H. Press, B. P. Flannery, S. A. Teukolsky, and W. T. Vetterling, "Numerical Recipes." Cambridge Univ. Press, Cambridge, 1989.
[79] See, for example, D. Gorenstein, this volume [14].

[14] ^{31}P NMR of DNA

By David G. Gorenstein

Introduction

^{31}P NMR has developed as a powerful probe of the structure and dynamics of DNA and DNA fragments in solution. Until recently ^{31}P NMR studies of nucleic acids were restricted to partial characterization of polymeric DNA and sequences less than six base pairs in length.[1] Limitations were due in part to the lack of availability of methods to conveniently chemically synthesize large quantities of DNA fragments, as well as NMR spectral methods that were unable to resolve signals of interest. However, synthetic schemes have now been developed where defined DNA duplexes can be routinely produced in multimilligram quantities.[2] The concomitant development of sequence-specific two-dimensional (2-D) ^{1}H/^{1}H and ^{1}H/^{31}P NMR assignment methodologies[3-13] and higher-field spectrometers has made the study of the ^{31}P NMR spectrum of modest size oligonucleotides [12–20 base pairs (bp)] possible.

The latest, routine, multinuclear Fourier transform (FT) NMR spectrometers (1.8–14.1 Tesla) have reduced if not eliminated the serious limitation to the widespread utilization of phosphorus NMR in biological systems, which is the low sensitivity of the phosphorus nucleus (6.6% at constant field compared to ^{1}H NMR). Today, routinely, millimolar (or

[1] D. G. Gorenstein, ed., in "Phosphorus-31 NMR: Principles and Applications," p. 1. Academic Press, Orlando, Florida, 1984.

[2] M. J. Gait, "Oligonucleotide Synthesis: A Practical Approach." IRL Press, Oxford, 1984.

[3] J. Feigon, W. Leupin, W. A. Denny, and D. R. Kearns, *Biochemistry* **22**, 5930, 5943 (1983).

[4] J. Feigon, W. Leupin, W. A. Denny, and D. R. Kearns, *Biochemistry* **22**, 5943 (1983).

[5] D. R. Hare, D. E. Wemmer, S. H. Chou, G. Drobny, and B. Reid, *J. Mol. Biol.* **171**, 319 (1983).

[6] S. A. Schroeder, J. M. Fu, C. R. Jones, and D. G. Gorenstein, *Biochemistry* **26**, 3812 (1987).

[7] D. R. Kearns, *Crit. Rev. Biochem.* **15**, 237 (1984).

[8] D. Frechet, D. M. Cheng, L.-S. Kan, and P. O. P. Ts'o, *Biochemistry* **22**, 5194 (1983).

[9] M. A. Broido, G. Zon, and T. L. James, *Biochem. Biophys. Res. Commun.* **119**, 663 (1984).

[10] R. M. Scheek, R. Boelens, N. Russo, J. H. van Boom, and R. Kaptein, *Biochemistry* **23**, 1371 (1984).

[11] S. Schroeder, C. Jones, J. Fu, and D. G. Gorenstein, *Bull. Magn. Reson.* **8**, 137 (1986).

[12] A. Pardi, R. Walker, H. Papoport, G. Wider, and K. Wüthrich, *J. Am. Chem. Soc.* **105**, 1652 (1983).

[13] K. Lai, D. O. Shah, E. Derose, and D. G. Gorenstein, *Biochem. Biophys. Res. Commun.* **121**, 1021 (1984).

lower) concentrations of phosphorus nuclei in as little as 0.3 ml of solution are conveniently monitored. The ^{31}P nucleus has other convenient NMR properties suitable for FT NMR: spin $\frac{1}{2}$ (which avoids problems associated with quadrupolar nuclei), 100% natural abundance, moderate relaxation times (providing relatively rapid signal averaging and sharp lines), and a wide range of chemical shifts (>600 ppm).

Conventional 2-D ^{1}H/^{31}P (^{31}P observe) heteronuclear correlation (HETCOR) techniques have been of limited value in resonance assignments for longer oligonucleotides,[12,13] and regiospecific labeling of the phosphorus residues by selective ^{18}O and ^{17}O labeling has proved quite useful.[6,14–17] Only oligonucleotide sequences six or fewer base pairs in length have been successfully assigned using the conventional HETCOR method[12,18] because this NMR technique is limited by both spectral resolution and sensitivity. However, newer indirect detection[19] and long-range, constant time HETCOR[20,21] methods have been successfully applied to the ^{31}P assignment problem. Because of these latter methods, assignment of the ^{31}P resonances of modest-size oligonucleotide duplexes is now quite feasible.[16,19,20,22] Thus, the assignment problems that initially held back the use of ^{31}P NMR have now been solved.

One of the main reasons for assigning ^{31}P resonances of oligonucleotides is to obtain information on the conformation of the phosphodiester backbone (Fig. 1).[23–26] The internucleotide linkage is defined by six torsional angles from one phosphate atom to the next along the DNA backbone. Theoretical studies have shown that the conformations of two of the six torsional angles (α: O3'–P–O5'–C5' and ζ: C3'–O3'–P–O5') appear to be most important in determining ^{31}P chemical shifts.[22,23,26] Although

[14] D. O. Shah, K. Lai, and D. G. Gorenstein, *J. Am. Chem. Soc.* **106,** 4302 (1984).

[15] M. Petersheim, S. Mehdi, and J. A. Gerlt, *J. Am. Chem. Soc.* **106,** 439 (1984).

[16] B. A. Connolly and F. Eckstein, *Biochemistry* **23,** 5523 (1984).

[17] A. P. Joseph and P. H. Bolton, *J. Am. Chem. Soc.* **106,** 437 (1984).

[18] D. M. Cheng, L.-S. Kan, P. S. Miller, E. E. Leutzinger, and P. O. P. Ts'o, *Biopolymers* **21,** 697 (1982).

[19] V. Sklenár, H. Miyashiro, G. Zon, H. T. Miles, and A. Bax, *FEBS Lett.* **208,** 94 (1986).

[20] J. M. Fu, S. A. Schroeder, C. R. Jones, R. Santini, and D. G. Gorenstein, *J. Magn. Reson.* **77,** 577 (1988).

[21] C. R. Jones, S. A. Schroeder, and D. G. Gorenstein, *J. Magn. Reson.* **80,** 370 (1988).

[22] D. G. Gorenstein, S. A. Schroeder, J. M. Fu, J. T. Metz, V. A. Roongta, and C. R. Jones, *Biochemistry* **27,** 7223 (1988).

[23] D. G. Gorenstein, *in* "Phosphorus-31 NMR: Principles and Applications" (D. G. Gorenstein, ed.) p. 7. Academic Press, Orlando, Florida, 1984.

[24] D. G. Gorenstein, *Annu. Rev. Biophys. Bioeng.* **10,** 355 (1981).

[25] D. G. Gorenstein, J. B. Findlay, R. K. Momii, B. A. Luxon, and D. Kar, *Biochemistry* **15,** 3796 (1976).

[26] D. G. Gorenstein, *Chem. Rev.* **87,** 1047 (1987).

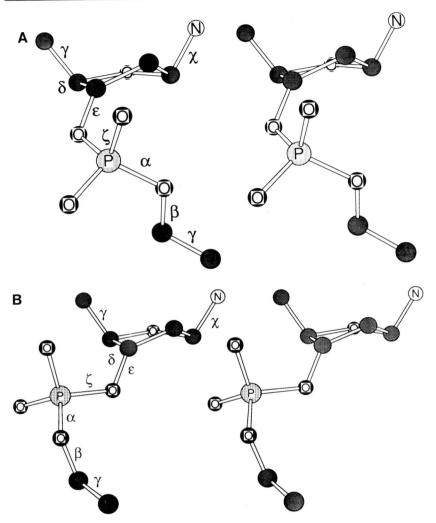

FIG. 1. Stereoview of the sugar–phosphate backbone (A) in the B_I conformational state and (B) in the B_{II} conformational state. α, β, ϵ, ζ: g^-, t, t, g^- (A); g^-, t, g^-, t (B). The B_I conformation represents the low-energy phosphate state normally observed in the crystal structures for many of the duplex phosphates, and the B_{II} conformation represents the other major sugar–phosphate state for B-DNA. Torsional angles are gauche(−) (g^- or −60°) or trans (180°). Crystal structures of duplex oligonucleotides show that these angles are only approximate, and indeed the ζ angle is generally closer to −90° for what is defined as g^-.

^{1}H/^{1}H distances derived from nuclear Overhauser effect spectroscopy (NOESY) provide important structural constraints, they give no direct information on the phosphate ester conformation, and NOESY distance-constrained structures have been suggested to be effectively disordered in this part of the structure.[27] However, as discussed in this chapter, ^{31}P chemical shifts and ^{1}H–^{31}P coupling constants can provide valuable information on the phosphodiester backbone conformation.

In duplex B-DNA, the gauche(−), gauche(−)(g^-, g^-; ζ, α) (or B$_I$) conformation about the P–O ester bonds in the sugar–phosphate backbone is energetically favored, and this conformation is associated with a more shielded ^{31}P resonance. In both duplex and single-stranded DNA the trans, gauche(−) (t,g^-; ζ,α) (or B$_{II}$) conformation is also significantly populated. The ^{31}P chemical shift difference between the B$_I$ and B$_{II}$ phosphate ester conformational states is estimated to be 1.5–1.6 ppm.[24,28] As the result of this sensitivity to the backbone conformational state, ^{31}P chemical shifts of duplex oligonucleotides have been shown to be dependent on both the sequence and position of the phosphate residue.[16,24,26,29]

With the tremendous increase in sensitivity and pulse sequence capabilities in modern FT NMR spectrometers, ^{31}P NMR spectroscopy of nucleic acids grew dramatically during the 1980s. Several monographs[1,30,31] entirely devoted to ^{31}P NMR, covering various aspects of biochemical applications, have been written. In this chapter I shall describe methods for obtaining one- and two-dimensional ^{31}P NMR spectra of DNA as well as the interpretation of ^{31}P chemical shifts and coupling constants and their application to nucleic acid structure. More comprehensive reviews of the literature on ^{31}P NMR studies of duplex polynucleotides[32] and on duplex oligonucleotides[33] and drug–DNA complexes[34] are recommended. Addi-

---

[27] F. J. M. van de Ven and C. W. Hilbers, *Eur. J. Biochem.* **178**, 1 (1988).

[28] V. A. Roongta, R. Powers, E. P. Nikonowicz, C. R. Jones, and D. G. Gorenstein, *Biochemistry* **29**, 5245 (1990).

[29] D. G. Gorenstein, *Prog. NMR Spectrosc.* **16**, 1 (1983).

[30] L. D. Quin and J. G. Verkade, "Phosphorus-31 NMR Spectroscopy in Stereochemical Analysis," Vol. 8. VCH Publ., Deerfield Beach, Florida, 1987.

[31] C. T. Burt, "Phosphorus NMR in Biology." CRC Press, Boca Raton, Florida, 1987.

[32] C.-W. Chen and J. S. Cohen, *in* "Phosphorus-31: NMR Principles and Applications" (D. G. Gorenstein, ed.), p. 233. Academic Press, Orlando, Florida, 1984.

[33] D. G. Gorenstein, R. P. Meadows, J. T. Metz, E. Nikonowicz, and C. P. Post, *in* "Advances in Biophysical Chemistry" (C. A. Bush, ed.), Vol. 1, pp. 47. JAI Press, Greenwich, Connecticut, 1990.

[34] D. G. Gorenstein and E. M. Goldfield, *in* "Phosphorus-31 NMR: Principles and Applications" (D. G. Gorenstein, ed.), p. 299. Academic Press, Orlando, Florida, 1984.

tional excellent discussions of solid state DNA studies by Tsai[35] and phosphorus relaxation methods by Hart[36] and James[37] are also available.

Experimental Considerations

Oligodeoxyribonucleotide Synthesis and Sample Preparation

Oligonucleotides may be synthesized by automated or manual modification of the phosphite triester method on a solid support by standard protocols.[2,6,38] The resulting products are purified by C_{18} reversed-phase high performance liquid chromatography (HPLC) with an acetonitrile–triethylammonium acetate (0.1 M solution at pH 7.2) gradient on a semipreparative reversed-phase column (Econosil C_{18}, Alltech, Deerfield, IL, is adequate). The sample is detritylated with 80% acetic acid for 25 min at room temperature followed by extraction with ether. Longer chain, (> 8 bp) palindromic duplexes are desalted by dialysis in a cellulose 1000 molecular weight cutoff dialysis tubing against double-distilled water. After lyophilization, the sample is run through a 1 cm column of Dowex ion-exchange resin to remove the remaining acetate. The resin should previously have been exchanged with KCl. For ^{31}P NMR, paramagnetic impurities must be reduced to a very low level in order to avoid significant broadening of the lines. Paramagnetic species are typically removed by exposing solutions to a chelating cation-exchange resin, such as Chelex 100 (Bio-Rad Laboratories, Richmond, CA), 200–400 mesh. In addition, at all times after Chelex treatment, it is important that all glassware be pretreated by washing in 6 M HCl or concentrated HNO_3 solutions and soaking in 0.1 mM EDTA to remove bound paramagnetic metal ion impurities. For 2-D NMR studies it is best to synthesize the oligonucleotide on a 10 μmol scale.

NMR Considerations

Commercial NMR spectrometers cover the ^{31}P frequency range from 24 to 243 MHz. Generally for small, phosphorus-containing compounds, high signal-to-noise and resolution requirements dictate the use of as high a

[35] M.-D. Tsai, in "Phosphorus-31 NMR: Principles and Applications" (D. G. Gorenstein, ed.), p. 175. Academic Press, Orlando, Florida, 1984.

[36] P. A. Hart, in "Phosphorus-31 NMR: Principles and Applications" (D. G. Gorenstein, ed.), p. 317. Academic Press, Orlando, Florida, 1984.

[37] T. L. James, in "Phosphorus-31 NMR: Principles and Applications" (D. G. Gorenstein, ed.), p. 349. Academic Press, Orlando, Florida, 1984.

[38] D. O. Shah, K. Lai, and D. G. Gorenstein, *Biochemistry* **23**, 6717 (1984).

magnetic field strength as possible since both sensitivity and chemical-shift dispersion increase at higher operating frequency (and field). However, consideration must be given to field-dependent relaxation mechanisms such as chemical shift anisotropy which can lead to substantial line broadening of the ³¹P NMR signal at high fields. Indeed, for moderate-sized DNA fragments (MW >10,000), sensitivity is often poorer at very high fields because of considerably increased linewidths. Because signal intensity is proportional to the number of nuclei in the region of the observing receiver coil of the probe, high concentration and large volume provide minimal acquisition times. Probes for 20 mm outer diameter (o.d.) sample tubes are available on wide-bore instruments, but the large sample volumes of 10–12 ml required often prohibit their use for any material of low availability, which is often the case in biological studies. More typical are ³¹P probes of 5 to 10 mm o.d., which require 0.5–2.5 ml sample volumes. Microcells can reduce the volume to less than 0.5 ml. The latest probe designs provide signal-to-noise ratios of 420:1 on a 1% trimethyl phosphite standard sample in a 10 mm tube in a single spectral acquisition (at 202 MHz) on a Varian VXR 500 spectrometer or 200:1 on a 0.0485 M triphenyl phosphate standard sample in a 5 mm tube in a single spectral acquisition (at 242 MHz) on a Varian VXR 600 spectrometer.

Because of the low sensitivity of the ³¹P experiment, high sample concentrations or larger sample volumes are generally required for obtaining an adequate signal-to-noise ratio (S/N) in reasonable total acquisition times. The 2-D NMR samples generally require 5–50 mg of oligonucleotide dissolved in 100–400 mM phosphate or Tris buffer, pH 7.0, 100 mM KCl, 1 mM EDTA, and 1 mM sodium azide in a total volume of 400–600 μl of 99.996% D_2O. Typical acquisition times for the ³¹P free induction decay (FID) following a 90° radiofrequency pulse are 1–8 sec depending on the required resolution (dictated by the linewidth of the signal, $1/\pi T_2^*$, where T_2^* is the time constant for the FID). Waiting longer than $2T_2^*$ will generally not improve S/N. Additional consideration for optimization of S/N must be given to the time it takes for the ³¹P spins to return to thermal equilibrium after a 90° radiofrequency pulse, which is roughly 3 times the spin-lattice relaxation time (T_1). If $T_1 \approx T_2^*$, as would be true for small phosphorus-containing molecules where magnetic field inhomogeneity and paramagnetic impurities do not lead to any additional line broadening, then a waiting period between pulses of $3T_2^*$ provides a good compromise between adequate resolution and signal sensitivity. If $T_1 > T_2^*$, as is often the case in duplex fragments, then waiting only $3T_2^*$ does not allow the magnetization to return to equilibrium, and an additional delay must generally be introduced so that the total time between pulses is approximately $3T_1$. This wait can be substantially shortened if the Ernst relationship[39] is used to set the pulse flip angles to less than 90°. At low field, 60–70° pulses, 4K to 8K data points and 2.0–5.0 sec recycle times are generally used. The spectra are generally broad-band ^1H decoupled.

The temperature of the sample can be controlled to within ±1° by commercial spectrometer temperature control units. Broad-band decoupling at higher superconducting fields can produce about 10–15° heating of the sample above the gas stream measured temperatures, even using a gated, two-level decoupling procedure.[25] This problem may be partially circumvented by using decoupling methods such as Waltz or MLEV decoupling. The temperature of the sample should be measured under the experimental decoupling conditions. It is incorrect to assume that the temperatures of the sample and of the gas used to regulate the temperature of the sample are the same because the decoupler can heat the solution substantially above the gas temperature. This is especially true for biological solutions of high ionic strength. It is therefore necessary to be able to measure the solution temperature directly using either a nonmagnetic thermocouple sensor or thermistor inserted into the NMR solution, or, better, a sample whose ^{31}P spectrum reflects the correct temperature. Several ^{31}P "thermometers" have been proposed for this purpose.[25,40,41]

The ^{31}P spectra are generally referenced to an external sample of 85% H_3PO_4 or trimethyl phosphate in D_2O, which is approximately 3.46 ppm downfield of 85% H_3PO_4. Note that throughout this chapter the IUPAC convention[42] is followed so that positive values are to high frequency (low field). One should cautiously interpret reported ^{31}P chemical shifts because the early literature (pre-1970s) and even many later papers use the opposite sign convention.

Routine ^{31}P one-dimensional spectra of small DNA fragments may be obtained with a sweep width of 800 Hz (at 200 MHz ^1H, 81 MHz ^{31}P) and 1600 data points. A total recycle time of 2–4 sec is generally used. The data are generally processed with 8K zero filling and a Gaussian apodization function to generate resolution enhancement. For broader, nonoverlapping signals an exponential line-broadening function may be applied.

Quantification of Peak Heights. At lower field, typically, 2000 to 4000 transients per 1-D spectrum are accumulated to achieve adequate signal to noise. Assuming a T_1 of about 5 sec, this accumulation requires 3 to 6 hr.

[39] M. L. Martin, J. J. Delduech, and G. J. Martin, "Practical NMR Spectroscopy," Heyden, London, 1980.
[40] F. L. Dickert and S. W. Hellman, *Anal. Chem.* **52**, 996 (1980).
[41] D. G. Gorenstein, this series, Vol. 177, p. 295.
[42] *Pure Appl. Chem.* **45**, 217 (1976).

The intensity of a resonance can be measured in several ways: (1) peak heights and areas obtained from the standard software supplied by the spectrometer manufacturer, (2) peak heights measured by hand, (3) peaks cut and weighed from the plotted spectrum, and (4) peaks fit to a Lorentzian lineshape. For flat baselines, intensity measurements are generally straightforward. However, in the event of curved baselines the measurements are somewhat uncertain, and manual measurements are generally more reliable than intensity values obtained from computer software.

It is often necessary that experiments be carried out without allowing time for full recovery of longitudinal magnetization between transients because of the limited availability of spectrometer time or because of the limited lifetime of the sample. Because of variations in T_1 between different phosphates and variation in the hetereonuclear NOE to nearby protons, care should be made in interpretation of peak areas and intensities. Addition of a recycle delay of at least $5T_1$ between pulses and use of gated decoupling only during the acquisition time to eliminate the $^1H/^{31}P$ NOE largely eliminate quantification problems.

^{17}O-Labeling Methodology for Assigning ^{31}P Signals of Oligonucleotides

A phosphoramidite ^{17}O-labeling methodology has been developed to assign the ^{31}P signals of the phosphates in a number of duplexes.[11,20] Using a manual solid-phase phosphoramidite method[2] ^{17}O or ^{18}O labels in the phosphoryl groups can be introduced[6,11,43] by replacing the I_2/H_2O in the oxidation step of the phosphite by $I_2/H_2^{17}O$ (40%) or $H_2^{18}O$. By synthesizing a mono-^{17}O-phosphoryl-labeled oligonucleotide (one specific phosphate is labeled along the strand), it is possible to identify the ^{31}P signal of that phosphate diester. This is because the quadrupolar ^{17}O nucleus (generally $\sim 40\%$ enriched) broadens the ^{31}P signal of the directly attached phosphorus to such an extent that only the high-resolution signal of the remaining 60% nonquadrupolar-broadened phosphate at the labeled site is observed.[15–17,38] In this way each synthesized oligonucleotide with a different monosubstituted ^{17}O-phosphoryl group allows identification of one specific ^{31}P signal, and the full series of monolabeled oligonucleotides gives the assignment of the entire ^{31}P NMR spectrum.

Figure 2 shows representative ^{31}P spectra[44] of an unlabeled and singly ^{17}O-labeled extrahelical adenosine 13-mer duplex, d(CGCAGAATT-

---

[43] D. G. Gorenstein, K. Lai, and D. O. Shah, *Biochemistry* **23**, 6717 (1984).
[44] E. Nikonowicz, V. Roongta, C. R. Jones, and D. G. Gorenstein, *Biochemistry* **28**, 8714 (1989).

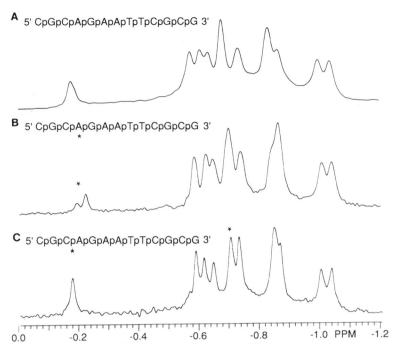

FIG. 2. (A) ^{31}P NMR spectrum of the unlabeled 13-mer, d(CGCAGAATTCGCG)$_2$. Examples of site-specific ^{17}O-labeling of two of the phosphates of the 13-mer at positions 4 and 3 (shown by an asterisk) are shown in (B) and (C), respectively. ^{31}P chemical shifts are reported relative to 85% phosphoric acid. (Reproduced with permission from Ref. 44.)

CGCG)$_2$. Figure 2A is a spectrum of the 13-mer, without any ^{17}O label, that can be used as a comparison between labeled spectra. As can be seen in Fig. 2B,C, a decrease in intensity of a single resonance is observed. Almost all of the resonances can clearly be distinguished, most integrating for a single phosphorus resonance. It is interesting to note that the resonance of the labeled phosphate is observed as two reduced intensity, resolved peaks associated with ^{16}O (unlabeled) and ^{18}O-labeled phosphorus resonances. (The H$_2$^{17}O sample also contains both H$_2$^{16}O and H$_2$^{18}O.) This can most easily be seen in Figure 2B where the ^{18}O-labeled phosphate ^{31}P signal is shifted slightly upfield relative to the remaining ^{16}O phosphorus resonance.[14] Joseph and Bolton[17] have also shown that enzymatic ^{17}O-labeling can also be used to identify ^{31}P signals of polynucleotides.

PAC Two-Dimensional $^{31}P/^1H$ Heteronuclear Correlated Spectra of Oligonucleotides

Whereas the most straightforward and unambiguous method for complete assignment of ^{31}P signals in oligonucleotides relies on site-specific labeling of the monesterified phosphoryl oxygens of the backbone with ^{17}O or ^{17}O/^{18}O, it suffers by being rather expensive and time consuming. For an oligomer n nucleotides in length, $n - 1$ separate syntheses of ^{17}O-labeled oligonucleotide are required. Two-dimensional ^{31}P–^{1}H NMR spectroscopy[12,18–20] can generally provide a convenient, inexpensive alternative for the assignment of ^{31}P chemical shifts in moderately sized oligonucleotides.

Conventional 2-D ^{31}P–^{1}H heteronuclear correlation (HETCOR) NMR spectroscopy has been applied with limited success to short oligonucleotides.[11,12,43] More recently a ^{31}P–^{1}H long-range COLOC experiment[20,21] and an indirect detection (^{1}H detection) HETCOR experiment[19] have been successfully used to assign ^{31}P signals in duplexes 6 to 14 bp long. For oligonucleotides longer than 4–5 nucleotide residues the simple heteronuclear correlation measurements suffer from poor sensitivity as well as poor resolution in both the ^{1}H and ^{31}P dimensions. The poor sensitivity is largely due to the fact that the ^{1}H–^{31}P scalar coupling constants are about the same size or smaller than the ^{1}H–^{1}H coupling constants. Sensitivity is substantially improved by using a heteronuclear version of the "constant time" coherence transfer technique, referred to as COLOC (correlation spectroscopy via long-range coupling) and originally proposed for ^{13}C–^{1}H correlations.[45] Our laboratory has developed a pure absorption phase, constant time *(PAC)* pulse sequence to emphasize ^{1}H–^{31}P correlations in oligonucleotides as well as to give rise to pure absorption phase spectra, thereby further improving resolution.[20] This pulse sequence incorporates both evolution of antiphase magnetization and chemical shift labeling in a single, "constant time" delay period, thereby improving the efficiency of coherence transfer and increasing sensitivity. The PAC sequence also gives homonuclear decoupling during t_1 and thus improves the resolution.

Superficially, the idea of producing pure absorption phase spectra using a "constant time" sequence seems quite discouraging. In conventional coherence transfer experiments the first-order phase correction needed in the t_1 dimension is normally small or even zero. This is due to the fact that the chemical shift labeling delay (t_1) for the first spectrum in the data field

[45] H. Kessler, C. Griesinger, J. Zarbock, and H. R. Loosli, *J. Magn. Reson.* **57**, 331 (1984).

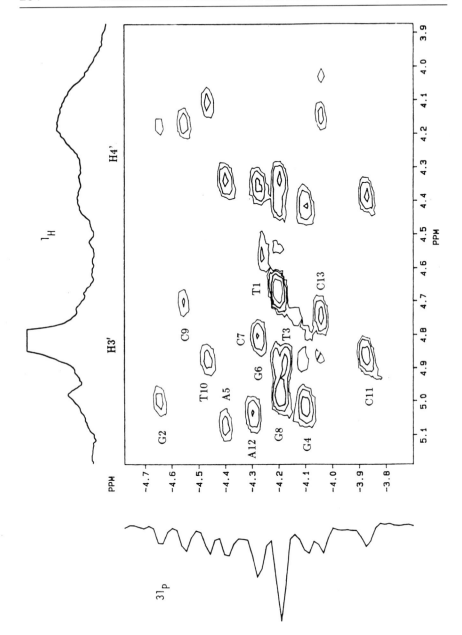

is small or zero. In contrast, the first spectrum in a PAC sequence is obtained at a large value of t_1. Consequently, the first-order phase correction needed in the t_1 dimension is large, namely, often many thousands of degrees. However, although this first-order phase correction is extremely large, it can be estimated rather accurately. In a PAC experiment the total first-order phase correction can be estimated as (spectral window in ω_1) × (constant delay) × 360°, assuming the entire constant delay is used for chemical shift labeling. This is the total first-order phase correction from one end of the ω_1 dimension to the other. A more accurate estimate could be made by including the finite lengths of the pulses, or these smaller phase corrections can be made interactively.

The PAC spectrum of the self-complementary 14-bp oligonucleotide duplex d(TGTGAGCGCTCACA)₂ is shown in Fig. 3. The cross-peaks represent scalar couplings between ³¹P nuclei of the backbone and the H3′ and H4′ deoxyribose protons. The chemical shifts of these protons are determined, as described above, by the sequential assignment methodology for B-DNA using the ¹H/¹H NOESY and correlation spectroscopy spectra. Assignment of the ³¹P signal of the *i*th phosphate is achieved through connectivities with both the 3′H(*i*) and 4′H(*i* + 1) deoxyribose protons. (The ¹H chemical shifts are first assigned by the 2-D ¹H/¹H sequential assignment methodology.) Although the 5′H(*i* + 1) and 5″H(*i* + 1) protons overlap with the 4′ protons, the intensities for the ³¹P–5′H and ³¹P–5″H PAC cross-peaks generally appear to be much weaker than the ³¹P–4′H cross-peaks. Similar behavior has been noted in ³¹P/¹H HETCOR studies on other oligonucleotides[6,12,13,19,46] and presumably reflects the

---

[46] M. H. Frey, W. Leupin, O. W. Sørensen, W. A. Denney, R. R. Ernst, and K. Wüthrich, *Biopolymers* **24**, 2371 (1985).

FIG. 3. Pure absorption phase ³¹P–¹H PAC spectrum of d(TGTGAGCGCTCACA)₂, at 200 MHz (¹H). ³¹P chemical shifts are reported relative to trimethyl phosphate, which is 3.456 ppm downfield from 85% phosphoric acid. (Reproduced with permission from Ref. 20.)

The pulse sequence and phase used for this spectrum were

¹H:	PA	90°	CD – t_1/2	180°	t_1/2	90°	RD/2	180°	RD/2	Decouple
³¹P:	PA		CD – t_1/2	180°	t_1/2	90°	RD/2	180°	RD/2	t_2

where the composite 180° pulse was a 90°, 180°, 90° sequence with phases x, y, x applied simultaneously to ¹H and ³¹P resonances; the preacquisition delay *PA* was 2 sec; the constant delay, *CD*, was 0.051 sec; and the refocusing delay *RD* was 0.035 sec. For one of the two states, data fields $\phi1 = \phi2 = x, -x, x, -x$, and for the second data field $\phi1 = y, -y, y, -y$ and $\phi2 = -y, y, -y, y$. A first-order phase correction of 11,520° was used in the ω_1 dimension. The spectrum was acquired on a Varian XL200A spectrometer (200 MHz ¹H).

strong and large 5' and 5" coupling. Variations in 3'H and 4'H coupling constants also explain the weaker intensities for some of these PAC cross-peaks.

It should be emphasized that sufficient resolution and sensitivity are achievable at 200 MHz (^{1}H) to allow the assignment of the ^{31}P signals of a fairly large DNA fragment (MW <9000). In contrast to the rather demanding spectrometer and probe performance required for the indirect detection HETCOR experiment at 500 MHz,[19] PAC can be readily implemented on most commercial, medium-field spectrometers.

DOC Two-Dimensional ^{31}P/^{1}H Heteronuclear Correlated Spectra of Oligonucleotides

Even with the development of the PAC and indirect detection heteronuclear correlation 2-D NMR methodologies, ^{31}P resonance assignments via heteronuclear correlation in duplex oligonucleotides represent a difficult 2-D NMR problem, largely because the resolution is often mediocre in both the ^{1}H and ^{31}P spectra, especially at lower fields.[20,47] Although the PAC method certainly works, the resolution in the proton dimension is digitally limited to that of the constant time delay. Acceptable values of this delay are sharply constrained by the evolution of proton–proton antiphase magnetization.

As noted above, the "constant time" PAC coherence transfer pulse sequences minimize this delay by using the same "constant time" for chemical shift labeling and the evolution of heteronuclear antiphase magnetization. This pulse sequence reduces the problem by limiting the time for the evolution of the parasitic proton–proton antiphase magnetization. Although PAC is a significant improvement over the normal HETCOR experiment, the sensitivity is still significantly reduced by the evolution of the parasitic proton–proton antiphase magnetizations.

A double constant time (DOC) sequence addresses the problem more directly by refocusing the proton–proton antiphase magnetization with a selective proton 180° pulse in between the two constant delays.[21] Nonselective 180° pulses move through each constant time period to chemical shift label the proton coherences. Thus the proton magnetization, which is antiphase with respect to the dilute spin, evolves for the entire constant delay period for protons in the region inverted by the semiselective 180° pulse. Likewise, the entire constant delay period is effectively used for chemical shift labeling. Proton–proton antiphase magnetization that evolves in the first constant time period is refocused in the second constant

[47] D. Williamson and A. Bax, *J. Magn. Reson.* **76**, 174 (1988).

time period, in the case of interest, where one proton is within the region inverted by the semiselective 180° pulse and the proton is outside of this region. This retains the sensitivity and resolution advantages of constant time pulse sequences and provides selectivity in the type of antiphase magnetization present at the end of the two constant time periods.

The DOC sequence is

^{1}H: 90° $CD - t_1/4$ 180° $t_1/4$ s180° $CD - t_1/4$ 180° $t_1/4$ 90° $RD/2$ s180° $RD/2$ decouple
X: 180°   180°   180°   90°   180°   ACQ

where s180° is a semiselective 180° pulse for protons within a selected region; CD is the length of one constant time delay so $2CD$ is the total evolution time; RD is a delay for refocusing the heteronuclear antiphase magnetization; t_1 is the delay which starts at 0 and is incremented up to $2CD$ to chemical shift label the proton coherences.

Acceptable values of the total evolution time are twice as large for DOC as compared to PAC.[21] This illustrates the well-known fact that proton–proton antiphase evolution is often the major limitation in obtaining the highest possible resolution in the proton dimension[47] and that DOC has largely or completely eliminated this limitation.[20] This should hold true up to the point where the proton T_2 drops to a value comparable to the time required to evolve significant proton–proton antiphase magnetization. It is quite clear that, as expected, twice the resolution is obtained in the proton dimension.

DOC can give not only much better resolution but also better sensitivity for the most difficult correlations. This sequence can thus compete effectively with indirect detection schemes.[19] The major limitation of the DOC sequence is that it may be necessary to collect the spectrum in parts, and, depending on the resolution required in the proton dimension, this can take more instrument time. However, if the highest resolution and the observation of the maximum number of correlations are required, then the DOC sequence is most likely the best choice.

Two-Dimensional Exchange ^{31}P NMR

Exchange of a phosphate between several different sites can lead to exchange-broadened lines[48] or even to separately resolved signals for the different sites. Especially during melting of duplex DNA or binding of a drug to DNA, various exchange processes may be observed.[49,50] Exchange

[48] E. P. Nikonowicz and D. G. Gorenstein, *Biochemistry* **29**, 8845 (1990).
[49] K. Lai and D. G. Gorenstein, *Biochemistry* **28**, 2804 (1989).
[50] M. Delepierre, C. Van Heijenoort, J. Igolen, J. Pothier, M. Le Bret, and B. P. Roques, *J. Biomol. Struct. Dyn.* **7**, 557 (1989).

processes can be detected by two-dimensional ^{31}P NMR in addition to the conventional one-dimensional methods. The two-dimensional exchange experiment (the same as NOESY) described by Ernst and co-workers[51,52] involves three 90° pulses. Nuclei are frequency labeled by a variable delay time (t_1) separating the first and second pulses, the mixing time is between the second and third pulses, and the detection of transverse magnetization as a function of time (t_2) follows the third pulse. During the mixing time, nuclei labeled in t_1 with a frequency corresponding to one site are converted by the exchange processes to a second site and evolve in t_2 with the frequency of the second site, giving rise to cross-peaks in the two-dimensional spectrum. Analysis of two-dimensional exchange data has been described[53] and differs from cross-relaxation data only in that the exchange rate matrix is not symmetrical.

Advantages in 2-D exchange experiments over the selective irradiation experiment[53] are as follows: (1) all exchange processes can be observed in a single experiment, (2) the nonselective irradiation problem is eliminated, and (3) there is no need for a selective pulse requiring either a low power radio frequency source or a DANTE pulse, which can be a limitation on some instruments. However, the quantification of cross-peak volumes is less precise than that of 1-D peak areas because of the practical limitation in the achievable signal-to-noise ratio. As such, rate constants obtained by 2-D ^{31}P exchange on large molecules would have large errors.

Measurement of ^1H–^{31}P Coupling Constants

The $J_{H3'-P}$ coupling constants in oligonucleotide duplexes can often be readily determined from the 2-D J-resolved long-range correlation spectrum (Fig. 4). Although other pulse experiments are available,[54] a Bax–Freeman selective 2-D J experiment with a DANTE sequence for a selective 180° pulse on the 3' proton signals allows one to measure phosphorus–H3' coupling constants, which can vary from approximately 1.5 to 8 Hz in duplexes as large as 14-mers. Measurement of the 4' or 5' (5″) coupling constants by this pulse sequence has been less successful.[48] The larger linewidths of longer duplexes limit measurement of the small coupling constants. Typically the selective 2-D J long-range correlation experiment is acquired [on a 200 MHz (^1H) spectrometer] with a sweep width of 50 Hz in the t_1 (^{31}P) dimension and about 500 Hz in the t_2 (^1H)

[51] J. Jeener, B. H. Meier, P. J. Bachmann, and R. R. Ernst, *J. Chem. Phys.* **71**, 4546 (1979).
[52] G. Bodenhausen and R. R. Ernst, *J. Am. Chem. Soc.* **104**, 1304 (1982).
[53] C. B. Post, W. J. Ray, Jr., and D. G. Gorenstein, *Biochemistry* **28**, 548 (1988).
[54] V. Sklenár and A. Bax, *J. Am. Chem. Soc.* **109**, 7525 (1987).

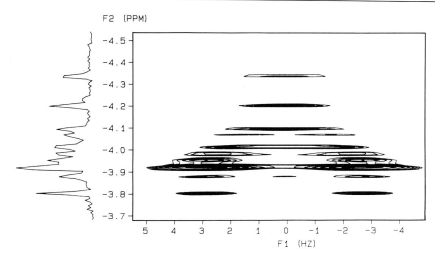

FIG. 4. Two-dimensional ³¹P/¹H J-resolved spectra at 200 MHz (¹H) of the duplex 14-mer d(TGTGAGCGCTCACA)$_2$. The 1-D decoupled ³¹P NMR spectrum is also shown along one axis, and the H3′ coupled doublets are shown along the second dimension. ³¹P chemical shifts are reported relative to trimethyl phosphate. (Reproduced with permission from Ref. 22.)

dimension. Selective excitation can easily be achieved with a DANTE pulse chain consisting of 20 pulses of an approximate length of 9° (total of 180°) in which the pulses are separated by a delay of 20 μsec. The data are processed with a Gaussian apodization plus a negative exponential function to give maximal resolution enhancement in both dimensions.

Notice as shown in Fig. 4 the very interesting pattern of monotonically decreasing coupling constants for the signals to higher field. This "Christmas tree" shaped pattern is usually observed in most other duplex oligonucleotides[22,28,55] and strongly supports a correlation between ³¹P chemical shifts and $J_{H3'-P}$ coupling constants (see below). This pattern collapses at higher temperatures, and at temperatures above the melting temperature of the duplex, all of the coupling constants are nearly identical ($J_{H-P} \approx$ 8.0–8.5 Hz).

The observed three-bond coupling constants are analyzed with a proton–phosphorus Karplus relationship to measure the H3′–C3′–O–P

[55] E. Ragg, R. Mondelli, A. Garbesi, F. P. Colonna, C. Battistini, and S. Vioglio, *Magn. Reson. Chem.* **27**, 640 (1989).

torsional angle θ from which the C4'–C3'–O–P torsional angle ϵ (= $-\theta - 120°$) may be calculated. The relationship

$$J = 15.3 \cos^2 (\theta) - 6.1 \cos(\theta) + 1.6$$

has been determined by Lankhorst *et al.*[56] As shown by Dickerson[57,58] there is a strong correlation ($R = -0.92$) between torsional angles C4'–C3'–O3'–P (ϵ) and C3'–O3'–P–O5' (ζ) in the crystal structures of various duplexes (ζ may be calculated from the relationship[57,58] $\zeta = -317 - 1.23\epsilon$). Thus both torsional angles ϵ and ζ can often be calculated from the measured P–H3' coupling constant.

As shown in Fig. 5, the Karplus relationship provides for four different torsional angle solutions for each value of the same coupling constant. Although all four values are shown in Fig. 5, the limb which includes ϵ values between 0 and $-270°$ is sterically inaccessible in nucleic acids.[59] Generally it is assumed that the coupling constants, which can vary only between 1 and 11 Hz, all correspond to ϵ torsional angles on a single limb nearest the crystallographically observed average for nucleic acids[59] ($\epsilon = -169 \pm 25°$). This need not always be true, especially since phosphate ester conformations corresponding to either of the other two solutions are observed crystallographically.[59] However, as shown in Fig. 5, nearly all of the phosphates for normal Watson–Crick duplexes fall along only a single limb of the Karplus curve. Exceptions include several phosphates adjacent to a GA mismatch in a decamer.[48] In addition, it is quite interesting that the measured ^{31}P chemical shifts and coupling constants for the two phosphates at the site of a tetramer duplex–actinomycin D drug complex do not fit on the main limb of the Karplus curve, although the values do nicely fit on the other two solutions of the curve (squares, Fig. 5). Indeed, crystal structures[59] of related intercalator–duplex complexes confirm that the phosphates often are constrained to conformations that best correspond to the other solutions. As described below, these intercalator drug–duplex complexes quite generally show large downfield shifts of the ^{31}P signals.[24,34,60,61]

In addition, perturbations in ^{31}P shifts can also arise from variations in the α (O3'–P–O5'–C5') torsion angle, which could also explain the

[56] P. P. Lankhorst, C. A. G. Haasnoot, C. Erkelens, and C. Altona, *J. Biomol. Struct. Dyn.* **1**, 1387 (1984).

[57] R. E. Dickerson, *J. Mol. Biol.* **166**, 419 (1983).

[58] R. E. Dickerson and H. R. Drew, *J. Mol. Biol.* **149**, 761 (1981).

[59] W. Saenger, "Principles of Nucleic Acid Structure." Springer-Verlag, New York, 1984.

[60] D. J. Patel, *Acc. Chem. Res.* **12**, 118 (1979).

[61] D. J. Patel, *Biochemistry* **13**, 2388 (1974).

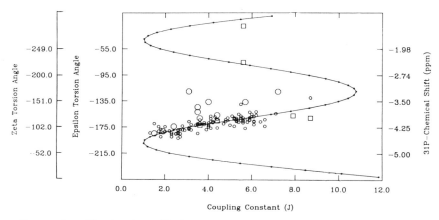

FIG. 5. Plot of ³¹P chemical shifts for the oligonucleotide sequences previously studied in our laboratory[22,28,44] (O) and the actinomycin D-bound d(CGCG)$_2$ tetramer complex[50] (□) with their measured $J_{H3'-P}$ coupling constants (the O data points correspond to phosphates in a tandem GA mismatch decamer duplex,[48] which shows unusual, slowly exchanging signals). Also shown are the theoretical ϵ and ζ torsion angles (solid curve) as a function of coupling constant derived from the Karplus relationship (ϵ) and the relationship $\zeta = -317 - 1.23\epsilon$. ³¹P chemical shifts are reported relative to trimethyl phosphate. (Reproduced with permission from Ref. 48.)

"anomalous" behavior of those ³¹P signals of phosphates which do not fall on the main limb of the Karplus curve. Whereas the ζ torsional angle is found to be the most variable one in the B form of the double helix, the other P–O torsional angle, α, is one of the most variable in the A form of the duplex.[59] The two points falling furthest off the main limb correspond to two phosphates in a 10-mer GA mismatch, d(CCAAGATTGG).[48] The unusual geometry of the tandem mismatch could allow significant variation of the α (O3'–P–O5'–C5') and β(P–O5'–C5'–C4') torsional angles. As previously noted, these angles have in practice been unobtainable even in duplexes of moderate length. Broad, overlapping lines at low temperature and the inability to stereospecifically assign H5' and H5" are significant barriers to deriving useful quantitative data, although qualitative measurements of $J_{H5'-P}$ may be obtained from the Bax–Freeman selective 2-D J-resolved spectrum.[48] There does appear to be a slight variation in these coupling constants, suggesting to some degree variation of the β torsional angles at the site of the mismatch and at residues 5' to that site (the two slowly exchanging phosphates that fall off the correlation).

It is important to recognize that ³¹P chemical shifts are perturbed by factors other than torsional angle changes alone. As has previously been shown, ³¹P chemical shifts are very sensitive to bond angle distortions as well.[23,41,62] It is quite reasonable to assume that backbone structural distortions as observed in mismatch duplexes, extrahelical base duplexes, or drug–duplex complexes also introduce some bond angle distortion as well. Widening of the ester O–P–O bond angle indeed is expected to produce an upfield shift[23,62] while narrowing of this bond angle causes a downfield shift, and it is possible that this bond angle effect could account for the anomalous shifts.

Ignoring the drug–duplex points that fall off the main limb of the Karplus curve and assuming a linear correlation between the coupling constants and ^{31}P chemical shifts (effectively a linear fit to the data on the main limb of the Karplus curve) the correlation coefficient is found to vary from -0.79 to -0.92 (at ambient temperature) for various oligonucleotide duplexes (the correlation coefficient for correlation of all of the main limb data is about -0.81).[22] Remarkably, a linear correlation is even observed up to a temperature of 80°.

Thus, for "normal" B-DNA geometry, there is an excellent correlation between the phosphate resonances and the observed torsional angle, although phosphates which are greatly distorted in their geometry must be more carefully analyzed. Importantly, knowledge of both the ^{31}P chemical shift and $J_{H3'-P}$ coupling constant can be very helpful in the proper interpretation of phosphate ester conformation.

^{31}P Melting Curves

The temperature dependence of ^{31}P chemical shifts and coupling constants provides an important monitor of both melting and premelting structural transitions in the phosphate ester backbone. Thus, as represented by the ^{31}P chemical shift melting curve (Fig. 6) for an extrahelical adenosine 13-mer,[44] there is a general downfield shift of many of the resonances with increasing temperature.[23,24] This is consistent with the change in the backbone conformation, generally g^- for P–O ester bonds in the duplex, to a random mix of g^- and nongauche conformations in the single-stranded random coil. However, it is interesting to note that one of the peaks actually moves considerably upfield beginning at about 30°. The inflection point in the curves for all phosphates occurs at approximately 40°, corresponding to the T_m of the duplex. The coupling constants generally parallel the variation in ^{31}P chemical shifts and show quite similar melting curves.[28,48] Thus ^{31}P chemical shifts (and coupling constants) follow the expected variation in the sugar–phosphate backbone conformation as the duplex melts into a random coil single strand.

[62] D. G. Gorenstein, *J. Am. Chem. Soc.* **97**, 898 (1975).

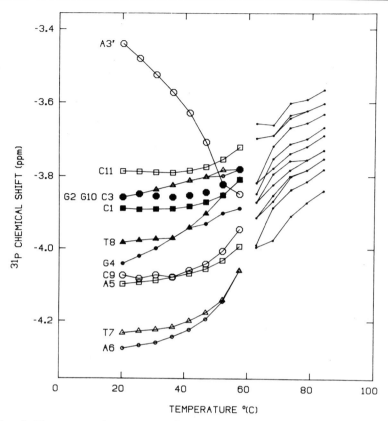

FIG. 6. Temperature dependence of ³¹P chemical shifts of an extrahelical adenosine duplex 13-mer, d(CGCAGAATTCGCG). ³¹P chemical shifts are reported relative to trimethyl phosphate. (Reproduced with permission from Ref. 44.)

³¹P NMR of Phosphoryl-Modified Analog of Oligonucleotides

With the demonstration that oligonucleotides and various analogs of oligonucleotides exhibit antiviral and antioncogene activity, considerable interest has developed around the synthesis and characterization of these "antisense" agents.[63] Various oligodeoxyribonucleotides and oligodeoxy-

---

[63] E. Uhlmann and A. Peyman, *Chem. Rev.* **90,** 543 (1990).

nucleoside methyl phosphonate, phosphorothioate, phosphoroamidate, and phosphorodithioate analog, among others, have been studied.[63-66] Whereas unmodified oligodeoxynucleotides are susceptible to nuclease digestion,[67] the phosphate-modified oligodeoxynucleoside methylphosphonate, phosphorothioate,[66] and phosphorodithioate[68-70] analog are generally nuclease resistant, and they have recently received great attention.

In the monothiophosphate and methyl phosphonate oligonucleotide analogs, the phosphorus center is chiral, and thus each diastereomeric phosphorus center generally gives rise to separate ^{31}P signals for each R_p and S_p diastereomer. Complicated ^{31}P NMR spectra are observed for modified oligonucleotides containing even just a few achiral phosphorus centers. In contrast, the ^{31}P NMR spectra of achiral phosphorodithioates are considerably simplified relative to the other analogs consisting of mixtures of unresolved diastereomers.

Synthesis and purification of dithiophosphate oligonucleotide analogs may be accomplished using thiophosphoramidite chemistry.[68-72] The large downfield ^{31}P shift (~ 110 ppm) of the dithiophosphate groups relative to the phosphate moiety (~ -4 ppm) readily identifies the substitution. (The ^{31}P signals of monothiophosphates resonate at $\sim 50-60$ ppm.) As shown in the ^{31}P spectrum of d(CGCTpS$_2^-$TpS$_2^-$AAGCG) (Fig. 7), multiple dithiophosphoryl substitutions show the expected reduction in the number of phosphoryl oligonucleotide signals with replacement by new dithiophosphoryl signals.[73] A pure absorption phase, constant time ^{1}H/^{31}P heteronuclear correlated spectrum can also be used to assign the resonances.[20] Interestingly, in contrast to the parent palindromic decamer sequence,[74] which has been shown to exist entirely in the duplex B-DNA conformation under comparable conditions (100 mM KCl, 10 mM strand concentra-

[64] P. S. Miller, M. P. Reddy, A. Murakami, K. R. Blake, S.-B. Lin, and C. H. Agris, *Biochemistry* 25, 5092 (1986).

[65] P. C. Zamecnik and M. L. Stephenson, *Proc. Natl. Acad. Sci. U.S.A.* 75, 280 (1978).

[66] M. Matsukura, K. Shinozuka, H. Mitsuya, M. Reitz, J. Cohen, and S. Broder, *Proc. Natl. Acad. Sci. U.S.A.* 84, 7706 (1987).

[67] E. Wickstrom, *J. Biochem. Biophys. Methods* 13, 97 (1986).

[68] W. K. D. Brill, J.-Y. Tang, Y.-X. Ma, and M. H. Caruthers, *J. Am. Chem. Soc.* 111, 2321 (1989).

[69] W. K. D. Brill, J. Nielsen, and M. H. Caruthers, *Tetrahedron Lett.* 29, 5517 (1988).

[70] N. Farschtschi and D. G. Gorenstein, *Tetrahedron Lett.* 29, 6843 (1988).

[71] B. H. Dahl, K. Bjergàrde, V. B. Sommer, and O. Dahl, *Nucleosides Nucleotides* 8, 1023 (1989).

[72] D. G. Gorenstein and N. Farschtschi, U.S. Patent Appl. P-88045 (1989).

[73] M. E. Piotto, J. N. Granger, Y. Cho, N. Farschtschi, and D. G. Gorenstein, *in* "Tetrahedron, Symposium in Print on Bioorganic Chemistry" (D. Boger, ed.), 47, 2449, 1991.

[74] R. Powers, C. R. Jones, and D. G. Gorenstein, *J. Biomol. Struct. Dyn.* 8, 253 (1990).

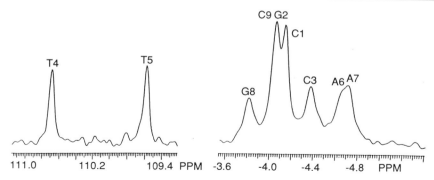

FIG. 7. ³¹P NMR spectra and phosphate assignments of the dithiophosphate decamer d(CGCTpS₂⁻TpS₂⁻AAGCG). Numbering corresponds to phosphate position from the 5′ end of the strand. ³¹P chemical shifts are reported relative to trimethyl phosphate. (Reproduced with permission from Ref. 73.)

tion), the dithiophosphate analog forms a hairpin, even at twice the strand concentration.

³¹P NMR of Polynucleic Acids

³¹P NMR may usefully be applied to the study of polynucleic acids as well. As described above, the range of ³¹P chemical shifts for individual phosphates in duplex fragments is less than 0.9 ppm. It is not surprising, therefore, that individually resolved signals are not observed for longer (> 100 bp) nucleic acid fragments, for which much broader lines are expected. Under favorable circumstances, however, separate signals may be resolved.

Sonicated poly(A)·poly(U) ³¹P spectra show two equal intensity signals separated by approximately 0.4 ppm (−1.2 and −1.6 ppm relative to 85% H_3PO_4). As shown by Alderfer and Hazel,[75] these signals are associated with the duplex. The chemical shifts of the triplex are +0.07, −0.38, and −0.90 ppm[75] referenced to 85% phosphoric acid. Joseph and Bolton[17] have shown by ¹⁷O labeling of the phosphates that the two main ³¹P signals are associated with phosphates in alternating conformations along the duplex poly(A)·poly(U) rather than the phosphates in the two different strands. The separation of the two signals could possibly be attributed to alternating helical twist values along the duplex (presumably an "alternating A-RNA" structure).

---

[75] J. L. Alderfer and G. L. Hazel, *J. Am. Chem. Soc.* **103**, 5925 (1981).

The ^{31}P spectrum of poly[d(AT)] also gives two signals separated by as much as 0.8 ppm, depending on salt conditions.[76] In low salt solution poly[d(AT)] shows a separation of ^{31}P signals of 0.24 ppm.[76] By thiophosphoryl labeling, Eckstein and Jovin[77] were able to establish that the deshielded ^{31}P signal arises from the TpA phosphates, which, based on an X-ray crystal model, are in a more extended translike phosphate ester conformation. The ^{31}P signal of the ApT phosphates is quite similar to that of normal B-DNA phosphates and indeed is in a g^-,g^- conformation.

^{31}P NMR of DNA – Drug Complexes

Patel[61,78] first reported that the intercalating drug actinomycin D (ActD) shifted several phosphate diester signals up to 2.6 ppm downfield from the double helical signal on binding to oligonucleotide duplexes containing dGdC base pairs. Thus, downfield shifts of 1.6 and 2.6 ppm in the dCdGdCdG – Act D (2:1) and 1.6 ppm in the pdGdC – Act D (2:1) complex at 8° have been observed. Reinhardt and Krugh[79] showed that at even lower temperature ($-18°$) in methanol – water the ^{31}P signals of the two phosphates in duplex pdGdC is split into two signals and shifted 1.7 and 2.4 ppm downfield on complexation with actinomycin D. Similar shifts are also observed in 2:1 Act D – oligonucleotide duplex complexes (see Ref. 80).

These shifts are consistent with the Sobell and Jain[81] model for these intercalated complexes: partial unwinding of a specific section of the double helix allows these planar, heterocyclic drugs such as actinomycin D to stack between two base pairs. The downfield-shifted peak in a d(AGCT) – Act D complex has been assigned to the dGdC base pair by the [17]O/^{18}O phosphoryl-labeling method[38] (see also Ref. 15). X-Ray studies on various intercalating drug – duplex complexes[59,82] suggest that the major backbone deformation of the nucleic acid on intercalation of the drug involves changes in both the P – O and C – O torsional angles.

Similar spectral changes have been observed for drug – polynucleic acid complexes. For example, titration of the poly(A)·poly(U) ^{31}P spectra with ethidium ion produces a new signal 1.8 – 2.2 ppm downfield from the

[76] H. Shindo, R. T. Simpson, and J. S. Cohen, *J. Biol. Chem.* **254**, 8125 (1979).

[77] F. Eckstein and T. M. Jovin, *Biochemistry* **22**, 4546 (1983).

[78] D. J. Patel and L. L. Canuel, *Proc. Natl. Acad. Sci. U.S.A.* **73**, 3343 (1976).

[79] C. G. Reinhardt and T. R. Krugh, *Biochemistry* **16**, 2890 (1977).

[80] E. V. Scott, R. L. Jones, D. L. Banville, G. Zon, L. G. Marzilli, and W. D. Wilson, *Biochemistry* **27**, 915 (1988).

[81] H. M. Sobell and S. C. Jain, *J. Mol. Biol.* **68**, 21 (1972).

[82] H.-S. Shieh, H. M. Berman, M. Dabrow, and S. Neidle, *Nucleic Acids Res.* **8**, 85 (1980).

duplex signal.[83] At saturating levels of the drug ([drug]/[DNA] ratios over 0.5), the relative intensities of the downfield signal relative to the largely unperturbed upfield signal is around 1:1.

This deshielding for around one-half of the phosphate signals in the poly (A)·poly(U)·drug complexes is entirely consistent with the intercalation perturbation of the phosphate ester geometry observed in the actinomycin D and related ethidium ion complexes. The chemical shift of this downfield peak is supportive of the intercalation mode of binding, since the purely electrostatic association between drugs and nucleic acid produces only small and generally upfield ³¹P shifts.[60,84] The largely unperturbed upfield signals between −1.2 and −1.6 ppm at 15° in these complexes likely represent undistorted phosphates in regions adjacent to the intercalation site. The approximately 1:1 relative intensity of the downfield and main upfield peaks at saturating drug ratios is consistent with the nearest-neighbor exclusion model for these drug–duplex complexes. Thus, at saturating drug ratios only every other base pair step can accommodate an intercalated drug. The phosphate ester backbone at the intercalation site is in an extended conformation (see below), whereas the phosphate ester backbone at the two neighboring sites is in a normal B-DNA conformation.

In contrast to the above RNA duplex results, Wilson et al.[84,85] and Lai and Gorenstein[49] did not observe a separate downfield peak in the binding of ethidium, quinacrine, daunomycin, and tetralysine to sonicated DNA. In ³¹P NMR studies on the intercalating drug–DNA complexes, increasing amounts of drugs only produce downfield shifts and line broadening. The difference between the duplex RNA and DNA systems is attributed to the rate of chemical exchange between the duplex–drug complex. In the drug–DNA complexes, the phosphates are undergoing fast chemical exchange between sites not involving the intercalated drug (nearest neighbor or further removed) and sites directly linking the intercalated base pairs.

Sequence-Specific Variation in ³¹P Chemical Shifts and Coupling Constants

To date, more than a dozen "modest" sized oligonucleotide sequences have had individual ³¹P resonances completely assigned. ³¹P chemical shifts of individual phosphates for similar sequences are quite similar, and

[83] E. M. Goldfield, B. A. Luxon, V. Bowie, and D. G. Gorenstein, *Biochemistry* **22**, 3336 (1983).
[84] W. D. Wilson and R. L. Jones, *Nucleic Acids Res.* **10**, 1399 (1982).
[85] W. D. Wilson, R. A. Keel, and Y. H. Mariam, *J. Am. Chem. Soc.* **103**, 6267 (1981).

perturbations in the shifts are largely localized near the sites of base pair substitution.

Although "complementary" phosphates (phosphates opposite each other on complementary strands) are chemically and magnetically none-quivalent, rather surprisingly, the ^{31}P chemical shifts at complementary phosphate positions generally follow the same pattern in both strands of the duplex regardless of base sequence or position, suggesting that the phosphate geometry is nearly the same in complementary positions along both strands.[86]

A very intriguing comparison[86] of the ^{31}P chemical shift of two 14-mers, d(TGTGACGCGTCACA)$_2$ and d(CACAGTATACTGTG)$_2$, is shown in Fig. 8. The two 14-mers are related by an "opposite" nucleotide sequence (a pyrimidine is replaced by the other pyrimidine and a purine is replaced by the other purine). This retains the purine–pyridimine base steps (and clashes) but obviously completely alters the sequence at every position. As shown in Fig. 8, remarkably, the pattern of variation of ^{31}P chemical shifts of the two 14-mers are quite similar. The observed variation of ^{31}P chemical shifts is largely a function only of the purine/pyrimidine sequence. The through space magnetic or electric chemical shielding effects of either a guanosine or adenosine base (i.e., through ring-current and electric charge effects) will not be identical at the phosphorus nucleus. Thus it is immaterial whether an A or G base is present at a particular purine position in the sequence.

As pointed out by Ott and Eckstein[87] and Schroeder et al.,[6,11] ^{31}P chemical shifts appear to vary in response to local, sequence-specific, and induced environmental distortions in the duplex geometry. Shown in Fig. 9 are the ^{31}P chemical shifts and helix twist values of the assigned phosphates of a number of oligodeoxyribonucleotides and drug–DNA complexes described in the previous section. Dickerson and co-workers[88] have shown that the helical distortions observed in the crystal structure of d(GCGCAATTGGC) could be quantitatively predicted through a series of simple "Calladine rule" sum function relationships.[89] Thus the global helical twist (t_g) can be calculated from $t_g = 35.6 + 2.1 \Sigma_1$, where Σ_1 is the helical twist sum function. Based on the ^{31}P assignments of over a dozen oligonucleotides, there does appear to be a modest correlation between ^{31}P

[86] S. A. Schroeder, V. Roongta, J. M. Fu, C. R. Jones, and D. G. Gorenstein, *Biochemistry* **28**, 8292 (1989).

[87] J. Ott and F. Eckstein, *Biochemistry* **24**, 2530 (1985).

[88] A. V. Fratini, M. L. Kopka, H. R. Drew, and R. E. Dickerson, *J. Biol. Chem.* **257**, 14686 (1982).

[89] C. R. Calladine, *J. Mol. Biol.* **161**, 343 (1982).

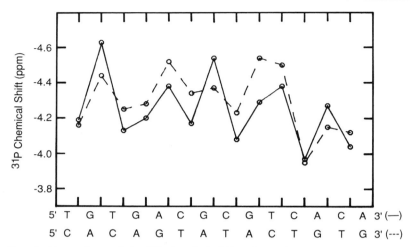

FIG. 8. Comparison of the ³¹P chemical shifts of pseudooperator 14-mers, d(TGT-GACGCGTCATA)₂ (solid line) and d(CACAGTATACTGTG)₂ (dashed line), versus sequence. ³¹P chemical shifts are reported relative to trimethyl phosphate. (Reproduced with permission from Ref. 86.)

chemical shifts and the helical twist sum function for a number of the oligonucleotides (Fig. 9).[6,11,22,49]

The correlation coefficient (r) between ³¹P shifts and helical twist is 0.71 if just the terminal phosphates and the ³¹P chemical shifts of the drug complexes described in Fig. 9 are included in the correlation. However, in the middle region of the duplex oligonucleotides, no correlation $(r = 0.05)$ is found between ³¹P chemical shifts and helix twist.[22] Either ³¹P chemical shifts behave differently in these two regions of the duplexes, or the sequence-specific Calladine rules derived from the X-ray crystal structures of the Dickerson 12-mer are to be questioned. Indeed, the current concern over the validity of the Calladine rules is most likely the explanation for these differences, and until more reliable "rules" are developed (if they even exist) further speculation is unwarranted. Note, however, that the plots of Figs. 8 and 9 and others not shown clearly demonstrate some sequence-specific variations in ³¹P chemical shifts.

The ³¹P chemical shifts of various drug complexes[49,84] appear to correlate with the observed degree of unwinding of the duplex DNA on drug binding. Indeed, these drug–duplex complex ³¹P chemical shifts also fit on the plot of Fig. 9. The helical twist values for various drug complexes that are used in Fig. 9 represent the difference between the normal B-DNA

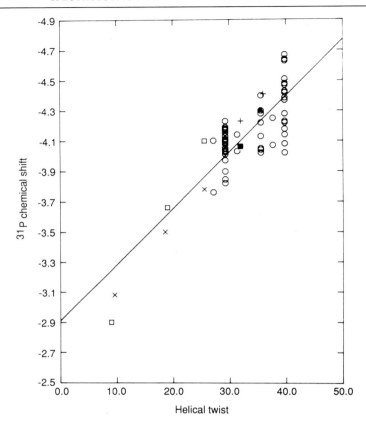

FIG. 9. Plot of ^{31}P chemical shift versus calculated helix twist, t_g [10–14 bp oligonucleo-tide phosphate shifts (○); t_g derived from $t_g = 35.6 + 2.1 \Sigma_I$]. Drug–DNA data (×), drug–calf thymus DNA data[49] (□), A-DNA (■), B-DNA (△), and alternating poly[d(AT)] (+) shifts are shown.[84] The helix twist for the drug–DNA complex is calculated from the difference between the helix twist of B-DNA ($t_g = 35.6°$) and the unwinding angle, ϕ, for the drug–DNA complex [for ethidium ($\phi = 26°$), quinacrine ($\phi = 17°$), and daunomycin ($\phi = 10°$)]. The ^{31}P chemical shifts of the drug–DNA complexes have been corrected for chemical shift averaging. ^{31}P chemical shifts are reported relative to trimethyl phosphate, which is 3.456 ppm downfield from 85% phosphoric acid. (Reproduced with permission from Ref. 49.)

helical twist angle of 35.6° and the observed helical unwinding angle resulting from binding of the drug. Because of the nearest-neighbor exclusion principle, at saturating concentrations of these drugs, only every other site has a bound drug. Because of fast chemical exchange averaging, only one ^{31}P signal is observed in these drug complexes, and this must represent an average of the ³¹P chemical shifts of phosphates at nonintercalative and intercalative sites. Correction for this chemical exchange averaging has been made in Fig. 9 by multiplying the observed perturbation of the ³¹P shifts by 2.

The possible basis for the correlation between helix unwinding, helix twist, and ³¹P chemical shifts in both RNA and DNA can be analyzed in terms of sugar–phosphate backbone distortions involved in duplex geometry changes,[22] as more fully described in the next section. Briefly, as the helix unwinds on binding of an intercalating drug (and the helix twist, t_g, decreases), the length of the sugar–phosphate backbone must increase to accommodate the additional heterocycle intercalated between the two stacked base pairs with a base-to-base separation of 6.7 Å. These local helical changes require changes in the sugar–phosphate backbone angles. This can be accomplished by the phosphate switching from the B_I ($\zeta = g^-$, $\alpha = g^-$) to the B_{II} conformation ($\zeta = t$, $\alpha = g^-$)[90] (Fig. 1). The B_I and B_{II} states represent ground-state minima with the P–O and C–O torsional angles in the staggered conformations.[91,92] Partially or fully eclipsed conformations which are not energy minima are only accessible through liberation in each of the staggered states or through transient passage during the rapid jumps between the two ground states.

Roongta et al.[28] have analyzed the variation in the ³¹P chemical shifts and $J_{H3'-P}$ coupling constants (Fig. 5) in terms of fractional populations of the two thermodynamically stable B_I and B_{II} states. From the crystal structure data, the average ϵ torsional angle value for the B_I state is $-190°$ and that for the B_{II} state is $-105°$.[57,58] These ϵ values approximate the maximum possible range of $J_{H3'-P}$ coupling constants (for $\epsilon = -105°$, $J_{H3'-P} = 10.0$ Hz, and for $\epsilon = -190°$, $J_{H3'-P} = 1.3$ Hz). In addition in this torsional angle range, the Karplus curve is nearly a single-valued function (actually it is only single valued in the range of $\epsilon \approx -200$ to $-120°$; between -105 and $-120°$, the coupling constant is in the range of 10.0– 10.8 Hz). The measured coupling constant represents the weighted average of the coupling constants in the two states, and, using the two extreme values of ϵ, the percentage B_I state can be estimated from the relationship

$$\%B_I = 100\% - \%B_{II} \approx \frac{J_{H3'-P}(\text{observed}) - J_{H3'-P}(-105°)}{J_{H3'-P}(-190°) - J_{H3'-P}(-105°)} \times 100\%$$

Under these assumptions and from the plot of Fig. 5, the ³¹P chemical shift and $J_{H3'-P}$ coupling constant of a phosphate in a purely B_I conforma-

[90] R. E. Dickerson, J. Mol. Biol. **166**, 491 (1983).
[91] P. Kollman, J. W. Keepers, and P. Weiner, Biopolymers **21**, 2345 (1982).
[92] D. G. Gorenstein, B. A. Luxon, and J. B. Findlay, Biochim. Biophys. Acta **475**, 184 (1977).

tional state are estimated to be approximately -4.6 ppm and 1.3 Hz, respectively. Similarly, the ^{31}P chemical shift and $J_{H3'-P}$ coupling constants of a phosphate in a purely B_{II} conformational state should be around -3.0 ppm and 10 Hz, respectively. The ^{31}P chemical shift difference between the B_I and B_{II} conformational states is thus estimated to be 1.6 ppm. Based on molecular orbital calculations of ^{31}P shielding, our laboratory had previously predicted that the ^{31}P chemical shift of a B_{II}-type phosphate ($\zeta = t$) is around 1.5 ppm downfield of a phosphate in a B_I conformation ($\zeta = g^-$).[23,24,93] It is thus extremely gratifying that the experimentally observed results fully support the hypothesis on the conformational dependence to ^{31}P chemical shifts. The largest coupling constant that is measured is approximately 7–8 Hz, which represents a ζ angle of about $-120°$. Analyzing the coupling constants in terms of the fractional populations of the B_I and B_{II} states, phosphates with $J_{H3'-P}$ values near this maximum thus have nearly equal populations of the B_I and B_{II} states. At 80°, based on the observed coupling constants, the B_I and B_{II} states are also approximately equally populated in the random coil single-stranded form of the oligonucleotides. This is quite reasonable because, based on molecular orbital calculations, the B_I phosphate ester conformation is estimated to be less than 1.0 kcal/mol lower in energy than a phosphate in the B_{II} conformation.[26,92]

The dispersion in the ^{31}P chemical shifts of oligonucleotides is thus likely attributable to different ratios of populations of the B_I and B_{II} states for each phosphate in the sequence. Assuming that only the staggered rotamers define stable conformations, the phosphate is assumed to make rapid jumps between the two states. Alternatively, the phosphate could be considered to be constrained to a single intermediate P–O ester conformation. The latter explanation appears to be ruled out by restrained molecular dynamics simulations[74] which clearly show that these intermediate conformational states are only transiently populated as the phosphate makes rapid jumps between the B_I and B_{II} states. Thus, the measured coupling constant appears to indeed reflect the populations of these two possible conformational states.

^{31}P NMR appears to be able to provide a convenient monitor of the phosphate ester backbone conformational changes on binding of various drugs to oligonucleic acid and polynucleic acid duplexes. The observed correlation between ^{31}P chemical shifts and helix twist in DNA fragments and duplex DNA–drug complexes is strongly supportive of the hypothesis that variations in phosphate ester torsional angles are a major contributor to variations in ^{31}P chemical shifts.

[93] D. G. Gorenstein and D. Kar, *Biochem. Biophys. Res. Commun.* **65,** 1073 (1975).

Stereoelectronic Effects on [31]P Spectra as Probe of
  DNA Structure

It is clear from the plots of Figs. 5 and 9 that ^{31}P chemical shifts and coupling constants provide a probe of the conformation of the phosphate ester backbone in nucleic acids and nucleic acid complexes. If ^{31}P chemical shifts are sensitive to phosphate ester conformations, they potentially provide information on two of the most important torsional angles that define the nucleic acid deoxyribose–phosphate backbone. One of these, the P–O3' torsional angle, is also the most variable one in the B form of the double helix, and the other P–O5' torsional angle is one of the most variable in the A form of the duplex.[59] Indeed, following the original suggestion of Sundaralingam,[94] and based on recent X-ray crystallographic studies of oligonucleotides, Saenger[59] has noted that the P–O bonds may be considered the "major pivots affecting polynucleotide structure."

As discussed above, one of the major contributing factors that determines ^{31}P chemical shifts is the main chain torsional angles of the individual phosphodiester groups along the oligonucleotide double helix. Phosphates located toward the middle of a B-DNA double helix assume the lower energy, stereoelectronically favored B_I, whereas phosphodiester linkages located toward the two ends of the double helix tend to adopt a mixture of B_I and B_{II} conformations, where increased flexibility of the helix is more likely to occur. Because the B_I conformation is responsible for a more upfield ^{31}P chemical shift, whereas a B_{II} conformation is associated with a lower field chemical shift, internal phosphates in oligonucleotides would be expected to be upfield of those nearer the ends. Although several exceptions have been observed, this positional relationship appears to be generally valid for oligonucleotides where ^{31}P chemical shift assignments have been determined.[33,87] Thus, position of the phosphorus (ends versus middle) within the oligonucleotide is one important factor responsible for variations in ^{31}P chemical shifts.

As described previously, Ott and Eckstein[87,95] noted that the occurrence of a 5'-pyrimidine–purine-3' base sequence (5'-PyPu-3') within the oligonucleotide has a more downfield than expected ^{31}P chemical shift is based solely on the phosphate positional relationship. They suggested that local helical distortions arise along the DNA chain owing to Calladine-type purine–purine steric clash on opposite strands of the double helix. As a result, 5'-PyPu-3' sequences within the oligonucleotide represent positions where the largest helical distortions occur. They have proposed, based on

[94] M. Sundaralingam, *Biopolymers* **7**, 821 (1969).
[95] J. Ott and F. Eckstein, *Nucleic Acids Res.* **13**, 6317 (1985).

the ^{31}P assignments of two dodecamers and an octamer, that a correlation exists between the helical roll angle parameter and ^{31}P chemical shifts. They have noted a considerably poorer correlation between ^{31}P chemical shifts of the oligonucleotides and other sequence-specific variations in duplex geometry (such as the helix twist). As shown in Fig. 9, our laboratory has found that there does appear to be a possible correlation between ^{31}P shifts and the Calladine–Dickerson rule calculated helical twist at a particular base step,[6,22] at least at the ends of the duplex.

Other factors, of course, can influence ^{31}P chemical shifts. Indeed, Lerner and Kearns[96] have shown that ^{31}P chemical shifts of phosphate esters are modestly sensitive to solvation effects, and they have argued that differential solvation of the nucleic acid phosphates could be responsible for the observed variation in the ^{31}P chemical shifts of nucleic acids. Costello et al.[97] have also noted a similar sensitivity of ^{31}P chemical shifts to salt. However, the strong correlation of ^{31}P chemical shifts and coupling constants demonstrates that these environmental effects on the ^{31}P chemical shifts of nucleic acids are smaller than the intrinsic conformational factors discussed above, assuming comparisons are made under similar solvation conditions. Likewise, other possible effects, such as ring-current shifts, are not likely to be responsible for shifts above 0.01 ppm.[23] The latter effects are so small because the phosphates are so far removed from the bases (> 10 Å).

Origin of [31]P Chemical Shift and Coupling Constant Variations: Backbone Torsional Angles and Local Helical Structure

As noted above, the possible basis for the correlation between Calladine rule-type structural variations and ^{31}P chemical shifts can be analyzed in terms of deoxyribose–phosphate backbone changes involved in local helical sequence-specific structural variations. As the helix unwinds in response to local helical distortions (and the helix twist t_g decreases), the length of the deoxyribose–phosphate backbone generally decreases.[22] These local helical changes require changes in the deoxyribose–phosphate backbone angles $\alpha - \zeta$ (Fig. 1). As the helix winds or unwinds, the distance between the adjacent C-4′ atoms of deoxyribose rings along an individual strand must change to reflect the stretching and contracting of the deoxyribose–phosphate backbone between the two stacked base pairs. To

[96] D. B. Lerner and D. R. Kearns, *J. Am. Chem. Soc.* **102**, 7611 (1980).
[97] A. J. R. Costello, T. Glonek, and J. R. Van Wazer, *J. Inorg. Chem.* **15**, 972 (1976).

a significant extent, these changes in the overall length of the deoxyribose–phosphate backbone "tether" are reflected in changes in the P–O ester (as well as other) torsional angles.[22] The sequence-specific variations in the P–O (and C–O) torsional angles may provide the linkage between the Calladine rule-type sequence-dependent structural variations in the duplex and ³¹P chemical shifts.

Analysis of the X-ray crystal structures[57–59] of B-form oligodeoxyribonucleotides has shown that torsional angles α, β, and γ on the 5′ side of the sugar are largely constrained to values g^- ($-60°$), t ($180°$), g ($+60°$), whereas significant variations are observed on the 3′ side of the deoxyribose–phosphate backbone. The greatest variation in backbone torsional angles is observed for ζ (P–O3′) followed by ϵ (C3′–O3′) and then δ (C4′–C3′). It is important to note that many of these torsional angle variations are correlated.[57–59]

When the P–O-3′ conformation is g^-, invariably the C–O-3′ conformation (ϵ) is found to be t. This ϵ (t), ζ (g^-) conformation is the most common backbone conformation. In this B_I (t,g) conformation, δ can vary considerably.[57,58] As δ increases to its maximum limit of about 160°, in order to relieve any additional local helical distortion, ζ now begins to vary from $-60°$ to $-120°$ while ϵ largely remains fixed at trans.[57,58] At a maximal value for δ the phosphate ester conformation can switch to the B_{II} state, thus providing a possible rationale for the different helical adjustments that appear to be responsible for the ³¹P chemical shift and coupling constant variations in different regions of the helix. It is largely this variation in δ, ϵ, and ζ that allows the sugar–phosphate backbone to "stretch" or "contract" to allow for sequence-specific variations in the local base pair geometry of B-DNA.

Conclusions

Through 2-D heteronuclear NMR or ¹⁷O-labeling experiments it is now possible to unambiguously assign the ³¹P signals of modest-sized duplex oligonucleotides. $J_{H3'-P}$ coupling constants [and hence the ϵ (C4′–C3′–O3′–P) torsional angles] can now also be measured. Correlations between experimentally measured coupling constants (and thus to P–O and C–O torsional angles) show that the sequence-specific variations in ³¹P chemical shifts are attributable to sequence-specific changes in the deoxyribose–phosphate backbone. Thus, ³¹P chemical shifts and ³¹P–¹H coupling constants serve as an important probe of the conformation and dynamics of nucleic acids, particularly the deoxyribose–phosphate backbone.

Acknowledgments

Research was supported by the National Institutes of Health (AI27744), the Purdue University Biochemical Magnetic Resonance Laboratory which is supported by the National Science Foundation National Biological Facilities Center on Biomolecular NMR, Structure and Design at Purdue (Grants BBS 8614177 and 8714258 from the Division of Biological Instrumentation), and the National AIDS Research Center at Purdue (AI727713). The contributions of Edward Nikonowicz, Vikram A. Roongta, Robert Powers, Claude R. Jones, Dean Carlson, Josepha Fu, Martial Piotto, Yesun Cho, Jill Granger, Stephen A. Schroeder, James T. Metz, Robert Powers, and Robert Santini are much appreciated.

[15] Furanose Sugar Conformations in DNA from NMR Coupling Constants

By JOHN VAN WIJK, BERNARD D. HUCKRIEDE, JOHANNES H. IPPEL, and CORNELIS ALTONA

Introduction

The flexible five-membered sugar ring plays a pivotal role in nucleic acid structure and dynamic behavior. Structural differences between A-, B-, and Z-type duplexes are intimately correlated with specific conformational ranges of individual (deoxy)riboses. Moreover, within the B-DNA family at least, each sugar responds to its surroundings (e.g., the base-stacking pattern) by an appropriate adaptation of its geometry. For this reason it is highly desirable to obtain a maximum of detailed quantitative information on the "atomic" level for each individual sugar in a DNA chain with the aid of existing experimental techniques. In this chapter we explore the information content embodied in sets of vicinal $^1H-^1H$ NMR coupling constants with the above aim in mind. Assignment procedures will only be touched on in rare cases where a peculiar behavior of the deoxyribose ring calls for caution. It is assumed that the reader is familiar with the basics of conformational analysis and NMR spectroscopy.

General Background

In current (bio)chemical literature two different D-ribose conformations are commonly distinguished: C3'-*endo* and C2'-*endo,* that is, C3' or C2' on the same side of the ring plane as C5', respectively (Scheme 1).[1,2]

[1] M. Sundaralingam, *Biopolymers* 7, 821 (1969).
[2] W. Saenger, "Principles of Nucleic Acid Structure." Springer-Verlag, Berlin, 1984.

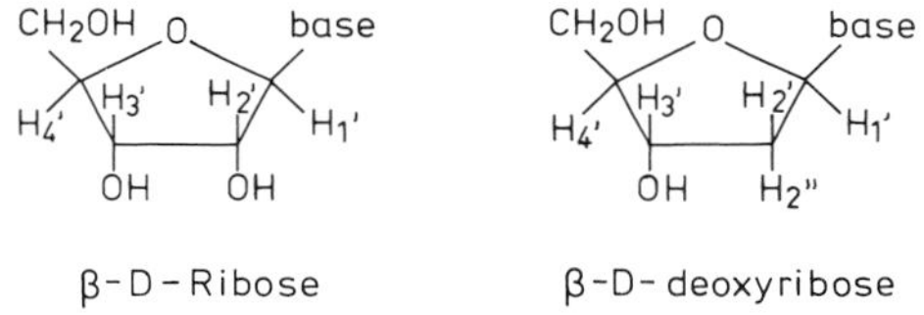

SCHEME 1. Stereochemistry of the sugar rings in RNA and DNA showing the numbering of the protons involved in vicinal spin–spin coupling.

This usage is too restrictive, however, since these terms carry a precise meaning: C3'-*endo* denotes a ^3E (envelope) conformer with a phase angle of pseudorotation[3] P of 18°, whereas C2'-*endo* stands for a ^2E form with P equal to 162°. For biochemical purposes such a notation is usually sufficient, but conformational analysis requires a more elaborate notation. It is known that actual sugar geometries in nucleic acids are determined by a complex interplay of many factors, for example, the substituent pattern of the ring, conformations adopted by the side chains, the stacking pattern of the attached nucleic acid base, and the conformational nature of the single or double helix at hand. The conformational distribution (in terms of a distribution of phase angles, see below) of riboses and deoxyriboses in nucleosides and nucleotides in the solid state is by now well documented.[2-5] X-Ray studies have shown that P values usually occur in two distinct ranges. In a conformational wheel representation[2-6] (Fig. 1), one range of forms occupies the "Northern" half of the circle, and these are conveniently denoted as belonging to the N-type family (P_N 0 ± 90°); the second range occurs in the "Southern" hemisphere, and these conformers are collected into the S-type family (P_S 180 ± 90°) (Fig. 2).

The five endocyclic torsion angles v_j of the furanose ring are mutually related on account of the simple fact that the ring must be closed. To a good approximation (0.4–0.7°) the torsion angles can be reproduced by a two-parameter pseudorotation equation:

$$v_j = \Phi_m \cos[P + 0.8\pi(j - 2)] \qquad (1)$$

for j equals 0–4. The torsion angles are numbered clockwise, starting with v_0(C4'–O4'–C1'–C2'); the amplitude of pucker is given by Φ_m. It is easily

---

[3] C. Altona and M. Sundaralingam, *J. Am. Chem. Soc.* **94**, 8205 (1972).
[4] H. P. M. de Leeuw, C. A. G. Haasnoot, and C. Altona, *Isr. J. Chem.* **20**, 108 (1980).
[5] M. Sundaralingam and P. T. Haromy, *in* "Landolt–Börnstein New Series VII/1a" (W. Saenger, ed.), p. 22. Springer-Verlag, Berlin and New York, 1989.
[6] IUPAC–IUB Joint Commission on Biochemical Nomenclature (JCBN), *Eur. J. Biochem.* **131**, 9 (1983).

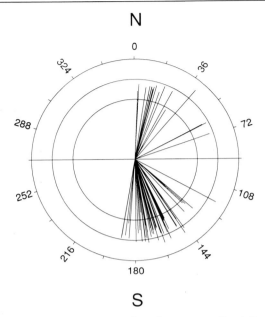

FIG. 1. Pseudorotational wheel representation of parameters P and Φ_m (in degrees) of 51 deoxyribonucleosides and deoxyribonucleotides, studied by single-crystal X-ray diffraction. The angle of each vector with the vertical axis represents the pseudorotation phase angle P, and the length of each vector represents the puckering amplitude Φ_m [Eq. (1)]. Raw data were taken from the Cambridge Crystallographic Data Base [F. B. Allen, O. Kennard, and R. Taylor, *Acc. Chem. Res.* **16,** 146 (1983)] and analyzed by us [B. D. Huckriede and C. Altona, unpublished results (1991)].

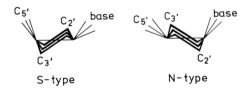

FIG. 2. Schematic representation of some of the possible puckering modes of the furanose ring in nucleic acids. The molecules are viewed through the C2'–C3' bond in the direction of ring oxygen O4'. The central C2'–C3' bars indicate the respective symmetrical twist conformations $\frac{2}{3}T$ (S) and $\frac{3}{2}T$ (N). In nucleic acids the upper bars give a better representation of the common situation.

deduced that N-type conformers are characterized by a positive sign of $v_3(C1'-C2'-C3'-C4')$, whereas in S-type forms v_3 has a negative sign. More accurate expressions for v_j (as well as for the endocyclic bond angles τ_j) are available.[7,8] However, for our present purposes the accuracy provided by Eq. (1) suffices.

In crystal structures of nucleosides and nucleotides usually a single pure N- or S-type conformer is found, but not necessarily the one that predominates in aqueous solution. In some cases both N and S forms reside side by side in the same unit cell. Statistical analyses[4,5] of X-ray data make it clear that fine details of sugar geometry (P and Φ_m) of monomers are influenced by anisotropic crystal packing forces. The situation appears to be different in helical oligomers, where stacking forces may play a more predominant role. However, X-ray studies cannot provide answers to questions concerning dynamic behavior, namely, what is the energy difference — or equilibrium composition — in cases where N- and S-type conformers are interconverting in solution, and what is the effect of an increase in temperature? NMR spectroscopy is eminently suited to supply answers to these questions.

In recent years much use has been made of NMR cross-relaxation rates R measured by means of one- or two-dimensional nuclear Overhauser enhancement (NOE) spectroscopy as the initial NOE buildup rate. R values can be translated (with more or less sophistication) into H–H distances. Unfortunately, it is not often recognized that the very nature of the pseudorotation phenomenon, in combination with conformational heterogeneity, virtually precludes an unambiguous and precise determination of sugar geometry from these distances. For a thorough review of the problems encountered, we refer to van de Ven and Hilbers,[9] and figures in that review[9] explain the difficulties well. All but two of the intrasugar H–H distances vary less than 0.25 Å relative to their average in going from the N to the S region of sugar conformational space and thus are useless. The H1'–H4' distance varies by more than 1 Å, but the minimum distance lies in the O4'-*endo* (P 90°) region; thus, this distance cannot discriminate between N and S forms. Only the H2"–H4' distance serves as a marker as it changes from approximately 2.8 Å in the N region to about 3.6–4.0 Å in the S conformational space.

In addition, one may take NOEs involving base–sugar H–H distances into account. It is then indeed possible to formulate general rules by which

[7] E. Diez, A. Esteban, F. J. Bermejo, C. Altona, and F. A. A. M. de Leeuw, *J. Mol. Struct.* **125**, 49 (1984).
[8] F. A. A. M. de Leeuw, P. N. van Kampen, C. Altona, E. Diez, and A. L. Esteban, *J. Mol. Struct.* **125**, 67 (1984).
[9] F. J. M. van de Ven and C. W. Hilbers, *Eur. J. Biochem.* **178**, 1 (1988).

A-, B-, and Z-DNA can be discriminated, at least qualitatively.[10-12] However, even when three distance constraints (H2'–H6/H8, H2''–H6/H8, H3'–H6/H8) are introduced simultaneously in a B-DNA model, one finds[9] that the glycosyl torsion angle χ is well defined ($\pm 10°$) but that the sugar geometry is not; the three constraints are satisfied over the range P_S 110–200°, Φ_S fixed at 35°, and mole fraction X_S fixed at 1.0 (pure S). The problem is further complicated by the fact that in an intact B-DNA double helix Φ_S may range from 28° to 47° and X_S from 0.67 to 1.0. The S–N interconversion barrier is probably low[13] (8–13 kJ mol^{-1}) and as a result the lifetimes of the S and N conformational species are less than 1 nsec. This means a $1/r^3$ averaging[14] of the cross-relaxation rates instead of the usual $1/r^6$. The conclusion must be that NOE data should be supplemented with other experimental facts before a reliable picture of the behavior of an individual deoxyribose ring in a DNA duplex can be obtained.

In this chapter we propose to show that a careful analysis of spin–spin splitting patterns of deoxyribose protons provides the following information: (1) the conformational equilibrium constant, usually expressed as the mole fraction X_S; (2) in cases where both interconverting conformers contribute significantly ($\sim 50/50$) to the equilibrium composition, the pseudorotation parameters P_N and P_S; and (3) in cases where a major conformation ($> 70\%$) exists, usually the S-type form, the parameters P_S and Φ_S. The geometrical properties of the minor N form are then less well defined and can be replaced safely by ensemble average properties of the N-type family. The reverse situation obviously applies to the dG residues in Z-DNA, where the N form predominates.[12]

Summarizing, the complete conformational analysis of a given interconverting furanose in the ideal case involves the determination of five independent parameters: P_S, Φ_S, P_N, Φ_N, and X_S. This is done by the creation of a simultaneous and consistent fit of a set of experimental coupling constants (or sums of couplings) to predicted values with the aid of a graphical method or a computer least-squares fit. Use is made of the standard sum rule for a time-average physical property:

$$J_{obs} = X_S J_S + X_N J_N \qquad (2)$$

[10] K. Wüthrich, "NMR of Proteins and Nucleic Acids." Wiley, New York, 1986.
[11] C. A. G. Haasnoot, H. P. Westerink, G. A. van der Marel, and J. H. van Boom, *J. Biomol. Struct. Dyn.* **2**, 345 (1984).
[12] L. P. M. Orbons, G. A. van der Marel, J. H. van Boom, and C. Altona, *Eur. J. Biochem.* **160**, 131 (1986).
[13] C. Altona, *Recl. Trav. Chim. Pays-Bas* **101**, 413 (1982).
[14] G. Lipari and A. Szabo, *J. Am. Chem. Soc.* **104**, 4546 (1982).

In Eq. (2), $X_S + X_N = 1$; J_{obs} stands for any observed coupling or sum of couplings; J_N and J_S are the predicted limiting couplings, which are known functions of the P and Φ parameters. In the following sections these functions are described in some detail.

Geometrical Relation between Proton–Proton Torsion Angles and Pseudorotational State

Endocyclic furanose bond angles involving carbon as the central atom usually vary between 99° and 108°. From the known mathematical relation between valency angles and projected valency angles[15] (not shown), it follows that such large deviations from the tetrahedral value cause changes of up to 6° from the idealized (120°) projection symmetry of a normal Newman projection. As such large deviations are not acceptable, recourse was taken to empirical correlations determined from 178 crystal structures of ribo- and deoxyribonucleotides, in conjunction with neutron diffraction studies.[4,16,17]

The general correlation between Φ_{HH} torsion angles on the one hand and the pseudorotation parameters Φ_m and P on the other hand is given by

$$\Phi_{HH} = B + A\Phi_m \cos(P + \text{phase}) \tag{3}$$

The A, B, and phase values deduced for various β-D-furanose rings are collected in Table I. It should be kept in mind that the A and B parameters have been established in a direct fashion only for ribose and deoxyribose nucleosides and nucleotides. For the remaining sugars, arabinose, lyxose, and xylose,[17] the appropriate B parameter was simply shifted by 120°, because the required number of X-ray studies was unavailable.

For any sugar geometry, expressed in terms of P and Φ_m, all relevant Φ_{HH} values now can be deduced. In a similar vein the backbone angle $\delta(O5'-C5'-C4'-C3')$ in β-D-ribosides is correlated with P and Φ_m according to

$$\delta = 120.6 + 1.1\Phi_m \cos(P + 145.2) \tag{4}$$

The constants in Eq. (4) are based on an extended pseudorotation equation[8] and were derived from a least-squares fit to a large crystal structure data set.[4]

[15] J. B. Hendrickson, *J. Am. Chem. Soc.* **83,** 4537 (1961).
[16] C. A. G. Haasnoot, F. A. A. M. de Leeuw, H. P. M. de Leeuw, and C. Altona, *Org. Magn. Reson.* **15,** 43 (1981).
[17] F. A. A. M. de Leeuw and C. Altona, *J. Chem. Soc., Perkin Trans.* **2,** 375 (1982).

TABLE I
A, B, AND PHASE PARAMETERS[a]

Φ_{HH}	Deoxyribose		Ribose		Arabinose		Lyxose		Xylose		Phase
	B	A	B	A	B	A	B	A	B	A	
$\Phi_{1'2'}$	121.4	1.03	123.3	1.102	3.3	1.10	3.3	1.10	123.3	1.10	−144
$\Phi_{1'2''}$	0.9	1.02									−144
$\Phi_{2'3'}$	2.4	1.06	0.2	1.090	120.2	1.09	0.2	1.09	−119.8	1.09	0
$\Phi_{2''3'}$	122.9	1.06									0
$\Phi_{3'4'}$	−124.0	1.09	−124.9	1.095	−124.9	1.10	−4.9	1.10	−4.9	1.10	144

[a] For use in Eq. (3), see text. These parameters are valid for β-D configurations. In the case of the α-D anomers one subtracts 120° from the $\Phi_{1'2'}$ ($\Phi_{1'2''}$) B values. Mirror images (L series) are obtained by reversing the signs of both A and B. A, B, and phase values are given in degrees.

Generalized Karplus-Type Equation

In a number of papers, published by the Leiden[18] group since 1980, extensive use was made of the "Haasnoot" equation[19] (not shown). This equation accounts both for the electronegativity and orientation of substituents attached to the $H_A - C - C - H_B$ fragment in question. In the course of time, some drawbacks of this particular formalism came to light. Most problematic appeared to be its poor reproduction of certain measured small (< 1.2 Hz) coupling values at torsion angles close to 90°. Recent improvements[20-22] include (1) the development of an empirical (group) electronegativity (λ^e) scale, based on accurate (±0.02 Hz) rotationally averaged vicinal coupling constants of molecules of the type CH_3CH_2X and CH_3CHXY[21]; and (2) the introduction of cross-terms to account for the observed nonadditivity of substituent effects.[22]

The "best" equation and parameters[20] are given as Eqs. (5) and (6):

$$^3J_{HH} = C_0 + C_1 \cos \Phi + C_2 \cos 2\Phi + C_3 \cos 3\Phi + S_2 \sin 2\Phi \qquad (5)$$

[18] See references cited in L. J. Rinkel and C. Altona, *J. Biomol. Struct. Dyn.* **4**, 621 (1987).

[19] C. A. G. Haasnoot, F. A. A. M. de Leeuw, and C. Altona, *Tetrahedron* **36**, 2783 (1980).

[20] L. A. Donders, Ph.D. Thesis, Leiden, The Netherlands (1989); parameters later slightly modified: J. van Wijk and C. Altona, unpublished results (1990).

[21] C. Altona, J. H. Ippel, A. J. A. Westra Hoekzema, C. Erkelens, M. Groesbeek, and L. A. Donders, *Magn. Reson. Chem.* **27**, 564 (1989).

[22] L. A. Donders, F. A. A. M. de Leeuw, and C. Altona, *Magn. Reson. Chem.* **27**, 566 (1989).

with

$$C_0 = 6.97 - 0.58 \; \Sigma_i \, \lambda_i - 0.24(\lambda_1\lambda_2 + \lambda_3\lambda_4) \tag{6a}$$

$$C_1 = -1.06 \tag{6b}$$

$$C_2 = 6.55 - 0.82 \; \Sigma_i \, \lambda_i + 0.20(\lambda_1\lambda_4 + \lambda_2\lambda_3) \tag{6c}$$

$$C_3 = -0.54 \tag{6d}$$

$$S_2 = 0.68 \; \Sigma_i \, \xi_i\lambda_i^2 \tag{6e}$$

The substituent numbering is shown in Fig. 3. The orientation of substituent i has an effect in the asymmetrical part of Eq. (5) (sin 2Φ term) and is incorporated by a sign factor ξ_i, which stands for $+1$ or -1 according to the definition given by Haasnoot *et al.*[19] (see Fig. 3). At this point it should be stressed that this sign factor depends only on the orientation of a given substituent S_i with respect to its geminal coupling proton, seen along the coupling path. The "sign" is introduced solely for reasons of convenience.

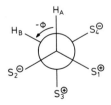

FIG. 3. Definition of substituent orientation factors ξ_i ($+1$ or -1) for use in the sin 2Φ part of the extended Karplus equation [Eq. (6e)]. Note that the orientation factor of a substituent S is defined with respect to the coupling hydrogen located on the same carbon atom and thus remains invariant on Φ rotation. The sign of the Φ angle is defined according to IUPAC–IUB rules [IUPAC–IUB Joint Commission on Biochemical Nomenclature (JCBN), *Eur. J. Biochem.* **131**, 9 (1983); see also C. A. G. Haasnoot, F. A. A. M. de Leeuw, H. P. M. de Leeuw, and C. Altona, *Org. Magn. Reson.* **15**, 43 (1981), and C. A. G. Haasnoot, F. A. A. M. de Leeuw, and C. Altona, *Tetrahedron* **36**, 2783 (1980)].

TABLE II
EXPERIMENTAL "GROUP ELECTRONEGATIVITY"
VALUES [a]

Group	$\lambda(D_2O)$	$\lambda(CDCl_3)$	$\lambda(DMSO)$
Adenine	0.57	0.57	0.61
Guanine[b]	0.57	0.57	0.61
Uracil	0.57	0.57	0.56
Thymine	0.57	0.57	0.56
Cytosine	0.58	0.59	0.57
OH	1.26	1.33	1.37
OR	1.27	1.40	1.40
CH_2OR	c	0.68	c
NH_2	1.11	1.19	1.23
Cl	c	0.94	0.99
F	c	1.37	c

[a] For use in Eq. (5), measured in three different solvents. From R. Francke, J. H. Ippel, and C. Altona, unpublished results (1990).

[b] λ values inferred from the close analogy between the couplings to methyl in 9-ethyladenine and 9-ethylguanine.

[c] λ values not measured in the solvent specified.

Its use can be avoided by the more cumbersome introduction of a second torsion angle Ψ.[23]

The original λ values[21] for use in Eq. (5) were deduced from NMR spectra taken under standard solvent conditions [3–5% (w/v) solutions in $CDCl_3$]. More recently we found that λ values of potential hydrogen bond accepting groups, notably —OR and —NR_2, are affected by hydrogen bond donating solvents, namely, H_2O and D_2O. The λ values of the common nitrogen bases are not so affected. Table II lists the λ values of substituents that are relevant to the present work, measured in three different solvents.[24] From Eqs. (5) and (6) it follows that a change in λ of 0.1 unit is accompanied by a change in J_{calc} of maximally 0.14 Hz in the case of a trans coupling (Φ_{HH} 180°). Equations (3), (5), and (6) and the parameters given in Tables I and II are incorporated in version 5.4 of our least-squares minimization program PSEUROT.

[23] K. G. R. Pachler, *J. Chem. Soc, Perkin Trans.* 2, 1936 (1972).

[24] R. Francke, J. H. Ippel, and C. Altona, unpublished results (1990).

Measurement of Vicinal Couplings and Sums of Couplings

The first step in the conformational analysis of a DNA fragment is the NMR assignment procedure. At present, assignments are usually based on NOESY (NOE spectroscopy) spectra and further aided by COSY (correlation spectroscopy) or HOHAHA (homonuclear Hartmann–Hahn) techniques.[25] A study of the one-dimensional NMR spectrum at maximum resolution (500–600 MHz) should not be forsaken, however, since such a spectrum yields maximum accuracy for nonoverlapping signals.

In the following discussion it is assumed that the ^{31}P nucleus has been decoupled, symbolized by $\{^{31}$P$\}$. Now one may ask under what conditions it is safe to apply first-order spectral analysis. An answer to this question is supplied by the study of computer-simulated spectra.[26] Given Eqs. (3), (5), and (6), the limiting couplings can be accurately predicted for any chosen furanose conformer, characterized by P, Φ_m. Via Eq. (2) these limiting couplings can be utilized to computer-simulate the splitting pattern of H1′, H2′, H2″, H3′$\{^{31}$P$\}$, and H4′ resonances for any chosen conformational mixture of N- and S-type forms, given a set of chemical shift values. Of course, the H4′ resonance has extra splittings because of the exocyclic H5′/H5″ protons and therefore will be left out of the present discussion.

Sums of couplings carry the same information content and are often easier to determine than individual couplings. We define

$$\Sigma\ 1' = J_{1'2'} + J_{1'2''}$$

$$\Sigma\ 2' = J_{1'2'} + J_{2'3'} + J_{2'2''}$$

$$\Sigma\ 2'' = J_{1'2''} + J_{2''3'} + J_{2'2''}$$

$$\Sigma\ 3'\{^{31}\text{P}\} = J_{2'3'} + J_{2''3'} + J_{3'4'}$$

Note that $\Sigma\ 2'$ and $\Sigma\ 2''$ include the absolute value of the geminal coupling $J_{2'2''}$, set at 14.0 Hz in the present work. The sums correspond to the distance in hertz measured between the outer peaks of the H1′, H2′, H2″, and H3′$\{^{31}$P$\}$ resonance patterns. For $\Sigma\ 2'$ and $\Sigma\ 2''$, the above statement is only strictly true when first-order conditions apply. However, sets of computer-calculated spectra,[26] computed for different values of the chemical shift difference $\Delta\delta2'2''$, revealed that first-order measurements of $J_{1'2'}$, $J_{1'2''}$, $\Sigma\ 2'$, and $\Sigma\ 2''$ can be trusted under condition that $\Delta\delta2'2''$ exceeds 40 Hz. This means that the nearest lines of the H2′ and the H2″ resonance

[25] R. R. Ernst, G. Bodenhausen, and A. Wokaun, "Principles of Nuclear Magnetic Resonance in One and Two Dimensions." Clarendon, Oxford, 1987; A. Bax and D. G. Davis, *J. Magn. Reson.* **65**, 335 (1985); A. Bax, this series, Vol. 176, p. 151.

[26] L. J. Rinkel and C. Altona, *J. Biomol. Struct. Dyn.* **4**, 621 (1987).

multiplets are separated from each other by at least 15 Hz. Significant discrepancies (>0.2 Hz) occur when $\Delta\delta2'2''$ lies between 40 and 4 Hz, and recourse must be taken to a computer simulation of the spectrum. The sums $\Sigma\ 1'$ and $\Sigma\ 3'$ remain independent of $\Delta\delta2'2''$. For (near-)isochronous H2′/H2″ signals ($\Delta\delta2'2'' <4$ Hz) "deceptively simple" splitting patterns are seen: a triplet for the H1′ resonance and a quartet for H2′/H2″. Nevertheless, these simple patterns still contain useful information, as is explained in the next section.

We conclude the present section by stressing the point that conformational conclusions, based on splitting patterns, are valid only when the spectra have been interpreted correctly, that is, when first-order conditions apply or when appropriate computer simulations of one-, two-, or three-dimensional spectra have been carried out whenever necessary.

Geometry of Deoxyriboses in Solutions: A Graphical Method

In this section we discuss the influence of the geometry (P, Φ_m) and of the relative populations (X_S) on the calculated sums of couplings and on $J_{1'2'}$ and $J_{1'2''}$. Knowledge of the latter individual J's allows a more refined analysis, and these values can often be determined with little effort, even from spectra of relatively large DNA helices. Table III lists sets of calculated couplings for sugar conformations occurring along the complete pseudorotational cycle (Φ_m 28°, 36°, 44°) (see also Fig. 4). In Table IV calculated sums of couplings are collected for Φ_m 28°, 36°, and 44°; these values represent the range observed so far in NMR studies of deoxyribose rings in DNA duplexes. Finally, in Table V we present calculated sums of couplings (P_N 18°, P_S 108°, 153°, and 180°; $\Phi_N = \Phi_S$ 36°) for equilibrium mixtures N–S (see also Fig. 5). Graphs of the type shown in Fig. 5 are easily prepared from the information collected in Tables III and IV and are extremely helpful in the simultaneous analysis of the coupling constants.

Figure 6 shows computer-generated splitting patterns for H1′, H2′, H2″, and H3′ resonances of hypothetical deoxyribofuranoses in nucleosides and nucleotides engaged in a progressively changing conformational equilibrium (X_S 1.0–0.5). Two extreme situations were selected in order to illustrate the fact that visual inspection may quickly lead to interesting conclusions. It is seen that the H1′ and H2″ signals are mainly sensitive to the molar composition. In contrast, the H2′ and H3′{^{31}P} patterns clearly reflect the pseudorotational state of the major (S-type) conformer; differences between a "low-S" ring (P_S 108°) and a "high-S" ring (P_s 180°) become apparent in the H3′ resonance at $X_S \geq 0.6$ and in the H2′ signal at $X_S \geq 0.8$.

Let us assume that one has obtained an experimental set of couplings of

TABLE III

CALCULATED COUPLING CONSTANTS FOR RANGE OF SUGAR GEOMETRIES OF DEOXYRIBOSE RING IN NUCLEOSIDES AND NUCLEOTIDES, ENCOUNTERED ALONG PSEUDOROTATION ITINERARY[a-c]

	$J_{1'2'}$ for Φ_m			$J_{1'2''}$ for Φ_m			$J_{2'3'}$ for Φ_m			$J_{2''3'}$ for Φ_m			$J_{3'4'}$ for Φ_m		
(°)	28°	36°	44°	28°	36°	44°	28°	36°	44°	28°	36°	44°	28°	36°	44°
0	1.3	0.8	0.7	7.9	7.5	6.8	7.5	6.6	5.5	8.6	10.0	11.0	7.0	8.0	8.7
9	1.5	1.1	0.8	8.1	7.7	7.3	7.5	6.6	5.6	8.5	9.9	10.9	7.4	8.3	9.0
18	1.9	1.4	1.0	8.2	8.0	7.7	7.6	6.8	5.8	8.3	9.7	10.8	7.6	8.6	9.2
27	2.4	1.9	1.5	8.3	8.2	8.1	7.7	7.1	6.2	8.0	9.3	10.4	7.8	8.7	9.3
36	3.0	2.6	2.2	8.4	8.3	8.3	7.9	7.4	6.6	7.5	8.8	9.9	7.8	8.8	9.3
45	3.7	3.4	3.2	8.3	8.3	8.3	8.1	7.7	7.2	6.9	8.1	9.2	7.8	8.7	9.3
54	4.4	4.4	4.4	8.2	8.2	8.2	8.3	8.0	7.7	6.2	7.2	8.2	7.6	8.6	9.2
63	5.3	5.5	5.7	8.0	8.0	7.9	8.4	8.3	8.1	5.4	6.2	7.0	7.4	8.3	9.0
72	6.1	6.6	7.1	7.8	7.6	7.4	8.5	8.5	8.4	4.6	5.1	5.6	7.0	8.0	8.7
81	6.9	7.6	8.3	7.5	7.2	6.8	8.5	8.5	8.5	3.8	4.0	4.2	6.5	7.4	8.2
90	7.6	8.5	9.3	7.2	6.7	6.2	8.4	8.4	8.4	3.0	3.0	3.0	6.0	6.8	7.5
99	8.2	9.2	10.1	6.9	6.3	5.6	8.2	8.2	8.1	2.3	2.1	1.9	5.3	6.0	6.6
08	8.7	9.8	10.6	6.6	5.8	5.0	8.0	7.8	7.6	1.7	1.4	1.2	4.6	5.0	5.5
17	9.1	10.2	11.0	6.3	5.5	4.6	7.7	7.4	7.0	1.3	1.0	0.8	3.9	4.1	4.3
26	9.4	10.5	11.2	6.1	5.2	4.2	7.4	6.9	6.4	1.0	0.7	0.6	3.2	3.2	3.2
35	9.6	10.6	11.3	6.0	5.1	4.0	7.0	6.4	5.7	0.8	0.6	0.7	2.5	2.3	2.2
44	9.6	10.7	11.3	6.0	5.0	4.0	6.7	6.0	5.1	0.7	0.6	0.9	1.9	1.6	1.4
53	9.6	10.6	11.3	6.0	5.1	4.0	6.5	5.6	4.6	0.6	0.7	1.1	1.5	1.1	0.9
62	9.4	10.5	11.2	6.1	5.2	4.2	6.3	5.3	4.3	0.6	0.8	1.4	1.1	0.8	0.6
71	9.1	10.2	11.0	6.3	5.5	4.6	6.1	5.1	4.1	0.6	0.9	1.5	0.9	0.6	0.5
80	8.7	9.8	10.6	6.6	5.8	5.0	6.1	5.1	4.0	0.6	0.9	1.6	0.7	0.5	0.6
89	8.2	9.2	10.1	6.9	6.3	5.6	6.1	5.1	4.1	0.6	0.9	1.5	0.6	0.5	0.7
98	7.6	8.5	9.3	7.2	6.7	6.2	6.3	5.3	4.3	0.6	0.8	1.4	0.6	0.5	0.8
07	6.9	7.6	8.3	7.5	7.2	6.8	6.5	5.6	4.6	0.6	0.7	1.1	0.6	0.6	0.9
16	6.1	6.6	7.1	7.8	7.6	7.4	6.7	6.0	5.1	0.7	0.6	0.9	0.5	0.6	0.9
25	5.3	5.5	5.7	8.0	8.0	7.9	7.0	6.4	5.7	0.8	0.6	0.7	0.6	0.6	0.9
34	4.4	4.4	4.4	8.2	8.2	8.2	7.4	6.9	6.4	1.0	0.7	0.6	0.6	0.5	0.8
43	3.7	3.4	3.2	8.3	8.3	8.3	7.7	7.4	7.0	1.3	1.0	0.8	0.6	0.5	0.7
52	3.0	2.6	2.2	8.4	8.3	8.3	8.0	7.8	7.6	1.7	1.4	1.2	0.7	0.5	0.6
61	2.4	1.9	1.5	8.3	8.2	8.1	8.2	8.2	8.1	2.3	2.1	1.9	0.8	0.6	0.5
70	1.9	1.4	1.0	8.2	8.0	7.7	8.4	8.4	8.4	3.0	3.0	3.0	1.0	0.8	0.6
79	1.5	1.1	0.8	8.1	7.7	7.3	8.5	8.5	8.5	3.8	4.0	4.2	1.5	1.1	0.9
88	1.3	0.8	0.7	7.9	7.5	6.8	8.5	8.5	8.4	4.6	5.1	5.6	1.9	1.6	1.4
97	1.1	0.7	0.7	7.8	7.2	6.4	8.4	8.3	8.1	5.4	6.2	7.0	2.5	2.3	2.2
06	1.0	0.7	0.7	7.6	7.0	6.1	8.3	8.0	7.7	6.2	7.2	8.2	3.2	3.2	3.2
15	0.9	0.7	0.8	7.6	6.8	5.9	8.1	7.7	7.2	6.9	8.1	9.2	3.9	4.1	4.3
24	0.9	0.7	0.8	7.5	6.8	5.8	7.9	7.4	6.6	7.5	8.8	9.9	4.6	5.0	5.5
33	0.9	0.7	0.8	7.6	6.8	5.9	7.7	7.1	6.2	8.0	9.3	10.4	5.3	6.0	6.6
42	1.0	0.7	0.7	7.6	7.0	6.1	7.6	6.8	5.8	8.3	9.7	10.8	6.0	6.8	7.5
51	1.1	0.7	0.7	7.8	7.2	6.4	7.5	6.6	5.6	8.5	9.9	10.9	6.5	7.4	8.2

[a] From Eqs. (3), (5), and (6).
[b] Amplitudes of pucker Φ_m were selected to cover the range thus far observed.
[c] Calculated by the authors.

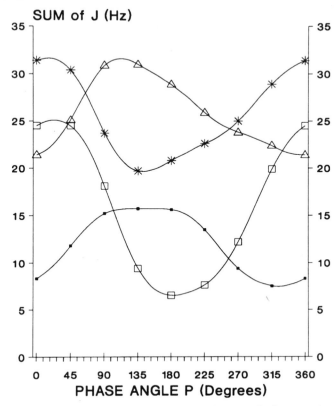

FIG. 4. Sums of calculated coupling constants (Hz) in deoxyribofuranose rings in nucleosides and nucleotides plotted versus the phase angle of pseudorotation P (°); Φ_m 36°. See also Table IV. $\Sigma\ 1'$ (■), $\Sigma\ 2'$ (△), $\Sigma\ 2''$ (*), $\Sigma\ 3'$ (□).

a given nucleotide residue in a DNA molecule, be it single strand, double helix, hairpin loop, or cruciform. The analysis of such a set should be worked out in successive stages. At the outset of the description of these stages we wish to remind the reader of the fact that the geometry of a minor conformer cannot be determined with precision. For this reason it is proper to introduce one or more constraints in the calculations. For example, an analysis[27] of 14 published X-ray structures of N-type deoxyriboses gave P_N(av) equal to 22°. The total collection of X-ray data (65 examples) revealed no significant difference between Φ_N(av) and Φ_S(av); we find

[27] B. D. Huckriede and C. Altona, unpublished results (1990).

TABLE IV

CALCULATED SUMS OF COUPLING CONSTANTS FOR RANGE OF SUGAR GEOMETRIES OF DEOXYRIBOSE RING IN NUCLEOSIDES AND NUCLEOTIDES, ENCOUNTERED ALONG PSEUDOROTATION ITINERARY[a-c]

| | Σ 1' for Φ_m | | | Σ 2' for Φ_m | | | Σ 2" for Φ_m | | | Σ 3' for Φ_m | | |
P (°)	28°	36°	44°	28°	36°	44°	28°	36°	44°	28°	36°	44°
0	9.2	8.3	7.5	22.7	21.4	20.2	30.5	31.4	31.8	23.1	24.5	25.3
9	9.6	8.8	8.0	23.0	21.7	20.4	30.6	31.6	32.2	23.4	24.9	25.6
18	10.1	9.4	8.7	23.5	22.2	20.9	30.5	31.7	32.5	23.5	25.1	25.8
27	10.7	10.1	9.6	24.1	23.0	21.7	30.3	31.5	32.5	23.5	25.1	25.9
36	11.3	10.9	10.5	24.9	24.0	22.9	29.9	31.1	32.2	23.3	24.9	25.9
45	12.0	11.8	11.6	25.8	25.1	24.4	29.2	30.4	31.5	22.8	24.5	25.6
54	12.7	12.7	12.7	26.7	26.5	26.1	28.4	29.4	30.4	22.1	23.8	25.1
63	13.3	13.5	13.6	27.7	27.8	27.8	27.5	28.2	28.9	21.2	22.8	24.1
72	13.9	14.2	14.5	28.6	29.0	29.5	26.4	26.7	27.0	20.1	21.5	22.7
81	14.4	14.8	15.1	29.4	30.1	30.8	25.3	25.2	25.1	18.8	19.9	21.0
90	14.8	15.2	15.5	30.0	30.9	31.7	24.2	23.7	23.2	17.4	18.1	18.9
99	15.1	15.5	15.7	30.5	31.4	32.2	23.1	22.4	21.5	15.8	16.2	16.6
108	15.3	15.6	15.7	30.7	31.6	32.2	22.3	21.3	20.2	14.3	14.3	14.3
117	15.5	15.7	15.6	30.8	31.6	32.0	21.6	20.5	19.3	12.8	12.4	12.1
126	15.5	15.7	15.4	30.8	31.4	31.5	21.1	19.9	18.9	11.5	10.8	10.1
135	15.6	15.7	15.3	30.6	31.0	31.0	20.8	19.7	18.7	10.3	9.4	8.6
144	15.6	15.7	15.3	30.4	30.6	30.4	20.6	19.6	18.9	9.3	8.3	7.4
153	15.6	15.7	15.3	30.0	30.2	29.9	20.6	19.8	19.2	8.6	7.4	6.7
162	15.5	15.7	15.4	29.7	29.8	29.5	20.8	20.0	19.6	8.0	6.9	6.2
171	15.5	15.7	15.6	29.3	29.3	29.0	20.9	20.4	20.1	7.6	6.6	6.1
180	15.3	15.6	15.7	28.8	28.9	28.6	21.2	20.8	20.6	7.5	6.5	6.1
189	15.1	15.5	15.7	28.4	28.3	28.1	21.5	21.2	21.1	7.4	6.5	6.3
198	14.8	15.2	15.5	27.9	27.8	27.6	21.8	21.5	21.6	7.5	6.7	6.4
207	14.4	14.8	15.1	27.4	27.2	26.9	22.1	21.9	22.0	7.6	6.9	6.7
216	13.9	14.2	14.5	26.8	26.5	26.2	22.5	22.3	22.3	7.9	7.2	6.9
225	13.3	13.5	13.6	26.3	25.9	25.5	22.8	22.6	22.6	8.4	7.6	7.3
234	12.7	12.7	12.7	25.8	25.3	24.8	23.2	22.9	22.8	8.9	8.1	7.8
243	12.0	11.8	11.6	25.3	24.8	24.3	23.6	23.3	23.1	9.6	8.9	8.4
252	11.3	10.9	10.5	24.9	24.4	23.9	24.1	23.8	23.5	10.4	9.8	9.4
261	10.7	10.1	9.6	24.6	24.1	23.6	24.6	24.3	24.0	11.4	10.9	10.5
270	10.1	9.4	8.7	24.3	23.8	23.4	25.2	25.0	24.7	12.5	12.2	12.0
279	9.6	8.8	8.0	24.0	23.6	23.3	25.8	25.7	25.5	13.7	13.6	13.6
288	9.2	8.3	7.5	23.8	23.3	23.1	26.5	26.5	26.4	15.0	15.2	15.4
297	8.9	7.9	7.1	23.5	23.0	22.8	27.2	27.4	27.4	16.4	16.8	17.2
306	8.6	7.7	6.8	23.3	22.7	22.4	27.9	28.2	28.3	17.7	18.4	19.0
315	8.5	7.5	6.7	23.0	22.4	21.9	28.5	28.9	29.1	18.9	19.9	20.6
324	8.4	7.4	6.6	22.8	22.0	21.4	29.1	29.6	29.8	20.0	21.2	22.0
333	8.5	7.5	6.7	22.7	21.7	20.9	29.6	30.2	30.3	21.0	22.3	23.2
342	8.6	7.7	6.8	22.6	21.5	20.5	30.0	30.7	30.9	21.9	23.3	24.1
351	8.9	7.9	7.1	22.6	21.4	20.3	30.3	31.1	31.4	22.6	24.0	24.8

[a] See footnotes to Table III.

[b] Round-off errors occasionally may cause a 0.1 Hz discrepancy with data calculated from Table III.

[c] Σ 2' and Σ 2" incorporate a value of $|J_{2'2''}| = 14.0$ Hz.

TABLE V

CALCULATED SUMS OF COUPLINGS FOR VARIOUS CONFORMATIONAL EQUILIBRIUM COMPOSITIONS AND S-TYPE GEOMETRIES[a,b]

X_S^c	P_S 108°				P_S 153°				P_S 180°			
	Σ 1'	Σ 2'	Σ 2''	Σ 3'	Σ 1'	Σ 2'	Σ 2''	Σ 3'	Σ 1'	Σ 2'	Σ 2''	Σ .
0	9.4	22.2	31.7	25.1	9.4	22.2	31.7	25.1	9.4	22.2	31.7	25
5	9.7	22.7	31.2	24.6	9.7	22.6	31.1	24.2	9.7	22.5	31.1	24
10	10.0	23.1	30.7	24.0	10.0	23.0	30.5	23.3	10.0	22.9	30.6	23
15	10.3	23.6	30.1	23.5	10.3	23.4	29.9	22.4	10.3	23.2	30.1	22
20	10.6	24.1	29.6	22.9	10.7	23.8	29.3	21.6	10.6	23.5	29.5	21
25	11.0	24.6	29.1	22.4	11.0	24.2	28.7	20.7	11.0	23.9	29.0	20
30	11.3	25.0	28.6	21.8	11.3	24.6	28.1	19.8	11.3	24.2	28.4	19
35	11.6	25.5	28.0	21.3	11.6	25.0	27.5	18.9	11.6	24.5	27.9	18
40	11.9	26.0	27.5	20.8	11.9	25.4	26.9	18.0	11.9	24.9	27.3	17
45	12.2	26.4	27.0	20.2	12.2	25.8	26.3	17.2	12.2	25.2	26.8	16
50	12.5	26.9	26.5	19.7	12.5	26.2	25.7	16.3	12.5	25.5	26.2	15
55	12.8	27.4	26.0	19.1	12.9	26.6	25.1	15.4	12.8	25.9	25.7	14
60	13.1	27.8	25.4	18.6	13.2	27.0	24.5	14.5	13.1	26.2	25.1	13
65	13.5	28.3	24.9	18.1	13.5	27.4	23.9	13.6	13.5	26.5	24.6	13
70	13.8	28.8	24.4	17.5	13.8	27.8	23.4	12.7	13.8	26.9	24.0	12
75	14.1	29.3	23.9	17.0	14.1	28.2	22.8	11.9	14.1	27.2	23.5	11
80	14.4	29.7	23.4	16.4	14.4	28.6	22.2	11.0	14.4	27.5	22.9	10
85	14.7	30.2	22.8	15.9	14.7	29.0	21.6	10.1	14.7	27.9	22.4	9
90	15.0	30.7	22.3	15.4	15.0	29.4	21.0	9.2	15.0	28.2	21.9	8
95	15.3	31.1	21.8	14.8	15.4	29.8	20.4	8.3	15.3	28.5	21.3	7
100	15.6	31.6	21.3	14.3	15.7	30.2	19.8	7.4	15.6	28.9	20.8	6

[a] See footnotes to Tables III and IV.
[b] Pseudorotation parameters selected were P_N 18°; Φ_N 36°; P_S 108°, 153°, and 180°; Φ_S 36°.
[c] Molefraction in %.

Φ_m(av) 36.4°. Moreover, the available couplings $J_{1'2'}$, $J_{1'2''}$, Σ 2', and Σ 2'' of the residues dG(2) and dG(4) in d(m⁵CG)₃, Z-DNA form,[12] are consistent with P_N 18–27°, Φ_N 36°, with perhaps a minute amount of S-type conformer mixed in. Thus, an educated estimate of N-form parameters brings us to assume P_N 18°, Φ_N 36°. These values are implied throughout the following analysis. Note that matters are less straightforward when the two conformers are present in approximately equal amounts ($0.35 \leq X_S \leq 0.65$). In such a case one would be advised to introduce the constraint $\Phi_N = \Phi_S = 36°$.

Step 1. The first marker to use is Σ 1'. A value in excess of 13.2 Hz indicates a predominance of the S-type conformer ($X_S \geq 65 \pm 5\%$). If that is the case a relatively safe conclusion follows: of the two H2'/H2'' multi-

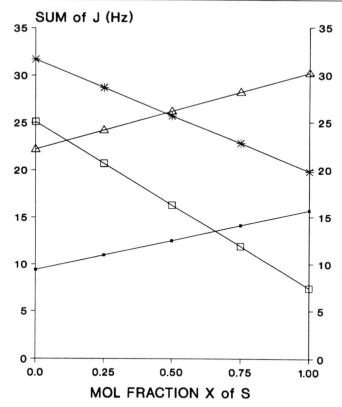

FIG. 5. Predicted sums of coupling constants plotted versus mole fraction of S-type conformer for deoxyribofuranose rings in nucleosides and nucleotides that occur in a fast conformational equilibrium between an N-type conformer (P_N 18°, Φ_N 36°) and an S-type conformer (P_S 153°, Φ_S 36°). See also Table V. Σ 1′ (■), Σ 2′ (△), Σ 2″ (*), Σ 3′ (□).

plets the one with the largest separation between the outer lines (typically 26–32 Hz) can be identified with H2′; the signal with the smaller separation (typically 19–25 Hz) is then assigned to H2″. In the great majority of cases an S/N ratio no less than 65/35 also implies that the larger of the two couplings present in the H1′ multiplet belongs to $J_{1'2'}$. However, in theory a reversal of the relative magnitudes of $J_{1'2'}$ and $J_{1'2''}$ can occur at the extremes of the normal conformational range. For this reason it is advisable to postpone a relative assignment of $J_{1'2'}$ and $J_{1'2''}$ until further details (approximate P_S and Φ_S) have been obtained.

The value of Σ 1′ is not very sensitive to variations in P_S and Φ_S and is

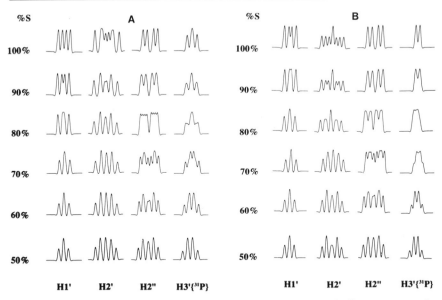

FIG. 6. Predicted splitting patterns of the H1′, H2′, H2″, and H3′{^{31}P} resonances for different values of the mole fraction X_S, expressed as %S. Limiting couplings for the N-type conformer were calculated for P_N 18°, Φ_N 37°. For the calculation of the limiting couplings of the S-type form two extreme situations were selected: (A) "low S", P_S 108°, Φ_S 41°; (B) "high S", P_S 180°, Φ_S 36°. Chemical shift values were chosen in accordance with those commonly observed for a pyrimidine residue in an ^{1}H NMR spectrum: δH1′ 2.8 ppm, δH2′ −1.5 ppm, δH2″ −0.7 ppm, δH3′ 1.5 ppm, and δH4′ 1.0 ppm (500 MHz), relative to the methyl resonance of Me$_4$NCl.

used to determine the approximate fraction of S-type conformer ($X_S \pm$ 0.05) whenever the S form predominates [Eq. (7)]:

$$X_S = (\Sigma\ 1' - 9.4)/(15.7 - 9.4) \tag{7}$$

Step 2. Once X_S is known to a first approximation a graphical or computerized refinement of the procedure can be attempted. In our experience large-scale graphs of the type shown in Fig. 5 are extremely helpful. The limiting (sums of) couplings for P_N 18°, Φ_N 36° are taken from Tables III and/or IV, and the limiting couplings of the unknown S-type form at hand are first taken for P_S 108°, 153°, and 180°, Φ_S 36°. Usually the observed sums will not fit any of the three sets of predicted lines at $X_S \pm 0.05$ simultaneously. However, with some experience one is able to narrow the possible P_S range of solutions or even obtain an acceptable solution [root mean square (rms) agreement ≤ 0.3 Hz]. At this stage one usually can assign $J_{1'2'}$ and $J_{1'2''}$ and bring these couplings into the procedure.

Step 3. Lack of satisfactory agreement between observed and predicted (sums of) couplings prompts one to pay special attention to $\Sigma\,2'$ and $\Sigma\,2''$ with a view to decide whether Φ_S should be chosen smaller or larger than 36°. Here Table IV comes into play. This step may require interpolation between the columns of Table IV. With the new value of Φ_S, steps 1–3 are repeated. Iteration can be regarded as successfully completed when no further improvement occurs. In the majority of cases, at least where experimental (sums of) couplings can be determined with good accuracy (± 0.1 Hz), one can expect a final rms deviation in $J_{obs}-J_{calc}$ of the same order of magnitude (≤ 0.2 Hz). In cases where the rms agreement remains above 0.3 Hz, one should be suspicious concerning the correctness of the assignment and/or doubt the accuracy of the experimental data.

The above recipe rests on two basic premises: (1) at least three out of four signals H1′ to H3′ are free from overlap, and (2) the chemical shift difference $\Delta\delta2'2''$ is sufficiently large (>4 Hz) to allow the extraction of $\Sigma\,2'$ and $\Sigma\,2''$ via spectral simulation or first-order analysis. When solely the splitting pattern of H1′ is available, only X_S can be obtained. An unambiguous solution of the conformational problem is then out of the question. Because of the symmetry of the individual couplings $J_{1'2'}$ and $J_{1'2''}$ about P 144°, two different but mathematically equivalent solutions usually exist.

In case H2′/H2″ signals display a "deceptively simple" quartet pattern, information concerning the geminal $J_{2'2''}$ is lost and $\Sigma\,2'$ and $\Sigma\,2''$ from Tables IV and V no longer apply. The first-order distance between the outer peaks of the isochronous H2′/H2″ quartet, multiplied by 2, corresponds to $\Sigma\,1' + J_{2'3'} + J_{2''3'}$. Knowledge of $\Sigma\,1'$ then allows determination of $J_{2'3'} + J_{2''3'}$. This sum, symmetrical about P 180°, contains valuable information about P_S because it increases strongly in going from P_S 162° (6.1 Hz) to P 90° (11.4 Hz). Again, the graphical method can be applied with J_N set at 16.5 Hz.

The foregoing procedure (steps 1–3) is now illustrated by an analysis of published[28] couplings of residue T(6) in d(CCGAATTCGG)$_2$, (Table VI, entry A). From $\Sigma\,1' = 14.2$ Hz, one calculates [Eq. (7)] a first estimate of X_S (0.76 ± 0.05). From step 2, a large-scale graph of $\Sigma\,1'$, $\Sigma\,2'$, and $\Sigma\,2''$ (P_N 18°, P_S 108°, 153°, and 180°, Φ_S 36°) reveals that the observed values of $\Sigma\,2'$ and $\Sigma\,2''$ require a P_S value in the lower range, 108–135°. From step 3, it is worthwhile to note that $J_{2'3'}$ varies little with X_S. However, this

[28] L. J. Rinkel, G. A. van der Marel, J. H. van Boom, and C. Altona, *Eur. J. Biochem.* **163,** 275 (1987).

TABLE VI

ANALYSIS OF COUPLINGS OF RESIDUE T(6) IN THE SELF-COMPLEMENTARY DNA
DECAMER d(CCGAATTCGG)$_2$[a]

Entry	Property	$J_{1'2'}$	$J_{1'2''}$	$J_{2'3'}$	$J_{2''3'}$	$\Sigma 1'$	$\Sigma 2'$	$\Sigma 2''$	X_S
A	J_{obs}	8.2	6.0	6.8	3.4	14.2	29.0	23.4	
B[b]	J_N	1.4	8.0	6.8	9.7	9.4	22.2	31.7	
	J_S	10.5	5.2	6.9	0.8	15.7	31.4	19.9	
	X_S	0.75	(0.67)	—	0.71	0.76	0.74	0.70	
C[c]	J_{calc}	8.0	6.0	6.9	3.2	14.0	28.9	23.1	0.73
	ΔJ	0.2	0.0	−0.1	0.2	0.2	0.1	0.3	

[a] See text.

[b] Predicted limiting couplings for pure N- and S-type conformers; P_N 18°, Φ_N 36°, P_S 126°, Φ_S 36°.

[c] Predicted couplings for a conformational mixture containing 73% S-type conformer. Pseudorotation parameters are the same as in footnote b (rms 0.17 Hz).

cis coupling is sensitive to P_S (and Φ_S). The observed value points to P_S ~ 126°; therefore, this is adopted as the "best guess." We now calculate X_S from each X_S-sensitive coupling in the conformational mixture characterized by P_N 18°, Φ_N 36°, P_S 126°, and Φ_S 36° (calculation B, Table VI). A mean X_S of 0.73 is obtained. The couplings calculated with the use of this value (J_{calc}, entry C in Table VI) give excellent agreement with the J_{obs} values (rms 0.17 Hz), and the procedure is terminated.

The major conformer (P_S 126°, Φ_S 36°) of the T(6) nucleotide has $_1E$ geometry, that is, an envelope with C-1′ pointing downward with respect to the plane formed by the remaining ring atoms. The five endocyclic torsion angles v_j are now calculated with the aid of Eq. (1). The five endocyclic bond angles follow from Eq. (8)[8]:

$$\tau_j = D_j + \Phi_m^2 \{E_j + F_j \cos(2P + 1.6\pi j)\} \tag{8}$$

for j equals 0–4. The bond angles are numbered clockwise, starting with $\tau_0(C4'-O4'-C1')$.[6] The D, E, and F parameters (not shown) can be found in Ref. 8 and were derived from a set of accurate X-ray data in a least-squares regression analysis as described elsewhere.[8] Bond lengths b are nearly independent of variations in P and Φ_m and can be taken from an X-ray data set.[4] Thus, we arrive at the following geometrical parameters for the deoxyribose ring in residue T(6) of our example:

$$v_0 - 34°(5), v_1\ 34°(5), v_2 - 21°(5), v_3\ 0°(5), v_4\ 21°(5)$$

$$\tau_0\ 108.2°(1.3), \tau_1\ 105.2°(0.6), \tau_2\ 102.4°(0.6), \tau_3\ 100.0°(0.6),$$

$$\tau_4\ 106.4°(0.6)$$

$$b_0(O4'-C1')\ 1.41\ \text{Å}, b_1\ 1.53\ \text{Å}, b_2\ 1.53\ \text{Å}, b_3\ 1.53\ \text{Å}, b_4\ 1.45\ \text{Å}$$

where values in parentheses pertain to the maximum expected errors that will arise when the true solution lies in the range P_S 117–135° and Φ_S 34–38°. As the positions of the five skeletal atoms in this ring are determined by nine independent parameters, the geometry is overdetermined, even without knowledge of the bond angles. In our least-squares computer program BUILDER (formerly called PSEU[8]) Cartesian coordinates of the ring atoms are generated from four selected bond distances, three bond angles, and two torsion angles. Positions of the exocyclic atoms are then calculated from known regressions,[8] completing our model building (Table VII). The backbone angle δ follows from Eq. (4): $\delta = 121°$ $(\pm 7°)$.

This example shows that the coupling constant method indeed provides us with a powerful entry into the detailed behavior of each individual sugar ring in a DNA duplex in aqueous solution, provided that at least four J_{obs} values can be determined with reasonable accuracy (0.1–0.2 Hz). In our opinion modeling of the sugar rings in a DNA duplex as outlined above provides an excellent basis for the application of distance geometry constraints in studies concerned with various helical properties of DNA in solution. Such an approach has the advantage that all experimentally available information is brought into play, not NOEs only.

TABLE VII

CARTESIAN COORDINATES[a] OF THE DEOXYRIBOSE
RING AND ATTACHED ATOMS OF RESIDUE T(6)
IN DECAMER DUPLEX d(CCGAATTCGG)$_2$ in
Aqueous Solution[a]

Atom	X	Y	Z
C-5′	−1.440	0.000	0.000
C-4′	−1.955	1.417	0.000
C-3′	−1.277	2.356	−0.992
C-2′	−0.674	3.454	−0.119
C-1′	−1.474	3.352	1.180
O-4′	−1.751	1.977	1.320
O-3′	−2.263	2.891	−1.873
N-1	−0.768	3.810	2.385
H4′	−3.039	1.399	−0.213
H3′	−0.490	1.828	−1.559
H2′	0.403	3.285	0.055
H2″	−0.785	4.449	−0.585
H1′	−2.422	3.909	1.074

[a] Deduced from the coupling constant analysis
and supplemented by correlations provided by
numerous X-ray studies (see text).

Results

What results can be expected from an analysis of coupling constant data of DNA sugars? A recent reinvestigation[29] of couplings of intact double-helical DNA fragments with the aid of our improved equations and parameterization [Eqs. (5) and (6), Table II, and PSEUROT version 5.4] fortunately revealed only minor differences with respect to the PSEUROT (version 4.0) analyses published earlier by us, except for a generally improved rms agreement between J_{obs} and J_{calc} when the new equations are used. Therefore, our earlier conclusions[13,16,18,30] remain essentially correct, except for minor numerical changes. The most important conclusions[29] are summarized as follows.

The sugar rings in an intact B-DNA duplex hardly ever display 100% S-type conformational purity, although purine residues come closer to this ideal than do pyrimidine residues. Excluding 3'-terminal ends, which traditionally enjoy conformational freedom compared to nonterminal residues,[31] one finds the following ranges: $X_S = 0.90 \pm 0.07$ for purines (13 residues), $X_S = 0.78 \pm 0.13$ for pyrimidines (22 residues). This means that the deoxyribose rings in an intact B-DNA double helix have retained their ability to flip from a major (S) form to a minor (N) one and vice versa.

In B-DNA duplexes the P_S values of purines (R) are generally greater than the P_S values of pyrimidines (Y), although the respective ranges overlap: P_S (R) $= 138-196°$, mean $162°$ (20 residues); P_S (Y) $= 87-156°$, mean $132°$ (22 residues). The variations in P_S appear to relate to the local base sequence and are outspoken in alternating sequences . . . R–Y–R–Y . . .. However, no definite rules for predictions of sugar behavior in an arbitrary B-DNA duplex have as yet been formulated.

The amplitudes of pucker Φ_S in B-DNA duplexes range from 28° to 47°, mean 36.2° (42 residues). No apparent distinction occurs between purine and pyrimidine residues. One notes with satisfaction that the NMR method independently yields a result that is identical with the mean X-ray value of nucleosides and nucleotides: Φ_S 36.4°, range $26-44°$ (51 residues).[27]

Acknowledgments

This research was supported in part by the Netherlands Foundation for Chemical Research (SON) with financial aid from the Netherlands Organization for Scientific Research (NWO).

[29] J. H. Ippel, J. van Wijk, and C. Altona, unpublished results (1991).
[30] C. Altona, *Nucleosides Nucleotides* **6**, 157 (1987).
[31] L. P. M. Orbons, G. A. van der Marel, J. H. van Boom, and C. Altona, *Eur. J. Biochem.* **170**, 225 (1987).

[16] Infrared Spectroscopy of DNA

By E. TAILLANDIER and J. LIQUIER

Introduction

Compared to other techniques some of the advantages of infrared spectroscopy in the study of nucleic acids are (1) the possibility of obtaining data from samples under an extremely wide variety of physical states (solutions, gels, hydrated fibers or films, and crystals); (2) no limitation introduced by the size of the investigated molecule, that is, one can work as well with short oligonucleotides as with long polynucleotides, DNA fragments obtained by enzymatic or chemical cleavage, or high molecular weight native DNAs. (3) Infrared spectroscopy is a nondestructive technique, which requires only small amounts of samples [a good Fourier transform infrared (FT-IR) spectrum can be obtained with 2 OD units of material]. (4) It gives vibrational information characteristic of the helical conformations of the nucleic acids. (5) The comparison of FT-IR spectra obtained in solution and in crystals permits correlations with precise structures known from X-ray crystal diffraction studies as well as NMR solution studies.

Different types of information can be obtained by IR spectroscopy. For example, it is possible to characterize molecular geometry using what is usually called IR "marker bands" which are conformation sensitive (wavenumber and/or relative intensity and/or polarization of the absorption). These bands can show the existence of a conformational transition when different factors such as temperature, hydration, concentration, nature, and amount of counterions are varied ($B \rightarrow A$, $B \rightarrow C$, $B \rightarrow Z$, helix $\rightarrow$ coil, etc.). In the case of mixtures of conformations, computer simulations based on "perfect" pure form spectra are used to evaluate quantitatively the amount of each geometric form in the sample, by reference with these marker bands. Another type of information which can be obtained from the FT-IR spectra concerns recognition of DNA sequences by a wide variety of molecules, such as oligonucleotides (triple-helical structures), drugs and proteins.

Although less commonly employed in studying nucleic acids than its "sister" technique, Raman spectroscopy, infrared spectroscopy has been successfully used since the early work of Fraser and Fraser[1] by several

---

[1] M. J. Fraser and D. B. Fraser, *Nature (London)* **167,** 759 (1951).

[2] H. Falk, K. A. Hartman, and R. C. Lord, *J. Am. Chem. Soc.* **85,** 391 (1963).

[3] H. T. Miles and J. Frasier, *Biochem. Biophys. Res. Commun.* **14,** 21 (1964).

groups.[2-7] In this chapter we shall briefly present a general survey of the FT-IR spectrum of a nucleic acid, describe the characterization of the different classic families of DNA conformations (A, B, and Z), and then consider how FT-IR spectroscopy can be used in the study of DNA recognition, particularly in the case of triple-stranded structures and DNA–drug interactions.

General Features of FT-IR Spectra of Nucleic Acids

Nucleic acid vibrations arise from different parts of the macromolecule, and the corresponding IR absorptions are detected in several regions of the spectrum. The main vibrations are observed in the spectral range between 1800 and 700 cm^{-1}, and four domains can be distinguished: between 1800 and 1500 cm^{-1} absorption bands due to the stretching vibrations of double bonds in the base planes; between 1500 and 1250 cm^{-1} bands due to the base–sugar entities, strongly dependent particularly on the glycosidic torsion angle; between 1250 and 1000 cm^{-1} strong absorptions of the phosphate groups and of the sugar; below 1000 cm^{-1} bands due to the vibrations of the phosphodiester chain coupled to vibrations of the sugar.

Figure 1a presents a DNA FT-IR spectrum recorded in H_2O solution. The enlarged area shows the spectral domain corresponding to the vibrations of the phosphate groups. On such a spectrum we can observe an important contribution of the solvent (H_2O), which makes it difficult to obtain data in the region around 1600 cm^{-1} and below 1000 cm^{-1}. To avoid this solvent problem, the spectrum of the same molecule is recorded in D_2O solution (Fig. 1b). Now we can easily study the FT-IR spectrum in the previously mentioned regions (see enlarged insets, Fig. 1). The combination of H_2O/D_2O solutions covers the complete spectral range necessary for nucleic acid studies.

If a DNA film is exposed to an atmosphere with controlled hydration obtained by using saturated solutions of known salts, the H_2O contribution is less important than in the case of an aqueous solution. Careful treatment of data allows one to obtain an infrared spectrum even in the 1800–1500 cm^{-1} region. When the film is exposed to D_2O vapor, H_2O/D_2O exchange occurs, and this can be followed on the IR spectrum. Important

---

[4] M. Tsuboi, *in* "Applied Spectroscopy Reviews" (E. G. Brame, ed.), Vol. 3, p. 45. Dekker, New York, 1969.

[5] J. Pilet and J. Brahms, *Biopolymers* **12**, 387 (1973).

[6] H. Fritzsche, H. Lang, and W. Pohle, *Biochim. Biophys. Acta* **432**, 409 (1976).

[7] E. Taillandier, J. Liquier, and J. A. Taboury, *in* "Advances in Infrared and Raman Spectroscopy" (R. J. H. Clark and R. E. Hester, eds.), Vol. 12, p. 65. Heyden, London, 1985.

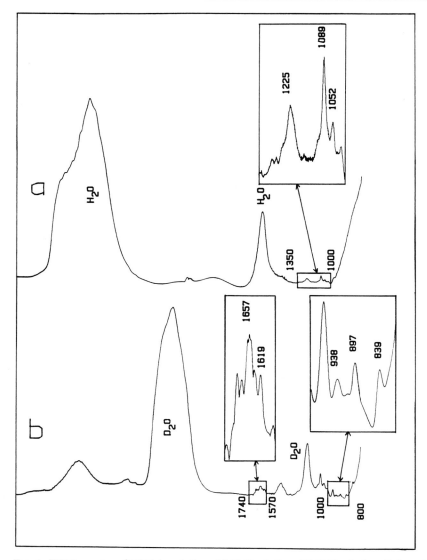

FIG. 1. FT-IR spectra of DNA in solution. (a) H_2O solution; (b) D_2O solution. Enlarged parts of the spectra present the absorptions involving mainly the vibrations of the phosphate groups (top), the double bonds of the bases in their plane (bottom left), and the sugar–phosphate backbone (bottom right).

modifications of the DNA spectrum can be observed that reflect the exchange of labile hydrogens of the bases, in particular the NH groups of thymines and the NH_2 groups of adenines, guanines, and cytosines. Exchange of hydrogens attached to C-8 of the purine base moieties is slow and observable only at elevated temperatures.

In the solid state, DNA can be studied by IR spectroscopy of microcrystals. Microscopes especially designed for IR spectroscopy that contain mirror all reflecting Cassegrain type optics are now commercially available and can be coupled to FT-IR spectrophotometers. Such systems need a more sensitive detector, usually a nitrogen-cooled MCT detector. An example is given in Fig. 2, which presents the micro-FT-IR spectrum of

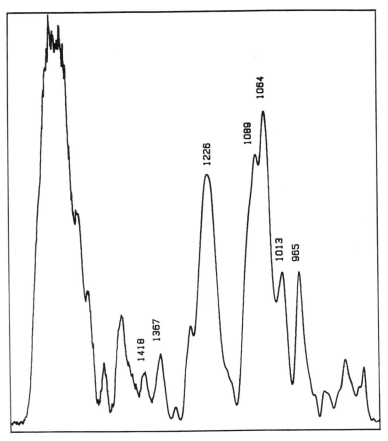

Fig. 2. Micro-FT-IR spectrum of a d(CGCGAATTCGCG) crystal.

d(CGCGAATTCGCG). In this case the crystal area analyzed by FT-IR microspectroscopy is about 50 × 50 μm.

Characterization of Nucleic Acid Secondary Structures

The main IR marker bands of A-, B-, and Z-DNA discussed below have been summarized in Table I. Polynucleotide and oligonucleotide conformations determined by IR spectroscopy are presented in Table II.

Right-Hand A- and B-DNA Conformations

In low ionic strength solutions, native DNAs and most synthetic double-strand poly- and oligonucleotides adopt a B-form conformation. Under the same conditions double-stranded RNAs adopt an A-type conformation. Figure 3 presents the FT-IR spectra obtained in H_2O solutions of two synthetic nucleic acids with regularly alternating A-U sequences, the first

TABLE I
MAIN INFRARED MARKER BANDS OF A-, B-, AND Z-DNA
CONFORMATIONS

Conformation			
A	B	Z	Assignment
1705	1715	1695	In-plane base double bond stretching
		1433	A-T bases
1418	1425	1408	Deoxyribose
1401			
1375	1375		dA · dG anti
		1355	dA · dG syn
1335	1344		dT
1335	1328		dA
		1320	dG
1275	1281		T
		1265	G
1240	1225	1215	Antisymmetric phosphate stretching
1188			Deoxyribose
		1065	Deoxyribose
899	894	929	
877			
864			
806	830–840		Exact positions of bands are within ±2 cm⁻¹ of these numbers, depending, in particular, on base sequence of DNA

TABLE II
Conformations Determined by Infrared Spectroscopy[a]

Polymer	Counterion	Physical state	Form	Refs
Polynucleotides				
Native DNA	Na^+	S, FH	B	b
		FL	A	b
	Li^+	FL	C	c
Poly[d(A-T)·poly[d(A-T)]	Na^+	S, FH	B	d,e
		FL	A	d,e
	Cs^+, Li^+	FL	C	f
	Ni^{2+}	FH	Z	d
Poly(dA)·poly(dT)	Na^+	S, FH	B	e,g
		FL	H	g
Poly(rA)·poly(dT)	Na^+	S, FH	H	h
		FL	A	h
Poly(dA)·poly(rU)	Na^+	S	H	h
		FL	A	h
Poly(rA)·poly(rU)	Na^+	S	A	h
Poly[d($2NH_2$A-T)]·poly[d($2NH_2$A-T)]	Na^+	S low salt	B	i
		S high salt	Z	i
Poly[d(A-C)]·poly[d(G-T)]	Na^+	S, FH	B	j
		FL	A	j
	Li^+	FH	C	k
	Ni^{2+}	F	Z	j
Poly[d(G-C)]·poly[d(G-C)]	Na^+	FH, S low salt	B	l,m,n
		FL, S high salt	Z	l,m,n
	Mg^{2+}, Co^{2+}, Ni^{2+}	FH	Z	o
Poly[d(m5C-G)]·poly[d(m5C-G)]	Na^+	F	Z	n
Poly(dG)·poly(dC)	Na^+	FH	B + A	p
		FL	A	
Poly[d(I-Br5C)·poly[d(I-Br5C)]	Na^+	S low salt	B	q
		S high salt	Z	q
Oligonucleotides				
d(CGGCCG)	Na^+	F	Z	r
d(CGm5CGCG)	Na^+	F	Z	s
d(CCGCGG)	Na^+	S low salt	B	t
		S high salt	Z	t
		C	B–Z	t
d(m5CGCGm5CG)	Na^+	F	Z	r
d(m5CGGCm5CG)	Na^+	F	Z	r
d(CBr8GCGCBr8G)	Na^+	F	Z	u
d(CBr8GGCCBr8G)	Na^+	F	Z	r
d(CCCGCGGG)	Na^+	FH	B	v
		FL	A	v
	Ni^{2+}	F	Z	v
d(C$2NH_2$ACGTG)	Na	FH	B	w
		FL	Z	w
d(m5CGCAm5CGTGCG)	Na	FH	B	w
		FL	Z	w
d(m5CGTAm5CG)	Na	FH	B	x
		FL	Z	x

Polymer	Counterion	Physical state	Form	Refs.
d(GGTATACC)	Mg^{2+}	C	B + A	*y*
d(CGCGAATTCGCG)	Mg^{2+}	C	B + A	*p*
d(ACATGT)	Na	F	B	*t,s*
	Ni^{2+}	F	Z	*t,s*
d(AAAAATTTTT)	Na^+	FH	B	*g*
		FL	H	*g*
r(A-U)$_6$, r(A-U)$_8$	Na^+	S	A	*h*

[a] S, Solution; FH, film high hydration; FL, film low hydration; F, film any hydration; C, crystal; and H, Heteronomous.

[b] J. Pilet and J. Brahms, *Biopolymers* **12,** 387 (1973); H. Fritzsche, A. Rupprecht, and M. Rupprecht, *Nucleic Acids Res.* **12,** 9165 (1984).

[c] J. Brahms, J. Pilet, P. L. Tran Thi, and R. L. Hill, *Proc. Natl. Acad. Sci. U.S.A.* **70,** 3352 (1973).

[d] S. Adam, J. Liquier, J. A. Taboury, and E. Taillandier, *Biochemistry* **25,** 3220 (1986).

[e] J. Pilet, J. Blicharski, and J. Brahms, *Biochemistry* **14,** 1869 (1975).

[f] S. Adam, P. Bourtayre, J. Liquier, J. A. Taboury, and E. Taillandier, *Biopolymers* **26,** 251 (1987).

[g] E. Taillandier, J. P. Ridoux, J. Liquier, W. Leupin, W. A. Denny, Y. Wang, and W. L. Peticolas, *Biochemistry* **26,** 3361 (1987).

[h] J. Liquier, A. Akhebat, E. Taillandier, F. Ceolin, T. Huynh Dinh, and J. Igolen, *Spectrochim. Acta* **47A,** No. 2, 177 (1991).

[i] F. B. Howard, C. W. Chen, J. S. Cohen, and H. T. Miles, *Biochem. Biophys. Res. Commun.* **3,** 848 (1984).

[j] E. Taillandier, J. A. Taboury, S. Adam, and J. Liquier, *Biochemistry* **23,** 5703 (1984).

[k] D. H. Loprete and K. A. Hartman, *J. Biomol. Struct. Dyn.* **7,** 347 (1989).

[l] E. Taillandier, J. A. Taboury, J. Liquier, P. Sautiere, and M. Couppez, *Biochimie* **63,** 895 (1981).

[m] J. Pilet and M. Leng, *Proc. Natl. Acad. Sci. U.S.A.* **79,** 26 (1982); P. B. Keller and K. A. Hartman, *Nucleic Acids Res.* **14,** 8167 (1986).

[n] J. A. Taboury, J. Liquier, and E. Taillandier, *Can. J. Chem.* **63,** 1904 (1985).

[o] J. A. Taboury, P. Boutayre, J. Liquier, and E. Taillandier, *Nucleic Acids Res.* **12,** 4247 (1984).

[p] E. Taillandier, *in* "Structure and Methods" Vol. 3 "DNA and RNA" (R. H. Sarma and M. H. Sarma, eds.), p. 73. Adenine Press, 1990.

[q] B. Hartman, J. Pilet, M. Ptak, J. Ramstein, B. Malfoy, and M. Leng, *Nucleic Acids Res.* **10,** 3261 (1982).

[r] S. Adam, J. A. Taboury, E. Taillandier, A. Popinel, T. Huynh Dinh, and J. Igolen, *J. Biomol. Struct. Dyn.* **3,** 873 (1986).

[s] S. Pochet, T. Huynh Dinh, J. M. Neumann, S. Tran Dinh, S. Adam, J. Taboury, E. Taillandier, and J. Igolen, *Nucleic Acids Res.* **14,** 1107 (1986).

[t] L. Urpi, J. P. Ridoux, J. Liquier, N. Verdaguer, I. Fita, J. A. Subirana, F. Iglesias, T. Huynh Dinh, J. Igolen, and E. Taillandier, *Nucleic Acids Res.* **17,** 6669 (1989).

[u] J. A. Taboury, E. Taillandier, T. Huynh Dinh, and J. Igolen, *J. Biomol. Struct. Dyn.* **2,** 1185 (1985).

[v] E. Taillandier, W. L. Peticolas, S. Adam, T. Huynh-Dinh, and J. Igolen, *Spectrochim. Acta* **46,** 107 (1990).

[w] J. A. Taboury, S. Adam, E. Taillandier, J. M. Neumann, S. Tran Dinh, T. Huynh Dinh, B. Langlois d'Estaintot, and J. Igolen, *Nucleic Acids Res.* **12,** 6291 (1984).

[x] E. Taillandier, S. Adam, J. P. Ridoux, and J. Liquier, *Nucleic Acids Res.* **16,** 5621 (1988).

[y] J. Liquier, E. Taillandier, W. L. Peticolas, and G. A. Thomas, *J. Biomol. Struct. Dyn.* **8,** 295 (1990).

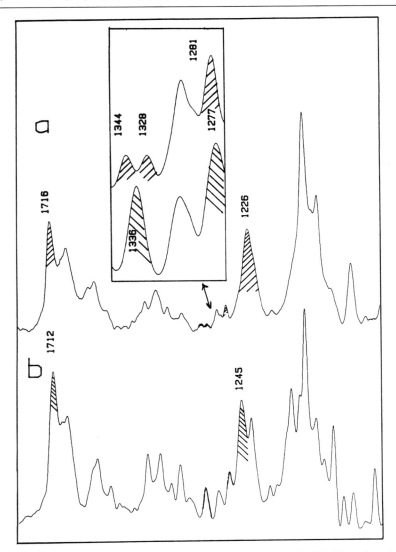

FIG. 3. FT-IR spectra of H_2O solutions of $d(A-U)_n$ (a) and $r(A-U)_8$ (b). The enlarged area between 1350 and 1270 cm^{-1} shows absorptions characteristic of A ($\backslash\backslash\backslash\backslash\backslash\backslash$) and B ($/////$) geometries.

one (Fig. 3a) with deoxy sugars, the second one (Fig. 3b) with ribo sugars: d(A-U)$_n$ and r(A-U)$_8$. Besides the important modifications induced on the IR spectrum by the O-2' deletion, differences between the spectra are due to the conformations adopted in solution by these two nucleic acids, namely, B and A. Some of the characteristic absorptions are shown hatched on Fig. 3, and the wavenumbers are given in Table I. In particular the phosphate antisymmetric stretching vibration is found at 1226 cm^{-1} in the B form and at 1245 cm^{-1} in the A form, and a band reflecting the base pairing of the nucleic acid is shifted from 1716 to 1712 cm^{-1}.

The A-DNA conformation can also be observed in DNAs by varying the water content of the nucleic acid sample. An easy way to achieve this is to decrease the relative humidity to which a DNA film is exposed. The decrease of the relative humidity (R.H.) from 100 to 58% R.H. will induce for some DNA sequences a B → A conformational transition, followed by subsequent disorganization of the double-helical structure if the hydration is further decreased. Some DNA sequences will not undergo this transition; for example, the A- conformation is not observed by decreasing the water content of a poly(dA)·poly(dT) sample. On the contrary, the alternating polymer poly[d(A-T)] readily adopts this conformation, as can be seen in Fig. 4. We present in Fig. 4 a series of FT-IR spectra of a poly [d(A-T)] film (1 Na$^+$ per phosphate counterion) exposed to decreasing relative humidities. The displayed spectral region (1000–800 cm^{-1}) contains several IR marker bands of the A and B geometries corresponding to vibrations of the phosphodiester chain coupled to deoxyribose vibrations. Both sets of characteristic bands have been shown hatched on Fig. 4. The B → A transition is reflected by the disappearance of the band located at 841 cm^{-1} and the emergence of a doublet at 880–865 cm^{-1} and of a band at 808 cm^{-1}.

The two A- and B-DNA conformations can also be characterized by a pattern of bands reflecting the vibrations of the bases coupled to vibrations of the sugar and located between 1250 and 1400 cm^{-1}. Figure 5 presents the FT-IR signatures of four double-stranded polynucleotides in this spectral region: from top to bottom poly(dA)·poly(dT), poly(rA)·poly(dT), poly(dA)·poly(rU), and poly(rA)·poly(rU). All spectra have been recorded in H$_2$O solutions, and the marker bands for the A-form strands (riboses) and B-form strands (deoxyribose) are shown hatched (Fig. 5). During the B → A transition we observe the shift of a band involving the N-3 H bending vibration of thymines (or uracils) from 1281 to 1275 cm^{-1} and the opposite displacements of an adenine band found at 1344 cm^{-1} and a thymine band found at 1328 cm^{-1} in the B form to lower and higher wavenumbers, respectively, around 1335 cm^{-1} in the A form.

Another possibility for characterizing DNA conformations is the use of

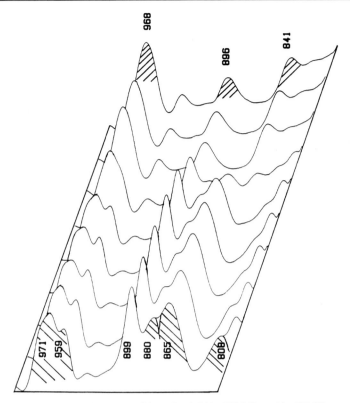

FIG. 4. B → A conformational transition of poly[d(A-T)] followed by FT-IR spectroscopy. The DNA film is exposed to decreasing relative humidities. (Top) B form, R.H. 100% (///). (Bottom) A form, R.H. 58% (\\\\). Bands characteristic of both geometries have been hatched.

polarized infrared radiation. If the sample presents a preferential orientation axis (e.g., in the case of stretched films or fibers) the infrared spectra obtained with an incident radiation having an electric field E polarized parallelly or perpendicularly to the orientation axis of the molecule will be different. These dichroic spectra can be used to compute orientations of transition dipole moments in the molecule and in simple cases give information concerning geometric parameters.[8,9] Such computations can be

[8] D. B. Fraser, *J. Chem. Phys.* **21**, 1511 (1953).
[9] R. L. Rill, *Biopolymers* **11**, 1929 (1972).

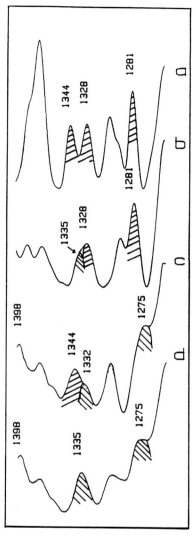

FIG. 5. FT-IR spectra between 1400 and 1250 cm⁻¹ of H₂O solutions. (a) Poly(dA)·poly(dT); (b) poly(rA)·poly(dT); (c) poly(dA)·poly(rU); (d) poly(rA)·poly(rU). (////) B form; (\\\\) A form.

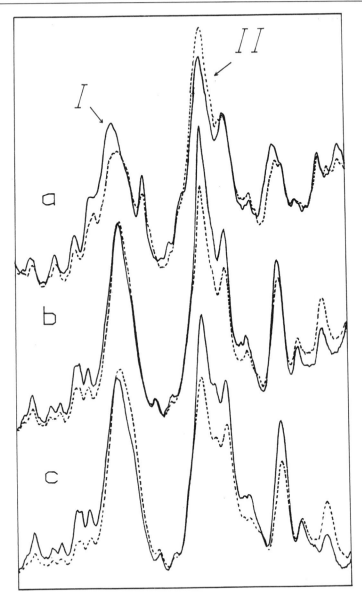

FIG. 6. Infrared polarized spectra of oriented films of poly[d(A-T)]. (a) 1 Na+ per nucleotide, R.H. 58%, A form; (b) 1 Na+ per nucleotide, R.H. 98%, B form; (c) 1 Cs+ per nucleotide, R.H. 76%, C form. (—) Electric vector of incident light perpendicular to orientation axis; (–––) electric vector of incident light parallel to orientation axis. I, Antisymmetric phosphate stretching vibration; II, symmetric phosphate stretching vibration.

TABLE III

POSITION AND POLARIZATION OF PHOSPHATE GROUP
VIBRATION ABSORPTION BANDS IN
DEOXYRIBOPOLYNUCLEOTIDES

Conformation	Antisymmetric stretching		Symmetric stretching	
	ν	Polarization	ν	Polarization
A	1241	⊥	1089	∥
B	1225	0	1089	⊥
C	1231	∥	1089	⊥
Z	1215	0	1090	⊥

performed on absorption bands involving the antisymmetric and symmetric motions of the phosphate groups.[5]

Figure 6 gives an example of such polarized infrared spectra in the case of a poly[d(A-T)] film.[10] The spectral region presented displays the strong absorptions of the phosphate groups. In Fig. 6 spectra (solid lines) have been obtained with polarization of the incident light perpendicular to the orientation axis, and spectra (dashed lines) have also been obtained with a parallelly polarized incident radiation. We can see the complete inversion of the polarization of the symmetric stretching vibration of phosphates (located around 1090 cm^{-1}) when the polymer undergoes the B → A transition (Figs. 6a and 6b). Simultaneously the antisymmetric stretching vibration located around 1230 cm^{-1} from nondichroic becomes perpendicularly polarized. In the case of the C form obtained in the presence of Li$^+$ or Cs$^+$ this band becomes parallelly polarized (Fig. 6c). The results concerning these bands are presented in Table III.

As previously stated in the introduction, these conformation-sensitive IR bands can be used to evaluate the amount of the two A and B helical geometries when both are present in the sample. Figure 7 presents the micro-FT-IR spectrum in the region of the phosphate–sugar backbone vibrations of a d(GGTATACC) crystal (Fig. 7b), along with two reference spectra of native DNA in the A (Fig. 7a) and B conformations (Fig. 7c). The crystal spectrum presents absorptions characteristic of both conformations (882, 861, and 812 cm^{-1} for the A form, 834 cm^{-1} for the B form). The amount of B conformation has been estimated to be 14% and inter-

[10] S. Adam, P. Bourtayre, J. Liquier, J. A. Taboury, and E. Taillandier, *Biopolymers* **26**, 251 (1987).

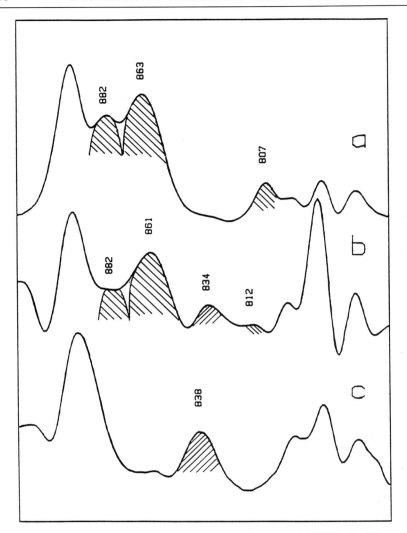

FIG. 7. FT-IR spectra of (a) A-form DNA (film at 58% R.H.); (b) d(GGTATACC) crystal; (c) B-form DNA (film at 98% R.H.). (\\\\) A-form bands; (////) B-form bands.

preted in agreement with a model in which one B-form octamer is trapped in a tunnel formed by six A-form octamers.[11,12]

[11] J. Doucet, J. P. Benoit, W. B. T Cruse, T. Prange, and O. Kennard, *Nature (London)* **337**, 190 (1989).
[12] J. Liquier, E. Taillandier, W. L. Peticolas, and G. A. Thomas, *J. Biomol. Struct. Dyn.* **8**, 295 (1990).

Left-Hand Z-Conformation

The first IR studies concerning left-hand conformations were devoted to the B → Z reorientation of poly[d(G-C)] in films.[13–15] In the presence of sodium counterions and in films this polymer undergoes a complete B → Z transition via decreasing the relative humidity. The marker bands of guanosines in a syn geometry were obtained first. Since then many other sequences have been shown to be able to adopt a Z conformation: regularly alternating purine–pyrimidine sequences, poly[d(A-C)·poly[d(G-T)][16] and poly[d(A-T)][17]; short oligomers containing AT base pairs, d(C2aminoACGTG), d(m⁵CGCAm⁵CGTGCG),[18] and d(m⁵CGTAm⁵CG)[19]; and sequences containing GC base pairs out of regular purine–pyrimidine alternation, d(m⁵CGGCm⁵CG), d(CB,8GGCB,8G), and d(CGCGGC).[20] A whole set of IR marker bands for GC and AT base pairs as well as the deoxyribose and the phosphate backbone in Z geometry has been obtained. The main IR Z-form marker bands are also summarized in Table I. Figure 8 presents the FT-IR spectra of the Z-conformations of three regularly alternating purine–pyrimidine polynucleotides. Main Z-form characteristic bands are dotted in Fig. 8.

The incorporation of AT base pairs in a regularly alternating GC sequence makes the Z form of the nucleic acid more difficult to observe. In fact, for sequences such as poly[d(A-C)·poly[d(G-T)] and poly[d(A-T)] a simple decrease of hydration in the case of a film does not induce, as in the case of regularly alternating GC sequences, a B → Z transition, but rather a B → A transition. The Z form of these polynucleotides has been obtained by interactions with divalent metal ions, Ni^{2+} and to a lesser extent Co^{2+}. We have shown by infrared spectroscopy that the hydrated nickel ions interact in a specific way with the N-7 groups of purines (guanines and/or adenines), stabilizing the syn geometry of the purine nucleosides and therefore the Z geometry of the DNA.[21] This can be done using polynucleotides selectively deuterated on the C-8 H of the purine. A vibrational

---

[13] E. Taillandier, J. A. Taboury, J. Liquier, P. Sautiere, and M. Couppez, *Biochimie* **63**, 895 (1981).

[14] J. Pilet and M. Leng, *Proc. Natl. Acad. Sci. U.S.A.* **79**, 26 (1982).

[15] J. A. Taboury, J. Liquier, and E. Taillandier, *Can. J. Chem.* **63**, 1904 (1985).

[16] E. Taillandier, J. A. Taboury, S. Adam, and J. Liquier, *Biochemistry* **23**, 5703 (1984).

[17] S. Adam, J. Liquier, J. A. Taboury, and E. Taillandier, *Biochemistry* **25**, 3220 (1986).

[18] J. A. Taboury, S. Adam, E. Taillandier, J. M. Neumann, S. Tran Dinh, T. Huynh Dinh, B. Langlois d'Estaintot, and J. Igolen, *Nucleic Acids Res.* **12**, 6291 (1984).

[19] E. Taillandier, S. Adam, J. P. Ridoux, and J. Liquier, *Nucleic Acids Res.* **16**, 5621 (1988).

[20] S. Adam, J. A. Taboury, E. Taillandier, A. Popinel, T. Huynh Dinh, and J. Igolen, *J. Biomol. Struct. Dyn.* **3**, 873 (1986).

[21] S. Adam, P. Bourtayre, J. Liquier, and E. Taillandier, *Nucleic Acids Res.* **14**, 3501 (1986).

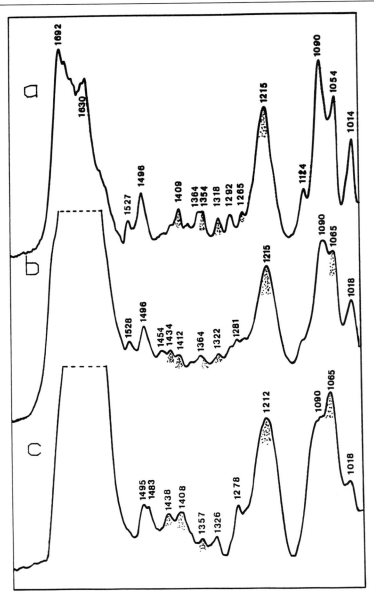

FIG. 8. IR spectra of the Z form of regularly alternating purine–pyrimidine polynucleotides. (a) poly[d(G-C)]; (b) poly[d(A-C)] · poly[d(G-T)]; (c) poly[d(A-T)].

bending mode of the N-7 C-8 D group is sensitive to interactions on the N-7 site. When nickel is added to the DNA sample, shifts in the corresponding IR band, located around 1465 cm^{-1}, reflect an interaction of the metal with the N-7 site.

The coexistence of right-hand and left-hand geometries in the same molecule, for instance, in the case of B–Z junctions, can be studied using the IR marker bands of both conformations.[22,23] With a whole set of A, B, and Z characteristic IR absorptions, it is possible to study the conformational flexibility of an oligonucleotide sequence. For example, the d(CCGCGG)[24] sequence has been studied by FT-IR in solution and in crystal. Figure 9 presents the spectrum of the crystal (Fig. 9b) and of the solution form at low ionic strength (Fig. 9a) and high ionic strength (Fig. 9c). We clearly detect in spectra (Fig. 9a,c) the marker bands of the B form (1420, 1372, phosphate bands at 1220 and 1090 cm^{-1}) and of the Z form (1410, 1360, 1320, 1264, phosphate bands at 1213, 1061, and 924 cm^{-1}). In solution an increase in the ionic strength induces a complete B $\rightarrow$ Z transition of the sequence. The spectrum of the crystal, however, is quite different. It reflects neither a B nor a Z conformation. The deoxyribose–base geometry seems to be of the B type (1378 cm^{-1}), whereas the phosphate–sugar backbone (924 cm^{-1}), sugar (1413 cm^{-1}), and phosphate group (1065 cm^{-1}) vibrations reflect a Z form. The oligonucleotide seems to be trapped in an intermediate conformation.

DNA Recognition

Recognition in Major Groove by Specific Nucleic Acid Sequences: Triple-Strand Structures

The specific recognition of a predetermined DNA sequence in a long-strand double-helical polymer can be achieved by an oligonucleotide which forms a triple-strand structure. This is of particular interest in the case of oligonucleotides equipped with efficient DNA cleaving moieties at the 5′ end. For example, metal chelates such as EDTA-Fe[25] or phenanthroline–

[22] S. Pochet, T. Huynh Dinh, J. M. Neumann, S. Tran Dinh, S. Adam, J. A. Taboury, E. Taillandier, and J. Igolen, *Nucleic Acids Res.* **14**, 1107 (1986).

[23] S. Adam, J. P. Ridoux, P. Bourtayre, E. Taillandier, S. Pochet, T. Huynh Dinh, and J. Igolen, *J. Biomol. Struct. Dyn.* **6**, 1 (1988).

[24] L. Urpi, J. P. Ridoux, J. Liquier, N. Verdaguer, I. Fita, J. A. Subirana, F. Iglesias, T. Huynh Dinh, J. Igolen, and E. Taillandier, *Nucleic Acids Res.* **17**, 6669 (1989).

[25] H. E. Moser and P. B. Dervan, *Science* **238**, 654 (1987).

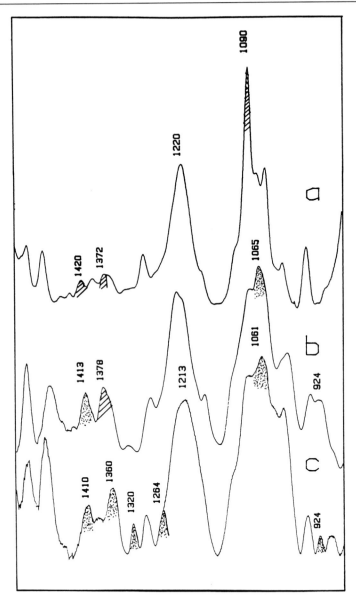

FIG. 9. FT-IR spectra of d(CCGCGG). (a) Solution, low ionic strength, B form; (b) crystal, intermediate form; (c) solution, high ionic strength, Z form. (///) B-form bands; (dotted) Z-form bands.

copper[26] or ellipticin derivatives[27] tethered to oligonucleotides might be used for sequence-specific recognition and cleavage of duplex DNA for chromosome analysis, gene mapping, and isolation via triple-helix formation. Interest in the study of the formation of triple helices also comes from the assumed existence of *in vivo* intramolecular triple-helical structures which may be formed by homopurine–homopyrimidine sequences. Such sequences are overrepresented in the eukaryotic genome and are often found within putative regulatory regions of genes and hot spots of recombination. A possible role for such triple helices in gene expression regulation has been proposed.[28]

Infrared spectroscopy can be successfully used to detect the formation of the triple-strand structures. As noted previously, the spectral region between 1800 and 1500 cm^{-1} mainly contains absorptions due to double bonds of the bases in their planes. The positions and relative intensities of these absorptions are extremely sensitive to base pairing. The formation of a triple helix deeply modifies the base–base interaction schemes and is reflected in the IR spectrum in this region. Several base triplets may theoretically be formed.

Figure 10 presents the characterization by FT-IR of three triplets, T-A-T on the left, C^+-G-C in the middle, and G-G-C on the right. All spectra have been obtained in D_2O solutions. For the T-A-T triplet, Fig. 10a presents the spectrum of double-strand poly(dA)·poly(dT), and Fig. 10d that of triple-strand poly(dT)·poly(dA)·poly(dT). The two spectra shown in Fig. 10b and 10c correspond to mixtures of poly(dA)·poly(dT) with increasing amounts of an oligonucleotide, dT_{10}. An absorption located in the poly(dA)·poly(dT) spectrum at 1622 cm^{-1} is assigned to an ND_2 bending vibration coupled to a pyrimidic ring vibration of adenine, whereas the band located at 1641 cm^{-1} corresponds to ring double-bond vibrations of thymines. When the amount of the third strand increases, these two bands progressively disappear and are no longer detected for the stoichiometric complex one double-strand poly(dA)·poly(dT)–one single-strand poly(dT), reflecting the formation of the triple helix.

In the case of G-C base pairs, a triple-helical structure can be obtained starting with double-helical poly(dG)·poly(dC) and a single poly(dC) strand, but only if the latter has been protonated. In this case a hydrogen bond can be established between the N-7 of the guanine initially involved

[26] J.-C. Francois, T. Saison-Behmoaras, C. Barbier, M. Chassignol, N. T. Thuong, and C. Hélène, *Proc. Natl. Acad. Sci. U.S.A.* **86**, 9702 (1989).
[27] L. Perrouault, U. Asseline, C. Rivalle, N. T. Thuong, E. Bisagni, C. Giovannangeli, T. Le Doan, and C. Hélène, *Nature (London)* **344**, 358 (1990).
[28] R. D. Wells, D. A. Collier, J. C. Hanvey, M. Shimizu, and F. Wohlrab, *FASEB J.* **2**, 2939 (1988).

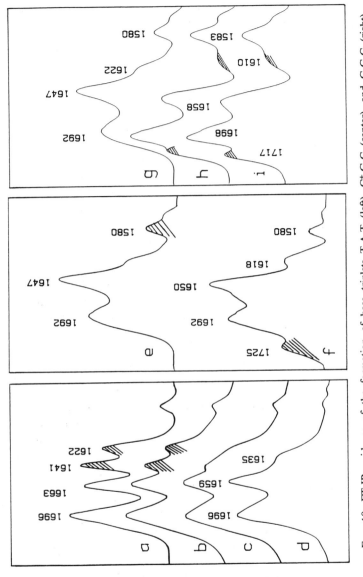

Fig. 10. FT-IR evidence of the formation of base triplets T-A-T (left), C⁺-G-C (center), and G-G-C (right). (a) Poly(dA)·poly(dT); (b) poly(dA)·poly(dT) + dT₁₀, 1/0.17 molar ratio; (c) poly(dA)·poly(dT) + dT₁₀, 1/0.33 molar ratio; (d) poly(dT)·poly(dA)·poly(dT); (e) poly(dG)·poly(dC); (f) poly(dC⁺)·poly(dG)·poly(dC); (g) poly(dG)·poly(dC); (h) poly(dG)·poly(dC)·poly(dG); (i) poly(dG)·poly(dC)·poly(dG) + 1.5 Na⁺ phosphate. Characteristic differences between double- and triple-strand geometries have been hatched.

in the G-C Watson–Crick base pair and the N-3 H of the protonated cytosine. Figure 10e,f presents the characterization of the C^+-G-C triplet. The upper spectrum (Fig. 10e) is that of poly(dG)·poly(dC), the bottom one (Fig. 10f) that of a stoichiometric mixture of poly(dG)·poly(dC) and poly(dC)$^+$ at pD 5. When the triple helix is formed, a decrease in the relative intensity of the absorption located at 1580 cm^{-1}, assigned to a vibration of C=N bonds of the guanine, shows that this base is involved in the formation of a triple-helical structure. A band also occurs at 1725 cm^{-1} which corresponds to a carbonyl group stretching vibration; this unusually high wavenumber is characteristic of the formation of a triple helix.

Among the possible purine–purine–pyrimidine triplets, the G-G-C triplet can be easily formed and has been characterized by UV circular dichroism.[29] Figure 10g–i shows the IR characterization of this triplet. Figure 10g is the spectrum of double-strand poly(dG)·poly(dC), whereas Fig. 10h, i correspond to an equimolecular mixture of poly(dG)·poly(dC) and poly(dG) in low and high salt. The addition of NaCl decreases the repulsions between the phosphate groups and favors the formation of the poly(dG)·poly(dG)·poly(dC) triplex. The characteristic features of the double to triple helical structure interconversion have been hatched on Fig. 10.

These infrared markers can be used to follow the formation of pyrimidine–purine–pyrimidine triple-helical structures containing iso-morphous T-A-T and C^+-G-C triplets. An example is given below. We have studied the associations formed by three oligomers:

22R: 5′-AAAGGAGGAGAAGAAGAAAAAA-3′

22Y: 5′-TTTTTTCTTCTTCTCCTCCTTT-3′

22T: 5′-TTTCCTCCTCTTCTTCTTTTTT-3′

The 22T oligomer contains a dTTTCCTCCTCT 11-mer. An ellipticine derivative tethered to this 11-mer was used to cleave DNA under photo-chemical activation by Hélène et al.[27] Figure 11 presents the results of the study of the self-association of these oligomers. In fact there is a competition between the formation of double and triple helices involving these sequences and the self-association of each oligomer.

Figure 11b is the spectrum of the oligopurine 22R strand, compared to reference spectra of dGmP (Fig. 11d), dAmP (Fig. 11c), and associated dG (Fig. 11a). No evidence of association is detected for the 22R oligomer. On the contrary, the homopyrimidine 22T or 22Y strand can be induced to

[29] B. L. Haas and W. Guschlbauer, *Nucleic Acids Res.* 3, 205 (1976).

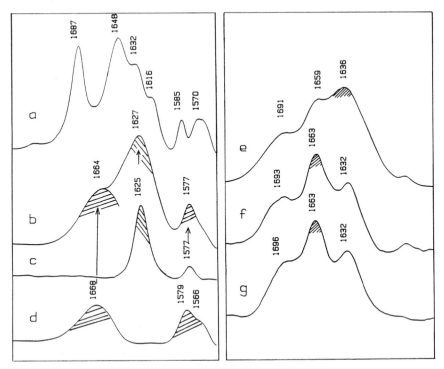

FIG. 11. Evidence for the self-association of 5'-TTTCCTCCTCTTCTTCTTTTTT-3' (right) and the absence of self-association of 5'-AAAGGAGGAGAAGAAGAAAAAA-3' (left). (a) dG autoassociated; (b) 5'-AAAGGAGGAGAAGAAGAAAAAA-3'; (c) dAmP; (d) dGmP; (e) 5'-TTTCCTCCTCTTCTTCTTTTTT-3', high concentration; (f) 5'-TTTCCTCCTCTTCTTCTTTTTT-3', low concentration; (g) dT$_{12}$.

self-associate by increasing the concentration or by decreasing the temperature. Figure 11f shows the spectrum of the nonassociated 22T oligomer (comparable to that of dT$_{12}$ known to be nonassociated); an increase of the concentration of the 22T oligomer leads to association, as seen on the spectrum in Fig. 11e. Therefore, precise concentration, temperature, and pD conditions were necessary to obtain double and triple helices with these oligomers.

Figure 12 (top) shows the spectrum of the 22R–22Y very unstable double helix which contains 15 A-T and 7 G-C base pairs. This spectrum can be satisfactorily compared to that of poly(dA)·poly(dT) (Fig. 10a). Figure 12 (bottom) shows the spectrum of the equimolecular mixture

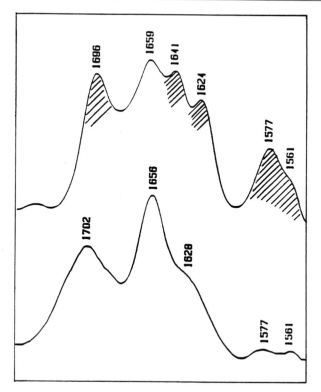

FIG. 12. FT-IR spectra in D_2O solutions of (top) double helix 22R–22Y and (bottom) triple helix 22R–22Y–22T.

22R–22Y–22T. We recognize on this spectrum the signatures previously given for the formation of the T-A-T and C^+-G-C triplets.

It is also possible to form triple helixes with unnatural oligonucleotides. For example, a triple-strand geometry can be obtained by adding to poly(dA)·poly(dT) the α anomer αdT_{10}. The IR spectrum of this triple helix presents the same characteristics as in the case of the triple helix formed with β anomers βdT_{10}. This may have significance in view of the resistance of the α anomers against native DNases in *in vivo* systems.

Infrared spectroscopy can also give information concerning the conformations of the deoxyribose in the triplexes. Figure 13 summarizes the results for poly(dT)·poly(dA)·poly(dT) (left), poly(dC$^+$)·poly(dG)· poly(dC), and poly(dG)·poly(dG)·poly(dC) (right). The first triplex was

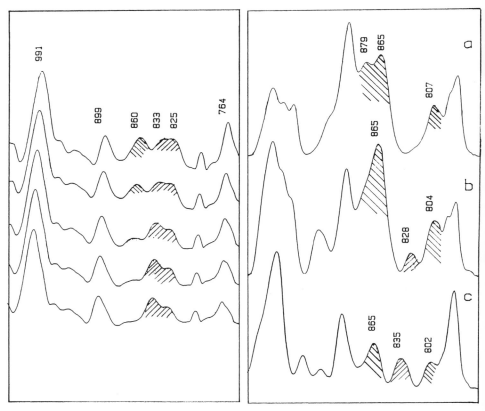

FIG. 13. Conformation of the deoxyribose and the phosphate–sugar backbone in triple helices. (Left) Poly(dT)·poly(dA)·poly(dT) hydrated film. Relative humidities from top to bottom: 47, 58, 76, 86, and 93%. (Right) Solutions in D_2O of (a) poly(dG)·poly(dC); (b) poly(dC+)·poly(dG)· poly(dC); (c) poly(dG)·poly(dG)·poly(dC). (\\\\\) A form; (/////) B form.

obtained starting with a B-form poly(dA)·poly(dT) duplex. $dA_n·dT_n$ sequences have never been observed in a canonical A form.[30,31]

As can be seen in Fig. 13 (left), the spectrum of the highly hydrated film of the poly(dT)·poly(dA)·poly(dT) triplex presents the characteristic ab-

[30] E. Taillandier, J. P. Ridoux, J. Liquier, W. Leupin, W. A. Denny, Y. Wang, G. A. Thomas, and W. L. Peticolas, *Biochemistry* **26**, 3361 (1987).
[31] B. Celda, H. Widmer, W. Leupin, W. J. Chazin, W. A. Denny, and K. Wüthrich, *Biochemistry* **28**, 1462 (1989).

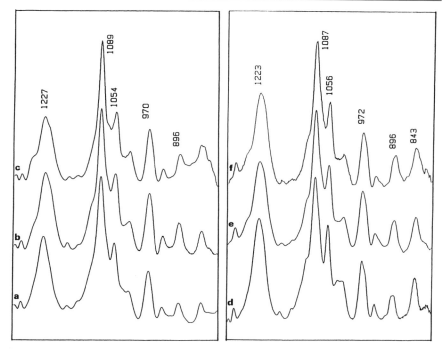

FIG. 14. FT-IR solution spectra in the region of the phosphate vibrations of (a) poly[d(A-T)]; (b) poly[d(A-T)] – netropsin, 1 drug per 4 base pairs; (c) poly[d(A-T)] – netropsin, 1 drug per 3 base pairs; (d) poly(dA)·poly(dT); (e) poly(dA)·poly(dT) – netropsin, 1 drug per 4 base pairs; (f) poly(dA)·poly(dT) – netropsin, 1 drug per 3 base pairs.

sorption of a B-form geometry of the sugar (840 cm⁻¹). If the hydration is decreased, the progressive emergence of a band at 860 cm⁻¹ reflects the presence of some deoxyriboses in the A form (Fig. 13, left, from bottom to top). In contrast both triplexes involving G-C base pairs have been obtained starting with the A form poly(dG)·poly(dC) (Fig. 13, right, a). This polymer is a very rare DNA sequence able to adopt an A form in solution. The solution spectra of both triplexes present a contribution around 830 cm⁻¹, showing that in the triple-strand structure B-form and A-form type sugars coexist. Infrared spectroscopy gives us the opportunity to study the conformation of sugars in the triple helices, which clearly are not all in a C3'-endo geometry, in agreement with the NMR study of d(TC)₄·d(AG)₄·d(TC)₄.[32]

[32] P. Rajagopal and J. Feigon, *Biochemistry* **28**, 7859 (1989).

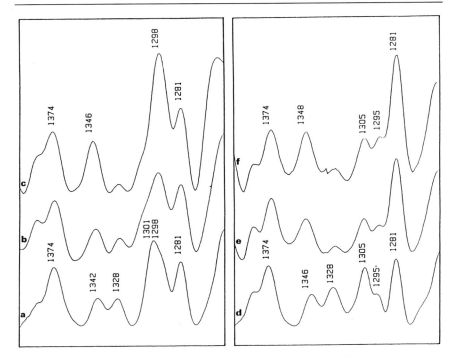

FIG. 15. Spectral region of base–sugar vibrations (as in Fig. 14).

Recognition in Minor Groove by Drugs

Among the numerous available drugs, the family of nonintercalating minor groove-binding drugs has been studied in particular because of their sequence-specific binding properties: netropsin, distamycin, and Hoechst 33258 bind preferentially to A-T base pairs, whereas several derived molecules have been synthesized to recognize G-C sequences.

The interaction of polymers containing A-T base pairs, either poly[d(A-T)] or poly(dA)·poly(dT), with netropsin has been studied in solution by FT-IR.[33] Infrared spectroscopy allows one to explore interactions in solution with polarized groups: phosphate groups, carbonyl groups of the bases, and deoxyriboses. Figures 14, 15, and 16 present the FT-IR

[33] J. Liquier, A. Mchami, and E. Taillandier, *J. Biomol. Struct. Dyn.* **7**, 119 (1989).

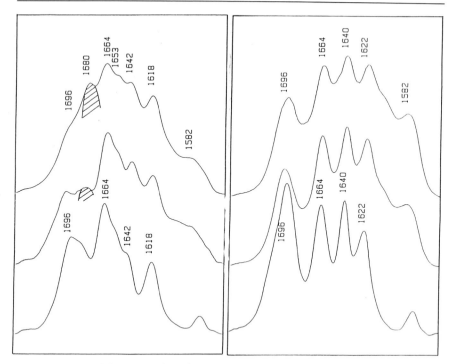

FIG. 16. Spectral region of base double bond vibrations (as in Fig. 14).

spectra of poly[d(A-T)]–netropsin (left-hand side of figures) and poly(dA)·poly(dT)–netropsin (right-hand side of figures) complexes in the spectral regions corresponding to these three putative interaction sites. In Figs. 14–16 the bottom spectra are those of the free polynucleotide, and the middle and top spectra are those of complexes with increasing drug/DNA ratios up to 1 netropsin molecule per 3 base pairs. In Fig. 14 we analyze the interaction with the phosphate groups. Both the antisymmetric stretching vibration band located around 1225 cm⁻¹ and the symmetric stretching vibration band located around 1088 cm⁻¹ remain unaffected by the progressive addition of netropsin. The possibility of an interaction between the phosphates and the drug can therefore be ruled out.

Figure 15 presents the spectra of the same complexes in the region where absorptions due to vibrations of the deoxyribose coupled to the bases are observed. The absorption located at 1328 cm⁻¹ involves vibra-

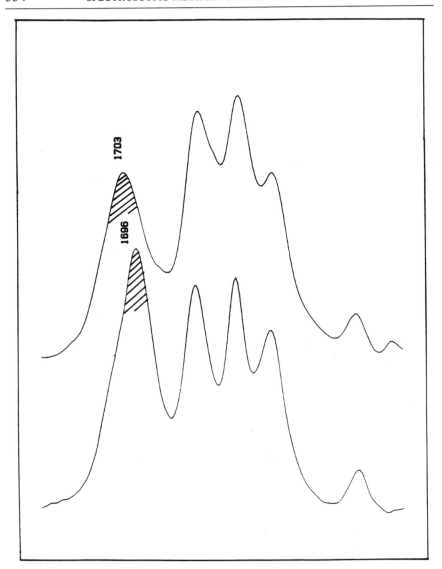

FIG. 17. FT-IR spectra of D$_2$O solutions of poly(dA)·poly(dT) (bottom); poly(dA)·poly(dT)–distamycin (top) (1 drug molecule per 5 base pairs).

tions of the deoxyribose of the dT entity. The important decrease of its relative intensity when the drug is added to either polymer is indicative that an interaction occurs at the level of the sugar.

The spectral region in which the carbonyl vibrations of the bases are detected is presented in Fig. 16. The shift of the C-2=O-2 carbonyl stretching vibration band located at 1696 cm^{-1} in poly[d(A-T)] reflects an interaction of the drug at this site in the narrow groove of the DNA. Infrared spectroscopy shows a different effect in the case of the homopolymer poly(dA)·poly(dT) (Fig. 16, right) than in the case of the alternating polymer poly[d(A-T)].

Figure 17 presents the FT-IR spectrum of a poly(dA)·poly(dT)]–distamycin complex with 1 drug per 5 bases pairs compared to that of the free polymer. We clearly observe the shift of the C-2=O-2 carbonyl band from 1696 to 1703 cm^{-1}, which shows an interaction at the level of the carbonyl group in the minor groove.

[17] Conformation of DNA *in Vitro* and *in Vivo* from Laser Raman Scattering

By WARNER L. PETICOLAS and ELISABETH EVERTSZ

Introduction

Raman spectroscopy is a method of obtaining the vibrational frequencies of a molecule. Some of the vibrational normal modes of large molecules such as DNA are conformationally sensitive. This can lead to changes in frequencies and/or intensities of the Raman bands with changes in conformation. Compared to other techniques, classic Raman spectroscopy has several advantages: minimal or no damage to the sample, relatively little interference from the water Raman signal in aqueous solutions or *in vivo*, a wide range of concentrations, and small sample volumes. In addition vibrational information is obtained that may be representative of one or more conformational states. Finally, the ability to compare Raman spectra from a sample *in vitro* or *in vivo* with Raman spectra from samples in the crystalline state where the precise structure is known from X-ray diffraction studies gives Raman spectroscopy a method of tying crystal structures to conformations in solutions or in living cells. From a comparison of the Raman spectra taken from different crystals of DNA of known structure, a correlation between the frequencies and intensities of Raman bands and the conformation of the DNA has been developed. Using this method, it is possible from the Raman spectrum to obtain the conformation of DNA in aqueous solution, in living cells, and also in crystals that are too small to give satisfactory X-ray diffraction.

To obtain a Raman spectrum of DNA, one places the sample of DNA at the focal point of a focused laser beam. Because a laser beam can be focused into a circle with a radius of approximately its wavelength, this allows for a spatial resolution of the order of 1 μm. The light scattered from the DNA sample in the focused laser beam is collected by means of a lens system and directed through a suitable monochromator to a photon detector. The intensity of the scattered light is measured as a function of its frequency. A plot of the intensity of the scattered light (photons per square centimeter per second) as a function of the frequency difference, Ω, between the incident laser frequency and the scattered light frequency, constitutes a Raman spectrum, that is, $\Omega = \omega_L - \omega_s$. When a band appears in the Raman spectrum at a particular frequency, Ω, it means that there is a normal mode of the molecule that vibrates at this frequency. It is a fundamental principle of the Raman effect that the spectrum is independent of the incident laser frequency. As we shall see later, in the resonance Raman effect, the frequency of the incident light falls within the absorption band of an excited electronic state. When this occurs, the intensity of some bands are enhanced while others are left weak. The rule is that the frequencies of the Raman bands are independent of the frequency of the incident light, but the intensities of the bands may vary.

There are three types of Raman spectrometers in general use today for the characterization of DNA structure or conformation: (1) the classic laser Raman spectrometer,[1-4] (2) the laser Raman microscope,[5,6] and (3) the ultraviolet resonance Raman spectrometer.[7-14] In the classic system the laser used is almost always an argon ion laser with prominent lines at

[1] B. Tomlinson and W. L. Peticolas, *J. Chem. Phys.* **52**, 2154 (1970).

[2] W. L. Peticolas, *in* "Procedures in Nucleic Acid Research" (G. L. Cantoni and D. R. Davies, eds.), Vol. 2, p. 94. Harper & Row, New York, 1971.

[3] E. W. Small and W. L. Peticolas, *Biopolymers* **10**, 69, 1377 (1971).

[4] K. A. Hartman, R. C. Lord, and G. J. Thomas, Jr., *in* "Physico-Chemical Properties of Nucleic Acids" (J. Duchesne, ed.), Vol. 2, p. 1. Academic Press, London, 1973.

[5] T. W. Patapoff, G. A. Thomas, Y. Wang, and W. L. Peticolas, *Biopolymers* **27**, 493 (1988).

[6] G. J. Puppels, F. F. M. de Mul, C. Otto, J. Greve, M. Robert-Nicoud, D. J. Arndt-Jovin, and T. M. Jovin, *Nature (London)* **347**, 301 (1990).

[7] D. Blazej and W. L. Peticolas, *Proc. Natl. Acad. Sci. U.S.A.* **74**, 2639 (1977).

[8] D. Blazej and W. L. Peticolas, *J. Chem. Phys.* **72**, 3134 (1980).

[9] L. Chinsky, P. Y. Turpin, and M. Duquesne, *Biopolymers* **17**, 1347 (1978).

[10] L. C. Ziegler, B. Hudson, D. P. Strommen, and W. L. Peticolas, *Biopolymers* **24**, 2067 (1984).

[11] B. Jolles, L. Chinsky, and A. Laigle, *J. Biomol. Struct. Dyn.* **1**, 1335 (1984).

514.5, 488.0, and 457.9 nm. The advantage of the classic Raman spectrometer is that the sample may be a fiber of DNA,[15-18] a solution of DNA,[19] or a crystal of a DNA oligomer.[20-24] Because these samples are transparent to the visible light, there is little or no heating of the sample by the laser beam. Classic Raman spectra of solutions of DNA are obtained from 5–50 μl samples in a glass capillary mounted in a copper block with the temperature controlled by circulating water from a water bath. The laser light is focused through the capillary, and the scattered laser light is collected at right angles by means of tunnels in the copper block that transmit the incident and scattered light. The concentration of the solution is characteristically 2.0–20 mg DNA per milliliter of solution, or about 30–70 mM in nucleotide.

Classic Raman Spectroscopy

The classic Raman spectrograph was used to obtain the marker bands for the A, B, C, and Z forms of DNA. The marker bands for A-, B-, and C-DNA were discovered in our laboratory in the 1970s[15,16] using fibers of DNA. This work was extended by Goodwin and Brahms.[17] The correlations between structure and spectra were confirmed by direct comparison of the X-ray diffraction patterns with the Raman spectra.[15-17] The marker bands for the Z conformation of DNA containing only alternating cytosine (C) and guanine (G) were discovered by Thamann et al.[20] A great amount

[12] W. L. Kubasek, B. Hudson, and W. L. Peticolas, *Proc. Natl. Acad. Sci. U.S.A.* **82**, 2369 (1985).

[13] S. P. A. Fodor, R. P. Rava, T. R. Hays, and T. G. Spiro, *J. Am. Chem. Soc.* **107**, 1520 (1985).

[14] M. Tsuboi, Y. Nishimura, A. Y. Hirakawa, and W. L. Peticolas, *in* "Biological Applications of Raman Spectrometry" (T. G. Spiro, ed.), Vol. 2, p. 109. John Wiley & Sons, New York, 1987.

[15] S. C. Erfurth, E. J. Kiser, and W. L. Peticolas, *Proc. Natl. Acad. Sci. U.S.A.* **69**, 938 (1972).

[16] S. C. Erfurth, P. J. Bond, and W. L. Peticolas, *Biopolymers* **14**, 1245 (1975).

[17] D. D. Goodwin and J. Brahms, *Nucleic Acids Res.* **5**, 835 (1978).

[18] J. C. Martin and R. M. Wartell, *Biopolymers* **21**, 499 (1982).

[19] S. C. Erfurth and W. L. Peticolas, *Biopolymers* **14**, 247 (1975).

[20] T. J. Thamann, R. C. Lord, H. H. T. Wang, and A. Rich, *Nucleic Acids Res.* **9**, 5443 (1981).

[21] J. M. Benevides, A. H.-J. Wang, G. A. van der Marel, J. H. van Boom, A. Rich, and G. J. Thomas, Jr., *Nucleic Acids Res.* **12**, 5913 (1984).

[22] Y. Yang, G. A. Thomas, and W. L. Peticolas, *J. Biomol. Struct. Dyn.* **5**, 249 (1987).

[23] W. L. Peticolas, Y. Yang, and G. A. Thomas, *Proc. Natl. Acad. Sci. U.S.A.* **85**, 2579 (1988).

[24] G. J. Thomas, Jr., and A. H.-J. Wang, *in* "Nucleic Acids and Molecular Biology" (F. Eckstein and D. M. J. Lilley, eds.), Vol. 2, p. 1. 1988.

of recent work has helped to clarify the assignments of these Raman spectra.[18,21-28] More recently it has been found that DNA polymers and oligomers containing only adenine (A) and thymine (T) can be induced into the Z form using Ni(II) ions in aqueous solution.[29,30]

Changes in the conformation of DNA lead to changes in the frequencies of some of the sugar–phosphate backbone vibrations because of the changes in the torsional angles. There are also changes in the frequencies of some of the base vibrations because the vibrations of the bases are coupled to some extent to those of the backbone. In addition there may be changes in intensity. In general, the base intensities decrease as the DNA goes from a melted or disordered conformation to a double helix. This phenomenon is called Raman hypochromism.[1] The change in the Raman intensity to some extent mirrors or amplifies the changes in the strength of the ultraviolet absorption, a well-known phenomenon called hypochromism.

Figure 1 shows the Raman spectra of hexameric oligonucleotides containing only cytidine (C) and guanine (G). The spectra are of r(CGCGCG) in the A form (top spectrum), d(CGCGCG) in the B form (middle), and d(CGCGCG) in the Z form (bottom). The A form is the natural form of ribonucleic acids (top spectrum), whereas the B-to-Z transition in the deoxyribose oligomer is induced by the change in salt concentration from 0.5 to 6 M. The bands that change with conformation are highlighted in black. These are the marker bands for guanine, cytosine, and the sugar–phosphate backbone as the DNA goes from the A to the B to the Z form.

Figure 2 shows the Raman spectra of pH 7, 0.5 M NaCl aqueous solutions of rA_{10} (top spectrum), dT_{10} (middle), and the duplex $(rA)_{10} \cdot (dT)_{10}$ (bottom). The rA strand is known to self-stack into an A-like structure.[2,3] The 816 cm^{-1} band is a marker band for the A form.[15-18] In the bottom spectrum of Fig. 2 we see the spectrum of a heteronomous duplex in which the rA strand is in the A form and the dT strand is in the B form.[31] A similar result with heteronomous poly(rA) and poly(dT) has also been reported.[24] Figure 3 shows the Raman spectrum of poly[d(A-T)] in the B form (top spectrum) and in the Z form (bottom). The spectrum of

[25] Y. Nishimura, M. Tsuboi, T. Nakano, S. Higuchi, T. Sato, T. Shida, S. Uesugi, E. Ohtsuka, and M. Ikehara, *Nucleic Acids Res.* **11**, 1579 (1983).

[26] Y. Nishimura, M. Tsuboi, and T. Sato, *Nucleic Acids Res.* **12**, 6901 (1984).

[27] J. M. Benevides and G. J. Thomas, Jr., *Nucleic Acids Res.* **11**, 5747 (1983).

[28] W. L. Peticolas, W. L. Kubasek, G. A. Thomas, and M. Tsuboi, *in* "Biological Applications of Raman Spectroscopy" (T. G. Spiro, ed.), Vol. 1, p. 81. John Wiley & Sons, New York, 1987.

[29] J. P. Ridoux, J. Liquier, and E. Taillandier, *Biochemistry* **27**, 3874 (1988).

[30] Z. Dai, G. A. Thomas, E. Evertsz, and W. L. Peticolas, *Biochemistry* **28**, 699 (1989).

[31] Y. Wang, Ph.D. Dissertation, University of Oregon, Eugene, Oregon (1987).

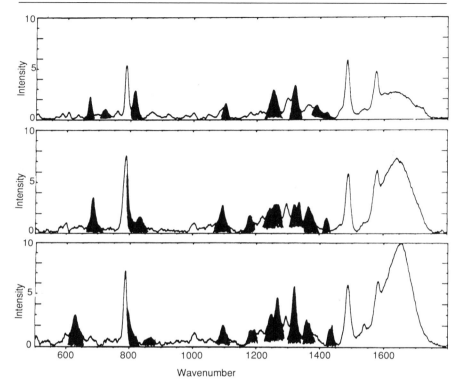

FIG. 1. Raman spectra of aqueous solutions of r(CGCGCG)·r(CGCGCG) in the A form (top), d(CGCGCG)·d(CGCGCG) in the B form (middle), and d(CGCGCG)·d(CGCGCG) in the Z form (bottom). The top and middle spectra were of samples dissolved in 0.5 M NaCl solution, whereas the bottom spectrum is from a sample dissolved in 6 M NaCl. The solutions are self-buffered at pH 7. The marker bands for the A, B, and Z forms are marked in black. All of the spectra were taken with the argon laser line at 514.5 nm (Spectra from Ref. 22.)

the Z form was taken in our laboratory using the recipe of Ridoux *et al.* with Ni(II) to induce the Z form.[29,30]

From spectra such as those in Figs. 1–3 it has been possible to make tables of the changes that occur in DNA when it goes from the standard B form to the A form or the Z form. Similarly, melting the DNA helices to form disordered structures also causes changes in the Raman intensities owing to Raman hypochromism[1–3] as well as changes in backbone and base frequencies. We have made a critical list of such changes, and they are presented in Tables I–III. Table I gives the spectroscopic changes for A,T-containing polymers going from the B form to either the A form or the

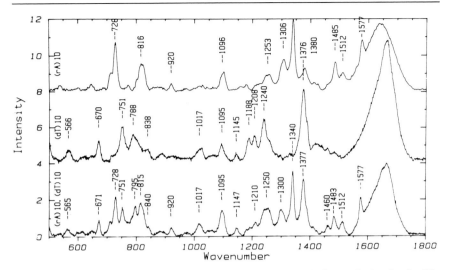

FIG. 2. Raman spectra of pH 7, 0.5 M NaCl aqueous solutions of rA$_{10}$ (top), dT$_{10}$ (middle), and the duplex (rA)$_{10} \cdot$(dT)$_{10}$ (bottom). At bottom we see the spectrum of a heteronomous duplex in which the rA strand is in the A form and the dT strand is in the B form.

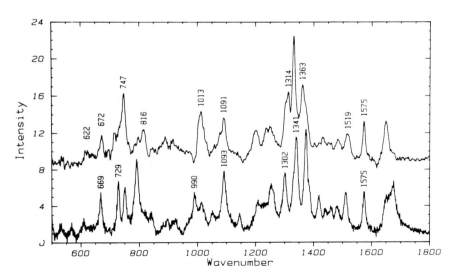

FIG. 3. The Raman spectrum of poly[d(A-T)] $\cdot$ poly[d(A-T)] in the B form (top) and in the Z form (bottom). The Z form was induced using Ni(II) according to the recipe of Ridoux et $al.$[29] The B form naturally occurs at 0.5 M NaCl and pH 7.

TABLE I
SPECTROSCOPIC CHANGES FOR A,T-CONTAINING POLYMERS[a]

Assignment	Disordered form (cm⁻¹)	Change B to disordered ←	B form (cm⁻¹)	Change B to A →	A form (cm⁻¹)
bk	—	—	644sh	inc	644
T	667	s.dec	670	dec	666
			n.r.	inc	706
A	728	inc	729	—	729
T	747	dec/shift	750	dec/shift	747
bk, T	794	s.dec	793	shift	779
C3′-endo bk	—	—	—	inc	807
C2′-endo bk	n.r.	dec	841	dec	n.r.
PO²⁻ bk	1096	—	1094	shift	1102
T	1186	inc	1186	—	—
T, A	1208	dec	1209	dec	n.r.
T	1237	inc	1240sh	inc	1239
T, A	1255sh	dec	1255	dec	1255sh
A	1307	s.inc	1303	s.inc	1301
A	1334	s.inc/shift	1341	inc/shift	1334
T, A	1374	No. change	1376	dec	1374
	—	—	n.r.	inc	1402
A, bk	1421	s.inc	1420	shift	(1415)
bk, A	1462		1462	inc	1462
A	1483	inc	1483	inc/shift	1478

[a] dec, Decrease in intensity; inc, increase in intensity; shift, change in band position; n.r., not resolvable; s., slight intensity change; sh, shoulder; A, adenine; T, thymine; and bk, deoxyribose phosphate backbone.

disordered form. Table II gives the spectroscopic changes for G,C-containing polymers going from the B to the A or the Z or the disordered form. Table III gives the more recent result on the effect of Ni(II) in inducing the Z form in poly[d(A-T)][29] or in oligomers with A-T base pairs.[30] Tables I–III can then be used to obtain subtle changes in conformation within the B form induced by interaction with proteins or to obtain changes from the crystal to the solution form.

Microscope Raman Spectroscopy

Microscope Raman spectroscopy has been used to determine the conformation of DNA in living sperm cells,[32] and in the bands of polytene

[32] W. L. Kubasek, Y. Wang, G. A. Thomas, T. W. Patapoff, K.-H. Schoenwaelder, J. H. Van de Sande, and W. L. Peticolas, Biochemistry 25, 7440 (1986).

TABLE II
SPECTROSCOPIC CHANGES FOR G,C-CONTAINING POLYMERS[a]

Assignment	Disordered form (cm⁻¹)	Change B to disordered ←	A form (cm⁻¹)	Change B to A ←	B form (cm⁻¹)	Change B to Z →	Z form (cm⁻
G¹	—	—	—	—	—	inc	625
G²	—	—	665	inc	—	—	—
G³	680	dec	—	dec	680	dec	—
bk, C	782	—	783	—	784	shift	784
C3'-endo	—	—	808	inc	—	inc	810
C2'-endo	n.r.	dec	—	dec	829	dec	n.r.
bk	—	—	852	inc	—	inc	855
PO²⁻ bk	1094	—	1100	—	1094	dec	1094
G, C	—	—	1180	—	1180	dec/split	1180
	—	—	—	—	—	—	1188
C	1220sh	s.dec	1218	dec	1218	dec	1213
C	1240	inc	1242	s.inc	1240sh	inc/shift	1246
C	1255	inc	1252	inc	1260	inc/shift	1265
C	1292	—	1295sh	s.dec	1293	—	1291
G	1320	inc	1314	inc/shift	1318	inc	1317
G	1333sh	s.dec	n.r.	dec	1334	—	n.r.
G	1361	s.dec	1361	s.dec	1362	dec/shift	1355
G	—	—	1388	appears	—	—	—
bk, G, C	1416	shift	1417	shift	1420	split	1417
bk, G, C	—	—	—	—	—	—	1426
G	1486	—	1482	shift	1489	—	1486
G	1528	inc	—	—	—	—	—
G	—	—	1574	shift	1578	—	1578

[a] dec, Decrease in intensity; inc, increase in intensity; shift, change in band position; n.r., n resolvable; s., slight intensity change; sh, shoulder; G, guanine; C, cytosine; bk, deoxyribose phc phate backbone; G¹, C3'-endo syn; G², C3'-endo anti; and G³, C2'-endo anti.[21]

chromosomes.[6] It has also been used to obtain the Raman spectra of microcrystals of deoxyoligonucleotides.[5,22,31] Figure 4 shows the schematic setup of a laser Raman microscope. The laser beam is sent through the epi-illuminator where it is reflected downward by means of a beam splitter and goes through the objective, where it is focused on the sample. The back-scattered light goes back through the objective, passes the beam splitter, and is focused through a small aperture between the microscope and the monochromator (not shown). The light from the aperture is focused into the monochromator with the proper f-number where the spectrum is recorded with a diode array. A modification of this design has been published with excellent spectra of polytene chromosomes.[6]

TABLE III
SPECTROSCOPIC CHANGES FOR A,T-CONTAINING
POLYMERS FOR THE B-TO-Z TRANSITION

B form	Change[a]	Z form
—	inc	622 cm^{-1} C3'-*endo*/*syn*
748 cm^{-1}	inc	748 cm^{-1}
790 cm^{-1}	shift	(748 cm^{-1})
835 ± 5 cm^{-1}	shift	815 cm^{-1}
1302 cm^{-1}	shift	1312 cm^{-1}
1342 cm^{-1}	shift	1332 cm^{-1}
1374 cm^{-1}	shift	1362 cm^{-1}

[a] inc, Increase in intensity; shift, change in band position.

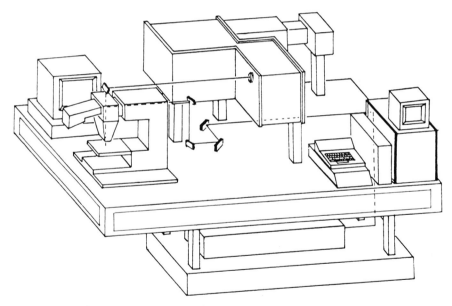

FIG. 4. Diagram of the laser Raman microscope used to take DNA spectra from samples 1 μl or so in volume. The argon laser beam is sent into the epi-illuminator of a Zeiss microscope where it is reflected by means of a beam splitter down through the objective to the same. The back-scattered laser light goes through the beam splitter and out the top of the microscope. The scattered light is then directed on the slit. A small aperture is placed at the focal point of the scattered light between the microscope and the entrance slit. The scattered Raman light is sent into the slit of a triple-grating monochromator so as to match as closely as possible the *f*-number of the monochromator.

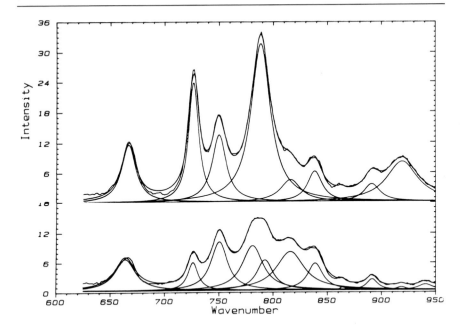

FIG. 5. Raman spectra of a single microcrystal of the duplex of d(AAAAATTTTT) taken with the polarization of the incident light perpendicular to the long axis of the crystal (top) and parallel to the long axis (bottom).

Figure 5 shows the microscope Raman spectra of a single microcrystal of d(AAAAATTTTT) taken with the long axis of the crystal perpendicular to the polarization direction of the incident laser light (top spectrum) and parallel to the laser polarization (bottom). From these spectra it is apparent that the helical axis of the d(AAAAATTTTT) duplex is parallel to the long axis of the crystal. In the top spectrum the polarization of the laser lies parallel to the plane of the bases. The in-plane stretching vibrations of the bases are active in this configuration. In the bottom spectrum, the direction of the laser polarization is parallel to the helical axis of the DNA and hence perpendicular to the bases. In this alignment the out-of-plane vibrations of the bases and the backbone are allowed. The backbone vibrations are allowed in both configurations because the furanose–phosphate chain is never aligned exactly parallel or perpendicular to the helical axis. This procedure has several advantages. It allows the identification of in-plane and out-of-plane vibrations. Furthermore, it allows the identification of the conformation of the DNA crystals of oligomers without the necessity of obtaining X-ray diffraction data.

Figure 6 shows the Raman spectrum of single microcrystals of four self-complementary oligonucleotides, all with the same base composition.

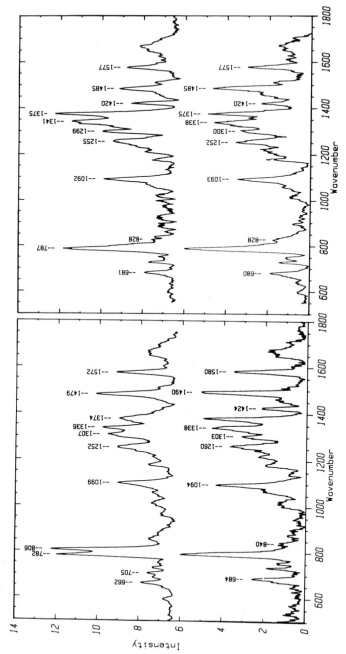

FIG. 6. Raman spectra of single microcrystals of d(GGTATACC) in the A form (top left), d(GGATATCC) in the B form (bottom left), d(GCTATAGC) in the B form (top right), and d(GCATATGC) in the B form (bottom right). Because all these oligomers have the same base composition, the differences in the Raman spectra are due to differences in base sequence.

The spectra were taken with the Raman microscope. Because each oligomer has a different base sequence, each one has a different Raman spectrum. Note that one of these oligomers is in the A form but the other three are in the B form. Slight differences are seen in the different oligomers in the B form. These differences, which can be interpreted by means of the data in Tables I–III, illustrate the point that B-DNA is a genus or family that allows some variation in its conformation.

Resonance Raman Spectroscopy

Resonance Raman spectroscopy of DNA requires the use of ultraviolet light since the absorption bands of DNA lie in the UV.[7–14] This is usually referred to as UV resonance Raman (UVRR) spectroscopy. Typically, light in the region 200–260 nm from pulsed laser sources is used. Because of the intense light used in the pulsed YAG (yttrium–aluminum–garnet) and dye laser systems, photochemical damage is a problem. However, frequency doubling of the continuous argon laser lines at 514.5 and 488.0 nm permits the obtaining of UVRR spectra using 257 and 244 nm without the concurrent photochemical damage.[33] The use of high powered, pulsed lasers permits the generation of far ultraviolet light by means of nonlinear frequency doubling using nonlinear optical crystals such as KDP. Additional use of stimulated Raman effects permits the shift of the laser frequency further into the UV by means of the stimulated anti-Stokes Raman effect in gaseous hydrogen.[10–13] The use of resonance Raman spectroscopy as a tool for conformational studies is just beginning but probably will have a bright future.

Interpretation of Spectra

The Raman bands of DNA come from vibrations that involve all of the atoms in the molecule. However, the displacement of many of the atoms is so small in a particular vibration as to be negligible, and this allows the identification of certain bands with the vibration of certain parts of the DNA. In general, the bands in a Raman spectrum may be assigned to vibrations that are predominately due to (1) the furanose–phosphate backbone or (2) one of the four bases, A, T, G, or C.

Raman Bands of DNA Backbone Vibrations

The original work on the correlation between DNA conformation and DNA Raman spectra concentrated on the bands due to the furanose–

---

[33] L. Chinsky, A. Laigle, W.L. Peticolas, and P.-Y. Turpin, *J. Chem Phys.* **76**, 1 (1982)

phosphate backbone.[2,3,15-17] There are three bands of great importance. The band at about 1100 cm^{-1} is due to the symmetric stretching of the two ionized phosphate oxygens in the diphosphate ester. This band changes from 1100 cm^{-1} in A-form DNA to 1095 cm^{-1} in B-form DNA. The intensity of this band appears to change very little with conformation so that it has been suggested as an internal intensity standard.[34] In RNA there is a sharp, strong band at 814 cm^{-1} that disappears when the double helix is melted.[2-4] This band also appears in the spectrum of A-form DNA although it is shifted to 809 ± 2 cm^{-1}.[15-17] This band was originally assigned to a marker band for the A-type conformations of nucleic acids and is still often loosely refered to as the A genus marker band. However, theoretical analyses of the normal modes of the backbone chain show plainly that this is a mode strongly dependent on the torsional angle delta (δ) about the C5'–C4'–C3'–O3' bond.[35,36] This in turn is related to the ring pucker of the furanose ring. This band at 807–814 cm^{-1} is assigned to the O–P–O symmetric stretching vibration of the X–O–P–O–C5 group of a mononucleotide in a DNA polymer, oligomer, or aggregate.[3] The X is the H atom in a mononucleotide aggregate at pH 4,[3] but in DNA X is the C-3' carbon of the previous nucleotide. In the B-form of DNA this band is weak, and a band at 835 ± 5 cm^{-1} is present.[15-17] The band at 835 ± 5 cm^{-1} is also a sugar furnanose vibration that is due to the antisymmetric O–P–O stretching vibration when the furanose ring is in the C2'-endo ring pucker.[35,36]

For guanine-containing polymers there is a guanine band in the region 625–680 cm^{-1} that is very sensitive to conformation. Although this is a guanine base vibration, it is so strongly coupled to the backbone conformation that we discuss it in this section. Nishimura et al.[25] first noted that the Raman spectra of various nucleic acids indicate that the guanine ring breathing frequency at about 650 cm^{-1} is sensitive to the internal rotation angle around the glycosidic bond and to the conformation of the five-membered ring of the furanose group to which it is directly connected. They suggested that the following correlation exists: 668 cm^{-1} (C3'-endo-anti); 625 cm^{-1} (C3'-endo-syn); 682 cm^{-1} (C2'-endo-anti). Noting the position of this band in A-, B-, and Z-DNA, Benevides and Thomas[27] reached the same conclusion.

Dai et al.[30] have used the 625–680 cm^{-1} bands to determine the fraction of CG base pairs in the Z form in B–Z conformational junctions.

[34] M. Tsuboi, S. Takahashi, S. Muraishi, T. Kaijiura, and S. Nishimura, *Science* **174,** 1142 (1971).
[35] E. B. Brown and W. L. Peticolas, *Biopolymers* **14,** 1259 (1975).
[36] K. C. Lu, E. W. Prohofsky, and L. van Zandt, *Biopolymers* **14,** 1259 (1977).

These investigators made use of the fact that the area of the 650 cm^{-1} band is constant as the conformation is changed even though the frequency changes from 680 to 625 cm^{-1}. By measuring the intensities (integrated areas) of the B-form band at 680 cm^{-1} and the Z-form band at 625 cm^{-1}, the fraction of CG base pairs in the Z form is given by $I_{625}/(I_{625} + I_{680})$. Using the rules for the relation between base sequence and conformation,[23] one can synthesize oligonucleotides of the form $d(T)_N(CG)_M(A)_N$ that are self-complementary. The number of CG base pairs that are in the Z form are obtained. These are reasonably assigned to the center of the sequence. The AT base pairs that flank the oligomer are in the B form and the CG base pairs that lie between the Z portion and the flanking AT portions are in the junction. This indicates that the junction is about 2–4 CG base pairs, that is, 2–4 CG base pairs are not in the Z form. They may be in the B form or in the junction or in a perturbed Z form.[30]

Raman Studies of Conformational Deviations within B-DNA Family

The data discussed above and listed in Tables I–III can now be applied to more subtle conformational variations of DNA that are totally within the B conformation. Such changes are distinctly biologically relevant because virtually all DNA *in vivo* is in the B form,[6,32] so that changes in conformation are subtle and made within the B family. Interesting conformational changes occur owing to the interaction of DNA with DNA-binding proteins. A noted characteristic of B-form DNA is that it is capable of a wide degree of conformational variation within B-form parameters.[37] These variations include B'- and C-DNA, both of which have been analyzed by Raman spectroscopy.[28,37a] This conformational variability may play an important role in protein–DNA recognition in addition to the primary mechanism of recognition by the array of hydrogen bond donors and acceptors presented by a sequence of bases.[38] This issue is difficult to address by crystallographic methods because of the difficulty crystallizing large pieces of DNA, as well as the uncertainty that small deviations are masked by crystal packing forces.

Another factor involved in protein–DNA interactions is that of DNA deformability. How much does a protein binding to a particular DNA sequence bend that DNA, and how does that bending contribute to the

[37] W. Saenger, "Principles of Nucleic Acid Structure." Springer-Verlag, Berlin and New York, 1983.
[37a] G. A. Thomas and W. L. Peticolas, *J. Am. Chem. Soc.* **105**, 986 (1983).
[38] P. H. von Hippel and O. C. Berg, *in* "NATO Advanced Study Institute Proceedings: DNA–Ligand Interactions" (W. Guschlbauer, ed.), p. 159. Plenum, New York, 1987.

stability of the complex? These issues have been addressed in three Raman spectrographic studies of DNA–protein interactions. The proteins studied were *Eco*RI,[39] the catabolite activator protein (CAP),[40] and most recently cro protein.[41] These studies are good examples of the use of Raman scattering to learn about subtle conformational changes that occur in DNA on binding to a DNA-binding protein.

We illustrate the use of Raman spectroscopy to study DNA–protein interaction using the data on the complex of cro protein and its 17-base pair (bp) DNA binding site.[41] The oligonucleotide is a consensus sequence of the six naturally occurring binding sites for cro in phage λ. The affinity of cro for this DNA is about $10^{12}\ M^{-1}$,[42] which is about 6 orders of magnitude higher than the affinity of cro for a random DNA sequence. This short sequence of DNA lends itself quite well to a Raman study because the outer bases are primarily A-T base pairs, whereas the center region of the binding site contains mainly G-C base pairs. Because many of the bands arising from base vibrations are spectroscopically resolvable by base composition, it becomes possible to correlate specific structural features to particular regions of the binding site DNA.

The spectrum of the free cro binding site shows some interesting deviations from the spectra of average B-form DNA. We have taken the spectra of the native DNA given by Wartell and Harrell[43] as our standard B-form DNA. Although the spectrum of the cro binding site is clearly that of B-DNA, the deviations are apparent in the 800–850 cm^{-1} sugar pucker region as well as in bands attributed to adenine base vibrations. In the sugar pucker region the binding site shows a higher than average 803 cm^{-1} band. As mentioned above, a band at 807 cm^{-1} has been assigned to furanose ring vibrations in the C3'-*endo* conformation.[19,35] We assume that the 803 cm^{-1} band in the cro binding site is the same as the usual 807 cm^{-1} band but shifted owing to a slight difference in conformation. It is estimated from the height of this band that the population of C3'-*endo* sugar pucker is about 25%. This is about 20% higher than estimated by Wartell and Harrell for native DNA in their study of eight naturally occurring and synthesized DNAs.[43] Other deviations from an average B-DNA spectrum include an upshifted 1423 cm^{-1} band and a downshifted 1460 cm^{-1} band. These bands normally appear at 1420 and 1463 cm^{-1}, respectively.[28] Both

[39] G. A. Thomas, W. L. Kubasek, W. L. Peticolas, P. Greene, J. Grable, and J. M. Rosenberg, *Biochemistry* **28,** 2001 (1989).
[40] H. De Grazia, D. Brown, S. Cheung, and R. M. Wartell, *Biochemistry* **27,** 6359 (1988).
[41] E. Evertsz, G. A. Thomas, and W. L. Peticolas, *Biochemistry* **30,** 1149 (1991).
[42] J. Kim, Y. Takeda, B. W. Mathews, and W. F. Anderson, *J. Mol. Biol.* **1,** 149 (1987).
[43] R. M. Wartell and J. T. Harrell, *Biochemistry* **25,** 2664 (1986).

of these bands are assigned to primarily adenine base vibrations, with the 1420 cm^{-1} band being coupled to sugar stretching modes.[44] These shifts are not consistent with any known structural variation, but indicate a novel conformational variation specific to this sequence.

The spectrum of the free 21-bp CAP binding site was compared with two 40- and 62-bp DNA fragments and showed mild deviations from average B-form DNA.[40] The deviations included the appearance of a small band at 867 cm^{-1} and a weaker than expected 682 cm^{-1} band. The 867 cm^{-1} band is assigned to backbone vibrations[45] and is characteristic of low humidity fibers in the C conformation.[15,17,46] As noted above, the 682 cm^{-1} band is assigned to guanine vibrations and is well known to be conformationally sensitive.[25,27] The effect of temperature on the conformation of the free binding site was also investigated by comparing the spectra at 25°, 35°, and 45°, all well below the melting temperature of the DNA. Changes that take place with increasing temperature include a 50% increase in the 497 cm^{-1} band assigned to thymine and guanine vibrations. This large change was interpreted as a conformational rearrangement of the guanine residues because it is too large an increase to be attributed to a simple melting transition. Another spectral change observed at increased temperature includes the appearance a small 705 cm^{-1} band, characteristic of a B-to-A transition in DNA of all compositions.[28] This study demonstrated that there are conformational variations in the CAP binding site that may indeed aid in recognition of the DNA by the protein.

Once the protein binds further deformation may occur. These deformations also may contribute to the specificity and stability of a protein–DNA complex. Spectra of the DNA bound to cro protein provide an excellent example of this conformational deviation. These spectra are obtained by subtracting the spectrum of free protein from the spectrum of the complex. This subtraction is valid in most regions of the spectrum if the protein retains its overall composition of α helix and β sheet. This is because most protein bands in this region are assigned to backbone vibrations that are sensitive to large global changes in conformation. Exceptions to this assumption include tyrosine stretching modes in the 810–870 cm^{-1} region and a strong phenylalanine band at 1004 cm^{-1}.[47] In this region the subtraction does not create a pure DNA spectrum, and so no conclusions

[44] R. Letellier, M. Ghomi, and E. Taillandier, *J. Biomol. Struct. Dyn.* **4**, 663 (1987).

[45] Y. Nishimura and M. Tsuboi, *in* "Spectroscopy of Biological Systems" (R. J. H. Clark and R. E. Hester, eds.), Vol. 16, p. 177. Wiley, New York, 1986.

[46] S. R. Fish, C. Y. Chen, and G. J. Thomas, *Biochemistry* **22**, 4751 (1983).

[47] M. M. Siamwiza, R. C. Lord, M. Chen, T. Takamatsu, I. Harada, and T. Shimanouchi, *Biochemistry* **14**, 1377 (1975).

about conformation can be drawn from bands in these regions. Enough of the spectrum is pure DNA, however, to see that conformational changes do occur when the protein binds.

Changes in the spectrum of the bound cro binding site include the appearance of a 705 cm^{-1} band and a sharp decrease in intensity and downshift in the 672 cm^{-1} band. The 705 cm^{-1} band is assigned to a purine-coupled sugar vibration and only appears in A-form DNA.[28] The change in the 672 cm^{-1} band is also consistent with the appearance of some A-like conformation. Here the band is assigned to thymine stretching modes, and so we can say that the AT-rich regions at the end of the binding site are undergoing some A-like conformational changes. Bands assigned to guanine and cytosine also change. These changes include an increase in relative intensity and downshift in the 685 cm^{-1} guanine band, as well as an increase in intensity in the 1262 cm^{-1} cytosine band. These changes indicate an overall destacking in the bases and can be correlated with an order–disorder transition as well as a B-to-A transition because stacking decreases in both cases. To distinguish between these two transitions we observe the positional shift in the small 1320 cm^{-1} guanine vibration: this band upshifts in a melting transition, whereas it downshifts in a B-to-A transition. Here, in the cro–DNA complex, the 1320 cm^{-1} band upshifts, indicating a destacking transition similar to an order–disorder transition. In this way we conclude that cro protein deforms the DNA such that the outer region adopts a slightly A-like conformation, whereas the central region destacks slightly in a manner resembling a simple melting transition.

The Raman study of the *Eco*RI–DNA cocrystal also reveals interesting changes that occur when the protein binds.[39] Here the sequence of the binding site has primarily A-T base pairs in the center and G-C base pairs at the ends. Spectra of the 15-bp *Eco*RI binding site free were compared with protein subtracted spectra of the protein–DNA complex. An increase in the 803 cm^{-1} C3-*endo* band was observed in the spectra of the bound DNA. This change in sugar pucker population was attributed to the center A-T region of the binding site because bands assigned to guanine showed no appreciable A-like character. For example, the 682 cm^{-1} band assigned to guanine-coupled sugar vibrations is very sensitive to changes in sugar pucker, and it does not change on protein binding. Bands which confirm that the A-T region was undergoing conformational changes include a 5 wavenumber downshift in the 750 cm^{-1} thymine band, as well as intensity decreases in the 1226 and 1380 cm^{-1} bands, both assigned to adenine and thymine. These changes have been correlated to melting phenomena as well as B-to-A transitions. Raman analyses of this protein–DNA system clearly show that protein binding causes some deformation in the DNA, demonstrating once again that deformability of particular sequence of DNA is a factor in recognition and specificity of protein–DNA interactions.

Raman spectroscopy can also facilitate the understanding of an unknown DNA conformation. This is the case in the Raman analysis of a parallel-stranded duplex DNA by Otto and co-workers.[48] Here spectra were compared between a conventional antiparallel sequence, containing only adenine and thymine, and a sequence of identical composition, but which allowed only for the formation of a parallel-stranded duplex. The spectra of the parallel-stranded duplex showed the characteristic band around 840 cm^{-1} indicating a basically B-like conformation. By analyzing the 1600–1700 cm^{-1} wavenumber region of the spectra, the authors were able to determine that the base pairing was reverse Watson–Crick in the parallel-stranded duplex. This region has been assigned to carbonyl stretching vibrations that are sensitive to the hydrogen bond environment of the base pairs. In Watson–Crick and Hoogsteen base pairing the C-2=O group of the adenine participates in the base pair. In the parallel-stranded duplex the carbonyl region showed strong differences in the 1661 and 1690 cm^{-1} bands from the corresponding bands in the antiparallel duplex. This demonstrated that the bands were either reverse Watson–Crick or reverse Hoogsteen, and the C-4=O was involved in the base pair. The possibility of reverse Hoogsteen base pairing was eliminated by observing that there was no change in the 1421 cm^{-1} band assigned to adenine base vibrations. This band is known to show a large decrease in intensity when the adenine participates in a Hoogsteen base pair in the triple-strand poly(dT)–poly(dA)–poly(dT) sequence.[49] This study demonstrates that with the existing basis set on the various conformations of DNA it is possible to determine important conformational parameters without a crystal structure.

[48] C. Otto, G. A. Thomas, K. Rippe, T. M. Jovin, and W. L. Peticolas, *Biochemistry* **30**, 3062 (1991).
[49] G. A. Thomas and W. L. Peticolas, *J. Am. Chem. Soc.* **105**, 993 (1983).

[18] Fluorescence Resonance Energy Transfer and Nucleic Acids

By ROBERT M. CLEGG

Introduction

Fluorescence resonance energy transfer (FRET) is a spectroscopic process by which energy is passed nonradiatively between molecules over long distances (10–100 Å). The "donor" molecule (D), which must be a fluorophore, absorbs a photon and transfer this energy nonradiatively to the "acceptor" molecule (A). A diagram of FRET is presented in Fig. 1. Energy can be transferred over distances on the order of common macromolecular dimensions (10–100 Å). This chapter will emphasize applications of FRET to nucleic acid molecules conjugated to dye molecules, but in general the donor–acceptor (D–A) pairs can be free in solution, one or both bound (covalently or noncovalently) to a macromolecule, or be an inherent part of the structure.

Interest in applications to biomolecules has existed since the first theoretical treatments of Perrin and Förster,[1-5] and FRET is frequently employed in polymer science.[6] The method has become popular in biochemistry[7-9] and has been used especially to study protein structures. The historical emphasis on protein structures and the relative scarcity of FRET applications to nucleic acids are due only partially to the particular research interests of the investigators. This predilection is also due to the relative ease with which certain proteins could be specifically labeled in the past, compared to nucleic acids. With the advent of automatic nucleic acid synthesizers, and the recent advances in solid-state chemical synthesis, it is now possible routinely to make oligonucleotides and to conjugate fluorescent probes at known nucleotide positions. Methods to label synthetic

[1] J. Perrin, *C. R. Acad. Sci. (Ser. 2)* **184,** 7 (1927).
[2] J. Perrin, *Ann. Phys. (Paris)* **17,** 283 (1932).
[3] T. Förster, *Ann. Phys.* **2,** 55 (1948).
[4] T. Förster, "Fluoreszenz Organischer Verbindungen." Vandenhoeck & Ruprecht, Göttingen, Germany, 1951.
[5] T. Förster, *in* "Modern Quantum Chemistry" (O. Sinanoglu, ed.), Part 3, p. 93. Academic Press, New York, 1966.
[6] B. Valeur, *in* "Fluorescent Biomolecules: Methodologies and Applications" (D. M. Jameson and G. D. Reinhart, eds.), p. 269. Plenum, New York, 1989.
[7] I. Z. Steinberg, *Annu. Rev. Biochem.* **40,** 83 (1971).
[8] L. Stryer, *Annu. Rev. Biochem.* **47,** 819 (1978).
[9] R. H. Fairclough and C. R. Cantor, this series, Vol. 48, p. 347.

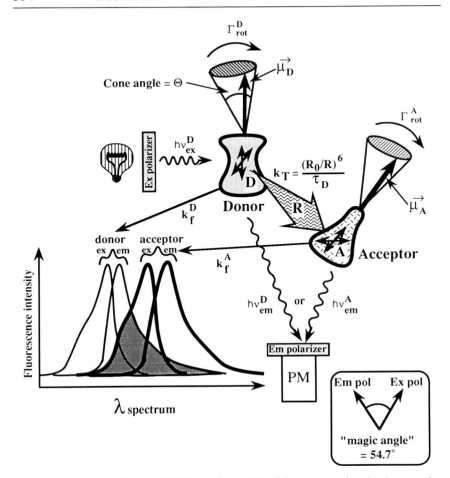

FIG. 1. Diagram depicting a FRET experiment. The light source excites the donor molecule (D, in solution) in the absorption band of the donor $h\nu_{ex}^D$ (the acceptor is also excited directly if the absorption spectra of the donor and acceptor overlap at this wavelength). The donor molecule loses this energy spontaneously by one of the processes shown in Fig. 2; here only the energy transfer k_T and photon emission k_f^P are shown. The cone above the donor depicts the solid angle of rotational freedom of the emission transition dipole moment $\vec{\mu}_D$ of the donor molecule; the angle Θ is the extent of this rotational movement, and Γ_{rot}^D expresses the rotational diffusion constant of the D chromophore (the molecules can rotate during their excited lifetime). R is the distance between the donor and acceptor. Γ_{rot}^A (with the corresponding cone) is the corresponding acceptor quantity, and μ_A is the absorption transition dipole of the acceptor. k_T and the two k_f's are defined in the legend to Fig. 2. τ_D is defined in Eq. (2) and R_0 in Eq. (6). In general, both the donor and the acceptor will emit (wavy lines with $h\nu_{em}^A$ and $h\nu_{em}^D$). The probability for energy transfer proportional to k_T decreases rapidly with the distance, $(R_0/R)^6$ [see Eq. (6)]. The fluorescence spectra indicate typical shapes of excitation and emission fluorescence patterns of donors and acceptors. The overlap of the emission of

DNA and RNA molecules are not discussed here, and space limitations prohibit a discussion of the merits of particular donor–acceptor pairs. The reader is referred to other sources for general information[10-12] and for recent specific examples.[13-15] The DNA synthesis methods open up new applications of steady-state as well as time-resolved fluorescence techniques (e.g., intensity measurements, quenching, polarization, photochemistry, energy transfer) to essentially any sequence of DNA that is covalently labeled at specific nucleotides with a fluorescence probe. In principle, the applications of these spectroscopic methods are independent of the particular molecular structure studied, at least as far as the basic theoretical description of the physical measurements are concerned.

Excellent discussions of FRET exist in the literature dealing with the theoretical background[3-5,16,17] and specifying the physical parameters that are available from a variety of fluorescence measurements.[18-22] Here some basic considerations and experimental procedures are described that are necessary to make reliable FRET measurements. The description assumes

[10] S. Agrawal, C. Christodoulou, and M. J. Gait, *Nucleic Acids Res.* **14,** 6227 (1986).
[11] M. S. Urdea, D. D. Warner, J. A. Running, M. Stempten, J. Clyne, and T. Horn, *Nucleic Acids Res.* **16,** 4937 (1988).
[12] R. P. Haugland, "Handbook of Fluorescent Probes and Research Chemicals." Molecular Probes, Eugene, Oregon, 1990 (yearly).
[13] R. A. Cardullo, S. Agrawal, C. Flores, P. C. Zamecnik, and D. E. Wolf, *Proc. Natl. Acad. Sci. U.S.A.* **85,** 8790 (1988).
[14] A. I. H. Murchie, R. M. Clegg, E. von Kitzing, D. R. Duckett, S. Diekmann, and D. M. J. Lilley, *Nature (London)* **341,** 763 (1989).
[15] J. P. Cooper and P. J. Hagerman, *Biochemistry* **29,** 9261 (1990).
[16] D. L. Dexter, *J. Chem. Phys.* **21,** 836 (1953).
[17] D. L. Dexter and J. H. Schulman, *J. Chem. Phys.* **22,** 836 (1954).
[18] T. Förster, *Naturwissenschaften* **33,** 166 (1946).
[19] T. Förster, *Discuss. Faraday Soc.* **27,** 7 (1959).
[20] F. Wilkinson, *Adv. Photochem.* **3,** 241 (1965).
[21] I. Z. Steinberg, E. Haas, and E. Katchalski-Katzir, *in* "Time-Resolved Spectroscopy in Biochemistry and Biology" (R. B. Cundall and R. E. Dale, eds.), p., 411. Plenum, New York, 1983.
[22] C. Bojarski, and K. Sienicki, *in* "Photochemistry and Photophysics" (J. F. Rabek, ed.), Vol. 1, p. 1. CRC Press, Boca Raton, Florida, 1990.

the donor and the absorption of the acceptor is illustrated by the shaded portion. Linear polarizers must be placed in front of the cuvette and before the photomultiplier (PM); artifacts in FRET measurements arising from linear polarization of the fluorescence are avoided if the axes of the two polarizers are placed at the "magic angle" relative to each other. This is especially important if the molecules have long tumbling rotational relaxations, such as DNA molecules.

no detailed previous knowledge by the reader, and emphasis is placed on basic ideas and methods so that the researcher can derive the necessary mathematical expressions and plan an experimental strategy for each particular situation. Some difficulties that can arise in the experimental proceedings and data analysis as well as considerations of sources of experimental error are pointed out. The main emphasis is on methods that are relatively simple and evident, and the presentation is biased toward eventual applications to nucleic acids (the literature will not be reviewed, but the interest reader should consult the classic FRET experiment on tRNA[23] and later related studies[24,25]). This chapter is restricted to singlet–singlet fluorescence energy transfer; however, the methods discussed are applicable to other energy transfer phenomena, such as triplet–singlet and triplet–triplet transfer.

Description of Spectroscopic Processes

There are many excellent general references on fluorescence,[26-30] and several reviews of fluorescence techniques[31-35] and energy transfer[7-9,36] are directed to a biochemical audience. Discussions dealing with general experimental techniques of measuring fluorescence can be found in some of these sources.

Fluorescence is a dynamic process; much of the information available from fluorescence measurements is derived ultimately from expressions relating the rates of the different conversions and the influence of the

[23] K. Beardsley and C. R. Cantor, *Proc. Natl. Acad. Sci. U.S.A.* **65,** 39 (1970).

[24] C.-H. Yang and D. Söll, *Proc. Natl. Acad. Sci. U.S.A.* **71,** 2838 (1974).

[25] R. F. Steiner and Y. Kubota, *in* "Excited States of Biopolymers" (R. F. Steiner, ed.), p. 203. Plenum, New York, 1983.

[26] D. M. Hercules, "Fluorescence and Phosphorescence Analysis: Principles and Applications." Wiley *(Interscience),* New York, 1966.

[27] C. A. Parker, "Photoluminescence of Solutions." Elsevier, Amsterdam, 1968.

[28] J. B. Birks, "Photophysics of Aromatic Molecules." Wiley, New York, 1970.

[29] R. S. Becker, "Theory and Interpretation of Fluorescence and Phosphorescence." Wiley, New York, 1969.

[30] J. R. Lackowicz, "Principles of Fluorescence Spectroscopy." Plenum, New York, 1983.

[31] F. W. J. Teale and G. Weber, *Biochem. J.* **65,** 476 (1957).

[32] J. R. Laurence, this series, Vol. 4, p. 174.

[33] L. Brand and B. Witholt, this series, Vol. 11, p. 776.

[34] C. R. Cantor and T. Tao, *in* "Proceedings in Nucleic Acid Research" (G. L. Cantoni and D. R. Davies, eds.), Vol. 2, p. 31. Harper & Row, New York, 1971.

[35] D. A. Harris and C. L. Bashford, eds., "Spectrophotometry and Spectrofluorimetry." IRL Press, Oxford, 1987.

[36] P. W. Shiller, *in* "Biochemical Fluorescence Concepts" (R. F. Chen and H. Edelhoch, eds.), Vol. 1, p. 285. Dekker, New York, 1975.

environment on these molecular spectroscopic rate constants. Figure 2A is an energy level diagram with the principal spectroscopic transitions important for this presentation of FRET. The first step of a FRET molecular process is the absorption of a quantum of energy (usually a photon) by the donor molecule.[37] D can also be excited into higher excited states S_n, but these will relax in approximately 10^{-12} sec to S_1 by internal conversion; the donor emission is almost always from S_1. Once the excited molecule is in thermal equilibrium with the surrounding medium (we assume that only the lowest vibrational level is occupied), it will return to the ground state by spontaneous radiative or nonradiative processes. The fate of each excited molecule now depends on the relative rates of several kinetic pathways away from this excited state. Figure 2A shows the deexcitation processes available to an excited donor molecule in the lowest vibrational level of the first excited singlet state S_1. All processes are nonradiative, except k_f. The process which we are concerned about is the rate k_T; however, the rates of all other processes affect the efficiency of long-range energy transfer.

Deactivation of D* by Fluorescence Resonance Energy Transfer

The deactivation process k_T (Fig. 2A) will become possible when an acceptor molecule A is close to the donor "sensitizer" molecule D. If the energy difference of the absorption transition in the nearby acceptor is the same as the energy difference of the deactivation process of the donor, both these processes can occur similtaneously provided that there is sufficient energetic coupling between the donor and acceptor. The spectra of polyatomic molecules in solution are broad, and the amount of spectral overlap is often sufficient for this "resonance transfer" to take place (Fig. 1). It is important to realize that the energy is transferred by inductive resonance; the process is nonradiative, and no photon is actually emitted from the donor. FRET can be viewed as a chemical reaction, as shown in Fig. 2B, where D* and A react over a long distance to form A* and D. Energy transfer from the excited triplet state to form either an excited singlet or triplet state in a nearby acceptor by the Förster mechanism is also possible.[20] Energy transfer from a long-lived donor has not yet found extensive use in biochemistry with doubly labeled macromolecules; however, diffusion-enhanced energy transfer[38,39] has been very useful, and there are specific problems where the long time of triplet decay (or decay from lanthanides) is a distinct advantage.

[37] Throughout the text, the donor and acceptor molecules in the ground and excited states, respectively, are referred to as D, D*, A, and A*.
[38] D. D. Thomas, W. F. Carlsen, and L. Stryer, *Proc. Natl. Acad. Sci. U.S.A.* **75**, 5746 (1979).
[39] L. Stryer, D. D. Thomas, and C. F. Meares, *Annu. Rev. Biophys. Bioeng.* **11**, 203 (1982).

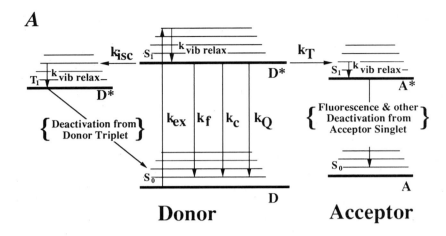

Donor   **Acceptor**

B   **Energy transfer reaction scheme**

$$D^* \cdots\cdots A \xrightarrow{\;k_T\;} D \cdots\cdots A^*$$

FIG. 2. (A) Energy level diagram (Jablonski diagram) showing the normal transitions which influence FRET measurements (in condensed systems). The thick lines are the lowest vibrational levels of the electronic molecular states, and the fine lines represent the higher vibrational levels. S_0, S_1, and T_1 refer to the ground and first excited singlet electronic states and the first excited triplet state. The donor is excited k_{ex} by a photon into higher vibrational levels of the first singlet state S_1 of the donor D^*. Following vibrational relaxation to the lowest vibrational state of D^*, the excitation energy is lost spontaneously by one of the following conversions: (1) internal conversion k_c ($\sim 10^8$ sec^{-1}), (2) intersystem crossing to a triplet state k_{isc} ($\sim 10^8$ sec^{-1}) with subsequent conversion of the triplet state to the ground singlet (which we do not discuss here), (3) collisional quenching reactions with an external quencher, $K_Q = k_q[Q]$, where Q is a quencher species (such as oxygen, halides, or heavy metals) and k_q is a bimolecular rate constant, usually diffusion controlled ($k_q = 10^9$–10^{10} M^{-1} sec^{-1}), (4) long-range resonance dipole energy transfer k_T {this rate is variable [Eq. (6)]; when it competes with fluorescence then $k_T \approx 10^8$ sec^{-1}} to an acceptor molecule (this can also be a nonfluorescent molecule) close to the excited molecule, and (5) radiative decay (fluorescence) to the ground electronic state. All processes, other than the last, are nonradiative. Equation (1) in the text shows the lifetime of the excited donor singlet state according to (A). The deactivation processes from the donor triplet state T_1 and from the singlet state of the excited acceptor are similar to those shown for the S_1 state of D^*, except that the transition from T_1 of D^* to S_0 of D is very slow. The latter transition is a spin nonallowed; the delayed emission from T_1 of D^* decays with a rate of around 10^0–10^6 sec^{-1}, depending on the concentration of external quencher, heavy atoms, or temperature. (B) Representation of long-range radiationless resonance energy transfer as a chemical reaction; the dotted lines indicate that the molecules are close to each other but do not contact.

Donor Decay Lifetimes in Presence and Absence of Acceptor

The decay rate of an excited molecule (in this case the donor) is the sum of the individual rates away from the excited state. With the inclusion of energy transfer (i.e., in the presence of an acceptor), the rate of decay (i.e., the reciprocal lifetime) from D* is (see Fig. 2A)

$$\tau_{DA}^{-1} = k_c + k_{isc} + k_Q + k_T + k_f \tag{1}$$

and without energy transfer (in the absence of an acceptor) the rate is

$$\tau_D^{-1} = k_c + k_{isc} + k_Q + k_f \tag{2}$$

The subscript D refers to the donor species alone (i.e., in the absence of the acceptor), and DA refers to the decay process of the donor in the presence of the acceptor. Equations (1) and (2) include the common kinetic decay mechanisms. Additional kinetic processes are included by simply adding the rate constant for this process ($k_{process}$) to both equations.

Definition of Efficiency of Energy Transfer E and determination k_T

The rate of energy transfer, k_T, is the fundamental characteristic describing energy transfer; however, in steady-state experiments a quantity called the efficiency of energy transfer E is usually determined. It is important to know the different ways E is defined, since E is the usual experimental variable determined in FRET measurements. E is defined below theoretically in three ways: from *(a)* the decrease in the decay times of the donor [Eq. (3)], *(b)* the decrease in the quantum yield of the donor [Eq. (4)], and *(c)* the enhancement of the fluorescence emission of the acceptor (if it is fluorescent) [Eq. (5)].

(a) The quantum yield of any spectroscopic process $\Phi_{process}$ is the ratio of the rate away from some selected state by way of this process, divided by the total rate of escape from this state; that is, $\Phi_{process} = k_{process}/$(total rate of escape). This is equivalent to defining $\Phi_{process}$ as a number equal to the quanta lost from some particular molecular state through this process, divided by the total quanta absorbed to create this particular state. For instance, according to Eq. (1), $\Phi_f^P = k_f/(k_c + k_{isc} + k_Q + k_T + k_f)$ is the quantum yield of donor fluorescence, $\Phi_{isc}^D = k_{isc}/(k_c + k_{isc} + k_Q + k_T + k_f)$ is the quantum yield of donor intersystem crossing (i.e., the rate at which the donor triplet state is populated from the donor excited singlet), and other quantum yields are defined accordingly. The "quantum yield of energy transfer" from the donor is

$$\Phi_{ET}^D = k_T/(k_c + k_{isc} + k_Q + k_T + k_f) = k_T\tau_{DA} \tag{3a}$$

By using Eqs. (1) and (2), Eq. (3a) can be written as

$$\Phi_{ET}^P = 1 - \tau_{DA}/\tau_D \equiv E \tag{3b}$$

where the symbol E is the efficiency of energy transfer.

The efficiency of resonance energy transfer E could just as well be called the quantum yield of energy transfer. Equation (3a) is the basic definition of E, and Eq. (3b) shows the direct connection to the lifetime of the excited state; E is the ratio of the number of quanta transferred per second from the excited donors to the acceptors to the total number of molecular conversions per second leading away from the excited donors (which is equal to the number of quanta absorbed by the donor), also in the presence of acceptors. E is the quantity that must be determined when using spectral, or quantum yield, methods to measure the extent of energy transfer. It is obvious from Eq. (3a) that this can be done only if the decay rates other than k_T are unaffected by the presence of the acceptor, and if the energy levels remain the same in both circumstances.

(b) Equation (3) can be rewritten by using Eqs. (1) and (2) together with the definitions given above for the quantum yields to give

$$E = 1 - (\Phi_f^{PA})/(\Phi_f^P) = (\Phi_f^P - \Phi_f^{PA})/\Phi_f^P \tag{4}$$

$(\Phi_f^{PA})/(\Phi_f^P)$ is the ratio of fluorescence quantum yields of the donor in the presence and absence of the acceptor; it is equal to the number of photons emitted from the donor (per unit time) in the presence of the acceptor, divided by the number of photons emitted from the same population of donor molecules in the absence of acceptor, assuming the same number of photons is absorbed in both cases. Consequently, Eq. (4) defines the efficiency of energy transfer E as the difference in the fluorescence quantum yields in the absence and presence of acceptor divided by the quantum yield without the acceptor. It is only possible to determine E from the definition of Eq. (4) if the adsorption and emission spectra of the donor are identical in the presence and absence of the acceptor (i.e., the interaction between the donor and acceptor cannot be so strong as to affect the molecular energy levels).

(c) The definition of E given in (b) can be reformulated in terms of acceptor emission if the acceptor molecule is fluorescent (acceptor fluorescence is not required in order to apply FRET). From the point of view of acceptor emission, E is equal to the ratio of the number of photons per second emitted from the acceptor molecules (that have been excited via energy transfer from the donor) to the photons per second that would be emitted by the acceptor (again only from those acceptors excited by energy transfer) if all the energy were to be transferred from the donor to the acceptor. That is, in terms of the acceptor emission

$$E = \frac{(\text{measured acceptor emission, excited via FRET})}{(\text{acceptor emission for 100\% transfer, excited via FRET})} \quad (5)$$

If $E = 1$ all the energy from the donor is lost to the acceptor, and the donor does not fluoresce at all.

The variables in Eqs. (4) and (5) cannot be measured directly [in contrast to Eq. (3), where the lifetimes are directly measured], and the remainder of this chapter is mainly concerned with methods to determine E from spectral data. The definitions in Eqs. (4) and (5) [which are derived from the definition in Eq. (3)] are used below to write expressions for the measured emission intensities (with contributions from both the donor and acceptor).

Fluorescence is not the only experimental variable to observe FRET; any process occurring through the excited donor state D^* will be affected by FRET. This is clearly seen from Eqs. (1), (2), and the definition of $\Phi_{process}$. For instance, if a photochemical reaction proceeds from D^*, the rate of this process must be added to both Eqs. (1) and (2). Thus, just as Φ_f is diminished by a significant k_T, $\Phi_{process}$ will also decrease relative to the case without energy transfer. If a photochemical reaction proceeds through the triplet state. FRET decreases the photochemical quantum yield by reducing Φ_{isc}, which builds up the triplet state. The action spectrum of thymine dimerization is affected by the presence of intercalated dyes; depending on the temperature and the particular dye, the decrease of the extent of the photochemical reaction has been attributed to singlet–singlet (room temperature), triplet–singlet (77 K), or triplet–triplet energy transfer (77 K).[40] This can be very useful; for instance, a photochemical reaction involving D^* will proceed slower the nearer the photoreactive donor is to an acceptor. If the donor is a free fluorescence dye which binds and photoreacts with the DNA (with a rate k_{PR}), there will be fewer photoproducts at those positions on the DNA molecule where an acceptor is present. If $R \ll R_0$, the photoreaction will be essentially stopped. If the rate of the reaction is $\gg 1/\tau_D$ (where $1\tau_D$ is the decay rate in the absence of the photoreaction), then donor fluorescence will not be observed; energy transfer will compete with the photoreaction when $k_T \approx k_{PR}$. Of course if the acceptor can react photochemically on excitation, this also provides a way to observe resonance energy transfer. To the knowledge of the author, such photochemical applications of energy transfer have not been employed yet with specific DNA sequences conjugated to acceptor (or donor)

[40] B. M. Sutherland and J. C. Sutherland, *Biophys. J.* **9**, 1045 (1969).

molecules; coupled with electrophoretic footprinting it could prove to be a useful technique.

A related novel technique to determine E has been developed recently[41,42] that takes advantage of donor photobleaching, a photochemical process that depletes the total pool of D molecules (via the D* excited state). The rate of donor photobleaching of a sample is measured in the presence and absence of an acceptor, and E is calculated by comparing the two bleaching rates. The efficiency is calculated from $E = 1 - \tau_D^{bl}/\tau_{DA}^{bl}$, where τ_D^{bl} is the lifetime of the photobleaching reaction in the absence of acceptor and τ_{DA}^{bl} is the photobleach lifetime in the presence of the acceptor (the sample photobleaching rate is much smaller, by about 12 orders of magnitude, than the rate of fluorescence). The rate of photobleaching is decreased in the presence of the acceptor, because energy transfer to the acceptor provides an additional pathway of escape from the D* state. This unique method promises to be useful in the light microscope, where it is more difficult to make precise spectral measurements and lifetime determinations.

Förster's Theoretical Expression for k_T in Terms of Spectroscopic Parameters, Relating Rate of Transfer to Distance between Donor and Acceptor

According to Eq. (3) the value of k_T determines the efficiency of transfer. To interpret FRET measurements on a molecular scale a theoretical expression for k_T in terms of molecular, solvent, and spectroscopic parameters is required. Förster[3] proposed a nonradiative mechanism for FRET which arises from long-range dipole–dipole interactions between the donor and acceptor (separated by the distance R). According to this, the transfer arises from mutual resonance dipole perturbations between the donor and acceptor (the Förster mechanism does not involve trivial reabsorption by the acceptor of a photon emitted by the donor). Dipole–dipole interaction energy decreases as the third power of the distance between the dipoles, and it is dependent on the relative orientation of the two dipoles (this is the factor κ in Fig. 3).[43] The probability of energy transfer (i.e., rate k_T) is proportional to the square of this interaction energy.

Subsequent to a classic description by Perrin[1,2] which predicted energy transfer over longer distances than observed, Förster[3-5] derived quantum

[41] T. Hirshfeld, *Appl. Opt.* **15**, 3135 (1976).
[42] T. M. Jovin and D. J. Arndt-Jovin, *in* "Cell Structure and Function by Microspectrofluorometry" (E. Kohen and J. G. Hirshberg, eds.), p. 99. Academic Press, New York, 1989.
[43] J. D. Jackson, "Classical Electrodynamics," Chap. 4. Wiley, New York, 1962.

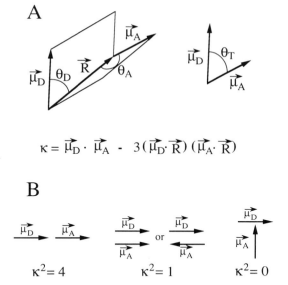

$$\kappa = \vec{\mu}_D \cdot \vec{\mu}_A - 3(\vec{\mu}_D \cdot \vec{R})(\vec{\mu}_A \cdot \vec{R})$$

FIG. 3. (A) Geometrical definition of κ from Eq. (6). κ is the orientation factor for dipole–dipole coupling, where μ_D is the emission transition dipole vector for the donor, μ_A is the absorption transition dipole of the acceptor, $\mathbf{R}$ is the separation vector pointing from the donor to the acceptor, θ_T is the angle between the donor and acceptor dipoles, and θ_A and θ_D are the angles between the dipoles and a vector joining the D–A pair. The mathematical expression for κ (a scalar) is given as a vector dot product. According to this definition, $\kappa = \cos(\theta_T) - 3\cos(\theta_A)\cos(\theta_D)$. The square of κ enters into Eq. (6); the possible values for κ are $0 < \kappa^2 < 4$. The interaction energy between two dipoles is proportional to $\kappa/(R)^3$. (B) Illustration of the orientations of μ_D and μ_A relative to each other, and of each dipole relative to the line separating them, that give $\kappa^2 = 4$, 1, and 0.

mechanically the following expression for the rate of energy transfer:

$$k_T = \left(\frac{9(\ln 10)\Phi^D \kappa^2 J(v)}{128\pi^5 N_A n^4 \tau_D R^6}\right) = \left(\frac{R_0}{R}\right)^6 \Big/ \tau_D \qquad (6)$$

where Φ^D and τ_D are the donor emission quantum yield and lifetime in the absence of the acceptor; $J(v)$ is the spectral overlap integral, $J(v) = \int_0^\infty (\varepsilon^A(v)\phi(v)/v^4)\, dv$ (the overlap region is indicated by the shaded portion of the spectra in Fig. 1), $\varepsilon^A(v)$ is the molar absorbance of the acceptor at the wavenumber v, $\phi(v)$ is the normalized flluorescence spectral shape function (its integral is normalized to 1); N_A is Avogadro's number; R is the scalar D–A separation (see Figs. 1 and 3); κ is defined and illustrated in Fig. 3 (see below for discussions of κ); and n is the refractive index of the medium between the donor and acceptor (for nucleic acids this has been suggested to be 1.4; it does not vary more than $\sim 7\%$ from water to 80% sucrose[34]). R_0 is the distance between the donor and acceptor when $k_T = 1/\tau_D$, and it is defined according to Eq. (6) to be

$$R_0^6 = (8.785 \times 10^{-25})\Phi^D\kappa^2 n^{-4}J(\nu) \qquad cm^6 \qquad (7)$$

R_0 is between 10 and 70 Å for most D–A pairs; the actual value depends on the relative angles between the dye transition moments and the actual quantum yield of the conjugated donor [see Eq. (7)]. Fairclough and Cantor[9] have a short table of useful D–A R_0 values assuming $\kappa^2 = 2/3$, a large compilation for aromatic fluorophores can be found in Berlman,[44] and much useful information can be found in handbooks.[12]

For a particular configuration of the D–A pair the efficiency of energy transfer E [Eq. (3b)] is related to the k_T for this D–A configuration:

$$E = \frac{k_T}{(\tau_D^{-1} + k_T)} \qquad (8)$$

By using the theoretical expression of Eq. (6) for k_T, the experimentally determined value of E can be related very simply to the ratio R/R_0:

$$E = \frac{1}{[1 + (R/R_0)^6]} \qquad (9)$$

From Eqs. (8) and (9) the association between lifetimes and R can be derived:

$$\frac{\tau_D}{\tau_{DA}} = 1 + (R_0/R)^6 \qquad (10)$$

Equation (6) is the pivotal connection between theory and experiment in FRET. When $R = R_0$, $E = 0.5$, and k_T is equal to the donor fluorescence decay time in the absence of the acceptor (i.e., $1/\tau_D$). R_0 can be calculated for a particular D–A pair if the parameters in Eq. (7) are known. R_0 is often measured in model systems and then assumed to apply to the FRET measurement in question. The above general references on energy transfer should be consulted for these procedures. It is important to keep in mind that R_0 may vary for different configurations of the D–A pair on the macromolecule, leading to a variety of values for E. In this case an average value of E will be measured. In addition, even if R_0 is the same for all D–A pairs, R may vary, and the value of E will be an average over the ensemble of D–A pairs (see below).

44 I. B. Berlman, "Energy Transfer Parameters of Aromatic Compounds." Academic Press, New York, 1973.

General Expressions for Fluorescence Intensities in Presence of Fluorescence Resonance Energy Transfer

For the remainder of this chapter the specific application in Fig. 4 is used to illustrate FRET measurements on nucleic acid structures. The system outlined in Fig. 4 encompasses many attributes of future FRET measurements on nucleic acids. Because DNA and RNA with the normal

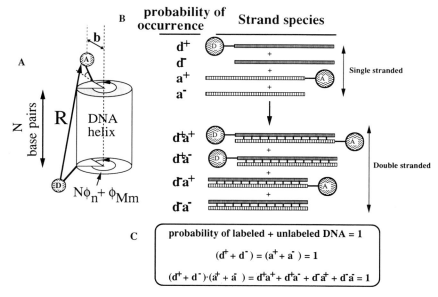

FIG. 4. (A) Typical application of FRET to a DNA molecule with the donor and acceptor conjugated to the ends of the helix. The maximum measurable distance for FRET will not extend over three DNA helical turns (this will be ~$2R_0$ for the D–A pairs with the largest R_0 values). R_0 is between about 10 and 60 Å for most dye pairs; thus, it may be necessary to consider the helical geometry shown in (A) when writing expressions for E in terms of R and κ^2. If the distance between the donor and acceptor varies as shown, and if b is the same for D and A,

$$R = [(Nh)^2 + 2b \sin[(n\phi_n + \phi_{Mm})/2]^2]^{1/2}$$

ϕ_{Mm} is the phase angle between the two labeling positions on the helix (when N theoretically is zero). Extensions including the angles of the tethers are obvious. (B) Single-stranded DNA molecules are labeled with the donor or acceptor on complementary strands. d^+ and a^+ are the percent donor or acceptor labeling of the singly labeled donor or acceptor single strands. d^- and a^- are the corresponding percentages of the single strands which are not labeled. (C) Expressions demonstrating how the probabilities are determined. The double letters (d^+a^+, etc.) are the probability of forming the double-stranded structures in (B). These probabilities are needed to write Eq. (11) (see text).

bases essentially do not fluoresce at room temperature, except for some bases of certain tRNA species, only examples with conjugated external labels are discussed. The fluorescence of dyes complexed to DNA has been reviewed recently.[25] Complementary singly labeled, single-strand synthetic nucleic acid oligomers that can form double-stranded DNA structures have been used for FRET measurements,[13-15] and FRET promises to yield much more interesting structural information about nucleic acids and their complexes with proteins in the future.

Appropriate expressions for specific cases can be derived easily by following the procedures outlined below. As shown in Fig. 4 the percentage of fully reconstituted duplexes that are doubly labeled will be set by the percent labeling of the single-strand oligomers that constitute the final structure. It is assumed that there is no preference for the association of the labeled versus unlabeled single strands during the process of annealing. The variables d^+, d^-, a^+, and a^- are defined in Fig. 4B,C. Similar considerations for partial labeling have been presented[45] for DNA duplexes labeled with a single dye. We assume that the donor and acceptor are well removed from each other (i.e., essentially not in contact) to ensure a minimum interaction between the dyes, and so that energy transfer by an exchange mechanism[16] is not possible.

Care should be taken either to purify essentially 100% labeled single-stranded molecules or to know the value of d^+. If a double-stranded structure is measured, duplexes should be cleanly separated from the single-stranded labeled strands. Usually high-performance liquid chromatography (HPLC) and electrophoresis are the methods of choice for these purifications. If similar molecules within a series are to be compared, comparisons of the spectroscopic parameters of the separate chromophores should be made: plots of all combinations of absorbance (also dye as well as DNA absorbance) and fluorescence measurements of the different doubly labeled macromolecules against each other give invaluable information whether d^+ and a^+ are the same for all samples. If the absorption and fluorescence spectra of individual chromophore species cannot be isolated for this purpose because the spectra overlap, a spectral component analysis similar to that described below will discriminate the components.

Combining Fluorescence and Absorption Measurements to Determine E

There are many ways to combine steady-state fluorescence and absorption measurements to determine E. The methods presented here involve

---

[45] H. Eshaghpour, A. E. Dieterich, C. R. Cantor, and D. M. Crothers, *Biochemistry* **19**, 1797 (1980).

essentially only fluorescence measurements made on a single sample containing both donor and acceptor. The reader is referred to the excellent article by Fairclough and Cantor[9] in this series; the specific methods discussed there require measurements on samples labeled with donor and acceptor, donor alone, and acceptor alone. The methodology presented by them[9] is closely related, but complementary, to this presentation, and the method of choice is dictated by the experiment and the data analysis options available to the experimenter. It is often not required to use multiple samples, and in order to compare a series of samples it may not be desired. The number of measurements on separate solutions with different molecules should be kept to a minimum, not only to avoid generating unnecessary measurement noise, but also to avoid systematic, or random, unknown differences between the samples. The methods presented here are not only useful for the usual objective of evaluating E, but they also provide a measurement to compare small differences between several related samples. This is very useful especially for applications involving labeled nucleic acids. It is now becoming possible, and will soon be routine, to label DNA molecules almost anywhere within their sequence, and structural information from FRET (for DNA, RNA, and protein–nucleic acid complexes) can be provided at least partially by "walking" the probe (donor or acceptor) along the DNA helix. In this way the stereochemistry of a macromolecule (or complex) may become apparent by comparing a series of measurements rather than trying to calculate an exact distance.[14]

Procedures for determining E from fluorescence spectral measurements are straightforward. The first step is to write an explicit expression for the fluorescence intensity, including direct excitation of the donor and acceptor and the energy transfer. The expression should take partial labeling into account and should pertain to any pair of excitation and emission frequencies. Depending on the D–A pair, one should check for possible wavelengths where only the donor, or only the acceptor, is either excited or emits (this simplifies the analysis). Because the fluorescence spectra (excitation and emission) of the donor and acceptor are known, it usually is possible to extract the individual contributions of the donor or acceptor from the fluorescence spectra containing emissions from both species. Then by using these data it is possible to combine fluorescence measurements at different wavelengths, or to combine fluorescence with absorbance data, to determine E in terms of known absorption constants and quantum yields.

The following calculations refer to the particular example in Fig. 4B, but the generalization to other specific cases is clear. The total emission $F(v,v')$, excited at a frequency of v' and emitted at v, measured from a solution containing complementary single-strand oligomers (the concentrations of the two strands do not have to be equal) labeled with either donor or acceptor is proportional to

$$F(v,v') \propto [S_{d^+a^+}]E\varepsilon^D(v')\Phi^A(v) + \{[S_{d^-a^+}] + [S_{d^+a^+}]\}\varepsilon^A(v')\Phi^A(v)$$
$$+ \{(1 - E)[S_{d^+a^+}] + [S_{d^+a^-}]\}\varepsilon^D(v')\Phi^D(v)$$
$$+ \text{fluorescence from single strand in excess} \qquad (11a)$$

Equation (11a) pertains even when the donor and acceptor strands are not 100% labeled, and when both single- and double-stranded DNA molecules are present. The excitation light intensity and the instrumentation factors are not given, but they are assumed to be the same for all measurements [these factors simply multiply the entire right side of Eq. (11a); they may be also v dependent]. The superscripts D and A refer to the donor and acceptor, $\varepsilon^D(v')$ is the molar absorption coefficient of the donor, $\Phi^D(v)$ is the emission shape factor for the fluorescence quantum yield of the donor, and these parameters are assumed to be the same for doubly and singly labeled molecules. $\varepsilon^A(v')$ and $\Phi^A(v)$ are the corresponding parameters for the acceptor. $[S_{d^+a^+}]$ is the concentration of double-stranded duplex which is doubly labeled, $[S_{d^+a^-}]$ is the duplex concentration which is labeled only with the donor, etc. We define the strand concentration of the excess single-stranded moiety (if it is present) to be [SS]. The last term of Eq. (11a) is then $[SS]d^+\varepsilon^D(v')\Phi^D(v)$ or $[SS]a^+\varepsilon^A(v')\Phi^A(v)$, depending whether the excess strand is labeled with donor or acceptor, respectively. As is evident from Fig. 4, if the total concentration of the duplex substrate is [S], then $[S_{d^+a^+}] = [S]d^+a^+$, $[S_{d^+a^-}] = [S]d^+a^-$, etc. Because one usually can separate experimentally double- and single-stranded moieties, the last term of Eq. (11a) is dropped to simplify the following presentation; however, it is easy to retain this additional term if needed. Then Eq. (11a) can be simplified to (see Fig. 4B)

$$F(v,v') \propto [S]\{\varepsilon^D(v')\Phi^A(v)Ed^+a^+ + \varepsilon^A(v')\Phi^A(v)a^+$$
$$+ \varepsilon^D(v')\Phi^D(v)d^+[(1 - E)a^+ + a^-]\}$$
$$= F^A(v,v') + F^D(v,v') \qquad (11b)$$

The emission from the acceptor $F^A(v,v')$ is derived from both energy transfer from the donor and direct excitation. $F^D(v,v')$ is the contribution of the donor to the total fluorescence. The inclusion of the factors d^+,a^+, and a^- conveniently express the effect of the extents of labeling of the donor and acceptor on the total fluorescence signal and on the derived efficiency of energy transfer, E, as shown below. For instance, if all the molecules have a donor but not all have an acceptor, the percentage of donor fluorescence will be higher than if there were complete 1 : 1 labeling. However, if every molecule has an acceptor, but not all have a donor, E

could be underestimated. If these labeling fractions are not constant it will be difficult to compare the samples with different values of R unless d^+ and a^+ are known. To account for this, these labeling fractions are retained throughout the equations below. Expressions for standard deviations of the derived values can be calculated easily from the equations using standard formulas[46] for propagation of errors.

Decomposition of Fluorescence

One of the major experimental complications in FRET is the decomposition of the measured fluorescence signal in terms of the separate components of Eqs. (11a) and (11b). If there are spectral regions where the donor or the acceptor can be observed without exciting the complement of the D–A pair, and if D or A emission can be observed alone without any contribution from the other fluorophore, then several of the terms of Eq. (11b) can be observed directly. However, this is often not the case, and even if this can be done, problems may arise with the normalization of the fluorescence values. It is advisable to take complete excitation and emission spectra, and not simply to measure the fluorescence intensity at singular wavelengths (or integrated over extended wavelength regions); this leads to more reliable results and a better identification of the individual contributions. This is especially important when the steady-state fluorescence spectra of the donor and the acceptor overlap.

The precision with which E can be determined depends on how well the components of the donor and acceptor can be separated from each other. Figure 5A shows four examples of fluorescence spectra (in Fig. 5 they are normalized to 1 for plotting convenience): the excitation spectra of the donor $F_{ex}^D(v,v')$ and acceptor $F_{ex}^A(v,v')$, and the emission spectra of the donor $F_{em}^D(v,v')$ and acceptor $F_{em}^A(v,v')$. These terms are defined in Fig. 5B, along with the corresponding emission processes. The actual measured excitation or emission fluorescence intensity from a sample containing both D and A for any pair of frequencies v,v' is a sum from three emission processes (provided that A fluoresces): (1) enhanced emission from the acceptor due to FRET, (2) acceptor emission from direct excitation of the acceptor, and (3) donor emission from direct excitation of the donor [the three terms of Eq. (11b)]. This fluorescence signal is dispersed on either an excitation or emission frequency (or wavelength) axis to produce a spectrum.

The excitation and emission spectra can be decomposed into characteristic donor and acceptor spectra. With only one donor and one acceptor

[46] P. R. Bevington, "Data Reduction and Error Analysis for the Physical Sciences." McGraw-Hill, New York, 1969.

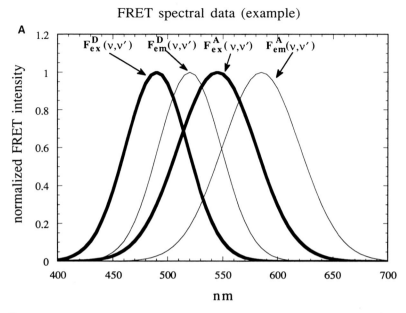

FIG. 5. (A) The separate donor and acceptor spectral components presented as excitation and emission spectra; all the spectra are normalized to 1 in order to plot them on the same ordinate. The measured fluorescence signal is the sum of the corresponding terms shown in (B); the emission spectrum intensity (constant v', varying v) is proportional to $F_{em}(v,v') = F_{em}^D(v,v') + F_{em}^A(v,v')$, and the excitation spectrum intensity (constant v, varying v') is proportional to $F_{ex}(v,v') = F_{ex}^D(v,v') + F_{ex}^A(v,v')$. Of course, the measured spectra are not nor-

species, but without FRET, this decomposition will give two spectral components for both spectra, such that $F_{ex}^D(v,v') = F_{em}^D(v,v')$ and $F_{ex}^A(v,v') = F_{em}^A(v,v')$ (after taking instrumentation differences of the excitation and emission spectra into account); that is, the magnitude of these components relate directly to the number of molecules in the observed volume. The tabulation in Fig. 5B shows how these individual spectral contributions to the total fluorescence are projected onto the characteristic donor and acceptor spectral axes in the presence of FRET. Of course, the total signal at a particular v,v' is the same whether this data point is acquired with an excitation or an emission spectrum; however, the spectral decomposition depends on the type of spectrum made. Such a decomposition is easy to write down; the excitation spectrum is a dispersion according to $\varepsilon(v')$ at a constant emission frequency v, and the emission spectrum is a dispersion according to $q(v)$ at a constant excitation frequency v'. Such a distinction of the excitation and emission spectral components is only possible when fluorescence readings are collected at multiple frequencies, preferably over an extended frequency span, so that the spectrum can be reliably decomposed by numerical fitting processes.

Four Experimental Methods to Determine E

Methods to determine E include measuring the following: (1) enhanced fluorescence of the acceptor ($\propto E$; A must fluoresce for this method), (2) decreased fluorescence quantum yield of the donor $[\propto(1 - E)]$, (3) decrease in the donor fluorescence lifetime $[\propto(1 - E)]$, and (4) change in the fluorescence anisotropy of the donor and acceptor.

Measuring Enhanced Fluorescence of Acceptor

The efficiency of energy transfer can be measured from the enhanced emission of the acceptor when exciting the donor. Reference should be

---

malized to 1. $F_{ex}(v,v') = F_{em}(v,v')$ for any particular frequency pair (v,v'), if instrumentation factors are taken into account. (B) The upper tabulation shows the donor and acceptor spectral components of the excitation and emission spectrum according to the notation used in Fig. 4 and for Eq. (11b). The columns identify the type of spectrum, and the rows refer to the D or A components of each spectrum. The boxes in the lower tabulation (which correspond to the same boxes in the upper tabulation) define the labeling of the spectral contributions resulting from the spectral decomposition of the excitation and emission spectrum, as shown in (A). It is easily seen in the upper tabulation that the intensity contribution from the enhanced acceptor fluorescence $[S]\varepsilon^D(v')\Phi^D(v)Ed^+a^+$ is in the donor spectral component of the excitation spectrum $F_{ex}^D(v,v')$, but in the acceptor spectral component of the emission spectrum $F_{em}^A(v,v')$.

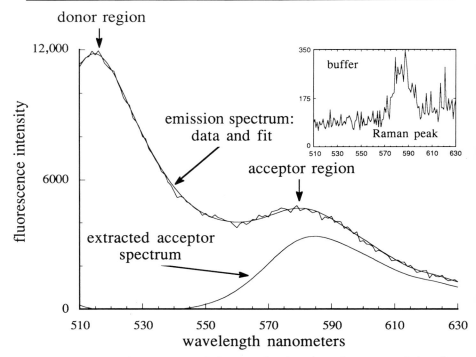

FIG. 6. Example of FRET data analysis, measuring the enhanced acceptor emission of a DNA duplex labeled with fluorescein (donor) and tetramethylrhodamine (acceptor) on the 5' ends of the complementary strands using a chemical C_6 linker (e.g., Fig. 4A). The emission spectrum [corresponding to Eq. (11b) in the text] is excited at 490 nm and fitted with a series of Chebyschev polynomials to smooth out random noise. Between 500 and 530 nm only fluorescein fluoresces; rhodamine has maximum emission at 585 nm. The "extracted acceptor emission" $F_{em}^A(v,v')$ of Fig. 5B is attained by fitting the spectrum to a standard fluorescein spectrum (conjugated singly to the same DNA) between 510 and 530 nm. The fluorescein (donor) contribution is then subtracted from the FRET data, giving the pure acceptor spectrum shown [see Eq. (12)]. An excitation spectrum [$F_{em}^A(v,v'')$ of Fig. 5B] of the same sample is taken from 400 to 590 nm, with emission in the range 580–600 nm. This excitation spectrum is fitted with polynomials similarly to the emission spectrum, and the peak [λ_{max} ~ 565 nm of $F_{em}^A(v_2,v'')$ in Eq. (13), where only rhodamine absorbs] is divided into the pure acceptor emission spectrum shown here. This normalizes the acceptor fluorescence to give the ratio of Eq. (13) in the text, and the analysis can then proceed according to Eq. (14). The buffer signal (inset), which contains the Raman emission from water, must be subtracted from every fluorescence spectrum. The spectra here were taken on an SLM 8000S spectrofluorimeter (Urbana, Illinois).

made to the tabulation in Fig. 5B and the spectra in Figs. 5A and 6 when reading this section. There are two main ways to do this: here a method where emission and excitation spectra are combined is explained; the complementary method (but not so advantageous) involving just two excitation spectra can be easily derived in a similar fashion. A derived emission spectrum containing only the emission from the acceptor, after subtracting the donor portion (see above), is

$$F'(v,v') = F_{em}^A(v,v') \propto [S] \, [\varepsilon^D(v')\Phi^A(v)Ed^+a^+ + \varepsilon^A(v')\Phi^A(v)a^+] \quad (12)$$

The subtraction is easily carried out if one knows the shape of the fluorescence spectrum of the donor, and if the donor fluorescence can be observed at certain wavelengths where the acceptor does not emit. Then, by subtracting the appropriate fraction of the donor fluorescence from the total fluorescence signal measured at (v,v'), the resulting signal corresponds to Eq. (12). This is readily done by taking an emission spectrum over a wavelength range covering both donor and acceptor emission peaks (i.e., varying v). If there is a region of the spectrum containing only donor fluorescence, it can be fitted to a standard donor spectrum, and this fitted curve is then subtracted over the entire sample emission spectrum (see Fig. 6) to yield Eq. (12). If it is not possible to measure donor emission without interference from the acceptor, the sample spectrum [corresponding to Eq. (11b)] can be fitted simultaneously to both the donor and acceptor standard spectra (see Fig. 5). This is not difficult on a computer, and it will lead directly to an acceptor spectra corresponding to Eq. (12).

One must check whether the shape of the resulting experimental curve of Eq. (12) corresponds to a standard acceptor spectrum. If this is not the case, the donor and the acceptor are too strongly coupled; that is, the presence of donor affects the energy levels of the acceptor, and FRET cannot be applied. The Förster mechanism [Eq. (6)] applies only to very weak coupling[5]; energy transfer can still take place in the presence of strong coupling, but the application of Eq. (6) becomes doubtful, and other mechanisms of energy transfer should be considered.[16] For the same reason the emission spectrum of the donor in the presence of an acceptor should be identical to a donor standard. Both standard spectra should be made in the same macromolecular environments as the samples which are doubly labeled to avoid errors in case the probe spectra change on conjugation to the nucleic acid. For the example of Fig. 4, the standards would be singly labeled duplexes. A word of caution is appropriate here. Dyes conjugated to even specific sites of macromolecular structures may exist in a variety of conformations with different solvent environments and different interactions of the probes with the macromolecule. This may result in spectral changes of the probe compared to the free unconjugated state, and there may be an ensemble of dyes present with a distribution of spectral parameters. This does not invalidate the application of FRET as long as these spectral parameters are not affected by the presence of the conjugate dye of the D–A pair. However, complexities arising from multiple species with different values of R_0, and possibility R (see below), must be taken into account.

A fluorescence intensity is not an absolute number (as absorption is), so the measurement must be normalized by some standard. Fluorescence instruments are usually corrected for the lamp intensity, and they can be standardized additionally for day-to-day comparisons by measuring an external standard, such as a known solution of rhodamine B in ethanol. However, the signal expressed in Eq. (12) must still be normalized by a signal proportional to the concentration of macromolecules. Probably the least accurate determination of concentration is calculating dilutions from a stock solution. If the sample concentration is high enough it is sufficiently accurate to make an absorption measurement directly on the sample, but this would still leave a dependence of the signal on a^+. However, it is more convenient if the acceptor fluorescence can be measured for every individual sample without donor interference, that is, at certain wavelengths where only the second term of Eq. (11b) (direction excitation of the acceptor) is finite. If this is not possible, reference to Fig. 5B shows that a decomposition of an excitation spectrum into the D and A components will give $F_{ex}^A(v,v')$ directly. $F_{ex}^A(v,v')$ is then divided into the signal corresponding to Eq. (12). This has the advantage that the factor a^+ in Eq. (12) cancels in $(ratio)_A$. Note that we cannot normalize by the donor fluorescence of the doubly labeled sample, since the magnitude of the donor fluorescence depends on the efficiency of transfer.

To generalize, we assume that the measurement of acceptor fluorescence alone with an excitation at v'' is made with the emission at v_2 (emission spectrum), and that the measurement of the doubly labeled sample with an excitation at v' is made with an emission at v_1 (excitation spectrum). The following ratio is then formed:

$$(ratio)_A = \frac{F_{em}^A(v_1,v')}{F_{ex}^A(v_2,v'')} \tag{13}$$

This ratio is (see Fig. 5B)

$$(ratio)_A = \left\{ Ed^+ \left[\frac{\varepsilon^D(v')}{\varepsilon^A(v'')} + \frac{\varepsilon^A(v')}{\varepsilon^A(v'')} \right] \right\} \frac{\Phi^A(v_1)}{\Phi^A(v_2)} \tag{14}$$

From this ratio we can easily extract the efficiency of energy transfer on each doubly labeled molecule from the information in only two fluorescence spectra, or even from fluorescence measurements at only two appropriate wavelength pairs. All parameters in Eq. (14) can be accounted for. $\varepsilon^A(v')/\varepsilon^A(v'')$ is known; it can be directly measured from the excitation spectrum, or even better from the absorption spectrum (if the absorption and fluorescence excitation spectra are not the same, only the ε values corresponding to the excitation spectrum are used). $\varepsilon^D(v')/\varepsilon^A(v'')$ either is known or can be measured. d^+ often can be measured by adsorption; ideally, $d^+ = 1$ for fully labeled molecules, although d^+ can be varied purposely if desired (see Fig. 7). Fortunately ϵ values are not very sensitive to the environment, compared to Φ values. If $v_1 = v_2$ the last term of Eq. (14) is 1. Of course all changes in instrumentation constants, resulting from differences between the settings of the instrument for excitation and emission spectra [used for measuring the spectra for Eq. (13)], must be taken into account. Any such correction factors can be calculated by comparing the last term of Eq. (14) (which has only parameters of the acceptor) with measurements on macromolecules singly labeled with acceptor (see, e.g., Fig. 7). If the acceptor can be excited where the donor does not absorb, an emission spectrum can be taken, and then $F^A_{em}(v_2,v'')$ can be substituted for $F^A_{ex}(v_2,v'')$ in Eqs. (13) and (14) (see Fig. 5B).

If double-stranded DNA is labeled with the donor and acceptor on complementary strands, it is easy to vary d^+ (or a^+) by mixing with unlabeled strands during the annealing procedure to make the duplex; then the intercept of a $(ratio)_A$ versus d^+ plot also yields an accurate estimate of the last term in Eq. (14). Figure 7 is an example of this plot for small values of E, where the linearity of Eq. (14) with E is shown. Such a percent labeling titration can be used to provide more reliable parameters for small values of FRET, and the intercept should agree with the experimental determination of $(ratio)_A$ using molecules labeled with only A.

The values of $F^A_{em}(v_1,v')$ and $F^A_{ex}(v_2,v'')$ can be determined from fluorescence measurements at single wavelength pairs. In general the component values cannot be determined so reliably, and the spectral decomposition cannot be checked for the shapes of the component spectra. The method of Eq. (13) extends earlier analyses[8,34,47,48]; it has been found to be useful and robust for FRET measurements on complex DNA samples.[14] It is particularly useful for making precise difference measurements between several samples with very low values of E when comparing a series of related samples, and it is also the best method when labeling cannot be guaranteed to be 100%.

The advantages of this method are (1) all fluorescence measurements

[47] R. H. Conrad and L. Brand, *Biochemistry* 7, 777 (1968).
[48] J. Szöllõsi, L. Trón. S. Damjanovich, S. H. Helliwell, D. Arndt-Jovin, and T. Jovin, *Cytometry* 5, 210 (1984).

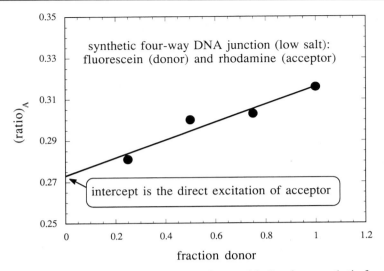

FIG. 7. Plot of *(ratio)*$_A$ versus the fraction of donor labeling for a synthetic four-way genetic junction (a model for the Holliday recombination junction). This DNA structure contains four DNA strands such that each single strand is base paired simultaneously to two other complementary partners in two helical arms of the structure. The four helical arms meet at a central "junction" where the strands exchange from one complementary single strand to the other. The helical arms extend from this junction, and the angular juxtaposition of the helical arms depends, among other parameters, on the nucleotide sequence at the junction and the ionic strength of the solution. The four-way junction folds from an extended form into a more compact structure (perhaps similar to an X) as the ionic strength of the solution is raised. Fluorescence resonance energy transfer has been used to correlate the relative efficiency of FRET when all possible pairs of arm ends are labeled with donor and acceptor, and the results have led to a proposal of an antiparallel X structure (two acute and two obtuse angles between neighboring arms) at 2–5 mM Mg^{2+}.[14] The plot shows the result of an experiment, using fluorescein and rhodamine as the donor and acceptor, where the percent labeling of the donor-labeled strand is varied (these are separate solutions, not a continuous titration) in one of the labeling isomers at low ionic strength (extended form of the junction). The efficiency of energy transfer is very small and corresponds to about 80 Å; this (together with other evidence) is expected for FRET across the ends of neighboring arms of an extended square structure (i.e., a + structure). Similar plots of junctions labeled on opposite arms of the expected + structure show a smaller slope, but the same intercept. The intercepts of all plots agree with the measurement of a singly labeled rhodamine (acceptor) molecule, and with a calculation of the second term of Eq. (14). The measurement precision is increased by fitting to a straight line, as opposed to measurements on singularly labeled molecules (R. M. Clegg, A. I. H. Murchie, A. Zechel, S. Diekmann, and D. M. J. Lilley, in preparation).

are taken on the same solution; (2) as shown in Eq. (14), the fluorescence quantum yield of the acceptor does not enter into the energy transfer ratio, since this parameter cancels out in the normalization; (3) the normalization procedure circumvents the necessity of knowing the concentration of the labeled molecule; (4) it is not even necessary to ensure 100% acceptor labeling since a^+ does not enter into Eq. (14), and this means one can measure E regardless of the percent labeling of the acceptor; (5) error due to the percent labeling of the donor (i.e., $\pm \Delta d^+$) will introduce this same percent error in the derived value of E since the factor Ed^+ enters directly into Eq. (14) [compare to Eq. (16), below], which means that the error $\pm \Delta d^+$ does not introduce a percent error in the determination of E based on the maximum value that E can attain (i.e., 1), but rather the percent error due to $\pm \Delta d^+$ is based on the real value of E which can be small; (6) the energy transfer ratio requires only two fluorescence measurements which are made on the same solution with the same instrument, and thus superfluous absorption measurements and concentration determinations that introduce error are not necessary for calculating $(ratio)_A$ [owing to the high fluorescence quantum yields possible and owing to point (8) below, very low concentrations can be employed where absorption measurements are impossible]; (7) the fluorescence contribution of the donor can be completely, and rigorously, removed; and (8) the quantum yield of the donor does not enter into Eq. (14). Only the ratio of the quantum yields of the acceptor at two wavelengths is involved, and this can be determined from an emission spectrum of the acceptor alone. $\Phi^D(v_1)$ only affects the actual value of E [through Eq. (6)], but not the measurement. This is an important consideration, since very low E values can be measured; on the contrary, if E is determined from the decrease in the donor fluorescence [e.g., according to Eq. (16)], the error can be much larger, especially since the donor quantum yield of different samples may vary widely unless the solution conditions or the macromolecular environment of the conjugated dyes are strictly the same.

Measuring Decreased Fluorescence Quantum Yield of Donor

The efficiency of transfer can also be determined by measuring the decrease in the donor quantum yield. With many D–A pairs there are emission wavelengths where only donor emission is observed, $F^D(v,v')$, with no contribution from the acceptor; for instance, with exciting at 490 nm and observing the emission between 500 and 530 nm, fluorescein emits, but rhodamine (a good acceptor for fluorescein) does not (see Fig. 6). As above, if the fluorescence of the donor and acceptor cannot be separated one can simply fit the signal numerically to a sum of singly labeled standard donor and acceptor emission spectra. Reference to Fig. 5 shows that the component $F_{em}^D(v,v')$ of the emission spectrum is the only term depending solely on the donor emission. Then the "donor only" fluorescence contribution can be determined as

$$F(v,v') \approx F^D(v,v') \propto F_{em}^D(v,v') = [S]\varepsilon^D(v')\Phi^D(v)d^+[(1-E)a^+ + a^-] \quad (15)$$

The unsubscripted $F^D(v,v')$ refers to a fluorescence signal containing only donor fluorescence, whether it is determined from a spectrum or not. If the donor fluorescence cannot be directly measured alone, it is best to determine the spectral components as described above. Figure 5 shows that a spectral decomposition of an emission spectrum will yield $F_{em}^D(v,v')$ which is present in Eq. (15). It is interesting to note that $F_{ex}^D(v,v')$ from the excitation spectrum also has a contribution from the acceptor. Therefore, the excitation spectrum does not provide a way to select a signal emanating from the donor alone [unless $\Phi^A(v) = 0$].

Equation (15) can be normalized by the same acceptor fluorescence measurement as in Eqs. (13) and (14)[49] to give

$$(ratio)_D = \frac{\varepsilon^D(v')}{\varepsilon^A(v'')} \frac{\Phi^D(v_1)}{\Phi^A(v_2)} \left(\frac{d^+}{a^+}\right) [(1-E)a^+ + a^-] \quad (16)$$

The disadvantages of Eq. (16) are common to all methods which determine E from the decrease of the donor quantum yield. Equation (16) shows that $(ratio)_D$ depends on both a^+ and d^+; therefore, error arising from uncertainty in the percent labeling affects $(ratio)_D$ of Eq. (16) more than $(ratio)_A$ of Eq. (14). In addition, an error in d^+ carries over directly to the determination of $(1-E)$, so that if E is small the percent error in the derived value of E can be very large.

Another disadvantage of Eq. (16) compared to Eq. (14) is that variations in the quantum yields of the donor and the acceptor do not cancel. Quantum yields are often strong functions of ionic strength, pH, and the polarity of the environment surrounding the probe; these solution variables are often varied to induce conformational and isomeric changes in nucleic acid structures. Unless the solution conditions, or the probe environments, of comparative samples are exactly identical, a large error may occur in calculating even relative values of $(1-E)$ from the donor fluorescence, because there is no way to eliminate the direct effect of $\Phi^D(v_1)$ in Eq. (15). One can measure a separate sample with no acceptor, but this will not eliminate inadvertent variations between samples. This is particularly crit-

---

[49] Of course one can normalize by any other convenient measure of [S], such as absorption measurements, or simply rely on the ability to make solutions with precise concentrations; but one must be cautious since these manipulations often involve error.

ical for small values of E; in this case it may be almost impossible even to determine reliable comparative values of E within a series of closely related samples by calculating *(ratio)*$_D$.

A major advantage of measuring E by observing the donor fluorescence, and normalizing Eq. (15) by another measurement of [S] other than the acceptor fluorescence, is that the acceptor must not fluoresce at all. Also, if $a^+ = 1$, or if a^+ is known, Eq. (16) gives a reliable estimate of $(1 - E)$ from fluorescence measurements on only one solution.

Measuring Decrease in Donor Fluorescence Lifetime

The rate of energy transfer can be determined directly from measurements of the fluorescence lifetime. By subtracting Eq. (2) from Eq. (1), and borrowing the last equality from Eq. (6), the following expression for k_T can be written:

$$1/\tau_{DA} - 1/\tau_D = (R_0/R)^6(1/\tau_D) \tag{17}$$

This is the most direct way to measure k_T and relate R to the measurement. One simply measures the fluorescence lifetime in the presence and absence of the acceptor and subtracts these inverse lifetimes to calculate k_T. The fluorescence lifetime of the D–A complex should be measured at an emission wavelength where only donor emission is present (e.g., 520 nm for fluorescein as a donor, rhodamine as acceptor, see Fig. 6), or the lifetimes of the donor and acceptor should be separated well enough so that a multicomponent analysis is reliable. A time-resolved measurement has the advantage that no extensive spectral data analysis is necessary. A disadvantage of this method is that the relatively complex instrumentation required is not always available, and considerable expertise is necessary to interpret complex decays. If R and R_0 are not the same for all D–A pairs, the donor fluorescence may not decay according to a single exponential function (or even as a series of two or three exponentials); the donor decay curve can be complex[50] (the reader is referred to the literature[21,51–54]). It is assumed below that the fluorescence of the donor has only one mode of

---

[50] The decay of fluorescence from a donor in the presence of a spherical random distribution of acceptors (static positions but random orientations) is approximately $r(t) = \exp[-(t/\tau) - \pi^{1/2}N(R_0/R_g)^3(t/\tau)^{1/2}]$, where N is the number of acceptor molecules surrounding a donor in the volume $(4/3)\pi R_g^3$.

[51] T. Förster, *Z. Naturforsch.* **4a**, 321 (1949).

[52] R. G. Bennett, *J. Chem. Phys.* **41**, 3037 (1964).

[53] K. B. Eisenthal and S. Siegel, *J. Chem. Phys.* **41**, 652 (1964).

[54] S. Georghiou, *J. Phys. B: At. Mol. Phys.* **2**, 1084 (1969).

decay (or at least only one major decay process). Multiple relaxation components complicate the analysis and interpretation. The following remarks apply to the simple case of single values of R and R_0.

Provided that τ_{DA} and τ_D are well separated, determining k_T by lifetime measurements has one special advantage: if the sample is a mixture of singly and doubly labeled molecules with unknown percent labeling, the calculation of k_T is still possible, since both lifetimes τ_{DA} and τ_D can be observed if the amplitudes are sufficient [see Eq. (17) below]. In this way errors arising from the percent labeling can be completely avoided (provided one is measuring transfer from one donor to one acceptor). Especially with recent developments in phase methods to determine fluorescence lifetimes, it is possible to measure many samples rapidly. Another advantage is that fluorescence lifetimes are in general very reproducible when measured against standards, so that measurements taken on different days, and even in different laboratories, can be compared.

There are, of course, limitations to this technique as well. The lifetime τ_{DA} will vary from approximately 1% to 99% of τ_D as the distance R between the donor and the acceptor varies in the range $0.5 \, R_0 < R < 2R_0$ [Eq. (10)]. If the lifetime measurement has a precision of about 100 psec and $\tau_D = 3-4$ nsec (e.g., fluorescein or rhodamine), we could just determine the difference between transfer and no transfer, provided the D–A distance was below $1.8R_0$. For $\tau_D = 3-4$ nsec the ability to measure lifetimes of 500 psec is required to measure $R = 0.75R_0$; if R is less than this, a further decrease in the lifetimes would not be detected without better time resolution. Owing to limitations on the precision of a lifetime measurement, donors with long lifetimes are preferred if lifetimes are to be measured (e.g., porphyrins, pyrene, ethidium bromide, and acridines). Because most fluorescence lifetimes are fairly short (<1 to 5 nsec), exceptionally accurate measurements of fluorescence lifetimes are required to determine accurate values of R/R_0 is R is much larger R_0.

When $a^+ \neq 1$ the time-dependent decay of the donor fluorescence signal is (each component amplitude is proportional to the concentration of the corresponding donor-containing species; see Fig. 4 for the labeling definitions)

$$F^D(t) \propto [S_{d^+a^+}] \, e^{-t/\tau_{DA}} + [S_{d^+a^-}] \, e^{-t/\tau_D}$$
$$= [S] \, [(d^+a^+) \, e^{-t/\tau_{DA}} + (d^+a^-) \, e^{-t/\tau_D}] \qquad (18)$$

$S_{d^+a^+}$, etc., are defined for Eqs. (11a) and (11b). If the lifetimes are measured by time decay pulse methods, the percent amplitude of the first term in Eq. (18) is the percent labeling of the acceptor; this is a direct way to measure a^+. Of course, the same information is available from the frequency-phase method of measuring lifetimes,[55] in which case the signal is proportional to the Fourier transform of Eq. (18),

$$F^D(\omega) \propto [S]\tau_D \left\{ (d^+a^+)\frac{\tau_{DA}}{\tau_D}\left[\frac{1}{1 + j\omega\tau_{DA}}\right] + (d^+a^-)\left[\frac{1}{1 + j\omega\tau_D}\right]\right\} \quad (19)$$

This expression has been written in complex notation, where $j = -1^{1/2}$, and $\omega = 2\pi f$ (f is the modulation frequency of the excitation light). Again, because we know τ_{DA}/τ_D from the lifetime measurement, we can calculate a^+. As a check of Eq. (18), we can derive the donor steady-state fluorescent value by integrating Eq. (18) from t equals 0 to infinity:

$$F^D_{\text{steady-state}} = \int_0^\infty F^D(t)\, dt \propto [S]\tau_D\left[(d^+a^+)\frac{\tau_{DA}}{\tau_D} + (d^+a^-)\right]$$

$$= [S]\tau_D[(d^+a^+)(1 - E) + (d^+a^-)] \quad (20)$$

which is equivalent to Eq. (15), as it must be. When $E \ll 1$, then $\tau_{DA} \approx \tau_D$, and it will be impossible to measure the separate times (here global fits according to a FRET model on a series of samples would extend the range of useful E values).

It is shown in Fig. 2B that energy transfer is essentially an excited state reaction. If A^* is fluorescent, and the time dependence of the fluorescence decay of the acceptor is observed while exciting the donor, a new exponential component of acceptor emission may be observed with a negative amplitude. The lifetime of this new component corresponds to the decay of the donor. Thus, it is also possible to establish the presence of FRET by observing the fluorescence lifetime of the acceptor, but this method has never been systematically applied.[36]

Measuring Change in Fluorescence Anisotropy of Donor and Acceptor

Fluorescence resonance energy transfer can be observed by measuring the anisotropy of the donor (and acceptor) fluorescence.[56-62] The anisot-

---

[55] J. Mugnier, B. Valeur, and E. Gratton, *Chem. Phys. Lett.* **119**, 217 (1985).
[56] G. Weber, *Trans. Faraday Soc.* **50**, 552 (1954).
[57] G. Weber, *Biochem. J.* **75**, 335 (1960).
[58] G. Weber, *Biochem. J.* **75**, 345 (1960).
[59] G. Weber, *in* "Fluorescence and Phosphorescence Analysis: Principles and Applications" (D. M. Hercules, ed.), Chap. 8, p. 217. Wiley *(Interscience),* New York, 1966.
[60] G. Weber and S. R. Anderson, *Biochemistry* **8**, 361 (1969).
[61] S. R. Anderson and G. Weber, *Biochemistry* **8**, 371 (1969).
[62] D. Genest and Ph. Wahl, *in* "Time-Resolved Fluorescence Spectroscopy in Biochemistry and Biology" (R. B. Cundall and R. E. Dale, eds.), p. 523. Plenum, New York, 1983.

ropy of a fluorescent molecule depends on the extent of rotational diffusion which the excited molecule undergoes before it emits a photon (see below to see how the anisotropy is measured). $\tau_{DA} < \tau_D$ [see Eqs. (1) and (2)], so that the extent of rotational diffusion within τ_{DA} will be less than that within τ_D; therefore, the fluorescence anisotropy of D usually increases in the presence of hetero-FRET, that is, when D and A are different chromophores. Energy transfer between like fluorophores (homo-FRET) leads to depolarization of the emission, since the deexcitation of one molecule leads to the excitation of the same emitting species with a rotated transition dipole; this was the original observation leading to the proposal of long-range energy transfer. This is the only method to observe FRET between like molecules ($\tau_D = \tau_{DA}$ and $\Phi^D = \Phi^{DA}$ since the acceptor is another donor molecule). If there is a distribution of D–A mutual orientations, the effect of FRET on the anisotropy of D* emission can be complex. The reader is referred to the above literature for excellent discussions of this method.

The excluded site binding mechanism of ethidium bromide to DNA has been verified[62] by observing the effect of energy transfer on the anisotropy decay of the dye fluorescence. This method has not been applied yet to dye molecules conjugated to nucleic acids, but it should be useful with defined length nucleic acid molecules. For instance, the fluorescence anisotropy from a labeled DNA molecule could be adjusted by admitting different amounts of energy transfer (double labeling at different distances) without affecting the rotational diffusion constants of the macromolecules.

The fluorescence anisotropy of the directly excited acceptor should not be affected by FRET (this is another check to be sure the interactions between D and A are very weak); however, the measured fluorescence anisotropy from acceptors excited by energy transfer is in general lower than those which are directly excited. The basic reason for this is that the FRET process excites molecules with transition dipole orientations distributed differently from the original distribution photoselected by the polarized excitation light.

Complexities in Interpretation

There are some well-known difficulties in determining accurate values of R from energy transfer measurements. R_0 must be known to calculate R, and Eq. (7) shows the parameters which are important for calculating $(R_0)^6$. The least certain parameters are Φ^D and κ^2, and the latter probably sets the limits of precision in determining R_0. R_0 can be measured independently on a model system, or it can be calculated by measuring the parameters defining it in Eq. (7). In the latter case a value for κ^2 must be assumed, since it is seldom known. It is not likely that only one configuration exists for many conjugated donors and acceptors; a distribution of orientations is more probable.

There is an extensive literature dealing with the so-called κ^2 problem.[63-65] The correct average value for κ^2 depends not only on the form of the orientational distribution, but also on the rate at which the different orientations are distributed throughout the ensemble. The general case is complex; however, a few simple cases exist. (1) $\kappa^2 = 0.476$ for random static orientations with a random R distribution of the donor and acceptor[66], and (2) $\kappa^2 = 2/3$ for rapid orientational randomization of the probes[4]; rapid randomization of probe orientations by rotational diffusion brings more D–A pairs into orientations preferential for energy transfer than does a random static distribution.

Limits can be set on the possible values for κ^2. By measuring the anisotropy[67] of the donor and acceptor (with D-, A-, and DA-labeled molecules, remember the anisotropy of D depends on E) for the same molecular situation as for the FRET measurements, estimates can be made for the dynamic rotational freedom of the dye molecules, and limits set on possible κ^2 values.[8,64] If the macromolecule is large enough not to rotate considerably during the lifetime of D*, and if the anisotropies of the donor and acceptor are less than about 0.2, it is probable that κ^2 is not much different from 2/3. The expected deviation of κ^2 from 2/3 is usually less than 20% if the cone angle (see Fig. 1) of rotational freedom for one of the probes is at least 30°.[8] This limit can be determined by measuring the fluorescence anisotropy r of the probe conjugated to the macromolecule (under conditions where the macromolecule does not rotate), comparing this to the anisotropy of the fully immobilized probe r_0, and applying the above analysis.[64,65,68,69] However, it should be noted that even if these anisotropy measurements are high (slow rotational diffusion), the static relative orientations of the D–A pair could still be random.

[63] R. E. Dale and J. Eisinger, *Biopolymers* **13**, 1573 (1974).

[64] R. E. Dale and J. Eisinger, *in* "Biochemical Fluorescence Concepts" (R. F. Chen and H. Edelhoch, eds.), Vol. 1, p. 115. Dekker, New York, 1975.

[65] R. E. Dale, J. Eisinger, and W. E. Blumberg, *Biophys. J.* **26**, 161 (1979); R. E. Dale, J. Eisinger, and W. E. Blumberg, *Biophys. J.* **30**, 365 (1980).

[66] I. Z. Steinberg, *J. Chem. Phys.* **48**, 2411 (1968).

[67] There are several ways to define fluorescence anistropy operatively. The most common (see Fig. 1) is $r = (F_{\parallel} - F_{\perp})/(F_{\parallel} + 2F_{\perp})$, where $F_{\parallel}$ is the fluorescence signal measured with vertical excitation and emission polarizers, and $F_{\perp}$ is the fluorescence measured with vertical excitation and horizontal emission polarizers, at right angles to the excitation beam (see the general references above).

[68] R. E. Dale, J. Eisinger, and W. E. Blumberg, *Biophys. J.* **26**, (1979); R. E. Dale, J. Eisinger, and W. E. Blumberg, *Biophys. J.* **30**, 365 (1980).

[69] One calculates the absolute limits of the cone angle (see Fig. 1) from $r/r_0 = \cos[(\Theta/2)^2][1 + \cos(\Theta/2)]^{2/4}$; if R' is the donor–acceptor distance calculated assuming $\kappa^2 = 2/3$, then the

Another phenomenon that can lead to κ^2 values close to 2/3 even if the donor and acceptor are oriented nonrandomly is the effect of mixed polarizations in the spectral bands of the chromophores. The overlap integral in Eq. (6) involves an integration over the entire frequency interval where the emission of the donor and the absorption of the acceptor overlap. If the electronic absorption and emission involve more than a single molecular transition, and if the different transitions have incoherent dipoles oriented differently in the molecule, the actual κ^2 value will be averaged over the overlap integral.[21,70] To check for this possibility of mixed polarizations, the fluorescence anisotropy of the donor and acceptor should be measured at many wavelengths over the entire emission frequency in a high viscosity solvent (so the probe does not rotate); if the anisotropy is not constant over the spectral band this effect is probable. Multiple transition moments in the main spectral bands with different directions in the molecules will lead to a natural depolarization. It is advantageous to use such chromophores for FRET, and further development of these type of probes for DNA would be very beneficial. In many cases the resulting effective κ^2 value is close to 2/3.

κ^2 should not always be viewed as an unwelcome problem. Certain fluorophores (e.g., Hoechst dyes and DAPI, see ref. 12 for dye acronyms) bind very tightly to double-stranded DNA in the minor grooves[71] in specific conformations, and intercalators[25] are rigidly held between the bases. These interactions can be accentuated if these molecules are conjugated to the DNA. Because the structure of double-stranded DNA is fairly uniform (much more so than for many proteins), it is possible that FRET measurements on DNA will supply interesting structural information based on variations in specific singular values of κ^2.[62] With respect to this, it is interesting that κ^2 is the only parameter in Eq. (6) which has never been rigorously verified. The effect of the mutual D–A transition dipole orientations on FRET can be appreciated qualitatively from the depictions in Fig. 3A,B.[72]

---

actual distance is $R = R'a = R'(1.5\kappa^2)^{1/6}$. Under the assumption of free rapid rotational diffusion within a cone with an angle of $(\Theta/2)$ (see Fig. 1), the following limits can be set on the value of a:

$$\{0.75 \,[1 + \cos^2(\Theta/2)]\}^{1/6} \le a \le [6 \cos^2(\Theta/2)]^{1/6}.$$

[70] E. Haas, E. Katchalski-Katzir, and I. Z. Steinberg, *Biochemistry* **17**, 5064 (1978).
[71] F. G. Loontiens, L. W. McLaughlin, S. Diekmann, and R. M. Clegg, *Biochemistry* **30**, 182 (1991).
[72] With respect to Fig. 3B, there is a misprint in a similar diagram in a popular fluorescence reference book.[30] The largest transfer ($\kappa^2 = 4)^4$ is expected for end-to-end dipoles, not for side-by-side dipoles. This is mentioned because approximate end-to-end dipole orientations pertain to some dyes which bind in the DNA grooves.

Static and dynamic averaging over distributions of donor and acceptor molecules must also be expected, especially if the polymer possesses a large degree of flexibility. So far we have discussed FRET assuming that every D–A pair in the solution has the same separation distance. Often this will not be the case, even for conjugated fluorophores; the molecular tether between the dye and DNA will allow some diffusional freedom, and the polymer itself may exist as an ensemble of configurations, with different R values (e.g., the flexible structure of single-strand regions of DNA and RNA, "bulges" and "kinks" in double-stranded structures). If E is measured the averaging can be expressed as $\langle E \rangle = \int [R_0^6/(R_0^6 + R^6)]\, g(R)\, dR$, where $g(R)$ is the equilibrium distribution function of R.[73] If the FRET can be measured over a range of R_0 values, it is theoretically possible to recover $g(R)$. It has been suggested[73] to use several D–A pairs with different R_0 values. External fluorescence quenching of the donor has lately been used to vary R_0,[74] and this method promises to be very useful in the future, also for labeled nucleic acid molecules.

If the decay kinetics of the donor (monoexponential without FRET) is observed, and the donor and acceptor relative positions do not change within the lifetime of the donor excited state, the observed fluorescence decay must be integrated over the R distribution, $\langle F(t) \rangle = \int g(R)\, e^{(t/\tau)[1 - (R_0/R)^6]}\, dR$.[75,76] Averaging can be over static ensembles, owing to the rotational and translational diffusion of the dyes (and the segmental motion of the polymers), and the diffusion may be in the time region of the fluorescence, especially if a long-lived donor is used. The problem can be quite complex, but it is important and informative; the reader is referred to the extensive literature.[66,76-84]

Such statistical distributions of donors and acceptors can easily result in empirical R dependencies other than that given in Eq. (9); even simple geometrical constraints of some molecules, such as the helical geometry of double-stranded DNA, could result in departures from Eq. (9) if the dis-

[73] C. R. Cantor and P. Pechukas, *Proc. Natl. Acad. Sci. U.S.A.* **68**, 2099 (1971).

[74] Gryczynski, W. Wiczk, M. L. Johnson, H. C. Cheung, C.-K. Wang, and J. R. Lakowicz, *Biophys. J.* **54**, 577 (1988).

[75] A. Grinvald, E. Haas, and I. Z. Steinberg, *Proc. Natl. Acad. Sci. U.S.A.* **69**, 2273 (1972).

[76] E. Haas and I. Z. Steinberg, *Biophys. J.* **46**, 429 (1984).

[77] Y. Elkana, J. Feitelson, and E. Katchalski, *J. Chem. Phys.* **48**, 2399 (1968).

[78] U. Gösele, M. Hauser, and U. K. A. Klein, *Chem. Phys. Lett.* **34**, 81 (1975).

[79] U. Gösele, M. Hauser, and K. A. Klein, *Z. Phys. Chem. (Wiesbaden)* **99**, 81 (1976).

[80] M. Hauser, U. K. A. Klein, and U. Gösele, *Z. Phys. Chem. (Wiesbaden)* **101**, 255 (1976).

[81] E. Haas, E. Katchalski-Katzir, and I. Z. Steinberg, *Biopolymers* **17**, 11 (1978).

[82] E. Katchalski-Katzir and I. Z. Steinberg, *Ann. N.Y. Acad. Sci.* **366**, 41 (1981).

[83] J. R. Lackowicz, M. L. Juhnson, W. Wiczk, A. Bhat, and R. F. Steiner, *Chem. Phys. Lett.* **138**, 587 (1987).

[84] S. Albaugh, J. Lan, and R. F. Steiner, *Biophys. J.* **33**, 71 (1989).

tance is assumed to be linear with the number of base pairs (see Fig. 4A). Just because some observations do not conform exactly to what is expressed in Eqs. (6)–(9), this should not be construed uncritically as an indication of non-Förster transfer; instead, this is probably an indication that the molecular situation is more complex than has been assumed in the data analysis. The Förster expression for k_T in Eq. (6) applies only to a particular relative configuration of the D–A pair. Equation (6) forms the theoretical basis for long-range dipole–dipole resonance transfer. Extensions to take into account D–A distributions in R, R_0, κ^2, and diffusion (rotational and translational) within the time scale of fluorescence decay can easily account for most situations which do not strictly adhere to Eq. (9), even though the energy transfer still occurs according to the resonance dipole–dipole Förster mechanism. One usually starts by assuming that the transfer takes place according to Eq. (9) and that $\kappa^2 = 2/3$; for many cases this is sufficient to account for the data and enable analysis of molecular structures. In general, however, there is much more information available in FRET measurements, and with improvements in the instrumentation (especially the time-resolved techniques) and increased sophistication in data analysis, future emphasis will be directed to these more intricate structural problems. Because of the ability to synthesize specific nucleic acid sequences and label them at specific sites, these applications will most certainly lead to interesting structural studies.

Another major complication with FRET measurements can occur if the quantum yield of the donor Φ^D is not the same for all the samples to be compared. The fluorescence quantum yield of the acceptor Φ^A directly affects FRET fluorescence spectra [Eq. (11)], even after subtracting the donor contribution [Eq. (12)]. The dependency on Φ^A is eliminated, however, by the ratios formed in Eqs. (13) and (14). If it is possible to subtract the donor contribution from the fluorescence data (as explained above), Φ_D will also not affect the determination of E. However, because Φ^D enters directly into calculations of R_0 [Eq. (7)], this must be taken into account when comparing E values within a series of measurements. Φ^D is often sensitive to the ionic strength, pH, and other solution parameters, which are convenient to vary during titrations. A change in Φ^D will not have a linear effect on the energy transfer ratio of Eq. (8), as can be seen from Eqs. (7) and (9), since R_0 will change between different experiments. This dependence can be calculated by iterating when fitting a series of titration measurements (e.g., ion titrations using FRET to follow a DNA structural transition), or by just comparing samples under different solution conditions. One must simply measure the relative Φ^D of a molecule labeled only with donor, for every solution condition. Then every value of E (or a signal proportional to E) can be corrected iteratively in a fitting procedure of the particular experiment. Because R_0^6 is directly proportional to the quantum yield of the donor [Eq. (7)], it is simpler to take account of Φ^D changes provided that E is very small [then $E \propto (R_0/R)^6$ from Eq. (9)].

Energy can be transferred by other mechanisms. Trivial readsorption of donor fluorescence radiated by the acceptor, and subsequent emission by the acceptor, will not be a problem in general, even for intramolecular energy transfer.[47] The extent of energy transfer by readsorption will occur mainly at very high concentrations of donor or acceptor (the donor of a doubly labeled molecule in Fig. 4 is not surrounded by acceptors, as is the case of D and A free in solution); it is dependent on the volume of the sample (FRET is not volume dependent), and its presence can be easily checked. Förster dipole–dipole transfer is the only other likely way to transfer energy (corresponding to optical transitions) over distances of 20–100 Å; exchange mechanisms[16] will not contribute until the distance between the donor and acceptor is approximately 5 Å or less. There have been reports of energy transfer over many base pairs, exciting where the DNA bases absorb.[40] Singlet–singlet transfer at room temperature probably follows a Förster mechanism,[40] as does the energy transfer from bases to dyes (triplet to singlet) at 77 K in a compact structure of poly(rA).[85,86] The possibility exists that one-dimensional exciton transport through the stacked base orbitals in the DNA helix can occur at 77 K in a well-stacked helix.[87]

It is an advantage to use a series of related molecules in order to distinguish stereochemical possibilities with energy transfer. Often this eliminates the necessity of accurately determining R_0. R is not a strong function of E; on the other hand, a small change in R will affect the measured value E by the sixth power of R, a very strong dependence. Because it is often impossible to determine R_0 rigorously for the exact molecular circumstances used for the energy transfer experiment, it is not advisable to deduce the symmetry of molecular structures solely by comparing molecular models with such ambiguous determinations of R. A more elaborate but more reliable approach is to construct a series of similar, doubly labeled molecules, judiciously chosen to test certain aspects of the molecular stereochemistry (e.g., labeled in different positions). The global alteration in E (resulting from differences in R), rather that attempting accurate individual estimates of R, may allow one to distinguish discordant molecular models; this avoids making structural conclusions based on possibly inaccurate estimates of R (even though the FRET measure-

[85] F. Van Nostrand and R. M. Pearlstein, *Chem. Phys. Lett.* **39,** 269 (1976).
[86] R. M. Pearlstein, F. Van Nostrand, and J. A. Nairn, *Biophys. J.* **26,** 61 (1979).
[87] F. Van Nostrand and R. M. Pearlstein, *Photochem. Photobiol.* **28,** 407 (1978).

ments may determine accurate values of E). Often it is sufficient to follow the variation of $(ratio)_A$ or $(ratio)_D$ of Eqs. (14) and (16). For instance, in this way it has been possible to deduce the probable overall stereochemistry of the four-way genetic junction,[14,88] a complex DNA structure with four strands. When comparing a series of labeled isomers, precautions must be taken that the value of R_0 is the same for all the samples to be compared, or the relative differences must be known.

Conclusion

The reader is reminded again that many different approaches exist to acquire and analyze FRET data, and the literature cited should be consulted to see how other situations have been handled. The methods which have been discussed in detail here have attributes advantageous for specific applications, and they have been found to be sensitive and to yield reliable, reproducible results, especially for experimental systems involving DNA structures (Fig. 4). By following analogous procedures outlined in the above sections [between Eqs. (11) and (20)] it should be possible to derive the appropriate experimental procedures and the corresponding analyses that are adapted to meet most particular experimental systems. Of course, the measurements cannot be better than the molecules that are being measured, so extreme care must be taken to ensure that the samples are well defined and pure. FRET with specifically labeled nucleic acids will surely become more popular in the near future; the method has the potential to distinguish many structural features and symmetries ranging over molecular distances less than 100 Å. The complexities discussed above portray not only difficulties that can interfere with simple interpretations of FRET results, even after reliable E values have been achieved, but also emphasize the rich variety of structural and dynamic information available from FRET measurements.

Acknowledgments

The author is indebted to Annelies Zechel for excellent technical assistance with energy transfer measurements, to NATO for a travel grant, and he would like to recognize the enjoyable collaborations with Prof. D. M. J. Lilley and Alastair I. H. Murchie.

[88] R. M. Clegg, A. I. H. Murchie, A. Zechel, C. Carlberg, S. Diekmann, and D. M. J. Lilley, *Biochemistry* (in press).

[19] Circular Dichroism Spectroscopy of DNA

By DONALD M. GRAY, ROBERT L. RATLIFF, and MARILYN R. VAUGHAN

Introduction

Circular dichroism (CD) measurements have been in use for over 20 years to study the conformations of nucleic acids in solution. The reliance on CD spectroscopy to study DNA conformations has stemmed from the sensitivity and ease of CD measurements, the nondestructive nature of such measurements, the fact that conformations can be studied in solution, and the requirement for relatively small amounts of material. Although detailed structural information, such as from X-ray crystallography or NMR spectroscopy, is not available from CD spectra, the CD spectrum of DNA in solution can provide a reliable determination of its overall conformational state when compared with the CD spectra of reference samples. Moreover, CD spectroscopy is applicable to a wide range of samples, including those that are difficult to crystallize or to obtain at high concentrations.

The information from CD spectra is complementary to that from absorption spectra. For example, whereas the percentage absorption change at 260 nm on denaturation of a DNA duplex is much larger than the percentage change of the long-wavelength CD bands, the different secondary structures of DNA can be more easily distinguished by CD spectra.

The value of CD spectra in characterizing the A and B forms of DNA was apparent in the work of Tunis-Schneider and Maestre,[1] and CD was the technique by which the Z conformation of DNA was first identified.[2] Subtle changes in secondary structures within the B-conformational family, for example the changes caused by modest dehydration, can readily be detected by CD spectroscopy.[1,3-6] CD spectroscopy showed that poly[d(A)·d(T)] can exist in a conformation different from mixed-sequence DNAs, since accurate first-neighbor analyses of DNA CD spectra could not be performed when the spectrum of this polymer was included

[1] M. J. B. Tunis-Schneider and M. F. Maestre, *J. Mol. Biol.* **52,** 521 (1970).
[2] F. M. Pohl and T. M. Jovin, *J. Mol. Biol.* **67,** 375 (1972).
[3] W. A. Basse and W. C. Johnson, Jr., *Nucl. Acids Res.* **6,** 797 (1979).
[4] S. B. Zimmerman and B. H. Pheiffer, *J. Mol. Biol.* **142,** 315 (1980).
[5] C. Y. Chen, B. H. Pheiffer, S. B. Zimmerman, and S. Hanlon, *Biochemistry* **22,** 4746 (1983).
[6] S. R. Fish, C. Y. Chen, G. J. Thomas, and S. Hanlon, *Biochemistry* **22,** 4751 (1983).

among the basis set of spectra.[7,8] It is now known that d(A)·d(T) segments can exist in an unusual conformation, denoted B_p, with bifurcated hydrogen bonds.[9] CD spectroscopy has also been valuable in the study of triple-stranded structures[10] and DNA structures containing protonated C·C+ base pairs.[11]

Because the CD spectrum above 210 nm is influenced by the nearest-neighbor composition of a nucleic acid, as well as by the nucleic acid secondary structure, both types of information are available from CD measurements. Of course, one must use caution when trying to assign CD spectral changes to a given conformational change if different sequences are involved, and, conversely, it is difficult to ascertain the nearest-neighbor frequencies of DNAs that have even slightly different secondary structures.

The chief practical application of CD spectroscopy to the study of DNA structures has been by making empirical comparisons with the CD spectra of known structures. Perhaps improvements in theoretical calculations, combined with the extension of CD spectra into the vacuum ultraviolet range, will some day allow a more absolute and detailed analysis of spectral data.[12-14] The quantitative comparisons of CD spectral data by curve-fitting methods is presented elsewhere in this series.[15] In this chapter, we provide examples from our laboratory of CD spectral changes in the near-ultraviolet range that illustrate the CD characteristics of structural transitions in natural and synthetic DNAs. Spectra of condensed forms of DNAs and spectra of complexes of DNAs with proteins or drugs are not discussed.

[7] D. M. Gray, F. D. Hamilton, and M. R. Vaughan, *Biopolymers* **17**, 85 (1978).

[8] F. S. Allen, D. M. Gray, and R. L. Ratliff, *Biopolymers* **23**, 2639 (1984).

[9] J. Aymami, M. Coll, C. A. Frederick, A. H.-J. Wang, and A. Rich, *Nucleic Acids Res.* **17**, 3229 (1989).

[10] V. P. Antao, D. M. Gray, and R. L. Ratliff, *Nucleic Acids Res.* **16**, 719 (1988).

[11] E. L. Edwards, M. H. Patrick, R. L. Ratliff, and D. M. Gray, *Biochemistry* **29**, 828 (1990).

[12] A. L. Williams, Jr., C. Cheong, I. Tinoco, Jr., and L. B. Clark, *Nucleic Acids Res.* **14**, 6649 (1986).

[13] J. H. Riazance, W. A. Baase, W. C. Johnson, Jr., K. Hall, P. Cruz, and I. Tinoco, Jr., *Nucleic Acids Res.* **13**, 4983 (1985).

[14] D. M. Gray, K. H. Johnson, M. R. Vaughan, P. A. Morris, J. C. Sutherland, and R. L. Ratliff, *Biopolymers* **29**, 317 (1990).

[15] W. C. Johnson, Jr., this series, Vol. 10, pp. 426–447.

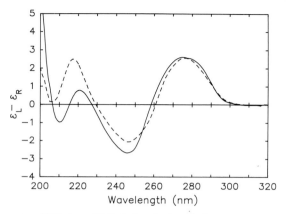

FIG. 1. CD spectra of T7 phage DNA before (—) and after (---) heat denaturation. The DNA was in 2 mM Na$^+$ (phosphate buffer), pH 7. The extinction coefficient at 260 nm was 6570 M^{-1} cm^{-1} for the native DNA. In all figures, data are plotted as $\varepsilon_L - \varepsilon_R$ in units of M^{-1} cm^{-1}, per molar concentration of nucleotide. Unless otherwise noted, spectra in this and subsequent figures were taken at 20°.

Transitions in Natural DNAs

Denaturation of Double-Stranded DNA

There are only modest CD changes above 210 nm on denaturation of double-stranded B-form DNA. T7 phage DNA is an excellent model natural DNA; it contains normal bases and has close to an average base composition, with 52–53 mol % A + T. Figure 1 shows the spectra of T7 phage DNA before and after heat denaturation, both spectra being obtained at 20°. (The heat-denatured DNA was quick-cooled after being denatured at 80°.[16] It may contain residual secondary structure.) Although the absorbance of the heat-denatured DNA increases 28–29% at 260 nm relative to the native DNA (both at 20°), the magnitudes of the major CD bands above 230 nm are not greatly changed on denaturation. There are CD changes at wavelengths below 230 nm, including an increase in the magnitude of the band near 220 nm. A shift of the long-wavelength crossover from 258 to 261 nm is a feature generally indicative of denaturation. The absence of major CD changes at long wavelengths during the helix–coil transition shows that CD bands arising from base pairing and base

[16] M. H. Patrick and D. M. Gray, *Photochem. Photobiol.* **24**, 507 (1976).

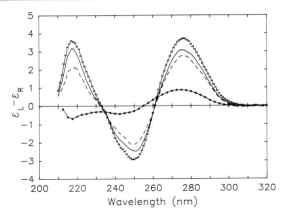

FIG. 2. CD spectra of single-stranded fd DNA at increasing temperatures of 0° (x–x–x), 40° (—), and 80° (---). The weighted average of the nucleotide spectra[18] at 20° (·–·–·) is shown for comparison. The fd DNA was in 2 mM Na$^+$ (phosphate buffer), pH 7. The extinction coefficient at 259 nm was 9470 M^{-1} cm^{-1} at 20°.

stacking in B-DNA, when compared with single-stranded DNA, are either small or tend to cancel.

For DNAs having unusual structures, the CD change on denaturation can be relatively much larger than the absorbance change (e.g., for the oligodeoxyguanylic acid self-complex; see below).[17]

Unstacking of Single-Stranded DNA

The effect of heating on the CD spectrum of fd phage DNA (a single-stranded DNA with negligible intrastrand base pairing at low salt concentration) is illustrated in Fig. 2. The major effect is that the magnitudes of the near-UV CD bands decrease as the temperature increases, owing to unstacking of the bases with a decrease in the interaction between neighboring bases. The long-wavelength crossover does not change if, as in this case, an isodichroic point lies near the baseline. Whereas the CD bands of this DNA decrease in magnitude on heating, the absorption at 260 nm increases by 15% as the temperature increased from 0 to 80°. The CD spectrum of the DNA at 80° still shows significant differences from the average spectrum of the mononucleotides.

[17] D. M. Gray and F. J. Bollum, *Biopolymers* **13**, 2187 (1974).
[18] C. R. Cantor, M. M. Warshaw, and H. Shapiro, *Biopolymers* **9**, 1059 (1970).

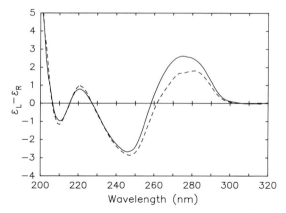

FIG. 3. CD spectra of T7 phage DNA at 0% ethanol (—) and 60% (w/w) ethanol (---). The DNA was in 2 mM Na$^+$ (phosphate buffer), pH 7, prior to the addition of ethanol, for this and Fig. 4.

Modified B Conformations

When natural DNAs are dehydrated by ethanol up to about 60% (w/w), the CD spectrum changes in a characteristic way, as shown in Fig. 3. The long-wavelength positive CD band decreases with a shift in the 258 nm crossover to longer wavelengths, but the CD spectrum is little changed at wavelengths below 250 nm. An increase in salt concentration similarly affects the long-wavelength positive CD band of duplex DNA,[19,20] whereas bands in the vacuum UV down to 180 nm are not significantly affected.[20] The introduction of negative superhelical turns has the effect of increasing the long-wavelength positive band.[21] Considerable effort has gone into identifying the structural changes that are related to this family of CD changes. The structural change was originally thought to be a type of B → C secondary structural change.[1] However, several lines of evidence have led to the conclusion that DNA with a decreased long-wavelength positive CD band remains in the B conformation.[3–6]

Flow-oriented DNA molecules do not show the type of change illustrated in Fig. 3 when the light beam is directed parallel to the direction of orientation.[22] This suggests that the observed reduction in the long-wave-

[19] V. I. Ivanov, L. E. Minchenkova, A. K. Schyolkina, and A. I. Poletayev, *Biopolymers* **12,** 89 (1973).
[20] C. A. Sprecher, W. A. Baase, and W. C. Johnson, Jr., *Biopolymers* **18,** 1009 (1979).
[21] M. F. Maestre and J. C. Wang, *Biopolymers* **10,** 1021 (1971).
[22] S.-Y. Chung and G. Holzwarth, *J. Mol. Biol.* **92,** 449 (1975).

length positive CD band is a consequence of a cancellation of two bands, a positive band from light absorbed parallel to the helix axis and a negative band, induced by dehydration, from light absorbed perpendicular to the helix axis.[22] An alternative possibility is that this type of CD change results, directly or indirectly, from a tertiary structural change that is perhaps not present during flow orientation, since similar CD changes are seen on compaction of the DNA into particles *in vitro,* into phage heads, or into nucleosomes.[23-25]

B → A Conformational Change

On dehydration of DNA in films[1] or by the addition of ethanol above 60% (w/w),[16,23,26] the CD spectrum undergoes a cooperative change to become like that of RNA.[27] As seen in Fig. 4, the CD spectrum of dehydrated T7 DNA has a large, broad positive band from 250 to 300 nm and a large, but narrower, negative CD band at 212 nm. X-Ray diffraction studies on similarly treated DNA samples leave little doubt that such CD spectra of dehydrated DNA samples correspond to the A-DNA conformation.[28,29] Denatured T7 DNA does not show a large CD change on dehydration by ethanol. Thus, the large CD bands for the DNA at 80% ethanol in Fig. 4 arise from the altered base–base and base–sugar interactions in the A conformation. CD measurements have shown that covalently closed PM2 phage DNA can also undergo a B → A conformational transition in solution.[23]

With larger CD bands, and presumably with a more restricted range of conformational variants, the CD spectra of the A family of DNA structures may be more suitable than spectra of the B family for analyses of the nearest-neighbor compositions of unknown sequences.

Sequence Dependence of DNA Circular Dichroism Spectra

Figure 5 illustrates the sequence dependence of DNA CD spectra. The three (A + T)-rich *Drosophila virilis* satellite DNAs whose spectra are shown in Fig. 5 are all repeating heptameric sequences. The sequences, all

[23] D. M. Gray, T. N. Taylor, and D. Lang, *Biopolymers* **17**, 145 (1978).
[24] B. P. Dorman and M. F. Maestre, *Proc. Natl. Acad. Sci. U.S.A.* **70**, 255 (1973).
[25] R. Mandel and G. D. Fasman, *Nucleic Acids Res.* **3**, 1839 (1976).
[26] V. I. Ivanov, L. E. Minchenkova, E. E. Minyat, M. D. Frank-Kamenetskii, and A. K. Schyolkina, *J. Mol. Biol.* **87**, 817 (1974).
[27] H. T. Steely, Jr., D. M. Gray, D. Lang, and M. F. Maestre, *Biopolymers* **25**, 91 (1986).
[28] D. M. Gray, S. P. Edmondson, D. Lang, M. Vaughan, and C. Nave, *Nucleic Acids Res.* **6**, 2089 (1979).
[29] S. B. Zimmerman and B. H. Pheiffer, *J. Mol. Biol.* **135**, 1023 (1979).

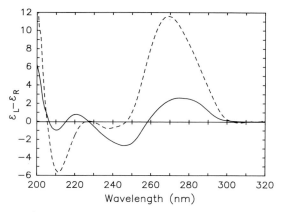

FIG. 4. CD spectra of T7 phage DNA at 0% ethanol (—) and at 80% (w/w) ethanol (---).

of which contain an AAA·TTT triplet, are AAACTAC, AAACTAT, and AAATTAC for satellite DNAs I, II, and III, respectively (these sequences are for one strand). It is obvious that a difference of one base pair out of seven, comparing the sequences of satellite DNAs II and III with the sequence for satellite DNA I, gives the spectra of these DNAs characteristic shapes that can be used to identify the sequence. The choice of reference spectra is important in trying to fit the spectra of such DNA sequences. In the case of satellite DNAs I and II, the spectral contribution of the

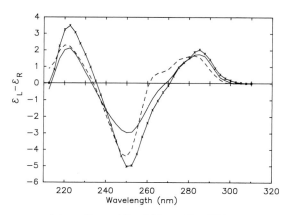

FIG. 5. CD spectra of three *Drosophila virilis* satellite DNAs, namely, satellite DNA I (—), satellite DNA II (---), satellite DNA III (x–x–x) (see text). Extinction coefficients at 260 nm were calculated to be 5890, 5910, and 6310 M^{-1} cm^{-1}, respectively, for satellite DNAs I, II, and III. The DNAs were in 10 mM Na$^+$ (phosphate buffer), pH 7.

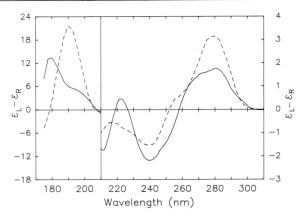

Fig. 6. CD spectra of poly[d(A-C)·d(G-T)] (——) and poly[d(A-G)·d(C-T)] (---). Extinction coefficients at 260 nm were 6500 and 5700 M^{-1} cm^{-1}, respectively. The DNAs were in 20 and 10 mM Na$^+$ (phosphate buffer), pH 7, respectively, for spectra above 210 nm. For spectra at short wavelengths, the polymers were in D$_2$O at pD 7–8, with 8–10 mM Na$^+$ counterions.

AAA·TTT triplet is better approximated by the spectrum of poly[d(A)·d(T)] than by the spectra of polymeric sequences that have an isolated AA·TT doublet.[30] The differences seen in Fig. 5 largely disappear when the DNAs are denatured by heating to 80°.

The CD spectra of sequence isomers poly[d(AC)·d(GT)] and poly[d(AG)·d(CT)] are shown in Fig. 6. There are significant differences in the spectra of these simple polymer sequences in both the near-UV and the vacuum-UV regions.[8,14] The CD bands in the vacuum-UV region below 200 nm are about 6 times greater in magnitude than the bands in the near-UV region. Such differences in CD spectra can potentially be used to estimate the numbers of different types of nearest-neighbor base pairs in a duplex DNA sequence. To date, analyses have been restricted to the region above 210 nm, because a complete set of basis spectra has been available only for this spectral region.[8]

At wavelengths above 210 nm, the CD spectrum of a nucleic acid polymer is, to a good approximation, a nearest-neighbor property.[31] If one has a basis set of spectra of eight sequences having different nearest-neighbor compositions, such that none of the eight sequences has a combination of nearest neighbors that is a linear combination of the nearest neighbors of the other seven sequences (i.e., the eight sequences have linearly indepen-

[30] D. M. Gray and J. G. Gall, *J. Mol. Biol.* **85**, 665 (1974).
[31] D. M. Gray, J.-J. Liu, R. L. Ratliff, and F. S. Allen, *Biopolymers* **20**, 1337 (1981).

dent combinations of nearest neighbors), it is theoretically possible to analyze the CD spectra of other sequences to derive their nearest-neighbor frequencies. Two conditions must be met: (1) the structures of the DNA sequences being compared all have to be identical, so that differences in CD spectra are caused only by differences in the nearest neighbors, and (2) the spectral components of the 10 different Watson–Crick nearest-neighbor base pairs have to be different enough so that the basis set of eight polymer spectra has eight significant components.[32] The first requirement is not met for some simple DNA sequences. In particular, the CD spectra of some DNAs, such as poly[d(A)·d(T)] and poly[d(G)·d(C)], have to be eliminated from the basis set of spectra for the most accurate analyses, presumably because these DNAs differ in structure from mixed DNA sequences.[7-9,33] Their spectra may also have significant contributions from non-nearest neighbors.[34]

Although the sequence information available from CD spectra of double-stranded DNAs is very limited and cannot compete with the information from chemical or enzymatic sequencing methods, CD measurements could be useful for comparing long repetitive DNA sequences, such as telomeric DNAs, that are difficult to sequence by other methods. The sequence dependence of CD spectra should be especially useful in the study of RNAs. Because RNAs in solution contain mixtures of sequences that are paired and unpaired, the analysis for those nearest neighbors that are base paired can help to distinguish among different models of secondary structure.[35]

Transitions in Synthetic DNAs

B → A Conformational Change

Some simple sequence DNAs can undergo an apparent B → A conformational change on dehydration. Figure 7 shows the CD spectra of poly[d(A-C)·d(G-T)] at 0% and 60% (w/w) ethanol. In 60% ethanol a large positive band appears at 260 nm and a large negative band at 210 nm; these bands are reminiscent of those found on dehydration of T7 DNA (Fig. 4). On dehydration, the spectrum of poly[d(A-C)·d(G-T)] becomes closer to the spectrum of the analogous synthetic RNA sequence poly[r(A-C)·r(G-U)], which does not change its major features during dehydration

[32] D. M. Gray and I. Tinoco, Jr., *Biopolymers* **9**, 223 (1970).
[33] C. Marck, D. Thiele, C. Schneider, and W. Guschlbauer, *Nucl. Acids Res.* **5**, 1979 (1978).
[34] S. R. Gudibande, S. D. Jayasena, and M. J. Behe, *Biopolymers* **27**, 1905 (1988).
[35] K. H. Johnson and D. M. Gray, *Biopolymers* **31**, 373, 385 (1991).

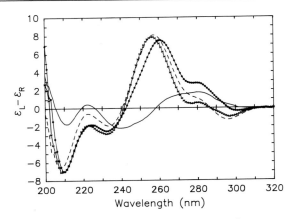

FIG. 7. CD spectra of the DNA sequence poly[d(A-C)·d(G-T)] at 0% ethanol (—) and at 60% (w/w) ethanol (·−·−·) and spectra of the RNA sequence poly[r(A-C)·r(G-U)] at 0% ethanol (---) and at 60% (w/w) ethanol (x–x–x). Extinction coefficients at 260 nm were 6500 and 6900 M^{-1} cm^{-1} for the DNA and RNA, respectively, in the absence of ethanol. The buffer was 20 mM Na$^+$ (phosphate buffer), pH 7, prior to the addition of ethanol.

(Fig. 7). The spectrum of this RNA sequence is undoubtedly that of the RNA A conformation, since the spectrum is consistent with those of other synthetic RNA sequences, as shown by nearest-neighbor calculations.[31] We conclude that the spectrum of poly[d(A-C)·d(G-T)] in 60% ethanol is that of the A conformation of this sequence. The analogous DNA–RNA hybrids also have CD spectra close to that of the duplex RNA when all are in 60% ethanol, except for a larger 280 nm band for the A·U-containing polymer poly[d(A-C)·r(G-U)] than for the A·T-containing polymer poly[r(A-C)·d(G-T)].[36]

B → Z Conformational Change

The CD spectral change for poly[d(G-C)·d(G-C)] that is generally acknowledged to be the result of a B → Z conformational change is shown in Fig. 8. As first reported by Pohl and Jovin,[2] the CD above 220 nm undergoes a change of sign (except near 270 nm) at high salt concentrations. The presence of ethanol or an increase in temperature can also similarly invert the CD of poly[d(G-C)·d(G-C)] at long wavelengths. The form with the inverted CD bands was determined by comparative Raman measurements of solutions and crystals to be the Z conformation of the polymer.[37] Such an inversion of CD bands is not by itself diagnostic of a

[36] D. M. Gray and R. L. Ratliff, *Biopolymers* **14**, 487 (1975).
[37] T. J. Thamann, R. C. Lord, A. H.-J. Wang, and A. Rich, *Nucleic Acids Res.* **9**, 5443 (1981).

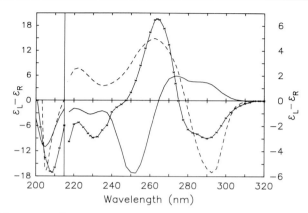

FIG. 8. CD spectra of poly[d(G-C)·d(G-C)] in 10 mM Na$^+$ (phosphate buffer), pH 7, before (——) and after adding NaCl to give 3.9 M Na$^+$ (– – –). The spectrum of poly[r(G-C)·r(G-C)] (x–x–x) was taken with the polymer in 2 mM Na$^+$ (phosphate buffer), pH 7. The extinction coefficients were 7100 M^{-1} cm^{-1} at 255 nm for the DNA in the absence of NaCl and 6560 M^{-1} cm^{-1} at 260 nm for the RNA.

shift from a right-handed B to a left-handed Z form. As shown in Fig. 8, the CD spectrum of poly[r(G-C)·r(G-C)], which is presumably right-handed, has bands above 250 nm with the same signs as in the spectrum of the Z form of poly[d(G-C)·d(G-C)]. Also, the large negative band at 205 nm is not inverted during a B → Z change. CD bands in the vacuum UV have been shown to be correlated with the handedness of the A, B, and Z forms of poly[d(G-C)·d(G-C)].[12,13,37a]

Mixtures of Homopolymers

The relation of CD bands in spectra of mixtures of polymers to bands in the spectra of isolated single strands can help to characterize the structures of duplexes and triplexes. A good example of such a relationship is provided by spectra of mixtures of poly[d(A)] and poly[d(T)]. Figure 9 shows spectra of poly[d(A)] at 0, 40, and 80°. The CD spectrum of poly[d(A)] has predominant bands at 217 and 250 nm. Although the 217 nm band decreases somewhat as the temperature increases, it is still very large at 80° The 250 nm band changes in only a minimal way over the same temperature range. A small band between 260 and 270 nm actually increases in magnitude as the temperature increases, so that the relative

---

[37a] J. C. Sutherland, K. P. Griffin, P. C. Keck, and P. Z. Takacs, *Proc. Natl. Acad. Sci. U.S.A.* **78,** 4801 (1978).

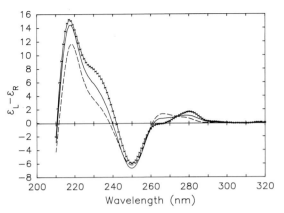

FIG. 9. CD spectra of poly[d(A)] at increasing temperatures of 0° (x–x–x), 40° (—), and 80° (- - -). The polymer was in 2 mM Na⁺ (phosphate buffer), pH 7, and the extinction coefficient at 260 nm was 9650 M^{-1} cm^{-1} at 20°.

magnitudes of the two bands above 260 nm change with temperature. Thus, single-stranded poly[d(A)] does not simply unstack as the temperature increases, with a decrease in the interaction between neighboring bases, as is mostly the case for fd DNA (Fig. 2). Along with a partial unstacking of the bases in poly[d(A)], there appears to be a shift between two related conformations as the temperature increases.

Figure 10 illustrates the information available from CD measurements

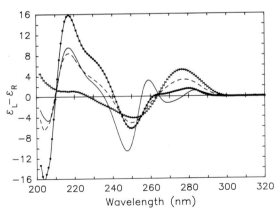

FIG. 10. CD spectra of the duplex poly[d(A)·d(T)] (—), single-stranded poly[d(A)] (·–·–·), single-stranded poly[d(T)] (x–x–x), and the average of the single strands (- - -). The polymers were in 100 mM Na⁺ (phosphate buffer), pH 7. Extinction coefficients at 260 nm were 6150, 9650, and 8140 M^{-1} cm^{-1} for the duplex, poly[d(A)], and poly[d(T)], respectively.

of homopolymer duplexes when spectra are available for the individual strands. Whereas the spectrum of single-stranded poly[d(A)] has a large band at 217 nm, the spectrum of poly[d(T)] is small in this wavelength region. The measured spectrum of the duplex poly[d(A)·d(T)] (solid curve) is closely matched by the average of the spectra of the single strands (dashed curve) in the region between 200 and 230 nm. This shows that base pairing does not add new CD bands to the 200–230 nm region and that the adenines interact in the duplex to give the same large 217 nm band that is present in the spectrum of poly[d(A)].

A comparison with the spectra of other homopolymer duplexes shows that the decrease in CD above 264 nm caused by base pairing in poly[d(A)·d(T)] is correlated with the type of base pair, being larger for an A·T pair than for an A·U pair.[38]

The formation of a poly[d(A)·d(T)] duplex causes a substantial increase in negative CD at about 247 nm. This negative CD band has been useful in monitoring the formation of neighboring A·T base pairs in mixed DNA sequences.[11,39,40]

Reported CD spectra of oligomers designed to contain parallel-stranded A·T base pairs are remarkably similar to the spectra of antiparallel-stranded A·T pairs in the 200–250 nm region.[41] The major influence of strand orientation on the CD spectrum occurs at long wavelengths, where base pairing and cross-strand interactions come into play.

A final example of the information attainable from spectra of homopolymer mixtures is given in Fig. 11, which shows the measured spectrum of triple-stranded poly[d(T)·d(A)·d(T)] (solid curve) compared with the weighted average of the spectra of duplex poly[d(A)·d(T)] and single-stranded poly[d(T)] (dashed curve). The major influence on the CD spectrum of the addition of the third strand is the apparent loss of the 217 nm CD band attributed to poly[d(A)]. However, there is still a negative CD band near 247 nm in the spectrum of the triplex, and we have found that a major characteristic band at 190 nm due to poly[d(A)] remains unchanged on triplex formation.[41a] Therefore, the apparent loss of the 217 nm band in Fig. 11 is more likely due to the addition of a new negative band from the second poly[d(T)] strand than to a major change in the structure of the poly[d(A)] strand.

[38] H. T. Steely, Jr., D. M. Gray, and R. L. Ratliff, *Nucleic Acids Res.* **14,** 10071 (1986).
[39] D. M. Gray, T. Cui, and R. L. Ratliff, *Nucleic Acids Res.* **12,** 7565 (1984).
[40] E. L. Edwards, R. L. Ratliff, and D. M. Gray, *Biochemistry* **27,** 5166 (1988).
[41] J. H. van de Sande, N. B. Ramsing, M. W. Germann, W. Elhorst, B. W. Kalisch, E. von Kitzing, R. T. Pon, R. C. Clegg, and T. M. Jovin, *Science* **241,** 551 (1988).
[41a] K. H. Johnson, D. M. Gray, and J. C. Sutherland, *Nucleic Acids Res.* **19,** 2275 (1991).

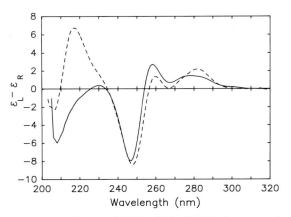

FIG. 11. CD spectrum of triplex poly[d(T)·d(A)·d(T)] (—) compared with a spectrum calculated as ($\frac{2}{3}$) × CD{duplex poly[d(A)·d(T)]} + ($\frac{1}{3}$) × CD{single-stranded poly[d(T)]} (---). The triplex was in 1.1 M Na$^+$ (phosphate buffer), pH 7, and had an extinction coefficient at 260 nm of 5140 M^{-1} cm^{-1}.

Self-Complexes of Simple Sequences

Large, characteristic CD bands appear when self-complexes are formed by poly[d(A)], poly[d(C)], or poly[d(G)]. Poly[d(A)] forms a protonated double helix with a pK of about 4.4 (0.1 M Na$^+$). The bases of protonated poly[d(A)] are presumably paired between parallel strands as in the case of protonated poly[r(A)].[42] The protonated form of poly[d(A)] has a CD spectrum that is very different from that of the stacked, single-stranded polymer (Fig. 12). The positive 217 nm band of the spectrum of the single-stranded polymer is lost, and a large composite positive CD band appears above 250 nm. The spectrum of the protonated polymer has the general shape of the DNA A-form spectrum (Fig. 4). It should be pointed out that the absorbance of deoxyadenosine undergoes only modest changes on protonation.[43]

Poly[d(C)] can also form a protonated double helix, which most likely has parallel strands. In this case, the bases are hemiprotonated, with one proton being taken up per two bases and probably being shared in a hydrogen bond between the cytosine N-3 atoms. With a pK above 7 (0.1 M Na$^+$), the poly[d(C)] self-complex is stable at neutral pH.[44] The absorption

[42] W. Saenger, "Principles of Nucleic Acid Structure," p. 305. Springer-Verlag, New York, 1984.

[43] G. D. Fasman, ed., "Handbook of Biochemistry and Molecular Biology, Volume 1: Nucleic Acids." Chemical Rubber Co., Cleveland, Ohio, 1975.

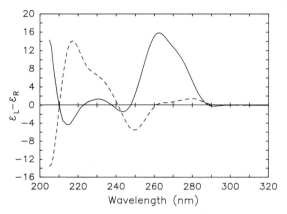

FIG. 12. CD spectra of poly[d(A)] at pH 4 (—) and pH 7 (---). The polymer was in 10 mM Na$^+$ (phosphate buffer), pH 7, and the pH was lowered by the addition of HCl.

spectrum of dCMP shifts by 9 nm to wavelengths longer than 300 nm on protonation,[44] and the CD spectrum of the poly[d(C)] self-complex is shifted to longer wavelengths relative to single-stranded poly[d(C)]. Thus, protonated poly[d(C)] has significant CD at wavelengths above 300 nm (see Fig. 13). Note that the poly[d(C)] self-complex has a negative band at 265 nm, whereas the poly[d(A)·d(T)] duplex has a negative band at about 247 nm (Fig. 10). These features enable one to monitor separately the formation of blocks of A·T and blocks of protonated C·C$^+$ base pairs in mixed-base sequences.[39,40] A structured form of poly[d(C-T)] containing C·C$^+$ base pairs at low pH was also first characterized by CD spectroscopy.[44]

Figure 14 shows a final set of spectra, which are of the single-stranded and self-complexed forms of oligo[d(G)$_5$], both spectra being taken at neutral pH. The propensity of guanine-containing sequences to form self-complexes is well-known and complicates the study of poly[d(G)] and duplexes such as poly[d(G)·d(C)].[17,33] The short d(G)$_5$ oligomer will aggregate when slowly frozen, and the resultant complex remains largely stable at 20°. As in the cases of the other DNA homopolymer self-complexes, significant CD changes occur as the d(G)$_5$ self-complex forms. Although there are large CD changes, there are only very minor changes in the absorption spectrum (<2% at 260 nm) on forming the d(G)$_5$ self-complex.[17]

[44] D. M. Gray, R. L. Ratliff, V. P. Antao, and C. W. Gray, in "Structure and Expression, Volume 2: DNA and Its Drug Complexes" (M. H. Sarma and R. H. Sarma, eds.), 147. Adenine Press, Schenectady, New York, 1988.

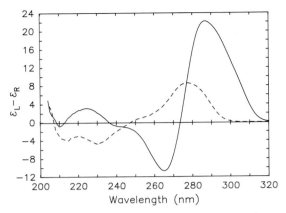

FIG. 13. CD spectra of poly[d(C)] at pH 7 (—) and pH 8.5 (---). The polymer was in 10 mM Na$^+$ (phosphate buffer), pH 8.5, and the pH was lowered by the addition of HCl. The extinction coefficient at 260 nm of the polymer at pH 7.0 was taken to be 5300 M^{-1} cm^{-1}.

The large positive CD band near 260 nm in the spectrum of the oligo [d(G)$_5$] self-complex approximates the spectral contribution of the poly[d(G)] strand of a poly[d(G)·d(C)] duplex near 260 nm, suggesting that the guanines are stacked similarly in the two cases. The oligo[d(G)$_5$] self-complex structure whose spectrum is shown in Fig. 14 may contain either a tetrameric parallel-stranded structure of paired guanine segments

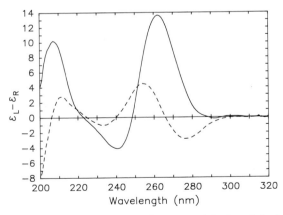

FIG. 14. CD spectra of oligo[d(pG)$_5$] in self-complexed (—) and single-stranded (---) forms. Both spectra were taken at 20°. The self-complexed form was obtained by freezing the sample at −20°. The oligomer was in 20 mM Na$^+$ (phosphate buffer), pH 7, and the extinction coefficient at 260 nm was 9530 M^{-1} cm^{-1}.

as described by Sen and Gilbert[45] or a tetramer of two antiparallel base-paired guanine segments, which may exist in telomeric DNA, as described by others.[46-48]

Structures of repeating dinucleotide sequences that have unusual base pairs have also been characterized by CD spectra. These include an acid form of poly[d(T-C)] with looped out thymines,[44] two acid forms of poly[d(G-C)],[44] two acid forms of poly[d(A-C)],[49] and self-complexes of poly[d(A-G)][10,50] and poly[d(G-T)].[51]

Summary

In general, the types of information available from the CD spectra of DNAs in solution (not including condensed forms of DNA or complexes with drugs and proteins) may be divided into three classes. First, the secondary conformations of DNAs containing Watson–Crick base pairs may be studied. These include the secondary A, B, and Z conformations, the triple-stranded H form (which has both Watson–Crick and Hoogsteen base pairs), and parallel-stranded structures. The interconversion of single-stranded and double-stranded forms can also be monitored. Second, the nearest-neighbor composition of a B-form DNA (or an A-form RNA) can be obtained from a CD spectral analysis. Third, the conditions under which unusual base pairs (e.g., $A^+ \cdot A^+$ and $C \cdot C^+$ base pairs) form can be investigated.

CD spectroscopy is probably one of the most sensitive techniques available for monitoring structural changes of DNAs in solution or for determining whether a new or unusual structure is formed by a particular polynucleotide sequence. Mixtures of structures, such as for samples of the polypurine–pyrimidine sequence poly[d(A-G)·d(C-T)] at various pH values and salt concentrations, can be readily analyzed by CD spectroscopy.[10] In the future, the empirical analysis of CD spectra of more complex sequences may be possible if measurements can be routinely extended to short wavelengths in the vacuum ultraviolet.

[45] D. Sen and W. Gilbert, *Nature (London)* **334**, 364 (1988).

[46] W. I. Sundquist and A. Klug, *Nature (London)* **342**, 825 (1989).

[47] I. G. Panyutin, O. I. Kovalsky, E. I. Budowsky, R. E. Dickerson, M. E. Rikhirev, and A. A. Lipanov, *Proc. Natl. Acad. Sci. U.S.A.* **87**, 867 (1990).

[48] J. R. Williamson, M. K. Raghuraman, and T. R. Cech, *Cell (Cambridge, Mass.)* **59**, 871 (1990).

[49] V. P. Antao, R. L. Ratliff, and D. M. Gray, *Nucleic Acids Res.* **18**, 4111 (1990).

[50] J. S. Lee, D. A. Johnson, and A. R. Morgan, *Nucleic Acids Res.* **6**, 3073 (1979).

[51] D. M. Gray and R. L. Ratliff, *Biopolymers* **16**, 1331 (1977).

Acknowledgments

We thank Dr. V. P. Antao for spectra of the acid forms of poly[d(A)] and poly[d(C)], and we thank Dr. H. T. Steely, Jr., for making available digitized values of homopolymers poly[d(A)] and poly[d(T)] and their mixtures, which were used for Figs. 10 and 11. We are grateful to Drs. C. W. Gray and V. P. Antao for critical readings of the manuscript. This work was supported by National Institutes of Health Research Grant GM19060, by BRSG Grant SO7 RR 07133 (Biomedical Research Support Grant Program, Division of Research Resources, NIH), and by Grant AT-503 from the Robert A. Welch Foundation. Work at the Los Alamos National Laboratory was supported in part by the U.S. Department of Energy.

Section IV
Structural Methods of DNA Analysis

[20] Crystallization of DNA

By YOURI TIMSIT and DINO MORAS

Introduction

The first step in a crystallographic study is the production of large, well-diffracting crystals. Although DNA oligonucleotides are easy to crystallize, good diffracting crystals are less common. To better understand the DNA crystallization processes it is necessary to consider the specific properties of a DNA molecule, to determine the physicochemical parameters which favor the three-dimensional assembly of the molecules, and to analyze the intermolecular contacts of the packing.

The intrinsic ability of DNA to be condensed or aggregated, as in the packaging of phage genomic DNA,[1] and its polyelectrolytic nature make possible its spontaneous organization in two-dimensional arrays by the addition of polycations or anionic polymers,[2-7] but these crystallike objects give limited diffraction information. The polymorphism and the plasticity of DNA structures are characteristic of this molecule.[8-10] It was also shown that several DNA forms can coexist in the same crystal,[11] and that different structures and crystal packing organizations can be found for the same oligonucleotide sequence.[12] In contrast with proteins, the packing environment stabilizes or induces important DNA structural alterations,[13,14] and crystallization conditions can induce structural transitions as, for example, in the high salt induction of B–Z transitions.[15,16] The DNA molecule is

[1] W. Earnshaw and S. Casjens, *Cell (Cambridge, Mass.)* **21**, 319 (1980).

[2] E. Dore, C. Frontali, and E. Gratton, *Biopolymers* **11**, 443 (1972).

[3] T. Maniatis, J. Venable, Jr., and L. Lerman, *J. Mol. Biol.* **84**, 37 (1974).

[4] L. Gosule and J. Schellman, *J. Mol. Biol.* **121**, 311 (1978).

[5] D. Chattoraj, L. Gosule, and J. Schellman, *J. Mol. Biol.* **121**, 327 (1978).

[6] R. Huey and S. Mohr, *Biopolymers* **20**, 2533 (1981).

[7] D. Porschke, *Biochemistry* **23**, 4821 (1984).

[8] R. Dickerson, H. Drew, B. Conner, R. Wing, A. Fratini, and M. Kopka, *Science* **216**, 475 (1982).

[9] Z. Shakked and D. Rabinovich, *Prog. Biophys. Mol. Biol.* **47**, 159 (1986).

[10] O. Kennard, *Q. Rev. Biophys* **22**, 327 (1989).

[11] J. Doucet, J. P. Benoit, W. Cruse, T. Prange, and O. Kennard, *Nature (London)* **337**, 190 (1989).

[12] Z. Shakked, G. Guerstein Guzikevich, M. Eisenstein, F. Frolow, and D. Rabinovich, *Nature (London)* **342**, 456 (1989).

[13] R. Dickerson, D. Goodsell, M. Kopka, and P. Pjura, *J. Biomol. Struct. Dyn.* **5**, 557 (1987).

[14] Y. Timsit, E. Westhof, R. Fuchs, and D. Moras, *Nature (London)* **341**, 459 (1989).

[15] F. Pohl and T. Jovin, *J. Mol. Biol.* **67**, 375 (1972).

[16] T. Thaman, R. Lord, A. Wang, and A. Rich, *Nucleic Acids Res.* **9**, 5443 (1981).

highly hydrated, and water molecules play an important role in the stabilization of its structure.[17]

Since DNA was first crystallized by Gianoni,[18] progress in DNA synthesis has been a determinant for the production of high-resolution diffracting crystals. A few dozen crystal structures are now known, and two types of information are available in literature: the description of the intermolecular contacts and the crystallization conditions. Unfortunately, only a few crystal packings were observed. In addition, published conditions result from an empirical selection from among a large number of assays, and important practical information such as the kinetic data or the geometry of the system used are seldom available. Nevertheless, the analysis of the intermolecular packing contacts found in the different structural forms of DNA can be a good starting point for understanding crystallization of DNA.[19] In this chapter, we search for correlations between the oligonucleotide sequences, the forces which maintain the molecular assembly in the crystal, and the crystallization conditions in cases of well-ordered crystals, whose structures are determined.

Crystal Packing Families

A Form

Good diffracting A-form crystals belong to two space groups: $P6_1$ and $P4_32_12$. In both space groups, the packing is directed by a common motif where the terminal base pair of one duplex stacks closely against the shallow minor groove of a symmetry-related molecule. With some variations in the intermolecular contacts this motif accommodates duplexes of different sequences and sizes. Repetition of this type of contacts produces spirals of double helices and creates large channels around the 6- or 4-fold screw axis. The structure of a 14-mer RNA has demonstrated a different manner of assembling A-form duplexes, using the 2'-OH of the riboses in the intermolecular contacts.[20]

Hexagonal Family ($P6_1$). All members of the hexagonal $P6_1$ family are octamers. In these crystals, the terminal base pairs of each octamer duplex stack against the minor groove of a symmetry-related helix (Fig. 1). As a result the minor groove of each duplex is completely filled by the ends of two neighboring molecules. The packing interaction is different for each

[17] E. Westhof, *Annu. Rev. Biophys. Chem.* **17**, 125 (1988).

[18] G. Gianoni, F. Padden, and D. Keith, *Proc. Natl. Acad. Sci. U.S.A.* **62**, 964 (1969).

[19] A. Wang and M. K. Teng, *J. Cryst. Growth* **90**, 295 (1988).

[20] A. C. Dock-Bregeon, B. Chevrier, A. Podjarny, J. Johnson, J. S. de Bear, G. R. Gough, P. T. Gilham, and D. Moras, *J. Mol. Biol.* **209**, 459 (1989).

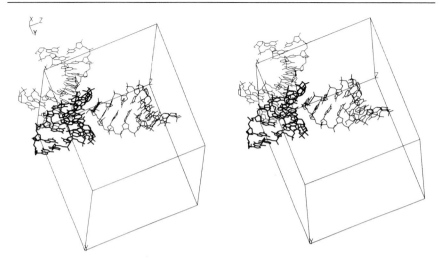

Fig. 1. Stereo view of packing of the d(GGGGCCCC) octamer duplex in the hexagonal cell. The terminal base of one duplex stacks closely against the minor groove of a symmetry-related molecule. This interaction is common for all A-DNA structures.

end of the molecule. The C8 extremity is in contact with a molecule related by the 6_1 operator, whereas the G1 end interacts with a molecule related by the 2_1 operator. Table I shows that, although maintaining the same space group, the 1–16 ends of the various isomorphous duplexes tolerate different hydrogen bonding partners. In the majority of cases, the 3'-hydroxyl of C16 is linked to the amino group of guanine 10. An N2 amino group in position 10 of the sequence is indeed common for all members of the family, but the group is involved in different intermolecular contacts. The N2(G1)–N3(G12) hydrogen bond which is found in the d(GGGGCCCC),[21] d(GGGGCTCC),[22] and d(GGIGCTCC)[23] parent octamer duplexes is replaced in d(GGTATACC)[24] by an N2(G1)–N3(A6) hydrogen bond since the guanine in position 12 is absent. N2(G1) is linked to O2(C6) in the d(GGGGTCCC)[25] mismatched duplex where G12 is still present.

[21] M. McCall, T. Brown, and O. Kennard, *J. Mol. Biol.* **183**, 385 (1985).
[22] W. Hunter, G. Kneale, T. Brown, D. Rabinovich, and O. Kennard, *J. Mol. Biol.* **190**, 605 (1986).
[23] W. Cruse, J. Aymami, O. Kennard, T. Brown, A. Jack, and G. Leonard, *Nucleic Acids Res.* **17**, 55 (1989).
[24] O. Kennard, W. Cruse, J. Nachman, T. Prange, Z. Shakked, and D. Rabinovich, *J. Biomol. Struct. Dyn.* **3**, 623 (1986).
[25] G. Kneale, T. Brown, O. Kennard, and D. Rabinovich, *J. Mol. Biol.* **186**, 805 (1985).

TABLE I
VARIATION OF INTERMOLECULAR CONTACTS WITHIN $P6_1$ SPACE
GROUP FAMILY[a]

Sequence	O3'(C16)	O2(C16)	N2(G1)	N3(G1)
CCATATGG GGTATACC	N2(G10)		N3(A6)	
GGGGCCCC CCCCGGGG	N2(G10)	N2(G11)	N3(G12)	N2(G12)
GGGGTCCC CCCTGGGG	N2(G10)		O2(C6)	b
GGGGCTCC CCTCGGGG	N2(G10)	N2(G11)	N3(G12)	N2(G12)
GGIGCTCC CCTCGIGG	N2(G10)	N2 absent	N3(G12)	N2(G12)

[a] Hydrogen bonding partners of the O3'(C16), O2(C16), N2(G1), and N3(G1) atoms at the 1–16 end of the duplexes for different isomorphous structures are shown.

[b] N2(G12) > O-1'(G2).

Tetragonal Family ($P4_32_12$ and $P4_3$). The tetragonal family presents the same motif but with the duplexes organized around a left-handed 4-fold screw axis and forming large oval channels (10–20 Å). Unlike the case in the $P6_1$ family, the contacts at the two ends of the helix are symmetric. The sequences which belong to this family are shown in Table III below. This family is unusual in that it includes oligonucleotides of different sizes, octamers, tetramers, and nonamers. In the case of tetramers, two stacked helices form an octamer in the asymmetric unit.

As in the $P6_1$ family, a great variety of intermolecular contacts participate in the minor groove stacking interactions. Different hydrogen-bonding schemes are encountered in related sequences: O5'(C1)–N3(A5) and O2(C1)–N2(G6) are reported for the d(CTCTAGAG)[26] octamer, O4'(G1)–N2(G5) and N3(G1)–N2(G6) in the d(GCCCGGGC).[27] The d(GGCCGGCC) octamer,[28] where no hydrogen bonds were identified in the contact region, represents the extreme situation.

[26] W. Hunter, B. Langlois d'Estaintot, and O. Kennard, *Biochemistry* **28**, 2444 (1989).
[27] U. Heineman, H. Lauble, R. Frank, and H. Blocker, *Nucleic Acids Res.* **15**, 9531 (1987).
[28] A. Wang, S. Fujii, J. van Boom, and A. Rich, *Proc. Natl. Acad. Sci. U.S.A.* **79**, 3968 (1982).

Z Form

The best ordered DNA crystals are obtained with the Z form, since the crystals diffract up to 0.9 Å resolution. The sequences which crystallize in the Z form are listed in Table IV. The $P2_12_12_1$ family is composed of hexamers. The $P6_5$ family contains octamer and tetramer duplexes. A common packing motif is observed in all Z-form crystals: end-to-end stacking of the helices to form infinite polymers along the c axis. Despite the missing phosphate, the stacking interaction is stable and specific. The lateral interactions between the parallel endless column stretches involve backbone–backbone contacts and a highly structured solvent network. In the $P2_12_12_1$ space group, the stacked left-hand hexamer duplexes d(CGCGCG)[19,29] are locked by four direct hydrogen bonds between the 3′-hydroxyl ends of G6 and G12 and the anionic oxygen of the phosphate group of C3 and G2, as represented in Fig. 2. In the analog brominated duplex,[30] several solvent intermolecular bridges where N7 atoms of the guanine bases play an important role are reported: sodium bridges link N7 to the anionic oxygen of the backbone, and ammonium ions bridge two guanine by their N7. Hexaaminecobalt bridges are also reported between N7 atoms of guanine bases and the phosphate group in the decamer d(CGTACGTACG).[31] Because the bases do not participate directly in the intermolecular contacts, there is no evidence of a direct influence of the sequence on the packing. Here, the principal sequence requirement to adopt this packing is the capacity to adopt the Z form.

B Form

The B form exhibits three principal packing types. Groove–groove interactions are observed in the dodecamer containing the EcoRI site d(CGCGAATTCGCG),[32] whereas backbone–backbone contacts are seen in the decamer d(CCAAGATTGG)[33] and the phosphorothioate hexamer d(GCGCGC).[34] Groove–backbone interactions are found in the dodecamer containing the NarI site d(ACCGGCGCCACA).[14]

[29] A. Wang, G. J. Quigley, F. Kolpak, J. Crawford, J. van Boom, G. van der Marel, and A. Rich, Nature (London) 282, 680 (1979).

[30] B. Chevrier, A. C. Dock, B. Hartman, M. Leng, D. Moras, M. Thuong, and E. Westhof, J. Mol. Biol. 188, 707 (1986).

[31] R. G. Brennan, E. Westhof, and M. Sundaralingam, J. Biomol. Struct. Dyn. 3, 649 (1986).

[32] R. Wing, H. Drew, T. Takano, C. Broka, S. Tanaka, K. Itakura, and R. Dickerson, Nature (London) 287, 755 (1980).

[33] G. Privé, U. Heineman, S. Chandrasegaran, L. K. M. Kopka, and R. Dickerson, Science 238, 498 (1987).

[34] W. Cruse, S. Salisbury, T. Brown, R. Cosstick, F. Eckstein, and O. Kennard, J. Mol. Biol. 192, 801 (1986).

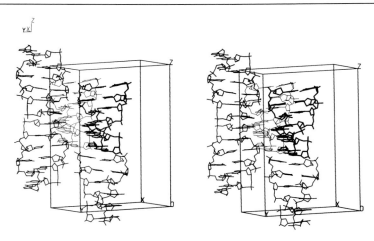

FIG. 2. Stereo view of end-to-end stacking of the Z hexamer duplexes d(CGCGCG) in the orthorhombic cell. The 3'-hydroxyl ends of G6 and G12 form hydrogen bonds with the anionic oxygen of the phosphate groups of C3 and G2.

Groove – Groove Contacts: $P2_12_12_1$ Family. The dodecamer d(CGCGAATTCGCG)[32] crystallizes in the orthorhombic space group $P2_12_12_1$ with a packing based on reciprocal minor groove interactions (Fig. 3a). The three terminal 5'-CGC . . . and . . . GCG-3' base pairs of the 2_1 related duplexes overlap, and hydrogen bonds connect N3 and N2 atoms of the guanine bases in the minor groove edge. In addition, the 3'-hydroxyl end forms a hydrogen bond with the N2 atom of G22. Lateral contacts resulting from the close approach of P2 and P7 (6.4 Å) and P10–P18 phosphates of symmetry-related duplexes act as pivots for the contrarotation of the helices.[13] The sequences which crystallize in an isomorphous packing are presented in Table V. All sequences exhibit the terminal CGC . . . GCG boxes with a conservation of the specific intermolecular contacts. The six central nucleotides can be changed without packing disruption (Fig. 5).

Backbone – Backbone Interactions: C2 Family. In the *C2* family, decamer duplexes are end-to-end stacked to form infinite columns aligned along the *c* axis. The end-to-end stacking step is identical to the intrahelical stacking. Adjacent parallel stacks are displaced relative to one another along their axis so that a phosphate backbone of one helix nests in front of the major or minor groove of its neighbor. Close intermolecular P–P phosphate group contacts and water intermolecular bridges block the columns. The absence of connectivity between two stacked helices is replaced by a hydrogen bond between the 5'-hydroxyl end of the C1 of one helix and the 3'-hydroxyl end of G10 of the following one. The lack of phos-

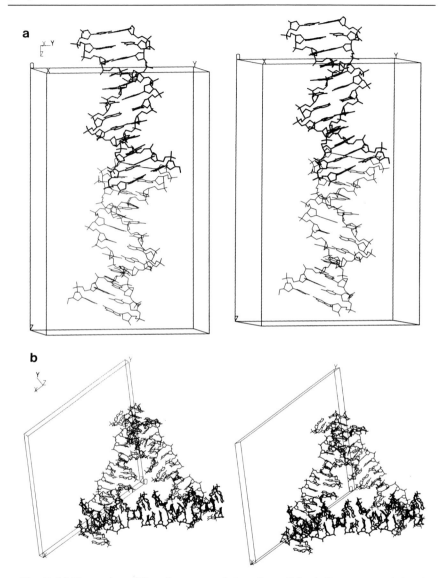

Fig. 3. (a) Stereo view of the minor groove interactions of the dodecamer containing the *Eco*RI restriction site. The three terminal base pairs of the symmetry-related duplexes overlap, and hydrogen bonds connect the N-3 and N-2 atoms of the guanine bases. (b) Crystal packing of the B dodecamer containing the *Nar*I site. Two types of interactions are shown: the duplexes are end-to-end stacked, and they are locked by groove–backbone interactions.

phate at this interhelical step is required to avoid the unacceptably short P–P distance (3.1 Å) that true infinite neighboring columns would form.[35] The sequence of the three decamers belonging to this family are reported in Table VI.

Groove–Backbone Interactions: R3 Family. Three dodecamers containing the *Nar*I site (Table VI) crystallize in a trigonal packing with two new types of interactions: end-to-end stacking of the duplex, with a negative twist at the interhelical base pair step, and two major groove–backbone interactions.[14] The helices are interlocked by the reciprocal fit of the backbone of a duplex in the major groove of a symmetry-related molecule (Fig. 3b). This close assembly of the duplexes is stabilized by the formation of four hydrogen bonds between the N4 amino group of the alternate cytosines C3/C21 and C6/C18 and the anionic oxygen atoms of the phosphate groups P17 and P20, respectively (Fig. 5).

Influence of Sequence on Space Group Distribution and Packing Stability of Oligonucleotide Crystals

Figure 4 summarizes the intricate relations between DNA sequence and crystallization conditions and their effects on the packing type, the space group, and the quality of the crystals. Both the sequence and the crystallization conditions act on the physical properties of the DNA in solution and determine the packing of the helices.

Direct Influence of Base and Backbone Substituents

Sequence specificity can influence the nature and stability of crystal packing through the presence of specific anchoring points which orient the formation of intermolecular hydrogen bonds. In the case of the B-form dodecamer duplexes, the terminal CGCxxxxxxGCG boxes lead to crystallization in the orthorhombic space group by favoring the formation of specific hydrogen bonds between the minor grooves of the staggered duplexes,[32] whereas the presence of cytosine in position $\frac{3}{21}$, $\frac{6}{18}$ of the dodecamer sequences allows a stable major groove–backbone interaction in the *R3* space group.[14] Consequently, under similar crystallization conditions, through these sequence specificities it is possible to orient the packing arrangement of the B dodecamers (Fig. 5). In the case of the A octamers where specific hydrogen bonds also contribute to the intermolecular contacts, the problem is much more complex, since they are able to tolerate different packing interactions while keeping the same global arrangement.

[35] G. Privé, K. Yanagi, and R. Dickerson, *J. Mol. Biol.* **217,** 177 (1991).

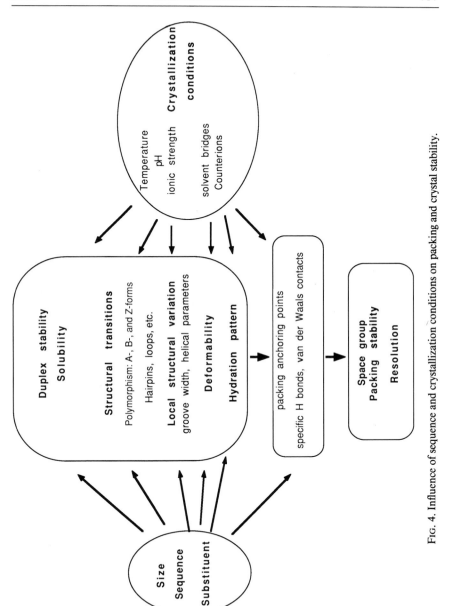

FIG. 4. Influence of sequence and crystallization conditions on packing and crystal stability.

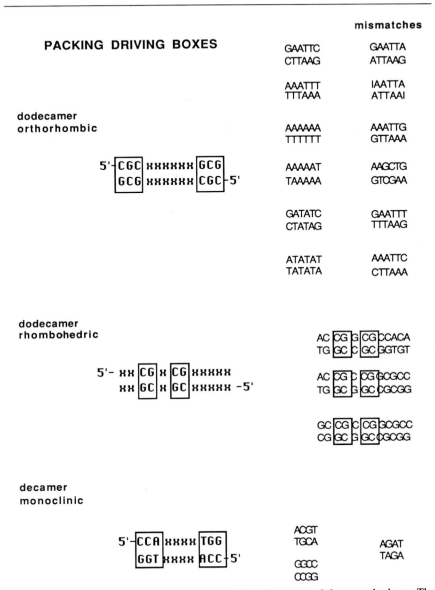

FIG. 5. Packing driving boxes in sequences of B dodecamer and decamer duplexes. The boxed sequences direct the intermolecular contacts and orient the crystallization of the duplexes in defined packings.

The presence of guanine bases at positions 1 and 10 seems to be a constant feature for the $P6_1$ family, whereas it seems not to be necessary for the $P4_32_12$ family.

Efficient van der Waals contacts can play an important role in the stability of crystals. The difference between crystal stability of the two mismatched octamers d(GGIGCTCC)[23] and d(GGGGCTCC)[22] is directly related to the presence of the N2 amino group of the mismatched base pair. The increased stability and better diffracting power (1.7 Å) of (inosine·thymine) octamer crystals could be the result of a more efficient van der Waals packing into the minor groove of both inosine bases. On the contrary, for the corresponding (guanine·thymine) octamer, the mismatched thymine is displaced in the major groove, thus reducing the van der Waals contacts. A similar effect is reported for the two mismatched Z-DNA hexamers d(CGCGFG)[36] and d(CGCGTG)[37] which diffract, respectively, to 1.5 and 1.0 Å. Differences in the intermolecular contacts involving fluorine atoms of 5′-fluorouracil (5′FU) and the methyl group of thymine are thought to be responsible for the decrease in diffraction power.

A good example of the importance of solvent bridges is given by the B-DNA d(GCGCGC)[34] phosphorothioate hexamer crystal. The coordination of cations to sulfur atoms of the backbone contribute to the formation of additional intermolecular solvent bridges and promote an ordered crystal growth, unlike the case with the unmodified hexamer which diffracts to only 8 Å.

Indirect Influence

The growth of left-handed hexamer crystals is strongly influenced by the ability of their sequences to adopt the Z conformation. Methylation or bromination of cytosines stabilizes the Z form and facilitates crystallization, whereas the presence of AT base pairs has a negative effect.

The sequence acts on the helical parameters and contributes indirectly to the adaptation of the helices to a given packing arrangement. For example, the majority of the octamers of the $P4_32_12$ family contain a central CG or pyrimidine–purine step (Table III) associated with a common structural feature, resulting in a wide open major groove, a partial unwinding at the central step, a low tilt, and large rise values. These typical features favor the tetragonal space group over the hexagonal one.[38]

[36] M. Coll, D. Saal, C. Frederick, J. Aymami, A. Rich, and A. Wang, *Nucleic Acids Res.* **17**, 911 (1989).

[37] P. Ho, C. Frederick, G. Quigley, G. van der Marel, J. van Boom, A. Wang, and A. Rich, *EMBO J.* **4**, 3617 (1985).

[38] D. Rabinovich, T. Haran, M. Eisenstein, and Z. Shakked, *J. Mol. Biol.* **200**, 151 (1988).

The monoclinic B decamers share a similar sequence constraint: CA steps with an extremely large twist value (50°) are required at positions 2 and 3.[35] A GA step in the same position in the d(CGATCGATCG) decamer favors an orthorhombic space group with different lateral intermolecular contacts.

Crystallization Conditions

Owing to recent developments in solid-phase automatic DNA synthesis by the phosphoramidite method, several milligrams of a defined oligonucleotide sequence can be readily prepared. The DNA is purified by reversed-phase or ion-exchange high-performance liquid chromatography (HPLC). In the majority of cases, oligonucleotides are crystallized by the vapor diffusion–hanging or sitting drop method in a manner similar to that used in protein crystallogenesis.[39] The hanging drop is the most widely used technique because it allows testing of a great number of different conditions with a small quantity of material. The crystallization box is usually kept in a temperature-controlled room. As a rule, the crystallization medium contains a cacodylate buffer (pH 6.5–7.5), an alcohol [2-propanol or methylpentanediol (MPD)], divalent cations (Mg^{2+}, Ba^{2+}, etc.), and polyamines (in most case spermine). An overview of published crystallization conditions is presented in tables II–VI.

Comparison of Crystallization Conditions Between Different Forms and Packing Families

Are crystallization conditions (spermine– and magnesium–base pair ratio, temperature, etc.) correlated with the crystal packing or the conformation of DNA in the crystal? Figure 6 shows that A-, Z-, and B-form duplexes which crystallize with spermine and magnesium can be differentiated by the range of spermine– and magnesium–base pair ratios used for crystallization. The Z form can be characterized by a high spermine to base pair ratio, whereas B-form dodecamers are found in the low spermine zone and higher magnesium. The A form (tetragonal space groups) is roughly confined to the left inferior corner, which corresponds to low magnesium– and spermine–base pair ratios.

Whereas B-form duplexes require MPD as a precipitant and crystallize at low temperature (Fig. 7), A-form and Z-form duplexes are less dependent on these parameters. A-form duplexes can in some cases crystallize

---

[39] A. McPherson, "Preparation and Analysis of Crystals." Wiley, San Francisco, California, 1982.

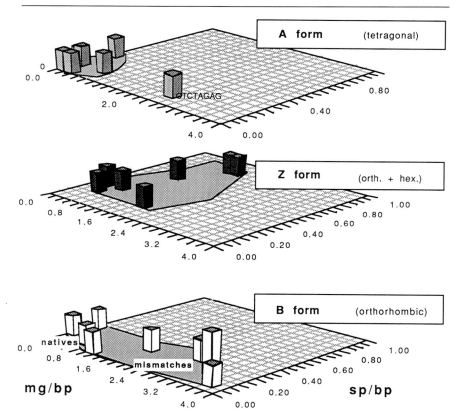

FIG. 6. Comparison of magnesium– and spermine–base pair ratios used in the crystallization of three space group families. Only crystals that are grown with both magnesium and spermine are reported here.

without any alcohol, and Z forms can crystallize with other precipitants at room temperature or higher (Fig. 7).

Z hexamers and B decamers yield high-resolution diffracting crystals (Fig. 7). They share similar packing motifs: end-to-end stacked helices with few lateral intermolecular contacts. The A-form octamer crystals exhibit a wide range of diffracting capacities (1.7–3 Å) and a significant variability in cell volume as a consequence of the great variety of the nature and number of intermolecular interactions within the same packing arrangement. Cell volume variations and limits of resolution are inversely correlated. The orthorhombic B dodecamers diffract homogeneously in the range of 2.2 Å resolution, and variation in cell volume is less important as a result of the conservation of the same intermolecular contacts within the family.

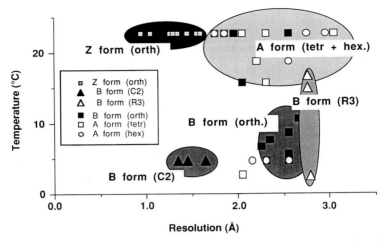

FIG. 7. Distribution of different DNA forms and crystal packing types as a function of crystallization temperature and diffracting power.

TABLE II

SEQUENCES AND CRYSTALLIZATION CONDITIONS USED FOR A-FORM OCTAMER DUPLEXES
BELONGING TO $P6_1$ SPACE GROUP

$P6_1$ A octamer	DNA (mM)	Spermine (mM)	Magnesium (mM)	MPD (%)	sp/bp	Mg/bp	Temperature	pH	Resolution (Å)	Re
GGTATACC	2	2	4 (Ba^{2+})	40	0.13	0.25	RT	6.9	1.8	a
GGBrUABrUACC	0.4	0.28	0.4	40	0.13	0.09	RT	7.0	1.7	b,
GGGGCCCC	0.8	—	3.8	—	—	0.59	18°	6.8	2.5	d
GGGGCTCC (G·T)	3	—	28	—	—	1.16	4°	6.5	2.25	e,
GGIGCTCC (I·T)	4	—	100	—	—	3.1	RT	6.5	1.7	g
GGGGTCCC (G·T)	0.4	—	60	—	—	18.75	4°	6.5	2.1	h
GGGATCCC	4	—	5	24	—	0.16	RT	7.5	2.7	i
GGGTACCC	1.5	0.1	5	10	0.01	0.41	4°	7.0	2.5	j
GGGTGCCC (G·T)	1.5	0.1	4.5	30	0.01	0.40	4°	7.0	2.5	k
GGGCGCCC	2	0.15	5	50	0.01	0.31	19°	7.0	2.9	l

[a] Z. Shakked, D. Rabinovich, O. Kennard, W. Cruse, S. Salisbury, and M. Viswamitra, *J. Mol. Biol.* **166**, 183 (1983).

[b] O. Kennard, W. Cruse, J. Nachman, T. Prange, Z. Shakked, and D. Rabinovich, *J. Biomol. Struct. Dyn.* **3**, 623 (1986).

[c] J. Doucet, J. P. Benoit, W. Cruse, T. Prange, and O. Kennard, *Nature (London)* **337**, 190 (1989).

[d] M. McCall, T. Brown, and O. Kennard, *J. Mol. Biol.* **183**, 385 (1985).

[e] T. Brown, O. Kennard, G. Kneale, and D. Rabinovich, *Nature (London)* **315**, 684 (1985).

[f] W. Hunter, G. Kneale, T. Brown, D. Rabinovich, and O. Kennard, *J. Mol. Biol.* **190**, 605 (1986).

[g] W. Cruse, J. Aymami, O. Kennard, T. Brown, A. Jack, and G. Leonard, *Nucleic Acids Res.* **17**, 55 (1989).

[h] G. Kneale, T. Brown, O. Kennard, and D. Rabinovich, *J. Mol. Biol.* **186**, 805 (1985).

[i] H. Lauble, R. Frank, H. Blocker, and H. Heineman, *Nucleic Acids Res.* **16**, 7799 (1988).

[j] M. Eisenstein, F. Frolow, Z. Shakked, and D. Rabinovich, *Nucleic Acids Res.* **18**, 3185 (1990).

[k] D. Rabinovich, T. Haran, M. Eisenstein, and Z. Shakked, *J. Mol. Biol.* **200**, 151 (1988).

[l] Z. Shakked, G. Guerstein Guzikevich, M. Eisenstein, F. Frolow, and D. Rabinovich, *Nature (London)* **342**, 456 (1989).

Variation of Crystallization Conditions Within Same Packing Families

A Form. Tables II and III show that for both the hexagonal and tetragonal families, crystallization conditions and sequence are correlated. For the two related octamers d(GGTATACC) and d(GGBrUABrUACC), the conditions are closely related: the crystals grow at the same spermine–base pair ratio, precipitant concentration, and temperature. The only differences are in the DNA concentration and the nature of the divalent cations. The octamer d(GGGGCCCC)[21] and its three mismatched analogs also crystallize under similar conditions. Crystals are grown by slow evaporation without spermine or precipitant. However, the magnesium–base

TABLE III

Sequences and Crystallization Conditions Used for A-Form Duplexes Belonging to Tetragonal Family $P4_32_12$ and $P4_3$

octamer	DNA (mM)	Spermine (mM)	Magnesium (mM)	MPD (%)	sp/bp	Mg/bp	Temperature	pH	Resolution (Å)	Ref.
$P4_32_12$										
GGCCGGCC	1.2	0.6	3	30	0.06	0.31	15°	7.0	2.25	a
CCCCGGGG	1.2	0.6	3	30	0.06	0.31	15°	7.0	2.25	a,b
GGmCCGGCC	1.2	3	1.5	20	0.31	0.15	15°	7.0	2.25	c
GCCCGGGC	2	—	10	30	—	0.63	RT	7.5	1.8	d
GGGCGCCC	2.8	0.1	7	40	0.005	0.31	RT	—	1.7	e
ATGCATGC		±							1.7/1.5	f
CTCTAGAG	1.2	1	25	50	0.10	2.6	18°	6.8	2.15	g
GTGTACAC	2	2	3	30	0.13	0.19	—	—	2.0	h,i
GTACGTAC	2	2	10	40	0.13	0.63	RT	7.0	2.25	j
CCGG	5.2	0.4	—	40 (2-propanol)	0.02	—	2°	7.5	2.0	k,l
GGATGGGAG	0.2	0.8	12	20	0.44	6.7	RT	6.5	3.0	m
$P4_3$										
ACCGGCCGGT	1	5	13	50	0.5	1.3	RT	6.5	2.0	n
r(GCG)d(TATACGC)	1.5	8	15	40	0.5	1.0		6.0	2.0	o

[a] A. Wang, S. Fujii, J. van Boom, and A. Rich, *Proc. Natl. Acad. Sci. U.S.A.* **79**, 3968 (1982).

[b] T. Haran, Z. Shakked, A. Wang, and A. Rich, *J. Biomol. Struct. Dyn.* **5**, 199 (1987).

[c] C. Frederick, D. Saal, G. van der Marel, J. van Boom, A. Wang, and A. Rich, *Biopolymers* **26**, 145 (1987).

[d] U. Heineman, H. Lauble, R. Frank, and H. Blocker, *Nucleic Acids Res.* **15**, 9531 (1987).

[e] D. Rabinovich, T. Haran, M. Eisenstein, and Z. Shakked, *J. Mol. Biol.* **200**, 151 (1988).

[f] G. Clark, D. Brown, M. Sanderson, T. Chwalinski, S. Neidle, J. Veal, R. Jones, D. Wilson, G. Zon, E. Garman, and D. Stuart, *Nucleic Acids Res.* **18**, 5521 (1990).

[g] W. Hunter, B. Langlois d'Estaintot, and O. Kennard, *Biochemistry* **28**, 2444 (1989).

[h] S. Jain, G. Zon, and M. Sundaralingam, *J. Mol. Biol.* **197**, 141 (1987).

[i] S. Jain, G. Zon, and M. Sundaralingam, *Biochemistry* **28**, 2364 (1989).

[j] F. Takusawaga, *J. Biomol. Struct. Dyn.* **7**, 795 (1990).

[k] B. Conner, T. Takano, S. Tanaka, K. Itakura, and R. Dickerson, *Nature (London)* **295**, 294 (1982).

[l] B. Conner, C. Yoon, J. Dickerson, and R. Dickerson, *J. Mol. Biol.* **174**, 663 (1984).

[m] M. McCall, T. Brown, W. Hunter, and O. Kennard, *Nature (London)* **332**, 661 (1986).

[n] C. Frederick, G. Quigley, M. K. Teng, M. Coll, G. van der Marel, J. van Boom, A. Rich, and A. Wang, *Eur. J. Biochem.* **181**, 295 (1989).

[o] A. Wang, S. Fujii, J. van Boom, G. van der Marel, S. van Boeckel, and A. Rich, *Nature (London)* **299**, 601 (1982).

pair ratio and the temperature of crystallization are different. The d(GGGGCTCC)[22] mismatched duplex is temperature sensitive (melting occurs at 6°). This sensitivity was attributed to the alteration of the crystal packing and not to the presence of a mismatched base pair, since the other mismatched sequences are stable at room temperature. Indeed, when the guanine of the mismatched base pair is replaced by an inosine base, the lack of the N-2 amino group in the minor groove allows more efficient van der Waals interactions and provides greater crystal stability and diffracting power.

In the $P4_32_12$ family (Table III), the majority of duplexes crystallize with MPD, magnesium, and spermine at room temperature. An interesting methyl substituent effect on the solubility and nucleation rate of the d(GGCCGGCC)[28,40] octamer has been reported: the methylated duplex crystallizes too rapidly under the conditions used for its native parent, and it was necessary to reduce both the magnesium ion and precipitant concentrations in order to decrease the nucleation rate and obtain sizable crystals.

Z Form. Whereas left-handed duplexes crystallize under a large range of conditions (Table IV), high spermine–base pair ratios are generally used. In some cases, however, the absence of spermine can be compensated by increasing the concentration of other cations (magnesium or cobalt hexaamine). All Z-form crystals can be grown at room temperature, and an interesting case of temperature-dependent diffracting power was reported for the brominated hexamer d(BrCGBrCGBrCG).[30]

In the $P2_12_12_1$ family, sequences containing exclusively cytosine/guanine alternating dinucleotides crystallize very easily, without the need of a "high salt" solution. Crystals of the hexamer d(CGCGCG) can be grown with small quantities of alcohol (5% 2-propanol, 10% MPD) or anionic polymers (3% polyethylene glycol 600).[29,41] When spermine is absent from the crystallization medium, a higher magnesium concentration or hexaaminecobalt is used to grow crystals.[42] Methylation or bromination of the cytosines greatly stabilizes the Z form and permits the use of lower spermine– and magnesium–base pair ratios. In the $P6_5$ family, unmethylated tetramers d(CGCG)[43] or octamers d(CGCGCGCG)[44] crystallize

[40] C. Frederick, D. Saal, G. van der Marel, J. van Boom, A. Wang, and A. Rich, *Biopolymers* **26**, 145 (1987).

[41] A. Wang, T. Hakoshima, G. van der Marel, J. van Boom, and A. Rich, *Cell (Cambridge, Mass.)* **37**, 321 (1984).

[42] R. Gessner, G. Quigley, A. Wang, G. van der Marel, J. van Boom, and A. Rich, *Biochemistry* **24**, 237 (1985).

[43] J. Crawford, F. Kolpak, A. Wang, G. Quigley, J. van Boom, G. van der Marel, and A. Rich, *Proc. Natl. Acad. Sci. U.S.A.* **77**, 4016 (1980).

[44] S. Fujii, A. Wang, G. Quigley, H. Westerink, G. van der Marcel, J. van Boom, and A. Rich, *Biopolymers* **24**, 243 (1985).

TABLE IV

SEQUENCES AND CRYSTALLIZATION CONDITIONS USED FOR Z-FORM DNA DUPLEXES

form	DNA (mM)	Spermine (mM)	Magnesium (mM)	MPD (%)	sp/bp	Mg/bp	Temperature	pH	Resolution (Å)	Ref.
$P2_12_12_1$ hexamers										
CGCGCG	2	10	15	5 (2-propanol)	0.83	1.25	RT	7.0	0.9	a
	2	7	10	3 (PEG)	0.58	0.83	RT	7.0	1.2	
	1.4	—	120	10	—	14.3	RT	7.0	1.25	b
mCGmCGmCG	2	3	4	10	0.25	0.33	RT	7.0	1.3	c
BrCGBrCGBrCG	0.3	—	200 [NaCl]	60	—	—	5°, 18°, 37°	6.5	—, 1.4, 14	d
mCGTAmCG	4	7	15	50	0.29	0.625	RT	7.0	1.2	e
BrCGTABrCG	3	3	10	30	0.16	0.55	RT	7.0	1.5	e
mCGATmCG	2.3	25 [Co(NH$_3$)$_6^{3+}$]	25 [Ca^{2+}]	25	—	1.8	RT	7.0	1.54	f
CACGTG	2	—	35	70	—	2.9	RT	6	2.5	g
CGCGTG (G·T)	2	—	100	25	—	8.3	—	7.0	1.0	h
CGCGFG (FU·G)	1	5.6	100	50	0.93	16.6	RT	7–9	1.5, 1.5	i
BrUGCGCG (BrU·G)	2	2	20	5 (2-propanol)	0.16	1.66	4°	8.2	2.25	j
$P6_5$ tetramers and octamers										
CGCG	2	—/10	15	5 (2-propanol)	—/1.25	1.8	RT	7.0	1.5	k
CGCGCGCG	1	—	20	10	—	2.5	RT	7.0	1.6	l
CGCATGCG	1	—	20	40 (2-propanol)	—	2.5	RT	7.0	1.9	l
CGTACGTACG	1.5	0.63	1.7 [Co(NH$_3$)$_6^{3+}$]	30 (2-propanol)	0.04	—	6°	—	1.5	m

[a] A. Wang, G. J. Quigley, F. Kolpak, J. Crawford, J. van Boom, G. van der Marel, and A. Rich, *Nature (London)* **282**, 680 (1979).

[b] R. Gessner, G. Quigley, A. Wang, G. van der Marel, J. van Boom, and A. Rich, *Biochemistry* **24**, 237 (1985).

[c] S. Fujii, A. Wang, G. van der Marel, J. van Boom, and A. Rich, *Nucleic Acids Res.* **10**, 7879 (1985).

[d] B. Chevrier, A. C. Dock, B. Hartman, M. Leng, D. Moras, M. Thuong, and E. Westhof, *J. Mol. Biol.* **188**, 707 (1986).

[e] A. Wang, T. Hakoshima, G. van der Marel, J. van Boom, and A. Rich, *Cell (Cambridge, Mass.)* **37**, 321 (1984).

[f] A. Wang, R. Gessner, G. van der Marel, J. van Boom, and A. Rich, *Proc. Natl. Acad. Sci. U.S.A.* **82**, 3611 (1985).

[g] M. Coll, I. Fita, J. Lloveras, J. Subirana, F. Bardella, T. Huynh-Dinh, and J. Igolen, *Nucleic Acids Res.* **17**, 8695 (1988).

[h] P. Ho, C. Frederick, G. Quigley, G. van der Marel, J. van Boom, A. Wang, and A. Rich, *EMBO J.* **4**, 3617 (1985).

[i] M. Coll, D. Saal, C. Frederick, J. Aymami, A. Rich, and A. Wang, *Nucleic Acids Res.* **17**, 911 (1989).

[j] T. Brown, G. Kneale, W. Hunter, and O. Kennard, *Nucleic Acids Res.* **14**, 1801 (1986).

[k] J. Crawford, F. Kolpak, A. Wang, G. Quigley, J. van Boom, G. van der Marel, and A. Rich, *Proc. Natl. Acad. Sci. U.S.A.* **77**, 4016 (1980).

[l] S. Fujii, A. Wang, G. Quigley, H. Westerink, G. van der Marel, J. van Boom, and A. Rich, *Biopolymers* **24**, 243 (1985).

[m] R. G. Brennan, E. Westhof, and M. Sundaralingam, *J. Biomol. Struct. Dyn.* **3**, 649 (1986).

under roughly the same conditions, with or without spermine, at low precipitant concentrations (Fig. 8).

Sequences containing AT base pairs, with or without the alternating pyrimidine–purine steps are more difficult to crystallize: the hexamers d(mCGTAmCG), d(BrCGTABrCG),[41] and d(mCGATmCG)[45] can only crystallize when the cytosines are methylated or brominated, in the presence of higher cation and precipitant concentrations. In both cases, crystals

[45] A. Wang, R. Gessner, G. van der Marel, J. van Boom, and A. Rich, *Proc. Natl. Acad. Sci. U.S.A.* **82**, 3611 (1985).

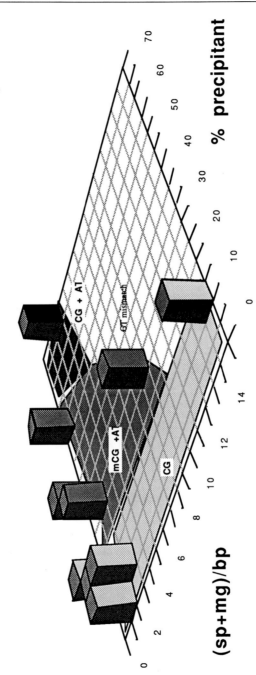

Fig. 8. Influence of AT content of sequences on crystallization conditions for the Z-DNA hexamers. It is necessary to increase the precipitant or the polycation concentrations when the sequence contains AT base pairs.

TABLE V

SEQUENCES AND CRYSTALLIZATION CONDITIONS USED FOR B-FORM DODECAMERS BELONGING TO $P2_12_12_1$ SPACE GROUP FAMILY

$P2_12_12_1$ B dodecamers	DNA (mM)	Spermine (mM)	Magnesium (mM)	MPD (%)	sp/bp	Mg/bp	Temperature	pH	Resolution (Å)	Ref.
CGCGAATTCGCG	1	1.5	2	30	0.13	0.17	RT	7.5	1.9	a
CGCGAATT*CGCG	0.5	0.2	6	35	0.03	1	7°	—	2.3	b
CGCGA*ATTCGCG	1.2	3	1.5	60	0.21	0.10	15°	7.0	2.0	c
CGCAAATTTGCG	1	1	8	50	0.08	0.66	RT	6.5	2.5	d
CGCAAAAAAGCG	0.2	0.5	10	45	0.2	1.6	4°	—	2.5	e
CGCAAAAATGCG	0.7	—	44	40	—	5.2	10°	7.0	2.6	f
CGCATATATGCG	0.5	0.4	22	40	0.07	3.66	6°	—	2.2	g
Mismatches										
CGCGAATTAGCG (G·A)	0.5	1	24	18	0.17	4	5°	7.4	2.5	h
CGCIAATTAGCG (I·A)	0.5	1	20	15	0.17	3.33	4°	7.4	2.5	i
CGCAAATTGGCG (A·G)	0.5	1	20	18	0.17	3.33	4°	6.6	2.25	j
CGCAAGCTGGCG (A·G)	0.75	2.5	25	32	0.27	2.77	5°	7.2	2.5	k
CGCGAATTTGCG (G·T)	0.5	1	25	20	0.17	4.16	8°	7.4	2.5	l
CGCAAATTCGCG (A·C)	0.5	1	24	18	0.17	4	4°	7.4	2.5	m

[a] R. Wing, H. Drew, T. Takano, C. Broka, S. Tanaka, K. Itakura, and R. Dickerson, *Nature (London)* **287**, 755 (1980).
[b] A. Fratini, M. Kopka, H. Drew, and R. Dickerson, *J. Biol. Chem.* **257**, 14687 (1982); M. Kopka, A. Fratini, H. Drew, and R. Dickerson, *J. Mol. Biol.* **163**, 129 (1983).
[c] C. Frederick, G. Quigley, G. van der Marel, J. van Boom, A. Wang, and A. Rich, *J. Biol. Chem.* **263**, 17872 (1988).
[d] M. Coll, C. Frederick, A. Wang, and A. Rich, *Proc. Natl. Acad. Sci. U.S.A.* **84**, 8385 (1987).
[e] H. Nelson, J. Finch, B. Luisi, and A. Klug, *Nature (London)* **330**, 221 (1987).
[f] A. DiGabrielle, M. Sanderson, and T. Steitz, *Proc. Natl. Acad. Sci. U.S.A.* **86**, 1816 (1989).
[g] C. Yoon, G. Privé, D. Goodsell, and R. Dickerson, *Proc. Natl. Acad. Sci. U.S.A.* **85**, 6332 (1988).
[h] T. Brown, W. Hunter, G. Kneale, and O. Kennard, *Proc. Natl. Acad. Sci. U.S.A.* **83**, 2402 (1986).
[i] P. Corfield, W. Hunter, T. Brown, P. Robinson, and O. Kennard, *Nucleic Acids Res.* **15**, 7935 (1987).
[j] T. Brown, G. Leonard, E. Booth, and J. Chambers, *J. Mol. Biol.* **207**, 455 (1989).
[k] G. Webster, M. Sanderson, J. Skelly, S. Neidle, P. Swann, B. Li, and J. Tickle, *Proc. Natl. Acad. Sci. U.S.A.* **87**, 6693 (1990).
[l] W. Hunter, T. Brown, G. Kneale, N. Anand, D. Rabinovich, and O. Kennard, *J. Biol. Chem.* **262**, 9962 (1987).
[m] W. Hunter, T. Brown, N. Anand, and O. Kennard, *Nature (London)* **320**, 552 (1986); W. Hunter, T. Brown, and O. Kennard, *Nucleic Acids Res.* **15**, 6589 (1987).

grow more slowly than the d(CGCGCG) crystals. Three unmodified sequences containing AT base pairs have been crystallized in the Z conformation with very high precipitant and cation concentrations: hexamer d(CACGTG)[46] crystals can grow for several months with or without spermine with a very high concentration of MPD (55 or 70%), the octamer d(CGCATGCG)[44] requires the addition of 400 mM magnesium chloride, and the decamer d(CGTACGTACG)[31] crystallizes in the presence of both spermine and hexaaminecobalt.

B Form. In the $P2_12_12_1$ family (Table V), the majority of dodecamers crystallize under relatively homogeneous conditions. Most crystals are grown at low temperature, around neutrally, and within a small range of

[46] M. Coll, I. Fita, J. Lloveras, J. Subirana, F. Bardella, T. Huynh-Dinh, and J. Igolen, *Nucleic Acids Res.* **17**, 8695 (1988).

TABLE VI

SEQUENCES AND CRYSTALLIZATION CONDITIONS USED FOR OTHER B-FORM DUPLEXES AND SOME UNUSUAL DNA STRUCTURES

Sequence	DNA (mM)	Spermine (mM)	Magnesium (mM)	MPD (%)	sp/bp	Mg/bp	Temperature	pH	Resolution (Å)	Ref.
C2 decamers										
CCAAGATTGG (G·A)	3	—	700	45	—	23.3	4°	—	1.3	a
CCAACGTTGG	1.6	—	760	60	—	47.5	4°	—	1.4	b
CCAGGCCTGG	3.3	—	250	25 (2-propanol), 32 (MPD)	—	7.5	4°	7.5	1.6	c
R3 dodecamers										
ACCGGCGCCACA	1	1.2	18	45	0.1	1.5	4	6	2.6	d
ACCGCCGGCGCC	1	—	24	40	—	1.5	15°	7	2.6	e
GCCGCCGGCGCC	1	—	24	40	—	1.5	15°	7	2.4	e
Other forms										
Phosphorothioate										
GCGCGC	1	1.25	0.75	10 (2-propanol)	0.21	0.125	14°	—	2.17	f
Looped out bases										
CGCAGAATTCGCG	0.6	2.4	15	20	0.30	1.9	4°	7	3	g
CGCGAAATTTACGCG	0.2	0.33	12	18	0.11	4	4°	7	3.6	h
Nicked dodecamer										
CGCGTT + TTCGCG	0.5	8.6 (Hoechst, 1.9 mM)	2.9	50	1.43	0.48	4°	5	3.0	i
Hairpin										
CGCGCGTTTTCGCGCG	1	20	5 (strontium)	40 (2-propanol)	2.5	—	7°	6	2.1	j

[a] G. Privé, U. Heineman, S. Chandrasegaran, L. K. M. Kopka, and R. Dickerson, Science 238, 498 (1987).
[b] G. Privé, K. Yanagi, and R. Dickerson, J. Mol. Biol. 217, 177 (1991).
[c] U. Heineman and C. Aling, J. Mol. Biol. 210, 369 (1989).
[d] Y. Timsit, E. Westhof, R. Fuchs, and D. Moras, Nature (London) 341, 459 (1989).
[e] Manuscript in preparation.
[f] W. Cruse, S. Salisbury, T. Brown, R. Costick, F. Eckstein, and O. Kennard, J. Mol. Biol. 192, 801 (1986).
[g] M. Saper, H. Edar, K. Mizuuchi, J. Nickol, E. Appela, and J. Sussman, J. Mol. Biol. 188, 111 (1986); L. Joshua-Tor, D. Rabinovich, H. Hope, F. Frolow, E. Appella, and J. Sussman, Nature (London) 334, 82 (1988).
[h] M. Miller, A. Wlodawer, E. Appela, and J. Sussman, J. Mol. Biol. 195, 967 (1987); M. Miller, R. Harrison, A. Wlodawer, E. Appella, and J. Sussman, Nature (London) 334, 85 (1988).
[i] J. Aymami, M. Coll, G. van der Marel, J. van Boom, A. Wang, and A. Rich, Proc. Natl. Acad. Sci. U.S.A. 87, 2526 (1990).
[j] R. Chattopadhyaya, S. Ikuta, K. Grzeskowiak, and R. Dickerson, Nature (London) 334, 175 (1988); R. Chattopadhyaya, S. Ikuta, K. Grzeskowiak, and R. Dickerson, J. Mol. Biol. 211, 189 (1990).

spermine–base pair ratios (0.033 to 0.27) except in the case of d(CGCAAAATCGC),[47] where the absence of spermine is compensated by a higher magnesium–base pair ratio. High magnesium–base pair ratios are also associated with the presence of a mismatched base pair in the duplex.

It is of interest to note that, despite their different packing, the conditions used for crystallization of the trigonal dodecamers (Table VI) are very close to those used for the orthorhombic crystals. In this case, the presence of anchoring points at specific positions in the sequence directs the molecular interactions. In the C2 decamer family (Table VI), crystal growth conditions are characterized by very high magnesium–base pair ratios without any spermine. As in the other B forms, crystallization occurs at low temperature. Although packing of the phosphorothioate hexamer d(GCGCGC)[34] is roughly identical to that of the C2 decamers, the hexamer crystallizes under conditions close to those used for R3 and $P2_12_12_1$ dodecamers.

Conclusion

An interesting observation which emerges from packing analysis is the importance of sequence for the organization of B-DNA molecules in the crystalline state. B-form DNA exhibits a larger variety of packings in which all parts of the molecule can participate in intermolecular contacts in a specific manner. In this case, it is possible to generate defined crystal organizations by designing packing driving boxes in the sequence. In contrast, A- and Z-form crystals are not directly sequence dependent, and each form of DNA exhibits only one packing arrangement. The different behavior of B-DNA is due to the existence of two easily accessible grooves and emphasizes its suitability for molecular recognition. It is clear that packing interactions can be good models of molecular recognition processes.

Spermine and magnesium are compounds that occur naturally in cells which could play important structural and regulatory functions. They are both used for the crystallization of oligonucleotides. Defined ranges of spermine– and magnesium–base pair ratios appear to be associated with each packing family, showing their importance in the specific stabilization of the tertiary arrangement of the DNA molecules. General predictive rules are difficult to find owing to the limited amount of available information.

[47] A. DiGabrielle, M. Sanderson, and T. Steitz, *Proc. Natl. Acad. Sci. U.S.A.* **86**, 1816 (1989).

[21] Dynamic Light Scattering for Study of Solution Conformation and Dynamics of Superhelical DNA

By Jörg Langowski, Werner Kremer, and Ulrike Kapp

Introduction

A flexible macromolecule like DNA, which can undergo complex transitions like local denaturation, changes from B-helix to unusual structures, and global shape fluctuations related to changes in superhelical density and branching, cannot be sufficiently described by one single conformation. Although DNAs shorter than the persistence length of 50 nm[1] behave in good approximation like a rigid rod, the time dependence of the conformation must be included in the description of longer DNAs. In addition, some methods frequently used to study the structure of long DNA molecules (e.g., electron microscopy or gel electrophoresis) may possibly distort the DNA conformation either in the course of binding the molecule to a support or by forcing the molecules through gel pores narrower than the average size of the molecule. It is therefore desirable to have methods available which can be used to study the structure of long DNA molecules in solution while disturbing the conformational equilibrium in the least possible way. Dynamic light scattering (DLS) is such a method. It yields information on the hydrodynamic properties of macromolecules: their diffusion, rotational motion, and internal motions arising from their flexibility.[2]

Dynamic Light Scattering

When radiation is scattered from a sample, the scattered intensity is related to the Fourier transform of the scattering power of the sample. For electromagnetic radiation, the scattered electric field amplitude is proportional to the amplitude of the Fourier component of the local excess polarizability in the direction of the scattering vector $\mathbf{K}$: $E_s(t) \propto \delta\alpha(\mathbf{K},t)$ (Fig. 1). In a solution of macromolecules, the local polarizability continuously fluctuates owing to Brownian motion; thus, the scattered intensity will vary on the same time scale as the motions of the particles in the solution. With DLS, one analyzes the fluctuations of scattered light in order to understand the Brownian motion in a sample.

[1] P. J. Hagerman, *Annu. Rev. Biophys. Biophys. Chem.* **17**, 265 (1988).
[2] K. S. Schmitz, "An Introduction to Dynamic Light Scattering by Macromolecules." Academic Press, San Diego, California, 1990.

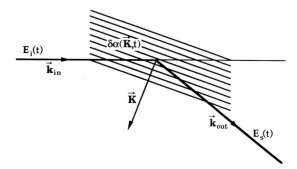

FIG. 1. Scattering of light with incident and observed wave vectors $\vec{k}_{in}$ and $\vec{k}_{out}$ and scattering vector $\vec{K}$. The scattered field amplitude is proportional to the Fourier component $\delta\alpha(\vec{K}, t)$ of the polarizability fluctuations in the medium. The magnitude of K is equal to $(4\pi n/\lambda) \sin(\Theta/2)$, where λ is the wavelength of the incident light, n the refractive index of the medium, and θ the scattering angle.

For particles small compared to the wavelength of the light (maximum dimension $\ll |K|^{-1}$) in dilute solution, one can approximate the excess polarizability by the concentration fluctuation, δc; thus, $E_s(t) \propto \delta c(K,t)$. In this case, it is easy to calculate the intensity fluctuations. Starting from the diffusion equation

$$\frac{\partial \delta c(\mathbf{r},t)}{\partial t} = D \nabla \delta c(\mathbf{r},t) \tag{1}$$

and expressing $c(\mathbf{r},t)$ through a Fourier transform, $c(\mathbf{r},t) = \int d^3\mathbf{r} \, e^{i\mathbf{K}\mathbf{r}} c(\mathbf{K},t)$, one obtains

$$\delta c(\mathbf{K},t + \tau) = \delta c(\mathbf{K},t) \, e^{-\mathbf{K}^2 D\tau} \tag{2}$$

For a system fluctuating around its equilibrium state, the initial condition $\delta c(\mathbf{K},t)$ is unknown. However, one may calculate suitable statistical averages. Multiplying both sides of Eq. (2) with $\delta c(\mathbf{K},t)$ and averaging yields

$$\langle \delta c(\mathbf{K},t + \tau)\delta c(\mathbf{K},t)\rangle_t = \langle \delta c(\mathbf{K},t)^2\rangle_t \, e^{-\mathbf{K}^2 D\tau} \tag{3}$$

where $\langle \ \rangle_t$ indicates the time average. The characteristic relaxation rate λ for the concentration fluctuations is related to the diffusion coefficient: $\lambda = K^2 D$. The scattered field amplitude is proportional to the concentration fluctuation; therefore,

$$\frac{\langle E_s(\mathbf{K},t + \tau)E_s(\mathbf{K},t)\rangle_t}{\langle E_s(\mathbf{K},t)^2\rangle_t} = e^{-\mathbf{K}^2 D\tau} = G_1(\tau) \tag{4}$$

where $G_1(\tau)$ is the first-order autocorrelation function of the scattered E field. This function can be shown to be related to the autocorrelation function of the scattered intensity[3]:

$$G_2(\tau) = \frac{\langle I_s(\mathbf{K},t+\tau)I_s(\mathbf{K},t)\rangle_t}{\langle I_s(\mathbf{K},t)\rangle_t^2} = 1 + [G_1(\tau)]^2 = 1 + e^{-2\mathbf{K}^2 D\tau} \tag{5}$$

$G_2(\tau)$ is usually obtained in real time from the photomultiplier output, using single-photon counting and a hardware correlator (see below). From the decay rate of the measured $G_2(\tau)$ autocorrelation function, one obtains the diffusion coefficient of the molecule under study. The diffusion coefficient can be expressed as the so-called Stokes radius, or hydrodynamic radius R_h,

$$R_h = \frac{k_B T}{6\pi\eta D} \tag{6}$$

where η is the viscosity of the medium, T the temperature, and k_B Boltzmann's constant. When the system becomes more complex, that is, the size of the molecule approaches the wavelength of the scattered radiation, or when one studies a mixture of particles, $G_2(\tau)$ cannot be described by a single-exponential decay function. If, for example, the sample contains a mixture of N species of diffusion coefficients D_i, $G_1(\tau)$ and $G_2(\tau)$ become a sum of single exponential decays with rate constants $\mathbf{K}^2 D_i$:

$$G_1(\tau) = \sum_{i=1}^{N} a_i \, e^{-\mathbf{K}^2 D_i \tau} \tag{7}$$

$$G_2(\tau) = A + \left[\sum_{i=1}^{N} a_i \, e^{-\mathbf{K}^2 D_i \tau}\right]^2 \tag{8}$$

A fundamental problem in analyzing DLS data from more complex systems is to decompose Eq. (7) or Eq. (8) into the individual exponential decay components, given a measured data set $G_1(\tau)$ or $G_2(\tau)$. This problem will not be discussed here; the reader is referred to selected references.[4-6]

Dynamic Light Scattering from DNA Solutions

When the dimensions of the particle under study are of the order of $|\mathbf{K}|^{-1}$ or greater, and if it is anisotropic, its rotation will also contribute to

[3] B. J. Berne and R. Pecora, "Dynamic Light Scattering." Wiley, New York, 1976.
[4] N. Ostrowski, D. Sornette, P. Parker and E. R. Pike, *Opt. Acta* **28**, 1059 (1981).
[5] S. W. Provencher, *Comput. Phys. Commun.* **27**, 213 (1982).
[6] R. Bryan, *Eur. Biophys. J.* **18**, 165 (1990).

fluctuations of the scattering intensity. Maeda and Fujime[7] showed that in the limit of $\mathbf{K} \to 0$, the first-order autocorrelation function of a rigid rod is given by

$$G_1(\tau) = P_0(\mathbf{K}) \, e^{-\mathbf{K}^2 D_0 \tau} + P_2(\mathbf{K}) \, e^{-\{\mathbf{K}^2[D_0 + (4/21)(D_3 - D_1)] + 6\Theta\}\tau} \tag{9}$$

where terms of higher order than $\mathbf{K}^2$ have been neglected. $P_0(\mathbf{K})$ and $P_2(\mathbf{K})$ are $\mathbf{K}$-dependent amplitudes, D_0 is the average translational diffusion coefficient, and D_1 and D_3 are the sideways and lengthwise diffusion coefficients of the rod. $G_2(\tau)$ is again related to $G_1(\tau)$ by Eq. (5). Using this relationship, one can evaluate the end-over-end rotational diffusion coefficient Θ of a rodlike molecule from a biexponential decomposition of $G_1(\tau)$ at the limit of small $\mathbf{K}$. Θ is obtained from the intercept of a plot of the fast relaxation rate against $\mathbf{K}^2$.

For flexible macromolecules, not only rotational motions, but also mutual motions of subparts of the molecule will give rise to intensity fluctuations. This is manifested by the appearance of faster decaying components in $G_1(\tau)$, whose amplitude increases with increasing $\mathbf{K}$. One can obtain an apparent diffusion coefficient D_{app} by approximating the autocorrelation function through a forced single-exponential fit, or by taking the initial slope of a semilogarithmic plot of $G_1(\tau)$ versus τ ("first cumulant"). In this case, an increase of D_{app} with increasing $\mathbf{K}$ is observed (Fig. 2). For the Rouse–Zimm model[8,9] of flexible polymer chains, the theoretical form of the autocorrelation function has been worked out by Lin and Schurr.[10]

The description of the autocorrelation function by one exponential decay component is, of course, only an approximation, and application of multiexponential data analysis procedures to DLS data from DNA shows two major exponential decay components.[11-13] The fast component may be related to an internal segmental diffusion coefficient of subparts of the molecule; as will be shown below, this quantity corresponds to the internal motion of DNA segments of approximately 300 base pairs.

In summary, DLS measurements on DNA will yield information on the translational diffusion coefficient, or hydrodynamic radius; the rotational diffusion coefficient; and the internal motions, expressed as a segmental diffusion coefficient.

[7] T. Maeda and S. Fujime, *Macromolecules* **17**, 1157 (1984).
[8] P. E. Rouse, *J. Chem. Phys.* **21**, 1272 (1953).
[9] B. H. Zimm, *J. Chem. Phys.* **24**, 269 (1956).
[10] S. C. Lin and J. M. Schurr, *Biopolymers* **17**, 425 (1978).
[11] J. Langowski, U. Giesen, and C. Lehmann, *Biophys. Chem.* **25**, 191 (1986).
[12] S. S. Sorlie and R. Pecora, *Macromolecules* **21**, 1437 (1988).
[13] J. Langowski, *Biophys. Chem.* **27**, 263 (1987).

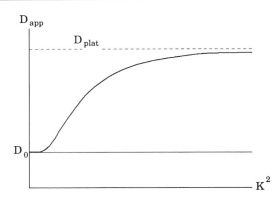

FIG. 2. Behavior of the apparent diffusion coefficient of a flexible polymer in a DLS experiment as a function of the square of the scattering vector. The low-K part of the curve approaches the translational diffusion coefficient D_0, while at the high-K end a plateau value is reached which corresponds to the motion of the smallest independent subunits of the chain.

Technique of Dynamic Light Scattering

A schematic of a typical DLS apparatus is shown in Fig. 3. The machine consists of a laser light source, a scattering cell in a thermostatted holder, a goniometer with a photomultiplier (PMT) to observe the scattered light from a defined angle, and the correlator.

Dynamic light scattering requires the incident light to be coherent over the length of the scattering volume. Also, because of the low scattering power of dilute DNA solutions, a high intensity is desired. The typical light source for a DLS experiment is therefore a laser. The most frequently used lasers in DLS are helium–neon (He–Ne) (λ 632.8 nm; 1–50 mW output power) and argon lasers (λ 514–351 nm; 0.5–2 W visible, 0.1–0.3 W UV output power). The large wavelength range and high power makes the argon laser the more versatile light source; the advantage of the He–Ne laser is its longer wavelength if small scattering vectors are desired, and its much lower cost.

The sample is usually contained in a cylindrical scattering cell made from optical grade glass or quartz, of typical diameter 1–3 cm. To minimize reflections at the glass–air interface and to thermostat the cell, it is kept in an index-matching bath, a cylindrical quartz vessel of 8–10 cm diameter filled with toluene, decaline, or water and kept at a constant temperature. For the higher powers employed with argon lasers, it is recommended to use water as an index-matching fluid; even though the refractive index of water does not match glass as closely as toluene or

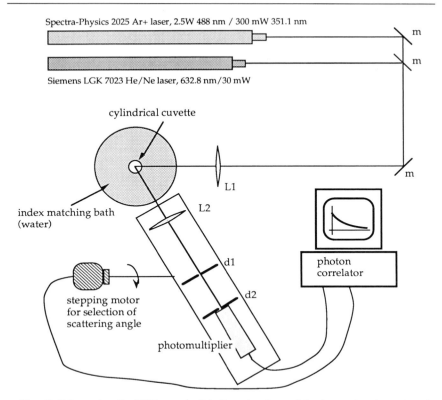

Spectra-Physics 2025 Ar+ laser, 2.5W 488 nm / 300 mW 351.1 nm

Siemens LGK 7023 He/Ne laser, 632.8 nm/30 mW

cylindrical cuvette

index matching bath
(water)

L1

L2

d1

d2

photon
correlator

stepping motor
for selection of
scattering angle

photomultiplier

m
m
m

FIG. 3. Schematic of a DLS system. L1, Focusing Lens; L2, observation lens; d1, d2, pinholes; m, mirrors. The observation optics together with the photomultiplier are mounted on a goniometer arm pivoting around the center of the scattering cell.

decaline do, the latter solvents give rise to thermal defocusing of the laser beam because of their highly temperature-dependent refractive index.

The observation optics consist of a lens which images the scattering volume on a first pinhole defining the observed section of the scattering volume, and a second pinhole defining the scattering solid angle (Fig. 3). The observed scattering solid angle determines the amplitude of the measured fluctuations; for maximum amplitude, a very small section should be observed across which the fluctuations are coherent. Usually, for the systems studied here, pinhole sizes around 0.1 mm for the first and 0.1 – 1 mm for the second one are used. More details on the quantitative relationship between the geometry of the optics and signal amplitude can be found in Ref. 14.

[14] N. C. Ford *In* "Dynamic Light Scattering" (R. Pecora, ed.), p. 7. Plenum, New York, 1985.

The light is detected by a photomultiplier tube which works in the photon-counting mode; the tube is selected for low dark count rate and low afterpulsing. The raw photon pulses are conditioned by discriminator electronics and then analyzed by the correlator, which computes the autocorrelation function of the scattered intensity [$G_2(\tau)$, Eq. (5)] directly from the PMT output. A variety of photon correlators is commercially available (see Ref. 14 for more detailed information); typically, $G_2(\tau)$ can be measured with a minimum sampling time $\Delta\tau$ of 0.05 to 0.1 μsec, and over a time range of several decades. A large dynamic range in τ is desirable for experiments on DNA, because of the multiexponential structure of $G_2(\tau)$.

Sample Concentration

The total scattered light intensity is at first approximation proportional to the molecular weight (MW). As an example, a decameric double-stranded DNA (with MW = 6600) will have approximately the same scattering intensity as the surrounding solvent (water, MW = 18), when its concentration is 18/6600, or approximately 3 mg/ml. For a plasmid DNA (MW $\approx 2 \times 10^6$), this point is reached at a concentration of 9 μg/ml. For reliable measurements, the concentration used should be 3–10 times above these limits, if possible. The volume of the scattering cell (typically 200 μl to several milliliters) determines the amount of sample required; for plasmid DNAs of several thousand base pairs, 10–500 μg is needed per experiment, for DNA fragments of hundreds of base pairs, 0.1 mg up to several milligrams, and for short oligonucleotides, 0.5–1 mg is the lowest amount on which reasonable quality data can be collected in a cylindrical cell of 1 cm diameter. If only very small sample amounts are available, rectangular fluorescence cuvettes can be used with inner dimensions of 3×3 mm; in that case, only 90° scattering will be measured. This is sufficient for clean preparations of small oligonucleotides [< 50 base pairs (bp)], where internal motions do not affect the scattering; for example, 514 nm light at 90° corresponds to a reciprocal scattering vector $\mathbf{K}^{-1}$ of 43 nm, for a total length of the DNA of 17 nm.

Sample Preparation

The requirements for DNA in DLS measurements are different from those in many other applications. First, light scattering measurements are extremely sensitive to the presence of dust and aggregates. For introducing a sample into the measuring cell, the solution is filtered through pore sizes of 0.2 to 0.6 μm depending on the hydrodynamic radius of the DNA; any particles smaller than this size will contaminate the sample. Because a solid dust particle of 0.2 μm diameter is much heavier than a plasmid DNA, its relative contribution to the scattered light intensity will exceed that of the DNA molecule by orders of magnitude. It is therefore necessary to have a preparation method available that excludes particulate contamination. Note that DNA eluted from agarose gels is highly contaminated by agarose particles in the size range of 0.2 μm and below. According to our experience, DNA prepared this way is useless for light scattering measurements.

Preparations using ethidium or other intercalators will leave some dye tightly bound to the DNA, unless extreme efforts are taken to remove remaining intercalators by dialysis against ion exchangers. For measurements in the near UV (370–350 nm), remaining dye can cause single-strand breaks owing to photoreactions. A preparation method which does not use intercalators to separate superhelical from linear and relaxed DNA is therefore highly desirable.

We have used two different high-performance liquid chromatography (HPLC) columns to prepare superhelical DNA in high purity for DLS experiments; one of them can also be used to separate topoisomers. Both methods start with a crude DNA preparation. The plasmid (pUC18 as an example) is grown in *Escherichia coli* HB 101 under ampicillin selection (50–100 μg/ml) at 37° in a 20-liter fermenter with 10 liters/min aeration and 400 rpm agitation to a turbidity (OD_{550}) of 1.0. Solid chloramphenicol is then added to a concentration of 150 μg/ml and incubation continued under the same conditions for 18 hr.[15] The cells are harvested by membrane filtration and centrifugation, then resuspended in TES buffer (50 mM Tris-HCl, pH 8.0, 50 mM EDTA, 15% w/v sucrose). Cell lysis is initiated by adding 500 mg of lysozyme, dissolved in the same buffer. The mixture is incubated at room temperature for 30 min without stirring. Ten percent (v/v) of 3 M sodium acetate and 5% (v/v) of 10% sodium dodecyl sulfate (SDS) solutions are mixed in by careful swirling, and incubation is continued for 30 min at room temperature. It is important at this step to handle the mixture gently, to avoid fragmentation of the chromosomal DNA.

The resulting viscous lysate is centrifuged at 100,000 g for 60 min to remove the chromosomal DNA. The clear supernatant is treated for 30 min at 37° with 100 μg/ml of RNase A (Sigma Chemical Co., St. Louis,

---

[15] HB 101 or another *recA*⁻ strain should be used to minimize the presence of dimers. Although dimeric plasmids in small amounts do not disturb gel electrophoretic analysis techniques (because the dimer can easily be distinguished from the rest), nor the study of linear DNA fragments (because restriction cutting will produce the same set of fragments for a head-to-tail dimer and a monomer), they are very undesirable in solution scattering techniques. Most of the time, monomer and dimer cannot be distinguished as two different species by DLS.

MO, stored in 10 mM Tris-HCl, pH 7.5, 15 mM NaCl, and preincubated for 20 min at 100° to remove DNases), then for 30 min at 37° with 10 μg/ml of proteinase K (Boehringer, Mannheim, Germany). NaCl is added from a 5 M stock solution to a final concentration of 0.5 M, then 20% (v/v) of a solution of 50% (w/v) polyethylene glycol (PEG) 10,000 in 0.5 M NaCl. This step precipitates the DNA, leaving short RNA fragments and most of the proteins in solution.

The solution is kept on ice for 1 hr. then centrifuged at 10,000 g. The supernatant is discarded, and the pellet is washed with 70% ethanol. The pellet is resuspended in 20 ml TE buffer (10 mM Tris, pH 8,0, 1 mM EDTA), and extracted 3 times with a phenol–chloroform mixture (1:1, v/v). It is then readjusted to 0.5 M NaCl and the PEG precipitation repeated. The precipitate is washed with 70% ethanol, resuspended in 10 ml TE buffer, 10% of a solution of 3 M sodium acetate added, and the DNA precipitated again by adding 0.8 volumes of 2-propanol. The precipitate is dried under reduced pressure.

At this stage, the DNA is 95% superhelical; very little contamination by chromosomal DNA and none by RNA can be detected (Fig. 4). Nevertheless, the purity is by no means sufficient for DLS studies, since the preparation still contains a large amount of non-UV-absorbing, high molecular weight contaminants (probably *E. coli* cell wall polysaccharides). Further purification is thus performed by HPLC. The pellet is resuspended in TE and mixed with 50% (v/v) of the initial buffer of the HPLC gradient.

Nucleogen Chromatography

Superhelical DNA can be separated from open circular and linear DNA by the wide-pore DEAE ion-exchanger Nucleogen 4000 (Diagen GmbH,

FIG. 4. Agarose (1%) gel electrophoresis of crude pUC18 DNA after lysis, RNase and proteinase treatment, and precipitation. The purity of the superhelical DNA is about 95%; no contamination by RNA or chromosomal DNA can be detected.

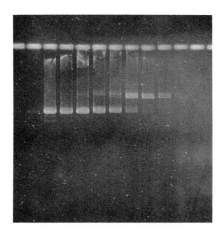

FIG. 5. Agarose (1%) gel electrophoresis of a HPLC separation of a 3400-bp plasmid DNA on a 12 × 1 cm DEAE-Nucleogen 4000 column. The gradient was 0.6–0.8 M NaCl in 30 mM sodium phosphate, pH 6.0, 6 M urea, 5 mM EDTA, 200 min at 1.5 ml/min. Fractions were taken every 2 min. The peak is centered around 0.7 M NaCl.

Düsseldorf, Germany). A 12 × 1 cm column is used with a 0.5 – 1.2 M KCl gradient in a buffer containing 40 mM sodium phosphate, pH 6.8, 5 mM EDTA, and 6 M urea. The typical flow rate is 2 ml/min, and the column capacity is approximately 5 mg of crude DNA. This method gives a satisfactory separation of superhelical DNA from the other species (Fig. 5), and separation of DNA fragments (data not shown). The DNA can be precipitated directly from the column fractions by adding 0.8 volume of 2-propanol.

RP-18 Chromatography

A method to separate superhelical DNA from linear and open circular forms on reversed-phase HPLC columns has been described.[16] We use a modification of the method to prepare superhelical DNA and to separate narrow populations of topoisomers in the amounts necessary for DLS. A Merck (Darmstadt, Germany) 25 × 1 cm RP-18 column is used in a buffer system employing 0.2 M ammonium acetate, pH 6.6, 10% dimethyl sulfoxide (DMSO), 90% water (buffer A) and 10% DMSO, 40% acetonitrile, 50% water (buffer B). The crude DNA is loaded onto the column equilibrated with buffer A, and a gradient of 0 – 100% buffer B is run at a flow

[16] S. Colote, C. Ferraz, and J. P. Liautard, *Anal. Biochem.* **154**, 19 (1986).

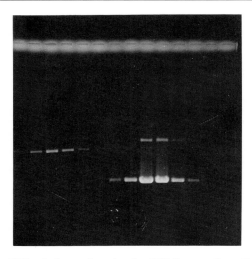

FIG. 6. Agarose (1%) gel electrophoresis of a HPLC separation of pUC18 DNA on a 25 × 1 cm Merck RP-18 column. The gradient was 0.2 M ammonium acetate, 10% (v/v) dimethyl sulfoxide in water (buffer A) to 5:4:1 water–acetonitrile–dimethyl sulfoxide (buffer B), 0–40% B in 90 min at 1.0 ml/min. Fractions were taken every 3 min. The peak is centered around 15% B.

rate of 1 ml/min for 100 min. Figure 6 shows that the superhelical, linear, and open circular species of pUC18 are well separated from each other. The DNA can be precipitated directly from the column fractions by adding 0.8 volume of 2-propanol.

Analysis of the fractions of the superhelical DNA peak on a gel containing 5 µg/ml chloroquine shows that the mean superhelical density of the fractions varies with their position in the peak. Thus, the RP-18 column also separates superhelical DNA on the basis of its superhelical density. The narrowest topoisomer distributions that we obtained contained only 2–3 topoisomers (Fig. 7).

Results

Translational and Rotational Diffusion of Superhelical DNAs

At low scattering vectors the scattered light autocorrelation function mainly contains contributions from translational and rotational diffusion. This has been shown, for example, for thin rods or semiflexible filaments.[5] The correlation function for a rigid rod is given by Eq. (9). For superhelical DNA, we could show that Eq. (9) describes the experimental data very

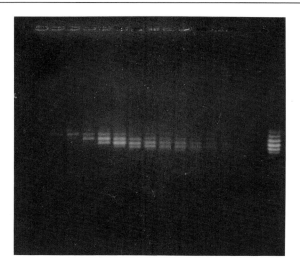

Fig. 7. Agarose (1%) gel electrophoresis of a HPLC separation of pUC18 topoisomers on a 12 × 0.4 cm Merck RP-18 column. The gradient was 0.2 M ammonium acetate, 10% (v/v) dimethyl sulfoxide, 10% methanol in water (buffer A) to 7:2:1 water–methanol–dimethyl sulfoxide (buffer B), 0–100% B in 120 min at 0.5 ml/min. Fractions were taken every 2 min. The peak is centered around 50% B. The rightmost lane contains the loaded sample. The gel and the gel buffer contained 5 μg/ml chloroquine.

well; the autocorrelation function is fit with a squared sum of two exponentials. Figure 8 shows the slow relaxation component, expressed in the form of a diffusion coefficient, for the superhelical and linear forms of pACL 29 (5400 bp). For K^2 below 3×10^{14} m^{-2} the slow component of the biexponential fit gives a **K**-independent diffusion coefficient, which we equate with the translational diffusion coefficient of the DNA. The fast relaxation rate is plotted in Fig. 9. For the superhelical form, the values extrapolate to a nonzero intercept as $\mathbf{K} \rightarrow 0$, as predicted for an anisotropic object [Eq. (9)], which is equal to the end-over-end rotational relaxation rate 6Θ of the superhelix.

The dependence of D_t and Θ of superhelical DNAs on molecular weight is shown in Figs. 10 and 11. D_t scales as MW$^{-0.6}$ and Θ as MW$^{-2.7}$. The scaling of the rotational diffusion suggests a rather extended structure; therefore, the interwound structure seems more likely for our DNAs than the toroid. However, comparing a stiff interwound with a stiff toroidal structure neglects completely the flexibility of DNA. To explain the measured D_t and Θ values in a quantitative manner, using known physical parameters of the DNA such as torsional and bending elasticity, we need a model that predicts the solution structure of a superhelical DNA more

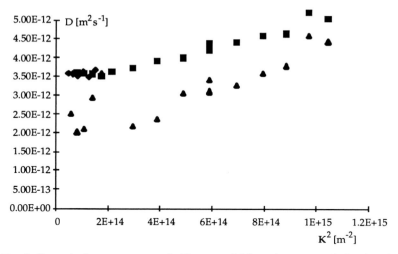

FIG. 8. Slow-relaxing components of a biexponential fit to the autocorrelation function of superhelical (■) and linear (▲) pACL29 DNA, expressed as diffusion coefficients [$D = (K^2\tau)^{-1}$, where τ is the slow relaxation time].

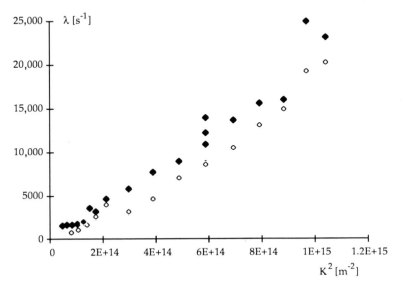

FIG. 9. Fast-relaxing components of a biexponential fit to the autocorrelation function of superhelical (◆) and linear (◇) pACL29 DNA, expressed as relaxation rates.

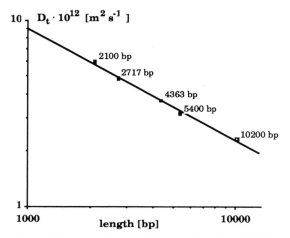

FIG. 10. Translational diffusion coefficients of superhelical plasmid DNAs as a function of their length.

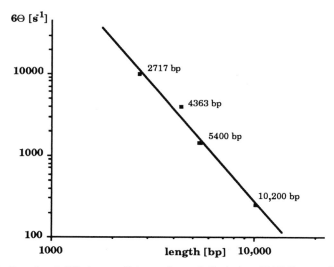

FIG. 11. Rotational diffusion coefficients of superhelical plasmid DNAs as a function of their length.

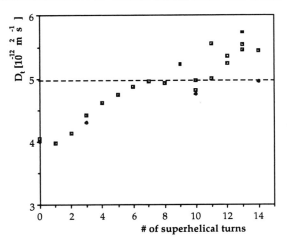

FIG. 12. Translational diffusion coefficients of superhelical DNA as a function of its linking number. (□) Simulated values from a Monte Carlo simulation of a 2700-bp DNA, using known values for hydrodynamic radius, bending, and torsional elasticity; (•) experimental data on pUC8 (2717 bp). The dashed line indicates the limiting experimental value for pUC8.

precisely than simple cylindrical or toroidal spirals can do. Such a model would allow a description of the dependence of the translational and rotational diffusion coefficient of superhelical DNA on DNA contour length, superhelical density, ionic strength (i.e., Debye screening length), and local sequence variations such as permanent curvature.

Recently, Frank-Kamenetskii and co-workers have developed a Monte Carlo procedure that predicts the structure of superhelical DNA very precisely. Details of that procedure can be found in the literature[17] (sec [23] in this volume). We are using this procedure to predict diffusion coefficients of a 2700-bp DNA (corresponding to the pUC series of plasmids) as a function of superhelical density.[18] For this purpose, we cover the superhelix contour generated by the Monte Carlo program with a continuous chain of beads of 1.65 nm radius; this bead size has been shown to predict the correct hydrodynamic radius for DNA.[19] The diffusion coefficient of the bead arrangement can then be calculated through a procedure described by de Haën et al.[20] The result is shown in Fig. 12.

[17] K. V. Klenin, A. V. Vologodskii, V. V. Anshelevich, A. M. Dykhne, and M. D. Frank-Kamenetskii, J. Mol. Biol. 217, 413 (1991).

[18] J. Langowski, U. Kapp, K. V. Klenin, and A. V. Vologodskii, manuscript in preparation.

[19] P. J. Hagerman and B. H. Zimm, Biopolymers 20, 1481 (1981).

[20] C. de Haën, R. A. Easterly, and D. Teller, Biopolymers 22, 1133 (1983).

We see from Fig. 12 that the diffusion coefficients of the relaxed form and several forms of intermediate superhelical density are predicted correctly. At the limit of very high superhelical densities, the program has the tendency to generate very compact structures with one or several branches in the interwound structure, which have a higher diffusion coefficient than measured for pUC8 or pUC18 of native superhelical density. The simulation converges only very slowly toward more extended structures with a lower diffusion coefficient. Work is in progress to accelerate the convergence of the simulation in order to compute correctly the equilibrium between branched and nonbranched forms at high superhelical density.

Internal Motions of Superhelical DNAs

For the linear form, and at higher K values for the superhelical form, the fast relaxation rate becomes proportional to K^2; it can therefore be expressed as a diffusion coefficient D_i. The average value of D_i is $(20 \pm 2.3) \times 10^{-12}$ m^2 sec^{-1} for the superhelical form and $(16 \pm 3) \times 10^{-12}$ m^2 sec^{-1} for the linear form. Similar D_i values are found for other plasmids.[21]

No theoretical description for the dynamic structure factor of semiflexible polymers exists so far that can include geometrical constraints like those in superhelical DNA. The dynamic structure factor of a 2311-bp linear DNA fragment has recently been measured and analyzed using CONTIN[22]; there it was shown that the data were consistent with a free-draining Rouse–Zimm model for the polymer chain.

To describe the internal dynamics of DNA in its various superhelical, open circular, and linear conformations, the DLS data from pACL29 (5400 bp) was subjected to multiexponential analysis using the maximum entropy method. Figure 13 shows a typical result for the superhelical and linear form. In both cases two peaks are obtained with mean values and peak areas close to the values obtained from the biexponential analysis.

The main result of our analysis is that the internal motions of intermediate size supercoiled and linear DNAs can be described by an internal diffusion coefficient in the range $D_i = (15-20) \times 10^{-12}$ m^2 sec^{-1}. This diffusion coefficient would correspond to the motion of segments of DNA approximately 300 bp long (2 persistence lengths).

The relative amplitudes of the fast-relaxing component for superhelical

[21] J. Langowski and U. Giesen, *Biophys. Chem.* **34**, 9 (1989).
[22] S. W. Provencher, CONTIN v.2 User's Manual, EMBL Technical Report DA07 (March, 1984).

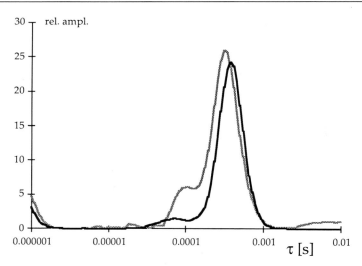

FIG. 13. Maximum entropy analysis (relative amplitude versus relaxation time in seconds) of the DLS data from pACL29 at $K^2 = 5.91 \times 10^{14}$ m^{-2}. Solid line, Superhelical form; gray line, *Eco*RI-linearized form.

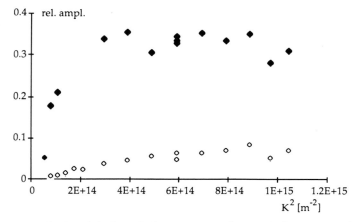

FIG. 14. Amplitudes of the fast-relaxing component for superhelical ($\diamond$) and linear ($\blacklozenge$) pACL29 DNA as a function of the scattering vector.

and linear pACL29 are plotted in Fig. 14 as a function of K^2. The internal motions become significant at much lower K for the linear than for the superhelical DNA. We may conclude that the increase in the apparent diffusion coefficient of the linear form relative to the superhelical form at high K is mainly caused by a change in the amplitude of an internal relaxation characterized by an internal diffusion coefficient.

Influence of Curved Sequences on Superhelix Conformation

Evidence that curved sequences can change the global conformation of superhelical DNA was first given by Laundon and Griffith, who showed that the relative positioning of curved sequences influenced the ratio of branched to nonbranched interwound superhelices.[23] The results were obtained by electron microscopy after binding the DNA to a carbon film.

Electron microscopy in vitreous ice on two plasmids constructed by Diekmann,[24] which differ in the curvature of an 80-bp insert (data not shown; Kremer *et al.*, in preparation) confirms that the proportion of branched to nonbranched conformations is 30% in the case of pK4A108 (strongly curved), but only 6% for pK1A108 (noncurved control). The reason why *one* curved insert causes this conformational change is probably due to the presence of a second, naturally curved sequence centered at approximately position 500.[23]

DLS can detect changes in the dynamics of superhelical DNA corresponding to the structural changes seen in electron microscopy. First experiments, both on plasmids containing curved inserts and for the binding of the DNA-bending CAP protein, show that the presence of strong curves decreases the amplitude of internal motions of superhelical DNA. This finding correlates with a theoretical prediction, using the Klenin/Vologodskii/Frank-Kamenetskii Monte-Carlo model of superhelical DNA (elsewhere in this volume), that permanent curves organize 1000 bp regions of a superhelical DNA into a more rigid configuration.

Conclusion

Dynamic light scattering is a powerful tool for studying the shape and internal motions of flexible biopolymers such as DNA. For superhelical DNA at small scattering angles, we can apply the rigid rod model with

[23] C. H. Laundon and J. D. Griffith, *Cell* (Cambridge, Mass.) **52**, 545 (1988).
[24] S. Diekmann, *FEBS Lett.* **195**, 53 (1986).
[25] G. Muzard, B. Theveny and B. Revet, *EMBO J.* **9**, 1289 (1990).

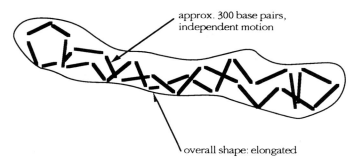

approx. 300 base pairs, independent motion

overall shape: elongated

FIG. 15. Proposed flexible model of superhelical DNA.

good success and obtain the translational and rotational diffusion coefficients of the molecule. The result leads to the conclusion that superhelical DNA molecules in solution have an elongated, interwound structure. This structure shows a high degree of internal motion which can be described by segmental diffusion with a diffusion coefficient corresponding to the independent motion of DNA pieces 300 bp long (Fig. 15). The overall elongated shape of the DNA is determined by the topological constraints arising from the circularity and internal torsional stress. A practical consequence of this model is that a time can be estimated that characterizes the mutual diffusion of distant parts of the DNA; in a 10,000-bp superhelical DNA, one piece of the DNA needs about 0.4 msec to move from one end of the structure to the other.

[22] Modeling DNA Structures: Molecular Mechanics and Molecular Dynamics

By Eberhard von Kitzing

Introduction

Understanding the biological function of biomolecules is one of the main reasons for studying their biophysical properties. Without detailed knowledge of the three-dimensional atomic structure of a molecule adequate insight into its biological function cannot be obtained. Experimental data are often interpreted in terms of molecular conformations; however, very few experimental methods for studying the biophysical properties of macromolecules provide detailed information regarding their three-dimensional atomic structure. Single-crystal DNA molecules exist only for very short sequences.[1,2] Because of the crystal structure of tRNA more conformational elements are known for RNA. Advances in biochemistry has made possible the direct synthesis of various molecules, and they can also be obtained from gene technological methods. However, the researcher must choose for study a few molecules out of an incredibly large number of possible candidates. For instance, in studying the curvature of DNA with a undecameric repeat by means of gel electrophoresis, there exists of the order of $4^{11}/11$ possible sequences (neglecting mispairs or modified base pairs). Thus, methods are needed that efficiently select a few highly significant sequences relevant for the biophysical effect under consideration. Detailed three-dimensional atomic models of the molecule studied experimentally can be used to simulate structural origins of the measured effect. The results of such simulations may provide information to discriminate relevant from irrelevant sequences for the effect under consideration.

During the last 30 years the powerful method of molecular mechanics and dynamics[3] has been developed. These methods use so-called force fields to calculate approximate internal energies of a given molecule as a function of the coordinates of its atoms. This internal energy may be used to discriminate between different conformations of the same molecule or to calculate statistical quantities.[4]

[1] R. Wing, H. Drew, T. Takano, C. Broka, S. Tanaka, K. Itakura, and R. E. Dickerson, *Nature (London)* **287,** 755 (1980).
[2] D. Moras and Y. Timsit, this volume [20].
[3] J. A. McCammon and S. C. Harvey, "Dynamics of Proteins and Nucleic Acids." Cambridge Univ. Press, London and New York, 1987.
[4] W. F. van Gunsteren and H. J. C. Berendsen, *Angew. Chem., Int. Ed. Engl.* **29,** 992 (1990).

There are many ways to use molecular mechanics or its variants. Molecular mechanics can be used as a computational tool in its own right. There are sophisticated force fields for calculating structural and thermodynamic properties of small molecules.[5] With Monte Carlo or molecular dynamics methods one can accurately simulate average properties of artificial hard sphere or the more realistic Leonard–Jones fluids.[6] With improvements in the method and the appearance of increasingly fast computers, larger molecules or systems of molecules can be studied.[4] The force field can be considered as some type of physicochemical knowledge base. It "knows" much about molecular geometries and their statistical relevance. It can therefore be used in structural model building studies. The three-dimensional model can be used to check geometric interpretations of structurally low resolution experimental data and to derive new or improved hypotheses.

In this chapter model building procedures will be presented that have been successfully applied to model curved DNA[7,8] and the DNA four-way junction.[9] Methods to obtain a set of starting coordinates are discussed. The central ideas regarding the formulation of a force field responsible for the calculation of the approximate internal energy are outlined. The advantages and shortcomings of various methods improving the initial estimate are discussed as well as the merits of local versus nonlocal optimization, Monte Carlo simulations, and molecular dynamics methods. The practical processs of using some of these methods to obtain refined structural models is described, and the final section concentrates on the interpretation of the large amount of data gathered in the model building study.

Obtaining a Starting Model

When modeling DNA structures, one should postulate what the target structure may look like. In the case of curved DNA the two strands are expected to form a double helix with the helix axis itself forming a circle or a superhelix. At the beginning it may not be clear whether the large-scale curvature is smoothly distributed over the double helix; it may also be located at specific junctions. In the case of the DNA four-way junction, one may expect four double helical stems connected by phosphate bridges.

[5] U. Burkert and N. L. Allinger, "Molecular Mechanics" (ASC Monograph 177). American Chemical Society, Washington D.C., 1982.
[6] I. I. Sheykhet and B. Ya. Simkin, *Comput. Phys. Rep.* **12**, 67 (1990).
[7] E. von Kitzing and S. Diekmann, *Eur. Biophys. J.* **15**, 13 (1987).
[8] S. Diekmann and E. von Kitzing, *in* "Structure and expression, Volume 3: DNA Bending and Curvature" (W. K. Olson, M. H. Sarma, R. H. Sarma, and M. Sundaralingam, eds.), p. 57. Adenine Press, Albany, New York, 1988.
[9] E. von Kitzing, D. M. J. Lilley, and S. Diekmann, *Nucleic Acids Res.* **18**, 2671 (1990).

A major motif in nearly all structures is the double helix; in the case of DNA or RNA one would generally assume double helices of the B- or A-type, respectively, unless experimental evidence leads to different expectations (Z-DNA for alternating GC sequences at high salt; RNA folding, including loops, etc.).

The earliest source for the coordinates of double helical structures came from fiber diffraction data.[10-16] Because of the low resolution of these data, they give only an averaged view of the double helix. Local sequence-specific information is generally not present. In recent years several short tracts of DNA molecules have been found to form single crystals, allowing the study of sequence-specific details at atomic resolution.[2] However, it is difficult to discern between the influence of the local sequence and crystal packing effects.[17,18] NMR gives many proton–proton distances in solution, but these distances do not have the precision to predict the local conformation accurately.[19]

Whereas the overall structure of double helical DNA and tRNA[20,21] is well known from experimental data, up to now the precise local conformation for arbitrary sequences could not be provided by purely experimental sources. Thus, initial estimates of the structure to be modeled were taken from idealized coordinates derived from fiber diffraction data. In some special cases, single-crystal structures may be found.

Starting from double helices the conformation of the studied molecule may be changed interactively and controlled by means of a molecular graphics program on a high resolution graphic screen. Because it is often very difficult to modify a structure by hand and at the same time retain a sterically reasonable structure, more advanced methods are used to change structure or to remove unfavorable steric interaction, as described in the next section.

[10] J. D. Watson and F. H. C. Crick, *Nature (London)* **171**, 737 (1953).
[11] F. H. C. Crick and J. D. Watson, *Proc. R. Soc. London, A* **223**, 80 (1954).
[12] S. Arnott and D. W. L. Hukins, *Biochem. Biophys. Res. Commun.* **47**, 1504 (1972).
[13] S. Arnott, R. Chandrasekaran, I. H. Hall, and L. C. Puigjaner, *Nucleic Acids Res.* **11**, 4141 (1983).
[14] A. Leslie, S. Arnott, R. Chandrasekaran, and R. Ratliff, *J. Mol. Biol.* **143**, 49 (1980).
[15] S. Premilat and G. Albiser, *J. Biol. Struct. Dyn.* **3**, 1033 (1986).
[16] H.-S. Park, S. Arnott, R. Chandrasekaran, and R. P. Millane, *J. Mol. Biol.* **197**, 513 (1987).
[17] R. E. Dickerson, D. S. Goodsell, M. L. Kopka, and P. E. Pjura, *J. Biol. Struct. Dyn.* **5**, 557 (1987).
[18] U. Heinemann and C. Alings, *EMBO J.* **10**, 35 (1991).
[19] See [13]–[15] in this volume.
[20] S. H. Kim, G. J. Quigley, F. L. Suddath, A. McPherson, D. Sneden, J. J. Kim, J. Weinzierl, and A. Rich, *Science* **179**, 285 (1973).
[21] J. D. Robertus, J. E. Ladner, J. T. Finch, D. Rhodes, R. D. Brown, B. F. C. Clark, and A. Klug, *Nature (London)* **250**, 546 (1974).

Weighting Different Estimates

In this section we present methods which discriminate between different structural estimates. Often the experimental information is not sufficient to determine a three-dimensional atomic model of the molecule uniquely. Thus it seems to be convenient to complete this missing information from a "structural data bank," based on a huge amount of physicochemical knowledge about possible molecular structures.

There exists a wealth of physicochemical knowledge about average bond distances, bond angles, preferred torsion angles, distances of closest approach, etc. The physics of the forces between atoms (at least in the gas phase) are well known. This knowledge has been sampled and condensed in so-called empirical force fields.[5,22-24] Compared to quantum mechanical *ab initio* calculations they describe the physics of the molecules in a highly simplified manner. These simplifications allow the calculation of physical properties of rather large molecules. There are force field calculations having various degrees of sophistication (for reviews see Refs. 3 and 4; for proteins see Ref. 25). In a typical force field the approximate internal energy of a molecule consists of contributions from bond stretching U_B, bond angle distorsion U_θ, dihedral angles U_ϕ, and nonbonded interactions U_{NB}:

$$U = U_B + U_\theta + U_\phi + U_{NB}$$

$$U_B = \sum_{ij \in \mathcal{M}_B} \frac{1}{2} k_{Bij}(r_{ij} - r_{0ij})^2$$

$$U_\theta = \sum_{ijk \in \mathcal{M}_A} \frac{1}{2} k_{\theta ijk}(\theta_{ijk} - \theta_{0ijk})^2$$

$$U_\phi = \sum_{\substack{ijkl \in \mathcal{M}_D \\ \mu}} V_{ijkl\mu} \cos(\mu \, \varphi_{ijkl} - \delta_{ijkl\mu}) \tag{1}$$

$$U_{NB} = U_{el} + U_{vdW} + U_{HB}$$

$$U_{Cou} = \sum_{ij \in \mathcal{M}_{el}} 332 \frac{q_i q_j}{\epsilon r_{ij}}$$

$$U_{vdW} = \sum_{ij \in \mathcal{M}_{vdW}} \left[\frac{A_{ij}}{r_{ij}^{12}} - \frac{B_{ij}}{r_{ij}^6} \right]$$

$$U_{HB} = \sum_{ij \in \mathcal{M}_{HB}} \left[\frac{C_{ij}}{r_{ij}^{12}} - \frac{D_{ij}}{r_{ij}^{10}} \right]$$

[22] A. J. Hopfinger and R. A. Pearlstein, *J. Comput. Chem.* 5, 486 (1984).

As long as no bond breaking chemical reactions are considered, one harmonic term suffices to account for the small deviations of the bond distances and angles from their average values. The energetic states U_ϕ stemming from the rotations around bonds are approximated by a short Fourier expansion. The interaction between nonbonded atoms U_{NB} is given by the electrostatic term U_{el}, the van der Waals term U_{vdW}, and possibly some hydrogen bond potential U_{HB}. One of the most difficult terms is the electrostatic energy.[4,26] The most simple form would be given by the Coulomb potential U_{Cou} with the atomic charge q_i and the effective dielectric constant ϵ. In this case interaction with the surrounding water and the counterions could be accounted for only via the dielectric constant. Some force fields also include the atomic polarization energy.[27]

In considering the results of force field calculations, a major concern is the reliability of the results. The reliability depends critically on the physical property of the molecule inspected and the force field chosen. Local geometries (e.g., bond distances, and bond angles) are the most reliable. Close contacts are avoided effectively by the van der Waals terms. In this respect the various force fields generally give similar results. Large-scale geometric properties, however, are much less reliable. The global conformation depends on complex physical properties of the molecule including hydration and counterion distribution. It is very difficult to incorporate these effects into the force fields on a sound physical basis. The problem of local minima may also affect the refinement procedure necessary to find a certain conformation.

Generating Improved Estimates

Various methods are available for the refinement of initially estimated structures; all use the force field as the "structural data bank" to supply the information lacking in the experimental data. If the experimental results can be directly related to structural properties of the molecule, this information may be introduced by constraints. In the case of X-ray data, the

[23] B. R. Brooks, R. E. Bruccoleri, B. D. Olafson, D. J. States, S. Swaminathan, and M. Karplus, *J. Comput. Chem.* **4,** 187 (1983).
[24] J. S. Weiner, P. A. Kollman, D. A. Case, U. C. Singh, C. Ohio, G. Alogerma, S. Profeta, Jr., and P. K. Weiner, *J. Am. Chem. Soc.* **106,** 765 (1984).
[25] G. Némethy and H. A. Scheraga, *FASEB J.* **4,** 3189 (1990).
[26] R. Klement, D. M. Soumpasis, E. von Kitzing, and T. M. Jovin, *Biopolymers* **29,** 1089 (1990).
[27] S. T. Russell and A. Warshel, *J. Mol. Biol.* **185,** 389 (1985).

atoms have to generate the measured electron densities; in the case of NMR the measured proton–proton distances have to be reproduced. Often experimental data are rather implicit with respect to atomic structural details. In this situation, the results may be used to exclude certain structures from the set of refined ones.

Several procedures can be used to generate new structures that are improved according to some weighted function (i.e., the appropriate internal energy) as compared to the initial ones. The choice of the appropriate procedure for refining the model is determined by the physical quantity under study. In many cases one would like to obtain the three-dimensional atomic structure of the molecule. In this case one may minimize the approximate internal energy of the molecules subject to the experimental constraints. The fast convergent optimization procedures converge only locally, that is, they do not surmount energetic barriers. The molecule may be trapped in a local minimum close to the starting point. Nonlocal methods have been proposed which generally are quite slow, especially in high dimensional spaces. Sometimes the calculation of a few static structures is not sufficient. For instance, there will be many structurally different hydration patterns with very similar energy. In this case statistical averages over many conformations are required. These conformations may be generated by the Metropolis–Monte Carlo procedure. Time-dependent information like base pair opening can be obtained from dynamics calculations.

Local Optimization

Various locally convergent optimization procedures will be surveyed, but only methods using the gradients are considered. These methods generally consist of two steps. At the beginning of cycle k a new search direction s_k is calculated. The various optimization procedures mainly differ in the way of calculating this direction s_k (see below). In the next step starting from the present set of coordinates x_k the minimum of the approximate internal energy function $f(x_k - \lambda s_k)$ is searched for by varying λ. Procedures to find the λ_k for which $f(x_k - \lambda_k s_k)$ is minimal are generally designated as line search routines (for details see Ref. 28). The new point x_{k+1} is given by $x_{k+1} = x_k - \lambda_k s_k$. Now a new search direction s_{k+1} has to be calculated. This loop is iterated as long as certain convergence criteria are not met. Possible criteria may be the number of iterations, the elapsed central processing unit (CPU) time, the norm of the instantaneous gradient $\|g_k\|$, the norm of the instantaneous step length $\|\lambda_k s_k\|$, etc.

[28] R. Fletcher, "Practical Methods of Optimization," Vol. 1. Wiley, Chichester, New York, Brisbane, and Toronto, 1980.

The most simple procedure for obtaining a search direction s_k is to equalize it with the instantaneous gradient $g_k = \nabla f(x_k)$. This method of steepest descent is known to converge rather slowly.[28] Another well-known method is the Newton–Raphson procedure. In this case the Hessian matrix $\mathcal{H}_k = \nabla \otimes \nabla f(x_k)$ (the matrix of second derivatives) of the internal energy function f is calculated, and s_k is obtained from solving the set of linear equations $g_k = \mathcal{H}_k s_k$. This method is known to locate a local minimum rather efficiently. Because the required computer space for storing the Hessian matrix $\mathcal{H}_k$ grows quadratically with the number of atoms, the Newton–Raphson procedure in its original form can be used only for relatively small molecules. Recently, truncated Newton–Raphson methods have been proposed, using only a small part of the total Hessian matrix $\mathcal{H}_k$.[29,30]

Another class of optimization procedures, designated variable metric methods, approximates the inverse Hessian matrix iteratively by means of the gradients. These methods require nearly the same amount of storage as the Newton–Raphson method. It has been shown[31] that these methods are rather inefficient in optimizing high dimensional functions like the energy function f.

A class of optimization procedures called conjugated gradients seems to be best suited for high dimensional energy functions f. These procedures require storage of the order of $3n$ (with n the number of atoms). In a series of test runs[31] the method of conjugated descent[28] has been shown to be a good choice in the case of large gradients (i.e., at the beginning of the optimization):

$$s_{k+1} = g_{k+1} + \frac{\langle g_{k+1} | g_{k+1} \rangle}{\langle g_k | s_k \rangle} s_k \qquad (2)$$

The Shanno method of conjugated gradients[32] has been shown to work very efficiently in the general case[31,33]:

$$s_{k+1} = \frac{\langle d_k | y_k \rangle}{\langle y_k | y_k \rangle} g_{k+1} + 2 \frac{\langle d_k | g_{k+1} \rangle}{\langle d_k | y_k \rangle} d_k - \frac{\langle y_k | g_{k+1} \rangle}{\langle y_k | y_k \rangle} d_k - \frac{\langle d_k | g_{k+1} \rangle}{\langle y_k | y_k \rangle} y_k \qquad (3)$$

with $d_k = x_{k+1} - x_k$ and $y_k = g_{k+1} - g_k$.

[29] J. W. Ponder and F. M. Richards, *J. Comput. Chem.* **8**, 1016 (1987).

[30] T. Schlick and M. Overton, *J. Comput. Chem.* **8**, 1025 (1987).

[31] E. von Kitzing, "Molekülsimulation mit Hilfe von Kraftfeldrechnungen am Beispiel der Aggregation von Nukleinsäuren verschiedener Konformation zu einem Komplex mit Übersetzungsfunktion." Edition herodot, Göttingen, Germany, 1986.

[32] D. F. Shanno, *SIAM J. Numer. Anal.* **15**, 1247 (1978).

[33] S. J. Watowich, E. S. Meyer, R. Hagstrom, and R. Josephs, *J. Comput. Chem.* **9**, 650 (1988).

Global Optimization

One of the most serious problems in the simulation of physical properties of large molecules is the enormous number of local minima, which grows exponentially with the number of atoms. So far there is no secure and efficient way to locate the global minimum definitively for functions of the form of the approximate internal energy [see Eq. (1)]. Because nearly all methods use an exhaustive search of different molecular conformations, successful applications of such procedures have been given only for relatively small molecules or for subsets of the total number of degrees of freedom.

Many procedures are variants of simulated annealing[34]: sampling many conformations at high "temperature" and then slowly cooling down the system.[35-38] The problem may be divided into subproblems and candidates estimated for the global minimum from the minima of the subproblems.[39] A few procedures modify the energy function in order to remove the barriers.[40-43] The Bremermann method,[44] which will be described below in more detail, has proved to be successful.[45] Unfortunately there are few comparisons between the various methods; generally they are given only for low dimensional problems. Therefore, a comparison of the merits of the different methods is difficult.

The Bremermann method has been used for the nonlocal optimization of the specific structure for the local sequence[8] and of the large-scale orientation of the stems of a DNA four-way junction.[9] At the beginning of the Bremermann optimization, a starting point has to be given, and its energy must be calculated. Then, the method works in an iterative manner. The starting point and its structural energy for a certain cycle are taken from the previous cycle. For n variables n random numbers are generated obeying a Gaussian distribution. These numbers represent a random direction in the space of the variables. Four additional points are generated along this direction in the neighborhood of the instantaneous values of the

[34] S. Kirkpatrick, C. D. Gelatt, Jr., and M. P. Vecchi, *Science* **220,** 671 (1983).

[35] J. A. Snyman and L. P. Fatti, *J. Optim. Theory Appl.* **54,** 121 (1987).

[36] Z. Li and H. A. Scheraga, *Proc. Natl. Acad. Sci. U.S.A.* **85,** 6611 (1987).

[37] M. J. Sabochick and D. L. Richlin, *Phys. Rev.* **B37,** 10846 (1988).

[38] G. Dueck and T. Scheuer, *J. Comput. Phys.* **90,** 161 (1990).

[39] K. D. Gibson and H. A. Scheraga, *J. Comput. Chem.* **8,** 826 (1987).

[40] G. M. Crippen, *J. Phys. Chem.* **91,** 6341 (1990).

[41] E. O. Purisima and H. A. Scheraga, *J. Mol. Biol.* **196,** 697 (1987).

[42] L. Piela, J. Kostrowicki, and H. A. Scheraga, *J. Phys. Chem.* **93,** 3339 (1989).

[43] D. R. Ripoll and H. A. Scheraga, *J. Protein Chem.* **8,** 263 (1989).

[44] H. Bremermann, *Math. Biosci.* **9,** 1 (1970).

[45] A. H. G. Rinnooy Kan, C. G. E. Boender, and G. Th. Timmeri, *in* "Computational Mathematical Programming" (K. Schittkowski, ed.), p. 281. Springer-Verlag, Berlin, 1985.

variables. The structural energies of these points are calculated. A fourth-order polynomial is determined to interpolate these four new energies together with the energy of the starting structure. This polynomial describes the variation of the function along the chosen line in the parameter space. The global minimum of this one-dimensional interpolation of the approximate internal energy is calculated by means of the Cardan formula. The structural energy of the DNA conformation corresponding to this minimum is determined; the energies of the starting structure, the four additional structures, and the expected global minimum structure are compared. The structure with the lowest energy serves as the starting point for the next Bremermann optimization cycle.

To find the global minimum or at least to escape the instantaneous local minimum toward a better one, an exhaustive search in the space of relevant degrees of freedom is required. A nonlocal optimization of all atomic degrees of freedom would thus lead to prohibitively long CPU times. This statement also holds for the Bremermann procedure.[46] To calculate an improved local conformation of a DNA double helix starting from an idealized set of atomic coordinates obtained from fiber diffraction data, only very few degrees of freedom will vary independently with a considerable amplitude. Those degrees of freedom are designated as *relevant degrees of freedom*. For instance, for a given stacked conformation of two adjacent bases, there are only a few sterically permitted sugar–phosphate–sugar conformations. For this reason it is useful to optimize nonlocally only very few independent degrees of freedom, while keeping the remaining ones at their energetic minimum by means of a fast local optimization procedure (see above). This method is generally designated as adiabatic relaxation. Thus, the slow nonlocal Bremermann optimization with random direction search is used only for a very small subspace, whereas the fast locally converging method of conjugated gradients is used for the remaining degrees of freedom. In our application of the Bremermann optimization method, a calculation of the structural energy means the variation of a small number of relevant variables, followed by a local optimization using the method of conjugated gradients, binding the atoms to the newly generated position by a harmonic force of 10 kcal/mol/Å^2. This allows the relaxation of distorted bond angles and close atomic contacts. The energy that results from this local optimization with constraints to the initial coordinates is taken as the structural energy in the Bremermann optimization. Further details are given by von Kitzing and Diekmann.[47]

[46] G. S. Rao, R. S. Tyagi, and R. K. Mishra, *Int. J. Quantum Chem.* **20**, 273 (1981).
[47] E. von Kitzing and S. Diekmann, in preparation.

Monte Carlo Method

In some cases statistical averages are required to calculate the physical properties under consideration. If the molecule has flexible groups, like side chains at the outer surface of proteins, the minimal energy configuration will only poorly represent the average structure. Statistical averages are also required for calculating free energy differences.[48] One method for obtaining canonical ensembles is the Metropolis–Monte Carlo procedure.[49] In this method the instantaneous set of coordinates x_k is modified in a random way to give x_k^*. The energy E_k^* of the new structure at x_k^* is compared to that of the instantaneous one E_k at x_k. If E_k^* is less than E_k the new structure x_k^* is accepted as x_{k+1}; in the other case it is accepted only with the probability of $\exp[(E_b - E_b^*)/k_B T]$, a function of the Boltzmann constant k_B and the absolute temperature T.[3,6]

The standard procedure to modify the instantaneous coordinates by adding an isotropic random direction vector or to change the coordinates of only one atom has proved to be very inefficient in the case of macromolecules. A careful weighting of the various directions by means of the eigenvectors and eigenvalues of the Hessian matrix of the energy function has been proposed.[50] Relaxing techniques have also been tested.[51]

Molecular Dynamics

Another method to obtain statistical ensembles is the molecular dynamics method. In contrast to the Monte Carlo procedure, in most cases it provides time-dependent information even for equilibrium systems.[4,52,53]

The most straightforward implementation is given by Newtonian dynamics. The term molecular dynamics is often used to indicate this type of dynamics. Newton's equations of motion are integrated with respect to time. In this case a microcanonical ensemble is generated. Several heuristic methods have been proposed to modify Newton's equations of motion to obtain canonical or grand canonical ensembles.[54-57] The major drawback

[48] W. F. van Gunsteren and P. K. Weiner, eds., "Computer Simulations of Biomolecular Systems," EXCOM, Leiden, The Netherlands, 1989.

[49] N. Metropolis, A. W. Rosenbluth, M. N. Rosenbluth, A. H. Teller, and E. Teller, *J. Chem. Phys.* **21**, 1087 (1953).

[50] T. Noguti and N. Go, *Biopolymers* **24**, 527 (1985).

[51] D. W. Heermann, P. Nielaba, and M. Rovere, *Comput. Phys. Commun.* **60**, 311 (1990).

[52] J. Kushick and B. J. Bern, *in* "Modern Theoretical Chemistry," Vol. 6. (B. J. Berne, ed.), p. 41. Plenum, New York, 1977.

[53] M. Karplus and G. A. Petsko, *Nature (London)* **347**, 631 (1990).

[54] H. J. C. Berendsen, J. P. M. Postma, W. F. van Gunsteren, A. DiNola, and J. R. Haak, *J. Chem. Phys.* **81**, 3684 (1984).

of these approaches results from the fact that by modifying the equations of motion, time-dependent information may be destroyed.[58] An alternative way to obtain canonical ensembles are provided by stochastic dynamics realized by means of Langevin's equations. In this case friction and a random force are introduced, simulating the coupling to an external bath (i.e., the surrounding solvent).[3,59]

Because the step length for integration must be lower than the fastest vibration within the molecule (i.e., the stretching modes of bonded hydrogens) this time step must be of the order of femtoseconds or even below. This implies that one needs a huge number of integration points to reach the time scale of nanoseconds or milliseconds. For this reason most dynamic simulations remain in the picosecond range. The step length can be increased, if the distance between bonded atoms is kept rigid.[4] An alternative may be the backward Euler method which automatically freezes all motions with time constants below the integration step.[60]

Summary

There are several tools to improve initial structural estimates of a molecule under study. The methods described in this section weight the structural models according to the approximate internal energy as a function of the atomic coordinates provided by a force field. The method of conjugated gradients is a powerful tool to find the way downhill toward a local minimum. Surmounting barriers and escaping local minima require nonlocal optimization procedures. Of the various possible choices the Bremermann method has been described in greatest detail. The central idea for the implementation of this method is its application to a small number of relevant degrees of freedom, allowing the remaining ones to relax to a local energetic minimum.

Monte Carlo methods may be used to obtain statistical averaged quantities. Again the convergence of this method may be considerably increased if the randomly chosen directions mainly concentrate on the relevant directions. Molecular dynamics calculations not only provide averaged quantities but also give information about time-dependent processes.

[55] S. Nosè, *Mol. Phys.* **57**, 187 (1986).
[56] C. H. Esparza and H. Kronmüller, *Mol. Phys.* **68**, 1341 (1989).
[57] H. J. M. Hanley and D. J. Evans, *Int. J. Thermophys.* **11**, 381 (1990).
[58] K. Hermannsson, G. C. Lie, and E. Clementi, *J. Comput. Chem.* **9**, 200 (1988).
[59] H. L. Friedman, "A Course in Statistical Mechanics." Prentice-Hall, Englewood Cliffs, New Jersey, 1985.
[60] C. S. Peskin and T. Schlick, *Commun. Pure Appl. Math.* **42**, 1001 (1989).

Building Improved Estimates

The previous section dealt with the technical aspects of energy minimization, Monte Carlo calculations, and molecular dynamics. In this section we examine the practical steps of structure refinement. First, problems are considered which often occur when starting to optimize structurally a built model. The design of stable building blocks is discussed, and some simple procedures are proposed for constructing larger units and for carefully optimizing their structure.

Instabilities in Initial Estimates

To study the local structural cause of DNA curvature, four turns of a double helical DNA were built from idealized atomic coordinates as provided by X-ray fiber diffraction.[12] The sequence was taken to agree with that of a molecule for which experimental data indicated curvature.[8] During the course of initial optimization the ideal helical structure became corrupted. Parts of the molecule wound up, generating kinks, etc. Such behavior was found by the author not to be exceptional.

The coordinates generated from a standard set or taken from some crystal structure have certain bond distances and angles. The force field used may have slightly different standard values. Because of the strong force constants for bond energy, a huge amount of energy is stored in the model. At the beginning the structure has the energy to surmount very high energy barriers and may push the molecule into unfavorable local minima with relative high energies. This initial high bond energy can be easily reduced with a short optimization by constraining the atoms to their initial coordinates with a force constant of about 1 kcal/mol/Å^2. This will allow a readjustment of bond distances, angles and close van der Waals contacts. At the same time the overall properties of the original structure are preserved.

Even at this stage an optimization of the relaxed structure will generally not lead to a curved molecule. The directions of motion related to a global bending or global twisting have very low related force constants, which are given by the eigenvalues of the Hessian matrix $\mathcal{H}_k$ for the respective eigenvectors. The "truncation noise" in the calculation prevents the optimization of the degrees of freedom in the direction of lowest force constants. Mostly these directions of low force constant are the most relevant ones for the large-scale conformation of the molecule.

Designing and Optimizing Building Blocks

One way to cope with large-scale conformational changes is to divide the total problem into subproblems. From the sequence of the molecule expected to be curved, other molecules may be designed that are expected not to be curved. This can be achieved by using a sequence repeat of one-half of a helical turn. The sequence d(GCTCGAAAAA)$_4$ · d(TTTTTCGAGC)$_4$ and closely related ones have been studied experimentally.[8] The molecule showed a strong gel migration anomaly which had been interpreted to result from intrinsic curvature. In a model building study to inspect the possible structural reasons for intrinsic curvature, this sequence was split into three substructures[7]: (1) 2d[C(GAGCTC)$_4$G], (2) d[A(GCGAA)$_4$] · d[C(TTCGC)$_4$], and (3) dA$_{12}$ · dT$_{12}$. Because in substructures (1) and (2) the sequence repeat is approximately one-half the helical repeat, these molecules are expected not to be curved. The effect of structural elements which in the case of proper phasing may give rise to curvature in these molecules will cancel within one turn. Thus, the changes in the structure induced by the force field will remain locally.

These types of calculations may be supported by means of symmetry constraints.[7] For instance, the rise and twist per base pair may be imposed on the molecule dA$_{12}$ · dT$_{12}$ via constraints during minimization. Because one effectively reduces the number of degrees of freedom, the convergence is accelerated. In a final calculation all such constraints have to be released. Symmetrical structures being optimized with symmetry constraints in an intermediate stage generally have a considerably lower final energy compared to the same molecule being optimized without such constraints.

As long as only local optimization procedures are involved one should not expect to achieve an appreciable reorientation of the bases. Because of the strong van der Waals forces, the bases stick together. It has been shown by Shakked and Rabinovich[61] that the van der Waals energy of single-crystal structures is considerably lower compared to the energy for the same sequence but with the coordinates taken from idealized X-ray fiber diffraction.[12]

To escape local minima the nonlocal Bremermann method has been used to optimize nonlocally the structure of a DNA four-way junction[9] and of double-stranded DNA[8,47] to study curvature. In the case of dA$_{12}$ · dT$_{12}$ the difference in final energies of the locally and nonlocally optimized structures is about 380 kcal/mol. The merits of the nonlocal optimization procedure are obvious. As explained above, the nonlocal optimization procedure is generally very slow in a large dimensional space. On average an infinitely extended regular double helix like dA$_n$ · dT$_n$ should have a repetitive structure. Thus, in principle only one base pair step has to be known to allow the reconstruction of the total helix by means of the helical symmetry operation, which is given by the rotation around and translation along the helix axis by the helical twist and rise, respectively.

[61] Z. Shakked and D. Rabinovich, *Prog. Biophys. Mol. Biol.* **47**, 159 (1986).

Because the deoxyribose and phosphate groups are quite flexible one can expect these groups to adjust easily. It seems reasonable, therefore, to assign at least 7 degrees of freedom to a base pair. These are three rotations and translations relative to the previous base pair and the internal propeller twist. These degrees have been used to optimize nonlocally the local conformation of molecules with several sequences.[8,9,47]

Reconstruction

The building blocks for the target molecule are designed and optimized. Now they must be put together to obtain the final molecule. In the case of the previously mentioned curved piece of DNA this may be obtained by means of sequential overlap[7]:

5'-GCTCGAAAAAGCTCGAAAAA-3'

1 1 1 1 **2** 2 3 3 3 **2** 2 1 1 1 **2** 2 3 3 3 3

3'-CGAGCTTTTTCGAGCTTTTT-5'

The local structure of the base pairs indicated by 1, 2, and 3 are from molecules (1), (2), and (3), respectively. At the base pairs indicated by boldface **2** the two structures are connected so that the molecule (1) or (2) has maximal base-pair overlap with the corresponding base pair of molecule (2) or (3), respectively. After this matching process, the doubly occurring atoms were omitted. The molecule was optimized twice by the method of conjugated gradients. During the first energy minimization process, the start coordinates were constrained to their initial coordinate values to relax mainly the slightly distorted bond distances and angles at the junction. During the second course of optimization no constraints were used. The atomic model of the molecule correctly predicts a shift of the proposed alternate DNA structure in dA tracts in the 3' direction of the dA strand[7] as found in hydroxyl radical probing.[62]

One might wish to optimize nonlocally the junctions of the curved piece of DNA obtained from sequential overlap. As degrees of freedom one may take three rotations and translations of one part of the molecule with respect to the rest. Another choice could be the use of pseudodihedrals,[63] rotations around pairs of nonbonded atoms. No bond distances are affected by this type of coordinate transformation. The distorted bond angles and some possible close van der Waals contact are easily adjusted in the local optimization for relaxing the structure. Possible axes could go through the O-5' and O-3' atoms between the reconnected adjacent DNA

[62] A. M. Burkhoff and T. D. Tullius, *Cell (Cambridge, Mass.)* **48**, 935 (1987).
[63] S. C. Harvey and J. A. McCammon, *Comput. Chem.* **6**, 173 (1982).

tracts (e.g., the dGA · dTC junction): (1) O-3'G – O-5'C, (2) O-5'A – O-3'T, (3) O-3'G – O-3'T, (4) O-5'A – O-5'C. This method has been applied to DNA sequences comparable to those tested experimentally by Hagerman.[64] The calculated curvatures were found to be in good agreement with experimental data.[8,47]

A similar approach has been used in the case of the DNA four-way junction. The sequence at the junction (the first two base pairs of each stem) of the optimized structure was chosen to be identical to that of junction 1 extensively studied experimentally.[65,66] The sequence of the stems was generated by taking the innermost two base pairs on each arm from junction 1 and repeating them twice. This repetitive sequence was required to allow for nonlocal optimization of the single double helical stems. The resulting junction contains six base pairs in each helical stem. An extended four-way junction was generated by means of an interactive program as the starting structure for the optimization.

The structures created during the optimization process were generated by rotations around the axes connecting pairs of oxygen atoms at the four-way junction (Fig. 1). The oxygens are indicated as 5' or 3' and by the label B, H, R, or X according to the stem they belong to. Twelve rotational axes have been introduced: (1) O-5'B – O-3'B, (2) O-5'H – O-3'H, (3) O-5'R – O-3'R, (4) O-5'X – O-3'X, (5) O-3'B – O-5'H, (6) O-3'H – O-5'R, (7) O-3'R – O-5'X, (8) O-3'X – O-5'B, (9) O-5'B – O-5'R, (10) O-3'B – O-3'R, (11) O-5'H – O-5'X, (12) O-3'H – O-3'X. Rotation angles around these axes were used as the variables for the nonlocal Bremermann optimization. Because these rotations do not commute, some axes were used up to 3 times (12 rotational angles used up to 3 times results in maximally 36 angles). The resulting structure was found to agree marvelously with the experimental data.[9]

Summary

Starting the structural local optimization from an idealized set of coordinates as obtained from X-ray fiber diffraction data often leads to corrupted structures. If one expects the large-scale conformation to be fairly correct, this problem may be reduced by using symmetry constraints. On the other hand, if a reorganization of the large-scale structure is expected, the design of building blocks may be useful. The sequence of the

[64] P. J. Hagerman, *Nature (London)* **321,** 449 (1986).
[65] D. R. Duckett, A. I. H. Murchie, S. Diekmann, E. von Kitzing, B. Kemper, and D. M. J. Lilley, *Cell (Cambridge, Mass.)* **55,** 79 (1988).
[66] A. I. H. Murchie, R. M. Clegg, E. von Kitzing, D. R. Duckett, S. Diekmann, and D. M. J. Lilley, *Nature (London)* **341,** 763 (1989).

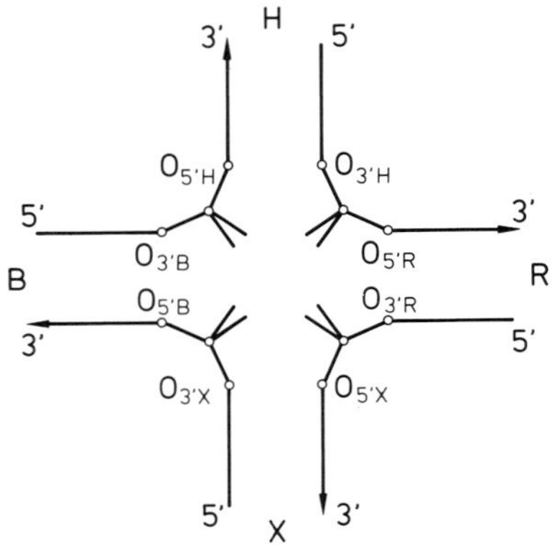

FIG. 1. Atoms used to define the rotational axes to optimize nonlocally a model of the DNA four-way junction.[9] The junction consists of four double helical stems designed X, B, H, and R. This nomenclature was taken from an experimental study of a DNA four-way junction[65,66] with a similar sequence close to the junction. The stems are connected via phosphate bridges; the terminal O-3′ atom of stem B is linked to the initial O-5′ atom of stem H, etc. The initial and terminal oxygens are used to define the pseudodihedrals (see text).

blocks should be chosen in such a way that the structural effects expected to lead to global deviations in the original sequence cancel within one or two turns of the building block sequences. To allow for nonlocal optimization the sequences should be kept as simple as possible. For instance, a dimeric repetitive sequence would require the nonlocal optimization of 14 degrees of freedom.[9,47] This number has been shown to be tractable for nonlocal optimization. At the end the blocks have to be built together to obtain the target structure. Further nonlocal optimization of the junction may be required. Finally, all selected structures have to be locally optimized without any constraints to obtain the final energy and to test the stability.

Interpretation of Theoretical Data

At the end of a modeling procedure including several stages of refinement, one or a few model structures for the target molecule have been obtained. They provide a wealth of information. What can we do with all these numbers? Can we blindly trust them? If not, are there hints for the reliability of certain aspects of the results?

Because the zero point of the energy in the force field is arbitrary, only energy differences of the same molecule in different conformations are comparable. If one has several candidates obeying the experimental restraints, the structure with lowest energy has to be taken as the best guess. Whereas the total energy is only one number for each structural model, a molecule with 1000 atoms has 3000 coordinates. This number reduces considerably only if the conformational angles are inspected; in the case of DNA or RNA the coordinates of 1000 atoms reduce approximately to 280 backbone angles ($\alpha - \zeta$, χ). Because of the many approximations involved in the force fields, we cannot expect all structural details to be realistic. On the other hand, there may be structural information which is insensitive to the instantaneous force field in use.

For instance, we inspected optimized $dA_{12} \cdot dT_{12}$ molecules with several conformations in search of properties of the adenine base responsible for the putative B′ structure.[7] Most resulting structures displayed special features in the stacking of adenine bases. This special stacking pattern was possible because of the lack of the exocyclic group at the 2 position in the adenines. Later the special importance of this position in the ring for the structural properties of longer adenine tracts was demonstrated experimentally.[67]

In the case of modeling the DNA four-way junction several aspects of the resulting structure are insensitive to the force field because they result from avoiding bad sterical contacts of groups relatively close in space but distant in sequence.[9] Experimental data on the DNA four-way junction[66] with Mg^{2+} ions in solution suggest the stacking of pairs of stems. In our model the DNA four-way junction consists of two quasi-continuous DNA double helices (see Fig. 2). At the junction two strands are exchanged. To avoid strain in the bonds of the backbone, the two helices have to come quite close to each other. A sterically feasible structure can be obtained by tilting the helix axes by 60° in such a way that the exchanging strands include the acute angle. With this angle the backbone of the continuous strand of one helix can easily be accommodated in the major groove of the other helix. A very similar packing of two helices has recently been observed in a single-crystal structure.[68] The opening of the minor grooves avoids short contacts of the exchanging strands. This model differs in many aspects from the previously accepted one.[69] On the other hand, experimental data[66] are only consistent with this newly proposed structure.

[67] S. Diekmann, E. von Kitzing, L. McLaughlin, J. Ott, and F. Eckstein, *Proc. Natl. Acad. Sci. U.S.A.* **84**, 8257 (1987).

[68] Y. Timsit, E. Westhof, R. P. P. Fuchs and D. Moras, *Nature (London)* **341**, 459 (1989).

[69] N. Sigal and B. Alberts, *J. Mol. Biol.* **71**, 789 (1972).

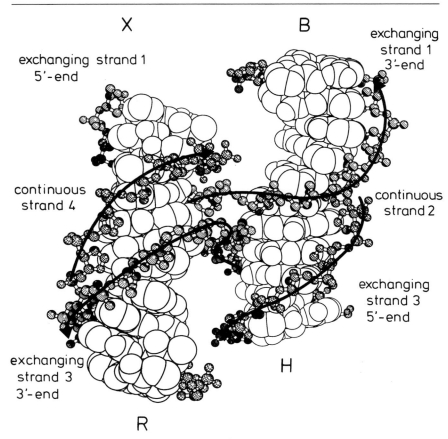

X

B

exchanging strand 1
5'-end

exchanging
strand 1
3'-end

continuous
strand 4

continuous
strand 2

exchanging
strand 3
5'-end

exchanging
strand 3
3'-end

H

R

FIG. 2. Atomic model of a DNA four-way junction as built by von Kitzing et al.[9] There are several structural properties which are insensitive to the empirical force field (see text) used for structure refinement. To avoid a steric clash between the two exchanging sugar–phosphate backbones, the minor grooves have to open at the junction. Starting at the 5' end of the continuous strand, the minor grooves open toward the junction and close immediately behind the junction. The angle between the two quasi-helices with stem R and B stacking on stem X and H, respectively, optimally allows for accommodating the backbone of the continuous strands 2 and 4 in the major grooves of stem R and B, respectively.

Summary and Conclusion

Model building studies may be used to supplement structurally low resolution experimental data with detailed three-dimensional hypothetical atomic models. Because of the strong relation between structure and function in biological molecules such models may give a consistent, integral view of a wealth of experimental data. In most cases such models will predict the outcome of certain experiments. The outcome of these experiments will often either confirm the model may be used for further refinement or even demand a major revision of the model.

Coordinates obtained from X-ray fiber diffraction data or in special cases single-crystal data may provide the elements for DNA or RNA model building. Local and nonlocal optimization may be used to refine these structures and to evaluate their statistical significance as estimated by a chosen force field. Appreciable progress using nonlocal optimization procedures can only be expected if the dimensionality of the problem can be reduced sufficiently to the relevant degrees of freedom. Taking advantage of structural symmetries may critically improve the convergence while refining the target molecule or its building blocks. Monte Carlo and molecular dynamics methods allow one to calculate averaged quantities. In addition, molecular dynamics provides time evolutions of certain averages.

During the simulation of certain physical properties of molecules a huge amount of data will be generated. They will provide many answers, but these answers may not always apply to the original question. So what type of questions will be reliably answered by a force field? Relatively safe answers concern the local geometry of the molecules. If a conformation leads to strong distortions of bond distances or angles or to close van der Waals contacts, this conformation can safely be rejected. Optimizing such unfavorable structures energetically may lead to structures showing how to avoid such distortions. More difficult are energetical questions: which of two conformers is more stable, or what is the free energy of the substrate in the active site?[48] One cannot always be sure that the force field provides the correct answer. Therefore, one should pose only those questions which can be checked experimentally. Because of the many possible answers, the experiment may benefit by starting with a choice proposed by the simulation. The application of this procedure to curved DNA[7,8] and the DNA four-way junction[9] was successful.

[23] Modeling Supercoiled DNA

By Alexander V. Vologodskii and Maxim D. Frank-Kamenetskii

Global DNA Structure

Numerous structural methods, such as X-ray crystallography and NMR, make it possible to elucidate the local DNA structure (see [5] and [13] in this volume). However, for DNA molecules comprising hundreds of base pairs, the global structure becomes of great importance. The global structure arises as a result of the accumulation of minute changes in local structure, and it cannot be detected by structural methods used to detect the local structure. Special methods, which are sensitive to the global structure, are used (see [21] and [26] in this volume).

Among the global DNA features the most significant is DNA supercoiling. Supercoiling entails the topological restrictions arising in closed circular DNA molecules. In closed circular DNA the two complementary strands are topologically linked to each other; therefore, the linking number, Lk, is a topological invariant, which cannot be changed without at least one of DNA strands being cut. Such notions as linking difference, ΔLk, and superhelical density, σ, are discussed in [8] in this volume.

The linking difference ΔLk is related to two geometrical characteristics of the DNA double helix:

$$\Delta Lk = \Delta Tw + Wr \tag{1}$$

The ΔTw value is the total twist of the double helix minus its reference value. The latter corresponds to an unconstrained double helix under given ambient conditions. Therefore, ΔTw is not a global quantity, but rather a trivial sum of local values. It is expressed in terms of a sum of changes in the winding angles between adjacent base pairs. By contrast, the writhing value, Wr, is a real global nonadditive quantity, which is determined by the path of the double helix axis in three dimensions. Therefore, the writhing value determines the global–structural contribution to DNA supercoiling.

We assume that the torsional rigidity constant does not depend on local DNA bending and vice versa. Available experimental data support this major assumption. Because twist is a simple additive quantity, its calculation creates no problem. The problem in theoretical treatment of DNA supercoiling lies in the treatment of the DNA global structure and in the calculation of the writhing number. Equation (1), in conjunction with the above assumption, makes it possible to treat independently the torsional and bending movements in the double helix. This means that twisting and writhing contributions to the linking number can be analyzed separately.

While studying DNA global structure, it is reasonable to neglect, within the first approximation, most of the details of local structure and to treat DNA within the framework of a simplified model. In so doing we treat the double helix as a homogeneous, isotropic elastic rod, which is characterized by three parameters: diameter d, the torsional rigidity constant C, and the bending rigidity constant, measured in terms of the Kuhn statistical length, b.

Adequate theoretical treatment of DNA supercoiling is essential for answering fundamental questions arising in the study of the phenomenon of supercoiling and its biological aspects. The key point is that globally a DNA molecule behaves as a polymer chain and exhibits an enormous variety of conformations. Its global behavior can by no means be interpreted in terms of one or a selected number of states. It may be treated adequately only in terms of statistical mechanics to allow for the entropic effects, which play the decisive role in its behavior.

A statistical–mechanical approach described in this chapter makes it possible to obtain, first of all, the images of superhelical DNA conformations in three dimensions. These images can be compared to the data obtained by electron microscopy and cryoelectron microscopy (see [26] in this volume). The approach is also very useful in interpreting numerous experimental data obtained by different indirect techniques, such as hydrodynamic methods and static and dynamic light scattering (see [21] in this volume). Among other things, it permits the calculation of the superhelix free energy as a function of superhelical density.[1]

Theoretical Background

Our procedure is based on the general method of Metropolis *et al.*,[2] which makes it possible to treat the elastic rod DNA model in terms of statistical mechanics.[3-5] To this end, a closed DNA molecule comprising n Kuhn statistical lengths is modeled as a spatial curve consisting of kn equal, straight elementary segments. Such a conformation has the energy

$$E = k_B T \alpha \sum_{i=1}^{kn} \theta_i^2 \tag{2}$$

where the summation is done over all the hinges connecting the elementary segments; k_B is the Boltzmann constant, T is the absolute temperature, θ_i is the angle (in radians) between two adjacent segments, and α is the bending rigidity constant. The latter parameter is defined in such a way that the Kuhn statistical length corresponds to k segments.[3] Within the framework of this model of the polymer chain, k may be expressed in the

[1] K. V. Klenin, A. V. Vologodskii, V. V. Anshelevich, A. M. Dykhne, and M. D. Frank-Kamenetskii, *J. Mol. Biol.* **217**, 413 (1991).

[2] N. Metropolis, A. W. Rosenbluth, M. N. Rosenbluth, A. H. Teller, and E. Teller, *J. Chem. Phys.* **21**, 1087 (1953).

[3] M. D. Frank-Kamenetskii, A. V. Lukashin, V. V. Anshelevich, and A. V. Vologodskii, *J. Biomol. Struct. Dyn.* **2**, 1005 (1985).

[4] J. Shimada and H. Yamakawa, *Biopolymers* **27**, 657 (1988).

[5] K. V. Klenin, A. V. Vologodskii, V. V. Anshelevich, V. Y. Klisko, A. M. Dykhne, and M. D. Frank-Kamenetskii, *J. Biomol. Struct. Dyn.* **6**, 707 (1989).

form

$$k = \frac{1 + \langle \cos \theta \rangle}{1 - \langle \cos \theta \rangle} \tag{3}$$

where

$$\langle \cos \theta \rangle = \frac{\int_0^\pi \cos \theta \sin \theta \exp(-\alpha \theta^2) \, d\theta}{\int_0^\pi \sin \theta \exp(-\alpha \theta^2) \, d\theta} \tag{4}$$

Substitution of the original wormlike chain, which is bendable at any site, by the chain consisting of kn unbendable straight segments connected with hinges assumes taking the limit at $k \to \infty$. In so doing one has to change the α value accordingly to satisfy Eqs. (3) and (4). In practice the k value is a calculation parameter, which is chosen in accordance with the guidelines described below.

The Metropolis–Monte Carlo procedure consists of consecutive steps which include displacements of separate parts of the chain. One may use different elementary displacements in the procedure. Usually[3-5] the elementary step consists of the rotation of an arbitrary number of adjacent segments by a random angle within the interval $(-\varphi_0, +\varphi_0)$ around the straight line connecting the first and the last hinges in the chosen set of segments (Fig. 1a). The φ_0 value is chosen in such a way that about one-half the steps would be successful. The starting conformation may be chosen arbitrarily, say, as a regular polygon. Within the framework of the above standard procedure[3-5] one obtains a set of chains, which correspond to the equilibrium distribution over different conformations of closed circular DNA.[1] This procedure completely neglects twisting and therefore actually treats nicked circular DNA. The distribution over the writhing values of the set of chains thus obtained will have the maximum at zero value of writhing. Chains with large absolute values of writing (Wr) will occur very seldomly.

In the case of closed circular DNA the actual variable, which determines the global properties of the molecule, is ΔLk [see Eq. (1)]. Therefore, to calculate the statistical–mechanical properties of supercoiled DNA one should use, instead of the energy in Eq. (1), the following value:[1,6]

$$E_g(\{\mathbf{r}\}) = E(\{\mathbf{r}\}) + 2\pi^2(C/L)[\Delta Lk - Wr(\{\mathbf{r}\})]^2 \tag{5}$$

where L is the total length of DNA under consideration. Equation (5) has a

[6] M.-H. Hao and W. K. Olson, *Macromolecules* **22**, 3292 (1989).

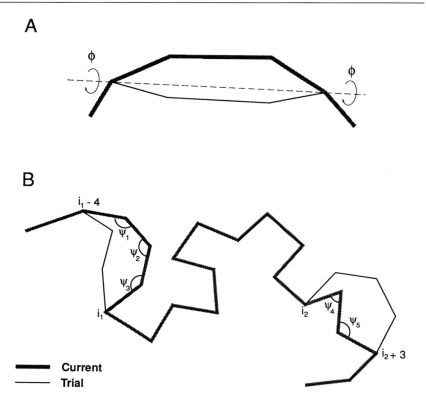

FIG. 1. Two modes of displacement in the course of the Metropolis–Monte Carlo procedure described in the text.

very simple physical meaning: it is the sum of the bending and torsional energy of DNA with given values of ΔLk and Wr. Using Eq. (5) as the energy function in the Metropolis–Monte Carlo procedure, it is possible to generate a set of conformations, which correspond to the equilibrium distribution of molecules with the given ΔLk value. Equation (5) is a special case of a more general approach to calculating conformational and thermodynamic properties of supercoiled molecules proposed by Klenin *et al.*[1]

Simulation Procedure

Based on the general simulation procedure outlined above, at each next step of the Metropolis procedure one should construct a new conformation of the chain, which is generated by an elementary displacement from the previous conformation. Then the generated conformation should be checked with respect to the excluded volume effects, knots, and energy.

Excluded Volume Effects

The excluded volume effects are taken into account by defining the DNA effective diameter, d. A special analysis shows that the statistical–mechanical properties actually depend on the DNA effective diameter value rather than on the shape of the repulsive potential between separated segments.[7] Within the framework of the Metropolis procedure a chain is assumed to have infinite energy if two nonadjacent segments in it are closer to each other than the d value. Therefore, for all pairs of elementary straight nonadjacent segments, the minimum distance between them is calculated and compared with the d value. If for at least one pair of segments the distance is less than d, the new conformation is not accepted.

Knot Checking

Usually closed DNA molecules are unknotted (see [6] in Volume 212 of this series). Therefore the starting conformation should be taken as unknotted (say, a polygon). However, this by no means guarantees that the chain will remain unknotted in the course of a Monte Carlo procedure because we actually deal with a phantom chain while deforming it: the separated parts of the chain can cross each other. Therefore, a knot-checking procedure is absolutely necessary even for short chains, because supercoiling favors compact conformations for which the probability of knotting sharply increases. If the next step entails knotting, it is not accepted. In principle, knotting could be avoided by forbidding crossings of separated segments in the course of the elementary Monte Carlo step (i.e., modeling a nonphantom chain). The two approaches are identical from a statistical mechanics viewpoint. The phantom-chain approach with subsequent knot checking, however, is much more practical in terms of computations. The knot-checking procedure is as follows.

Imagine that we have an arbitrary closed chain in three dimensions. We wish to know whether the chain is knotted or unknotted (the latter is often labeled as a trivial knot). First, one projects the knot on a plane along an arbitrarily chosen axis, while drawing breaks at the crossing points in the part of the curve that lies below the other part (Fig. 2). The projection of

---

[7] K. V. Klenin, A. V. Vologodskii, V. V. Anshelevich, A. M. Dykhne, and M. D. Frank-Kamenetskii, *J. Biomol. Struct. Dyn.* **5**, 1173 (1988).

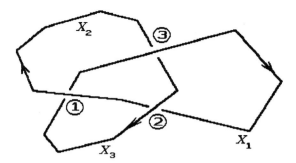

FIG. 2. Calculation of an Alexander polynomial for a knot. Here x_1, x_2, and x_3 are the generators, and 1, 2, and 3 are the crossing points in the projection of the chain. The simplest nontrivial knot ("trefoil") is shown, for which $|\Delta(-1)| = 3$.

the knot amounts to the set of curves, which are called the generators. Let us arbitrarily fix the direction of passage of the generators and number them, having arbitrarily selected the first generator. The crossing that separates the kth and $(k + 1)$th generators will be called the kth crossing. Now we can correlate the knot with a square Alexander matrix, in which the kth row corresponds to the kth crossing and which consists of N elements (N is the total number of crossings in the projection of the knot). Here all the elements except a_{kk}, $a_{k,k+1}$, and a_{ki} (i is the number of the overpassing generator) are zero.

To discriminate closed chains into two categories, knotted versus unknotted, it is sufficient to calculate the Alexander polynomial $\Delta(t)$ at one point $t = -1$. In this case the nonzero elements of the Alexander matrix for the kth row may be calculated according to the very simple rules that follow: (1) When $i = k$ or $i = k + 1$, then $a_{kk} = -1$ and $a_{k,k+1} = 1$; (2) when $i \neq k$, $i \neq k + 1$, then $a_{kk} = 1$, $a_{k,k+1} = 1$, and $a_{ki} = -2$. These relationships hold for all $k = 1, \ldots, N$ under the condition of the substitution $N + 1 \rightarrow 1$.

The value of Alexander polynomial, $\Delta(-1)$, which is an invariant of the knot, is obtained from the Alexander matrix by calculating any $(N - 1)$th order minor (the result does not depend on the choice of minor). The chain is considered unknotted only if $\Delta(-1)$ is equal to $+1$ or -1.

The knot-checking procedure is the most time consuming part of the whole calculation. Therefore this part of the computer program deserves the most attention in terms of its rationality. In this connection a very useful procedure consists of reducing the order of the Alexander matrix before calculating it by eliminating trivial intersections (for details, see Refs. 8–10).

Energy

If the new conformation is acceptable in terms of excluded volume effects and knotting, its energy is calculated using Eq. (5). To do this one has to calculate the writhing value of the new conformation. One can calculate writhing using its expression in terms of the Gauss integral (see, e.g., Ref 10). For straight segments this integral may be presented in the form of a double sum of simple terms. In the framework of our general procedure a more convenient method to calculate writhing is that proposed by Le Bret.[8] In this method, the total writhing value is presented in the form of two contributions, one of which is the directional writhing number, which can be calculated simultaneously with calculating the Alexander polynomial, virtually without additional computations. The directional writhing number in the z direction is the sum of -1 or $+1$ over all crossings. The sign of each term is determined by the type of crossing (Fig. 3).

The second contribution is a sum over all elementary segments r_m of the chain:

$$\sum_m \{\arcsin[\sin \xi_m \sin(\varphi_{m+1} - \chi_m)] - \arcsin[\sin \xi_m \sin(\varphi_m - \chi_m)]\}/2\pi$$

where φ_m is the angle of the projection of r_m on the x,y plane with the x axis. The perpendicular to the plane formed by vectors r_m and r_{m+1} makes the angle ξ_m ($0 < \xi_m < \pi/2$) with the z axis, and its projection on the x,y plane makes the angle χ_m with the x axis.

Metropolis–Monte Carlo Procedure

Whether the new conformation is accepted is determined by applying the classic rules of Metropolis *et al.*[2] (a) If the energy of the new conformation E_{new} is lower than the energy of the previous conformation, E_{old}, the new conformation is accepted. (b) If the energy of the new conformation is greater than the energy of the previous conformation, then the new conformation is accepted with the probability $p = \exp[(E_{old} - E_{new})/k_B T]$.

[8] M. Le Bret, *Biopolymers* **19**, 619 (1980).
[9] J. Des Cloizeaux and M. L. Metha, *J. Phys.* **40**, 665 (1979).
[10] M. D. Frank-Kamenetskii and A. V. Vologodskii, *Sov. Phys.-Usp. (Engl. Transl.)* **24**, 679 (1981).

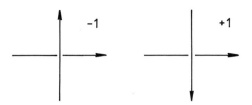

FIG. 3. Calculation of the directional writhing number, showing the two types of crossings and their contribution to the directional writhing number.

Equilibrium Conformations

The Metropolis–Monte Carlo scheme is a theoretically ideal procedure, which leads to, in the limit of an infinite number of elementary steps, reliable results in terms of statistical mechanics. In practice, however, the number of steps is limited by the affordable computer time. Within a limited number of steps no universal criterion exists which could guarantee that all essential conformations were reached in the course of calculations. The possibility that some regions of the configurational space cannot be reached by the chosen procedure for affordable time is the major problem in practical applications of the Metropolis–Monte Carlo method. Equilibrium in terms of different characteristics can be reached in quite different scales of the number of elementary steps of the Metropolis–Monte Carlo procedure.

Calculations performed with the aid of the above procedure showed that, for a chain consisting of 100 straight segments at a superhelical density around -0.05, after about 10^4 elementary steps the chain proves to be a branched interwound superhelix (Fig. 6). The writhing value and the superhelix energy do level off at some values, around which they slightly fluctuate (Fig. 4). Thus, in terms of these quantities the above procedure may be considered a satisfactory one. However, a special analysis has shown that equilibrium in terms of the number of branches of the chain cannot be reached within a realistic number of steps with the aid of the above procedure grounded on the type of displacements shown in Fig. 1a. Therefore, for such experimentally significant quantities as mean square radius of gyration or translational and rotational diffusion coefficients, equilibrium cannot be reached in practice, at least for highly supercoiled molecules (see [21] in this volume).

The fact that the deformations in Fig. 1a are very inefficient in the branch migration process is not unexpected. Indeed, these displacements favor rotations in the plane perpendicular to the general local direction of the chain. By contrast, the branch migration requires slithering movement

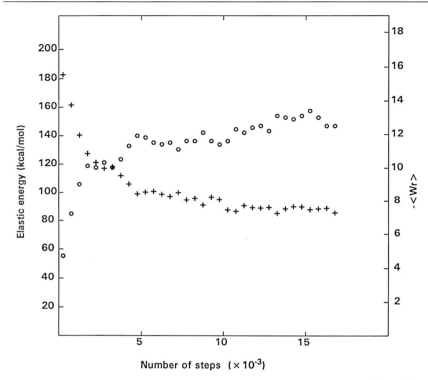

FIG. 4. Dependence of the calculated elastic energy (+) and the average writhing (O) on the number of steps of the Metropolis–Monte Carlo procedure. The chain contained 11.8 Kuhn statistical lengths [3500 base pairs (bp)], $\Delta Lk = -16.6$ ($\sigma = -0.05$), $k = 10$. Each point corresponds to the value averaged over 500 steps. The starting conformation was a closed, randomly coiled chain.

within the interwound superhelix. Therefore, an additional movement has been proposed (A. V. Vologodskii, *et al.*, 1992), which can be called the "internal reputation" (see Fig. 1b). This deformation includes the following operations: (a) One chooses two nodes i_1 and i_2 (Fig. 1b). The deformation is realized only if the distance between i_1 and $i_1 - 4$ is less than three segment lengths. (b) The four-segment chunk between i_1 and $i_1 - 4$ is deformed by changing angles ψ_1 ψ_2 and ψ_3 by an amount δ, so that this chain section could be incorporated between nodes i_2 and $i_2 + 3$. In so doing one conserves the dihedral angles determined by adjacent triplets of nodes between i_1 and $i_1 - 4$. Accordingly, the three-segment section between nodes i_2 and $i_2 + 3$ was deformed by changing angles ψ_4 and ψ_5 so that it would be incorporated between nodes i_1 and $i_1 - 4$. (c) After this one exchanges the two chunks. It is important how the changed sections of the chain are oriented about the axes spanning i_1 and $i_1 - 4$ and spanning i_2 and $i_2 + 3$. The acceptance probability of the deformation is greatly increased if one conserves the direction of vectors pointing from these axes to the centers of masses of the chunks.

The above procedure corresponds to the slithering out by one segment of the chain region situated between nodes i_1 and i_2 along its original conformation. Although such deformations are rarely accepted within the Metropolis–Monte Carlo algorithm, their incorporation significantly enhances the process of branch migration. The above procedure works well for the case of $k = 10$. In case of $k = 15$ the exchange of chunks consisting of three and two straight segments is much more efficient.

Our original modeling of DNA as a homogeneous isotopic elastic rod assumes DNA deformation in any site, that is, corresponds to the wormlike polymer chain. In practice, we generate chains consisting of finite numbers of straight segments having in mind the limit of $k \to \infty$, k being the number of straight segments in the Kuhn statistical length. It should be emphasized that the bending rigidity constant α in Eq. (2) is a function of k [see Eqs. (2) and (3)]. Of course, increasing k entails a dramatic increase in computer time. Therefore, for routine calculation one should choose an optimal value of k, large enough to ensure reliable results and small enough to make the calculations affordable. Figure 5 shows the dependence of the averaged writhing $\langle Wr \rangle$ for the chain comprising 6 Kuhn statistical seg-

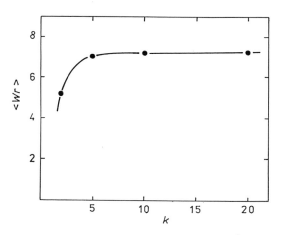

FIG. 5. Calculated writhing number, averaged over about 5×10^5 steps, as a function of k. The chain contained 6 Kuhn statistical lengths, $\Delta Lk = -9$.

ments as a function of k. One can see that even for highly supercoiled molecules ($\sigma = -0.054$) a k value equal to 10 is sufficient to obtain reliable results.

The described calculation scheme is rather time consuming and ideally needs a supercomputer for obtaining statistically reliable results. However, in many cases a powerful personal computer is sufficient. Using a computer, such as the IBM PC 386-25, one needs about 1 h for conducting 10^4 Monte Carlo steps for a chain comprising 100 straight segments. The number of Monte Carlo steps sufficient for calculating statistically reliable averaged quantities sharply increases with increasing the number of segments and degree of supercoiling; for 100 segments at $\sigma = -0.06$ it is about 1 million.

Parameter Values

Discussion of the Kuhn statistical length b (or persistence length $a = b/2$) and the torsional rigidity C can be found in Ref 11. We usually use the following values: $b = 100$ nm, $C = 3.0 \times 10^{-19}$ erg cm. The DNA effective diameter d strongly depends on ambient conditions, first of all ionic strength. We usually take it d as equal to 3.0 nm. According to the available data,[12-14] this value, corresponds to conditions close to physiological.

Shapes of Supercoiled Molecules

Figure 6 shows typical simulated conformations of a chain, corresponding to 3500 base pairs of DNA, for different superhelical densities. One can see that highly supercoiled DNA molecules adopt the conformation of an interwound superhelix, sometimes branched. As superhelical density increases, the conformations become more and more regular. Their shapes resemble electron microscopy patterns of supercoiled DNA molecules.[15-17] We never observed a conformation that would resemble a

[11] M. D. Frank-Kamenetskii, in "Lantold–Bornstein New Series, in Biophysics" (W. Saenger, ed.), Vol. 1c, 228. Springer-Verlag, Berlin, 1990.

[12] D. Stigter, *Biopolymers* **16**, 1435 (1977).

[13] A. A. Brian, H. L. Frish, and L. S. Lerman, *Biopolymers* **20**, 1305 (1981).

[14] E. G. Yarmola, M. I. Zarudnaya, and Yu. S. Lazurkin, *J. Biomol. Struct. Dyn.* **2**, 981 (1985).

[15] C. D. Boles, J. White, and N. R. Cozzarelli, *J. Mol. Biol.* **213**, 931 (1990).

[16] M. Adrian, B. ten Heggeler-Bordier, W. Wahli, A. Z. Stasiak, A. Stasiak, and J. Dubochet, *EMBO J.* **9**, 4551 (1990).

[17] A. V. Vologodskii, S. D. Levene, K. V. Klenin, M. D. Frank-Kamenetskii, and N. R. Cozzarelli, *J. Mol. Biol.*, in press.

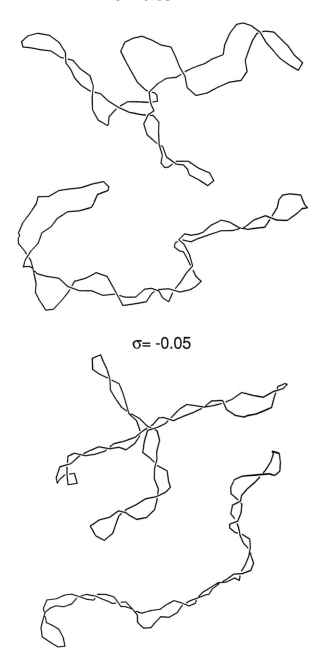

σ= -0.03

σ= -0.05

Fig. 6. Examples of simulated conformations corresponding to supercoiled DNA (3500 bp) at different superhelical densities (indicated) for $k = 10$.

$\sigma = -0.07$

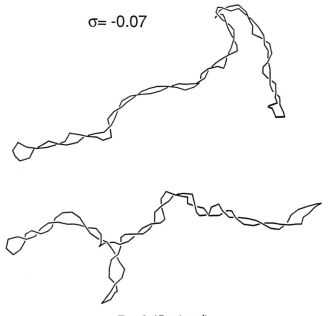

FIG. 6. *(Continued)*

chain wound around a torus. It was claimed, on the basis of very first simulations, that almost all chains are branched.[1] However, more recent calculations show that this question needs more careful analysis.[17]

Klenin *et al.*[1] used the above procedure to calculate the average writhing value $\langle Wr \rangle$ as a function of ΔLk. They arrived at the conclusion that for sufficiently long DNA molecules the $\langle Wr \rangle$ value is close to $0.7\Delta Lk$ within very wide range of ΔLk values. Boles *et al.*[15] and Adrian *et al.*[16] studied this problem experimentally and arrived at a remarkably similar quantitative result. Note that the writhing contribution decreases for shorter DNA chains.[3,5]

More recently, Vologodskii *et al.*[17] have applied the above procedure to calculate numerous characteristic of supercoiled DNA molecules, including the superhelix axis length, the superhelix diameter, the number of branches, etc. The theoretical results quantitatively agree with experimental data obtained by traditional and cryoelectron microscopy.[15,16]

[24] Electron Microscopic Visualization of DNA and DNA–Protein Complexes as Adjunct to Biochemical Studies

By RANDY THRESHER and JACK GRIFFITH

Introduction

Value of Electron Microscopy

Electron microscopy (EM) provides a rapid and unique method for analyzing DNA–protein complexes that are too large to be imaged by X-ray diffraction or two-dimensional (2-D) NMR methods, yet too small to be visualized by light microscopy. A vast majority of the nucleoprotein complexes engaged in DNA metabolism fall within this category. These complexes are usually large and irregular, and they may contain numerous, often different, proteins. Electron microscopy can be a powerful tool in dissecting the mechanism of protein action on DNA because reaction intermediates and products can be visualized throughout the course of the reaction. For example, in the studies of homologous recombination catalyzed by the RecA protein from *Escherichia coli,* the coupling of EM observations with biochemical analyses has provided the physical models for how these reactions occur. These RecA protein-driven recombination reactions occur within giant DNA–protein filaments that are easily seen by EM.[1] Similarly, the large, multiprotein complexes governing DNA replication[2] and repair[3] are amenable to EM analysis. Thus, there is a growing need for studies that integrate information from both EM and biochemistry.

There are many instances in which EM studies lay the foundation for subsequent biochemical work. For example, the formation of a strand invasion complex catalyzed a RecA-like protein might be easily detected by EM even if it occurred at a frequency of 1 in 500 molecules. With gel electrophoresis or filter binding assays, however, such low frequencies of reaction intermediates might well remain undetected. In a recent study,[4] EM was used to demonstrate that intercalating agents will induce the rapid binding of RecA protein to double-stranded DNA (dsDNA) and was used

[1] J. Griffith and L. D. Harris, *Crit. Rev. Biochem.* **23,** S43 (1988).

[2] B. E. Funnell, T. A. Baker, and A. Kornberg, *J. Biol. Chem.* **262,** 10327 (1987).

[3] S.-S. Su, M. Grilley, R. Thresher, J. Griffith, and P. Modrich, *Genome* **31,** 104 (1989).

[4] R. Thresher and J. Griffith, *Proc. Natl. Acad. Sci. U.S.A.* **87,** 5056 (1990).

to determine the kinetics of binding. Once these parameters were defined by EM, it was then possible to develop a gel retardation assay[5] that was used to screen a large number of intercalating compounds. Had EM not been available, the initial characterization of the drug-induced binding of RecA protein to dsDNA would have been much more difficult to define.

Requirements for Electron Microscopic Methods to Be Used as Adjunct of Biochemical Studies

For EM to be used in the manner illustrated above, the preparative methods must be rapid and easily accomplished, and the results unambiguous in their interpretation. Single-stranded regions must be distinguished from duplex regions, and proteins bound to the DNA should be visualized either directly or indirectly by using antibodies tagged with gold beads. Finally, it is essential that the methods be easy to carry out so that many molecules can be analyzed per experiment to yield statistically significant results. The EM preparative methods that require extensive computer enhancement or an averaging of many images, while useful for high resolution studies, will not be applicable as routine adjuncts to biochemical studies of DNA metabolism. This chapter will provide a guide to the simplest and most useful methods for using EM to follow DNA – protein interactions.

Methods for Visualizing DNA – Protein Complexes

It is useful to categorize the methods used to visualize DNA and DNA – protein complexes into three groups: surface spreading methods, direct mounting methods, and cryo methods. The oldest and simplest are the surface spreading methods that were initially developed by Kleinschmidt[6] and subsequently modified. These methods employ denatured basic protein films to spread out and thicken DNA at an air–fluid interface. Although often considered inappropriate for examination of DNA – protein complexes, these methods can be performed in most general EM laboratories and are, in fact, extremely versatile and useful for this purpose.

The direct mounting methods encompass techniques in which DNA – protein complexes are adsorbed to the surface of carbon support films and visualized after rotary shadow casting. Because these methods are more demanding and require absolute cleanliness of buffers, sample, glassware,

---

[5] W. Bonds and J. Griffith, in preparation.
[6] A. K. Kleinschmidt and R. K. Zahn, *Z. Naturforsch. B* **14**, 770 (1959).

water, etc., the surface spreading techniques, which are more forgiving in this respect, should be mastered first.

Cryo methods involve newer methodologies in which chemical fixation is circumvented by the use of very fast freezing to fix the sample physically. The sample is then examined directly in the frozen, hydrated state, or the water (as amorphous ice) is removed by sublimation in a high vacuum and the sample visualized following rotary shadow casting. These methods are quite demanding and require access to specialized equipment. They are also less applicable to the routine analysis of ongoing biochemical reactions and thus will not be described here.

Surface Spreading Methods

In the basic surface spreading methods, a film of denatured basic protein is formed at an air–fluid interface. When DNA contacts the denatured protein film, it is drawn into the film and coated with a sheath of denatured protein which, depending on the conditions, may be as much as 200 Å in diameter. This very thick coating about the DNA serves a number of functions. With the increase in the DNA thickness from 20 to 200 Å, the DNA–protein filament can be visualized by the most rudimentary shadow casting methods. The increase in thickness also allows DNA to be viewed at a much lower magnification with virtually any transmission electron microscope. Finally, the thick protein sheath greatly stiffens the DNA. This stiffening combined with the surface spreading forces act to reduce the otherwise high propensity of DNA to cross over itself, allowing long molecules to be followed with ease.

Because the surface spreading methods are much easier to establish in the average laboratory and are very rapid, there is a great advantage in exploiting them as much as possible for visualizing DNA and DNA–protein complexes. A major asset of the surface spreading procedures is that the simplest of the methods requires only a scant 5 ng of DNA.

Preparation of Supporting Films. Parlodion provides the best supporting film for surface spreading DNA because it is very hydrophobic and adsorbs the denatured protein-coated DNA better than Formvar or carbon-coated Formvar films. An important caution is that solid Parlodion is very hygroscopic and must be thoroughly dried in a vacuum (as small chips) prior to dissolving as a 2% (w/w) solution in amyl acetate. If the water is not completely removed the film will appear dark and may contain numerous small holes owing to the included water, and it will melt rapidly in the electron beam. The film is formed by dropping 50 μl of the 2% solution onto a water surface, allowing the film to form and then lowering the film onto 200- or 300-mesh copper grids.

Preparation of Samples. Before DNA–protein complexes are prepared by the surface spreading or direct mounting methods, the proteins must be fixed in place with either glutaraldehyde or a combination of formaldehyde and glutaraldehyde. Because primary amine buffers such as Tris will react with aldehydes, the reactions need to be buffered with, for example, HEPES or phosphate buffers. We recommend the following fixation protocols: (1) 1% formaldehyde for 5 min followed by 0.6% glutaraldehyde for 5 min more, or (2) 0.6% glutaraldehyde for 10 min. These are final fixative concentrations, and no single fixation regime will work equally well for all DNA–protein complexes. Thus, for any next complex to be examined, both fixation regimes should be tested at 4°, 20°, and 37° (using longer fixation times at the lower temperatures and shorter times at the higher temperatures). We have routinely observed that one of the two protocols at a particular temperature will provide the best structural preservation, when, for example, the DNA–protein associations are either labile or unusually stable at higher or lower temperatures.

It is often useful to purify the fixed samples by gel filtration to remove the fixatives and unbound proteins. If a Tris chromatography buffer is used, the fixation reaction will be quenched, allowing the purified sample to remain suitable for mounting for several days. Removal of the fixatives and unbound protein will also result in a cleaner background in the micrographs, even when surface spreading, and may be absolutely required for viewing samples by direct mounting methods. The disadvantages of gel filtration are that the amount of sample required increases by 10-fold and fractionation is time consuming. We prepare columns in 9-inch Pasteur pipettes with a glass wool plug and use chromatography media such as Sepharose 4B or BioGel A-5m. It is useful to practice collecting 7 to 10-drop fractions in microtiter wells using radiolabeled DNA to determine the fraction which will contain the peak of DNA–protein complexes. Generally, DNAs from 0.5 to 50 kilobase pairs, with or without bound proteins, will elute at approximately the same volume.

Aqueous Drop Spreading. The simplest method for surface spreading DNA is the "aqueous drop" method. Here DNA or DNA–protein complexes are diluted to a concentration of 0.20–0.5 μg/ml in a buffer of 10 mM Tris, pH 7.5. Alternatively, samples can be diluted into 0.25–0.5 M ammonium acetate (pH 7.6) instead of Tris. This is particularly helpful for spreading samples containing fixatives. Cytochrome c is then titrated to a concentration between 2 and 20 μg/ml. A 50-μl drop is placed on Parafilm and allowed to remain for a defined time. A Parlodion-coated grid is touched to the surface of the drop and then stained, dehydrated, and air-dried. Staining and dehydration are carried out in two steps: (1) 30 sec in 75% ethanol to which has been added a 1/100 volume of freshly prepared saturated uranyl acetate in water, and (2) 60 sec in 75% ethanol alone. The grids are blotted on filter paper, air-dried, and shadow cast as described below.

This method can provide superb detail for even very large DNAs if it is carefully optimized. A critical parameter is the length of time that the drop remains on the Parafilm. Because cytochrome c continuously diffuses to the drop surface and denatures, the protein film will steadily increase in thickness with time. There will be a window of time during which an optimal thickness of the protein film and protein coating of the DNA will be achieved. This window will be brief if a high cytochrome c concentration is used and longer if lower concentrations are employed. We allow the drop to remain on the parafilm for either 30 or 60 sec and then, for each series of experiments, titrate the amount of cytochrome c added to the DNA solution beginning with 2 μg/ml and increasing to 20 μg/ml. Note that with all the surface spreading methods described in this chapter, it is worthwhile to experiment with longer times and low concentrations of cytochrome c to improve visualization of DNA–protein complexes.

The aqueous drop method can be used for a great variety of samples from small plasmids to very large DNAs such as the 150-kilobase Epstein-Barr virus DNA. The small amount of DNA required is a major asset.

Aqueous Tray Spreading. In the aqueous tray spreading method, DNA or a fixed reaction sample is diluted to 0.1–0.5 μg/ml of DNA in 0.5 M ammonium acetate, pH 7.6 (the hyperphase). The diluted sample should be allowed to disperse for 1 min or more at room temperature. Cytochrome c is then added to 2–20 μg/ml, and 50 μl of the sample is immediately ramped down a glass surface onto the hypophase containing 0.25 M ammonium acetate, pH 7.6. As the sample drop meets the air–glass–fluid interface, the cytochrome c rapidly denatures and spreads out to form a film that traps nucleic acids as well as proteins, phages, etc. The film will spread rapidly and can remain for 30 sec to 2 min prior to collecting, but it will begin to break down with significantly longer times. To collect the sample, a Parlodion-coated grid is touched to the film 2–5 grid diameters from the glass–fluid interface. The edge of the grid can be bent slightly to facilitate a parallel aposition between the protein film and the Parlodion film. The grids are lifted straight up and then stained, dehydrated, and shadow cast as in the drop spreading method. It is vitally important that clean glassware be used in the tray spreading procedures. Glass microscope slides should be thoroughly cleansed in an organic solvent, such as ether, and stored in a strong acid, such as chromic acid. These must be rinsed in distilled water and air-dried completely prior to use.

The major advantage of tray spreading over the drop spread, which relies on microdiffusion, is that quantitative data regarding the fractional occurrence of a species of DNA or DNA-protein complex can be determined. For example, very precise DNA concentrations can be determined by spreading a mixture of DNA of unknown concentration with a DNA of known concentration and then counting the different species in the microscope. Note that the DNAs must be of sufficiently different size to make an accurate determination of concentration. Similarly, the percentage of intermediates and products in a reaction can be counted to derive kinetic data.

Extending Single-Stranded Regions with SSB Protein. In the aqueous spreading methods single-stranded DNA (ssDNA) remains in a complex secondary structure and appears as a collapsed bush of nucleic acid. The formamide methods described below were developed to extend single-stranded DNA. However, even with the best formamide methods, single-stranded and duplex regions are often difficult to distinguish, and short, single-stranded regions may be overlooked. Furthermore, the degree to which the ssDNA is extended will vary greatly, making precise length measurements impossible.

A useful approach to alleviate this problem is to bind the ssDNA with the SSB protein of *E. coli,* which allows these areas to be clearly visualized by aqueous drop or aqueous tray spreading. SSB protein binds to single-stranded DNA tightly, opens its secondary structure, and extends the ssDNA to 0.35–0.40 times the length of the equivalent sized duplex DNA (prepared for EM by the same method). The thickness of the SSB complex allows single-stranded DNA to be easily identified (Fig. 1). We add an amount of SSB protein that will achieve a ratio of 6 μg of SSB protein per microgram of single-stranded DNA, incubate on ice for 10 min, and then fix the complexes with 0.6% glutaraldehyde for 10 min on ice. The complexes can be used directly or chromatographed and then prepared for EM. SSB protein is commercially available and free of nucleases. The T4 gene 32 protein or the *E. coli* RecA protein could also be used, although the gene 32 protein does not thicken single-stranded DNA as much as SSB protein and is often contaminated with nucleases. RecA protein binding is much more problematic, and RecA may bind to duplex segments.

Formamide Drop Spreading. The inclusion of more than 30% formamide[7] in the spreading solution will disrupt the secondary structure of single-stranded DNA, allowing it to be extended and complexed with denatured protein. In this method, the DNA-protein complexes are diluted in 100 mM Tris, pH 8.4, and 10 mM EDTA to a final concentration of 0.1–0.25 μg/ml DNA. The spreading solution should contain 50% (v/v) spectral quality formamide, fresh from a frozen stock or recrystallized. The

---

[7] B. C. Westmoreland, W. Szybalski, and H. Ris, *Science* **163**, 1343 (1969).

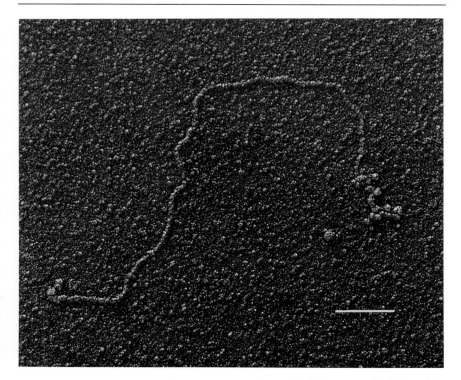

FIG. 1. Electron micrograph of double-stranded DNA with single-stranded tails bound by SSB protein. Tailed molecules were generated by incubating linear 7000-bp plasmid DNA with λ exonuclease to digest approximately one-half the length of the duplex molecules. DNA and protein were mixed at a ratio of 6 : 1 (w/w), incubated on ice for 10 min, and fixed with 0.6% glutaraldehyde for 10 min. The fixed sample was drop-spread in 0.5 M ammonium acetate (pH 7.6) and 2 μg/ml cytochrome c. The protein film was allowed to form for 5 min prior to collecting on a Parlodion-coated grid and then rotary shadow cast with platinum– palladium (80 : 20) at an angle of 10°. The smooth dsDNA in the middle of the molecule can be easily distinguished from the beaded structure of the SSB protein octamers (~ 144 kDa each) bound at the ends. Bar: 0.1 μm.

DNA is allowed to disperse at room temperature for 1 min or more in the formamide–buffer mixture, cytochrome c is added to a concentration of 2–20 μg/ml, and a 50-μl drop is formed on clean Parafilm. Samples are collected after the film has been allowed to build for 30 sec to 5 min (typically 1 min) by touching a Parlodion-coated grid to the surface of the drop. Staining, dehydrating, and rotary shadow casting are performed as described.

Formamide Tray Spreading. The use of formamide in a method similar to aqueous tray spreading has several advantages. By allowing single-stranded regions to be extended, this method can be applied to the mapping of insertions, deletions, inversion loops, and hybridizations between duplex DNA and homologous single-stranded DNA or RNA (D-loops and R-loops, respectively). Increasing concentrations of formamide in the hyperphase (containing DNA) effectively decreases the melting temperature of the DNA, thereby allowing regions with weak secondary structure of poor base pairing to remain separated as they are trapped in the denatured protein film. Components are mixed in 100 mM Tris, pH 8.4, and 10 mM EDTA to a final concentration of 0.5 μg/ml DNA and 40% formamide. Cytochrome c is added to 20–100 μg/ml and the sample ramped down a glass slide onto a hypophase containing 10 mM Tris, pH 8.4, 1 mM EDTA, and 10% formamide. It is critical that there be a 10-fold excess of salts in the hyperphase relative to the hypophase and that the formamide differential be maintained at 30%. Thus, if 80% formamide is required for melting out the nucleic acid, the formamide concentration of the hypophase must be increased to 50%. After spreading, the film will remain intact for up to 2 min before deteriorating. The sample is collected and processed as for aqueous tray spreading. Aggregates or flowerettes of DNA in the protein film can be indicative of buffers that are too low in pH.

Rotary Shadow Casting and Carbon Coating. Rotary shadow casting has been described in detail elsewhere.[8,9] For surface-spread samples we use an alloy of 80% platinum and 20% palladium (w/w). This metal coating provides high contrast, is stable indefinitely, and is easy to evaporate. It is suggested that the evaporator be equipped to deposit a thin layer of carbon over the metal coating. The carbon layer will greatly strengthen the Parlodion film against the thermal effects of the electron beam without adversely affecting visualization of the DNA–protein complexes. The Parlodion should *not* be carbon coated before use at this will inhibit its adsorption of the DNA.

Direct Mounting Methods

In the surface spreading methods, DNA is coated by a layer of denatured protein which will obscure all but the largest proteins bound to DNA. We estimate a 300-kDa globular protein complex bound to DNA to be near the lowest limit of unambiguous identification in the surface spreading methods. The direct mounting methods allow DNA to be directly visualized along with proteins of more moderate size ($\sim$50 kDa). This method has been described in detail[9] but will be briefly summarized here.

---

[8] J. Griffith, *in* "Methods in Cell Biology" (D. Prescott, ed.), p. 129. Academic Press, New York, 1973.

[9] J. D. Griffith and G. Christiansen, *Annu. Rev. Biophys. Bioeng.* **7**, 19 (1978).

All of the direct mounting methods involve three major steps: (1) preparation of a substrate and adsorption of the sample, (2) washing and dehydration, and (3) rotary shadow casting.

Preparation of Substrate and Adsorption. Supporting films of pure carbon are universally used for the direct mounting methods. The plastic and nitrocellulose films employed for surface spreading are too electron dense to allow a clear visualization of the thin DNA helix following metal coating, and these films are not stable to heating by the electron beam needed for the high magnification, high beam intensity viewing. The preparation of pure carbon films has been described.[8]

The major difference between the various direct mounting methods is the strategy employed to attach the DNA to the carbon film. Pure carbon is hydrophobic and does not bind DNA well. All of the approaches begin with a glow charging of the carbon supports in a vacuum of 200 mTorr using a high voltage discharge. Surface electron spectroscopy of glow-charged carbon supports[10] has revealed that this process creates a variety of surface oxide functionalities including hydroxyl, phenolic, carbonyl, and carboxyl groups. Although glow charging creates a hydrophilic carbon surface, additional treatments are required for the carbon to bind DNA efficiently. We advocate the use of 2 mM spermidine, 150 mM Na$^+$, and 1 mM Mg^{2+}. This combination provides efficient binding, and the monovalent and divalent salts keep the DNA from collapsing in the presence of spermidine. Magnesium alone[11,12] will facilitate binding to the glow-charged carbon surface, but not as well as when spermidine is included. Treatments described by others have included alcian blue[13] and polylysine.[14] These methods provide very efficient binding but also produce a relatively rough background.

Washing, Dehydration, and Shadow Casting. The washing and dehydration steps have been described in detail in Ref. 9. Rotary shadow casting, as opposed to unidirectional shadow casting, is preferable because it allows the DNA to be followed over very long distances. Metals used include gold, platinum, platinum–palladium (80:20), tantalum, and tungsten. Tungsten provides the finest resolution,[15] but it oxidizes rapidly in air and may form hydrates over the course of several hours. This combination of events results in the deterioration of fine details with time.

[10] R. Thresher and J. Griffith, unpublished data (1991).
[11] A. Stasiak, A. Z. Stasiak, and T. Koller, *Cold Spring Harbor Symp. Quant. Biol.* **49,** 561 (1984).
[12] C. Bortner and J. Griffith, *J. Mol. Biol.* **215,** 623 (1990).
[13] P. Labhort and T. Koller, *Eur. J. Cell Biol.* **24,** 309 (1981).
[14] R. Williams, *Proc. Natl. Acad. Sci. U.S.A.* **74,** 2311 (1977).
[15] J. Griffith, J. Huberman, and A. Kornberg, *J. Mol. Biol.* **55,** 201 (1971).

Thus, if tungsten is utilized, the samples must be viewed soon after shadow casting or kept under vacuum prior to viewing. Pure platinum, which is stable indefinitely and has relatively fine grain, is often used to avoid these problems. Resistance or vibrating crystal monitors for measuring the amount of metal deposited on the sample are extremely useful in reproducing high resolution mounts. The thickness of metal deposited is a critical factor in achieving the best detail in this method.

Conclusion

In this chapter we have provided an overview of the ways that DNA and DNA–protein complexes can be prepared for EM as an adjunct to parallel biochemical studies. The simple surface spreading methods, when used together with fixation and binding of single-stranded regions with proteins such as SSB, provide an extremely powerful and rapid method for examining reactions with only the most rudimentary vacuum evaporators and electron microscopes. Although there are many DNA–protein complexes that will require more advanced methods of preparation, the surface spreading methods must be established first in any laboratory endeavoring to utilize EM as an adjunct to biochemical studies of DNA metabolism.

[25] Scanning Tunneling Microscopy of Nucleic Acids

By PATRICIA G. ARSCOTT and VICTOR A. BLOOMFIELD

Introduction

The scanning tunneling microscope (STM) was invented in the early 1980s by Gerd Binnig and Heinrich Röhrer, at IBM, Zurich. Their phenomenal success in visualizing details as small as 0.01 nm in the surface structure of semiconductors raised new hopes of at last seeing the atoms in biological molecules. Their instrument was based on the concept of electron tunneling, which was first proposed in the 1920s.[1] It was realized that electrons can pass through a potential energy barrier between electrodes if the barrier is sufficiently thin, that is, equal to or less than about 1 nm. Thus, if a fine wire needle is brought within tunneling distance of a conducting surface, a current can be induced to flow between the tip of the needle and the surface by applying a bias of less than 100 mV. As illus-

[1] R. H. Fowler and L. Nordheim, *Proc. R. Soc. London, A* **119**, 173 (1928).

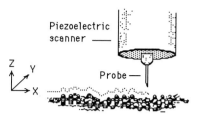

Fig. 1. Scanning mechanism. The STM probe scans the surface in a rastering motion, going back and forth across the specimen less than 1 nm from the surface. Movement up and down in all three directions is controlled by applying voltage to metallic strips embedded in the piezoelectric scanner. The one diagrammed here is typical of the tubular design employed in the most up-to-date instruments. It is made of a ceramic material, usually lead zirconium titanate, which expands, contracts, or bends a few tens of nanometers per volt. Voltage controlling movement in the z direction is supplied by way of a feedback loop. The flow of electrons between the tunneling tip and the surface is continuously monitored and adjusted to maintain either a constant current or a constant distance from the surface. Resolution in the x,y direction is a function of the size, shape, and singularity of the tunneling tip, whereas in the z direction it is a function of distance between the tip and the surface.

trated in Fig. 1, the STM probe is mounted at the base of a piezoelectric crystal which contracts, expands, and bends in response to the applied voltage. By scanning back and forth over the specimen and adjusting the voltage to maintain a constant current or distance from the surface, one can construct an image based on variations in the electronic properties of the surface.

The first STM image of DNA[2] was obtained very shortly after the microscope was introduced, but the goal of visualizing the atoms in the molecule remained elusive for several more years. Making an atomically sharp needle turned out to be surprisingly easy, but other problems, such as how to spread and immobilize the specimen on the substrate, have still not been satisfactorily resolved. Although many techniques have been adapted from electron microscopy, the higher resolution of the STM and the different environments in which it is operated require an entirely new methodology. The most recent images of DNA reveal both strands of the helix, individual bases, phosphates in the backbone, and atoms in the grooves. Reproducibility is an acknowledged problem, but there is ample evidence that progress is being made.

[2] G. Binnig and H. Rohrer, in "Trends in Physics" (J. Janta and J. Pantoflicek, eds.), p. 38, Eur. Phys. Soc., Petit-Lancey, Switzerland, 1984.

Equipment

Scanning tunneling microscopes and the associated hardware and software are available from a number of suppliers.[3] The system components include a piezoelectric scan head which holds the probe, a base to support the head and sample, a control unit containing the power source and amplifiers to drive the movement of the probe, and a computer workstation with a high resolution display screen. Almost all of the newer microscopes employ the tube scanner,[4] which allows faster scan rates than the original tripod design. Some instruments have temperature controls and are specially equipped for operation in ultra high vacuum or under water.

A separate light microscope ($\times 25$) and high intensity lamp are useful for coarse positioning of the probe. To maintain a scan less than 1 nm away from a surface, all parts of the system must be protected from power surges, and vibrations must be kept to an absolute minimum (≤ 1 Hz). The latter objective can be achieved most easily by isolating the STM on an air table or on a heavy platform suspended from the ceiling by bungee cords.

Probes

General Characteristics

The STM probe functions as an electrode and must therefore be conductive, resistant to oxidation, and able to withstand the Joule heating accompanying a high tunneling current. It is typically made of 0.25–0.5 mm tungsten,[5] gold, platinum, platinum–iridium (10–30% Ir),[5] or platinum–rhodium (40% Rh) wire. Tungsten oxidizes easily but produces a more uniform tip than some of the other metals, and it can be used for a limited time without problems.

The size and shape of the probe determine the quality and resolution of the image in the x,y dimensions. Atomic resolution requires an atomically sharp tip, that is, the tunneling tip must consist of a single atom to focus the flow of electrons on a single atom in the specimen or substrate. The force exerted by the probe is distributed over a much greater area than the tunneling current and is related to the overall radius of curvature at the end of the wire.[6–8] Long slender probes of the shape shown in Fig. 2 are

[3] Angstrom Technology, Mesa, AZ; Burleigh Instruments, Fishers, NY; Digital Instruments, Santa Barbara, CA; Park Scientific Instruments, Mountain View, CA; QuanScan, Inc., Pasadena, CA; VG Instruments, Stanford, CA; W. A. Technologies, Cambridge, U.K.

[4] G. Binnig and D. P. E. Smith, *Rev. Sci. Instrum.* **57**, 1688 (1986).

[5] Probes available from Digital Instruments, Santa Barbara, CA, and Materials Analytical Services, Raleigh, NC.

[6] C. Horie and H. Miyazaki, *Jpn. J. Appl. Phys.* **26**, L995 (1987).

[7] C. Horie and H. Miyazaki, *J. Phys. Paris* **48**, C6-85 (1987).

[8] S. Lindsay, O. Sankey, Y. Li, C. Herbst, and A. Rupprecht, *J. Phys. Chem.* **94**, 4655 (1990).

FIG. 2. Tip of an electrochemically etched probe. The size and shape of the probe depend on the composition of the wire, the age and composition of the etching solution, and the strength and duration of the applied potential. Probes with a long tapered neck, such as shown here, are preferred for scanning nucleic acids. More parabolic shapes are often obtained. A distinction must be made between the tip of the probe and the tunneling tip. The latter is defined as the electron donor or receptor and may or may not be located at the center of the probe tip.

preferred for nucleic acid work, not only to minimize the effects of compressive forces but to enable the narrow grooves in the helix to be closely tracked.

Fabrication

Cutting with Scissors. The first step in making the probe is to cut a 10 cm length of wire with scissors. Cutting at an oblique angle leaves numerous jagged points, which are frequently long and sharp enough to produce a good image. Problems arise, however, when instead of a single tunneling tip, there are two or more tips equally long or equally distant from the surface. Multiple tips produce multiple images, which are not always recognizable, especially when there is significant overlap.

Etching. Electrochemical etching gives the most control over tip geometry. The general procedure is to immerse the end of a wire in an electrolytic solution and apply enough voltage to get an initial current of about 1.5 A. The portion of the wire closest to the air–liquid interface etches faster than the rest, producing a long thin neck which eventually breaks in two. The voltage is turned off as soon as the wire breaks and the probe is quickly taken out of solution and rinsed with distilled water to prevent the tip from rounding. Comparative data relating size and shape of the tip to specific parameters such as type of wire and composition of the etching solution are presently lacking.

For tungsten probes, the procedure described by Michely *et al.*[9] requires only a small amount of etching solution. The wire is put into the needle of

[9] Th. Michely, K. H. Besocke, and M. Teske, *J. Microsc.* **152,** 77 (1988).

a syringe which is centered over a Pt loop holding a drop of 5% NaOH. Etching is initiated by sticking the end of the wire all the way through the drop and applying a 1.8 V ac potential to the loop. To prevent damage to the tip, a mound of foam is placed underneath to catch the wire when it breaks. It is retrieved and rinsed with distilled water as quickly as possible. Diameters vary from 10 to 20 nm, depending on the volume of the electrolytic solution.

It should be feasible to etch Pt tips by the same procedure as W tips by substituting an acidic solution and a tungsten loop. Saturated solutions of $CaCl_2 - H_2O - HCl$ (60:36:14, by volume) have been used to etch Pt–Ir wires under other conditions.[10] The procedure presently used is extremely hazardous. The etching solution consists of an unstirred mixture of 6 M KCN and 2 M NaOH. The entire procedure must be carried out in a fume hood, with strict attention to safety. The cyanide is present to facilitate dissolution of the wire by chelating metal ions, and the NaOH is added to inhibit formation of HCN. A Pt or Pt alloy wire is placed in the center of a circular electrode and immersed in the solution to a depth of about 2 mm. Heben and colleagues[11] use a wax-impregnated graphite electrode 3 inches in diameter. Sharp tips of the shape shown in Fig. 2 are obtained with Pt–Ir (30% Ir) wire in 6–7 min at 25 V root mean square (RMS) ac. Etching is terminated 5 sec after an abrupt decrease in the measured current signals a change in the length of the wire. The probe (upper portion of the wire) is then removed from solution and immediately rinsed with distilled water. Reproducible results are obtained only with fresh solutions, and no more than 3–4 tips can be etched per 100 ml.

Growing in a Scanning Electron Microscope. An elongated tip can be grown on the end of a probe by "electron deposition" in a scanning electron microscope (SEM).[12] The probe is glued to a specimen stub and positioned so that the electron beam is perpendicular to the tip. After a few minutes at 30 kV (100 μA), an amorphous deposit forms on the wire. It probably comes from hydrocarbon contaminants present in the column, but its exact composition has not been determined. The diameter of the deposit is proportional to the focal area on the tip and remains constant as the depth increases with time. Needlelike structures 400 nm long are produced in about 15 min.

[10] A. A. Gewirth, D. H. Craston, and A. J. Bard, *J. Electroanal. Chem.* **261**, 477 (1989).
[11] M. J. Heben, M. M. Dovek, N. S. Lewis, R. M. Penner, and C. F. Quate, *J. Microsc.* **152**, 651 (1988).
[12] Y. Akama, E. Nishimura, A. Sakai, and H. Murakami, *J. Vac. Sci. Technol. A* **8**, 429 (1990).

Insulation

The probes used in scanning under water must be insulated in order to minimize ionic conduction (Faradaic) currents which may be several orders of magnitude greater than the tunneling current (1 μA versus 1 nA). The smaller the area left exposed the better.[13-15] A variety of insulating materials have been tried, including glass,[16] poly(α-methylstyrene), silicone rubber, SiO_2, wax, epoxy,[16] and fingernail polish. Of these, the last three are by far the easiest to apply.

Apiezon Wax. Nagahara and colleagues[17] melt Apiezon wax[18] on a 1 cm² copper plate, which is designed to fit over the end of a soldering iron. It has a narrow slit extending from one outside edge to the center which allows the probe to be pushed into the molten wax from underneath. The tip is inserted about halfway between the edge of the plate and the center, held for a few seconds to reach thermal equilibrium, then pushed all the way through the wax and moved sideways out of the slit. Surface tension prevents the wax from adhering to the steep slopes of the tip. Tests of probes prepared by this method indicate negligible leakage (100 pA in 0.1 M $HClO_4$ at 100 mV).

Epoxylite Varnish. Gewirth *et al.*[19] dip the tip into a 5:1 solution of varnish (Epoxylite 6001) and solvent (Epoxylite 6001-s),[20] then turn the probe right side up so that the solution flows down the sides. As in the previous case, how much of the tip is left uncoated depends on its sharpness. It takes 1 hr at room temperature, followed by 2 hr at 80° and 1 hr at 170° for the epoxy to harden. An integrated circuit board makes a convenient holder for the probes during this process.

Nail Polish. The most efficient method of insulating a probe is to dip it into a bead of clear nail polish after it has already been mounted in the STM head. The head is afterward set upside down, so that the coating flows away from the tip. The sides of the probe and end of the probe holder are then painted with the nail polish brush. Drying takes less than 10 min at room temperature.[21]

[13] R. Sonnenfeld and P. K. Hansma, *Science* **232**, 211 (1986).
[14] S. M. Lindsay and B. Barris, *J. Vac. Sci. Technol. A* **6**, 544 (1988).
[15] S. M. Lindsay, T. Thundat, and L. Nagahara, *J. Microsc.* **152**, 213 (1988).
[16] Coated probes available from Longreach Scientific Resources, Orr's Island, ME.
[17] L. A. Nagahara, T. Thundat, and S. M. Lindsay, *Rev. Sci. Instrum.* **60**, 3128 (1989).
[18] Available from Angstrom Technology, Mesa, AZ.
[19] A. A. Gewirth, D. H. Craston, and A. J. Bard, *J. Electroanal. Chem.* **261**, 477 (1989).
[20] Available from Epoxylite Corp., Irvine, CA.
[21] Digital Instruments Instruction Manual, vers. 5, Digital Instruments, Inc., Santa Barbara, California (1990).

Substrates

General Characteristics

The primary requirement of an STM substrate is that it be conductive. It must also be able to withstand the considerable force exerted by the probe as it scans at very close range. In working with nucleic acids it is obviously advantageous to have a substrate that is both uniformly conductive and topographically smooth for hundreds of square nanometers. The surface should also be hydrophilic enough to permit the sample solution to spread evenly.

Highly Oriented Pyrolytic Graphite

Highly oriented pyrolytic graphite[22] (HOPG) is not an ideal substrate for biological molecules but offers several advantages. Its structural and electronic properties are well known,[23,24] and its appearance in the STM is by now familiar to most microscopists working with nucleic acids. The hexagonal symmetry and 0.246 nm lattice constant serve as calibration standards when testing probes and piezoelectric controls.

The weak bonding between layers (separated by ~ 0.34 nm) makes it easy to obtain a clean surface by sticking a piece of tape to the top layer and stripping it away. Problems arise when layers are not removed completely. The unsatisfied bonds at steps and cracks in the surface are reactive with some molecules and may contribute to the tendency of nucleic acids to accumulate and align at these sites. Twisted strips of graphite and debris left on the surface can be mistaken for sample material, and the rotation and sliding of large portions of the surface create moire patterns in the background image.[25]

More serious concerns relate to the deformability of the surface. HOPG has the lowest elastic modulus[23] (3.6×10^6 N/cm^2) of any of the substrates used with biological molecules. Giant corrugations, as much as 10 times greater than expected for atomically smooth surfaces, have been

---

[22] ZYA grade; Available from Union Carbide, Cleveland, OH.

[23] A. W. Moore, in "The Chemistry and Physics of Carbon" (P. L. Walker, Jr., and P. A. Thrower, eds.), Vol. 11, p. 69, Dekker, New York, 1973.

[24] A. W. Moore, in "The Chemistry and Physics of Carbon" (P. L. Walker, Jr., and P. A. Thrower, eds.), Vol. 17, p. 233, Dekker, New York, 1981.

[25] T. R. Albrecht, H. A. Mizes, J. Nogami, S.-I. Park, and C. F. Quate, *Appl. Phys. Lett.* **52**, 362 (1988).

reported[26-29] and are thought to be due to pressure-induced changes in conductivity. Estimates of the force exerted by the probe run as high as 10^8 N/cm^2.[8]

The hydrophobicity of the surface is not conducive to binding double-stranded nucleic acids, which are characterized by a high proportion of hydrophilic groups on the surface. Poly(dA),[30] and presumably other single-stranded forms that have bases exposed, can be expected to adsorb more strongly, but neither single- nor double-stranded forms adhere well enough for HOPG to be used as a substrate under water.[14]

Gold

Gold is currently the substrate of choice. Aside from being a good conductor, it is inert in air and water and does not have the disadvantages of softer materials. The elastic modulus of gold is 8.2×10^8 N/cm^2, two orders of magnitude greater than that of graphite.[23,31] The [100] surface is characterized by a square lattice 0.288 nm on a side which provides a smooth background for imaging the fine features of nucleic acids. The main disadvantages are that gold is hydrophobic in air[32] and easily contaminated. Lindsay and colleagues recommend storing gold films under argon.[33]

Gold Foil. Gold foil (0.1 mm thick) of the type used by jewelers is too rough to be used as a substrate for biological molecules, but it undergoes a surface reconstruction, which produces large flat areas, when heated to a temperature near its melting point (1336 K). The foil is cleaned in a 30% solution of HNO_3, then heated in an oxygen–acetylene flame until the

[26] H. J. Mamin, E. Ganz, D. W. Abraham, R. E. Thomson, and J. Clarke, *Phys. Rev. B* **34,** 9015 (1986).

[27] R. J. Colton, S. M. Baker, J. D. Baldeschwieler, and W. J. Kaiser, *Appl. Phys. Lett.* **51,** 305 (1987).

[28] H. Yamada, T. Fujii, and K. Nakayama, *J. Vac. Sci. Technol. A* **6,** 293 (1988).

[29] P. I. Oden, T. Thundat, L. A. Nagahara, G. B. Adams, S. M. Lindsay, and O. F. Sankey, *in* "Abstracts STM '90, Fifth Int. Conf. Scanning Tunneling Microscopy/Spectroscopy, and Nano I, First Int. Conf. Nanometer Scale Sci. and Technol.," p. 67, Amer. Vac. Soc., Baltimore, Maryland 1990.

[30] D. D. Dunlap and C. Bustamante, *Nature (London)* **342,** 204 (1989).

[31] U. Landman, W. D. Luedtke, N. A. Burnham, and R. J. Colton, *Science* **248,** 454 (1990).

[32] R. Emch, J. Nogami, M. M. Dovek, C. A. Lang, and C. F. Quate, *J. Microsc.* **152,** 129 (1988).

[33] J. Schneir, H. H. Harary, J. A. Dagata, P. K. Hansma, and R. Sonnenfeld, *Scanning Microsc.* **3,** 719 (1989).

color changes from gold to reddish yellow. A pair of forceps for holding the foil can be fashioned from tungsten wire (melting point 3669 K).[34]

Faceted Gold Spheres. Spheres approximately 2 mm in diameter such as used in reflection electron microscopy are produced by slowly feeding 4 cm of high purity (99.999%) gold wire, 0.55 mm in diameter, into an oxygen–acetylene flame. Flat facets, some of which are hundreds of nanometers across, form as the metal cools to room temperature.[14,15,33,35]

Gold Films Epitaxially Grown on Mica. Gold films require more effort to prepare than faceted spheres but are reportedly more reliable.[33] The method developed by Reichelt and Lutz[36] entails evaporating 250–300 nm of gold (99.999% pure gold wire) onto a piece of freshly cleaved mica under a vacuum of 5×10^{-7} to 10^{-6} Torr. Smooth surfaces are obtained by heating the mica to a high temperature (350–500°) and keeping the deposition rate below 0.1 nm/sec.[32] At least 12.5 to 15 nm is required to produce a continuous film.[37] The thickness is monitored and controlled with a quartz crystal oscillator.

Platinum–Carbon and Platinum–Iridium–Carbon on Mica

Carbon films as prepared for transmission electron microscopy (TEM) have too low a density and too high a resistivity to provide a good substrate for STM.[38,39] Pt–C films are less noisy, but the grain size of the metal, which averages 1.5–2.0 nm laterally and 0.6 nm vertically,[39] makes them unsuitable for imaging single molecules of nucleic acids. Amrein and co-workers[40] obtain finer grained films with Pt–Ir–C. A 1.5 mm Pt–Ir (25% Ir) wire is inserted into a 2 mm C rod and evaporated onto freshly cleaved mica under standard conditions at room temperature and 10^{-6} Torr. An average thickness of about 2 nm is sufficient for good coverage.

[34] Z. Wang, T. Hartmann, W. Baumeister, and R. Guckenberger, *Proc. Natl. Acad. Sci. U.S.A.* **87**, 9343 (1990).

[35] S. M. Lindsay, T. Thundat, and L. Nagahara, *in* "Biological and Artificial Intelligence Systems" (E. Clementi and S. Chin, eds.), p. 125. ESCOM Sci. Publ. B. V. Leiden, The Netherlands, 1988.

[36] K. Reichelt and H. O. Lutz, *J. Cry. Growth* **10**, 103 (1971).

[37] P. Echlin, *in* "Preparation of Biological Specimens for Scanning Electron Microscopy" (J. A. Murphy and G. M. Roomans, eds.), p. 137, Scanning Electron Microscopy, Inc., AMF, O'Hare, Illinois, 1984.

[38] R. Guckenberger, C. Kösslinger, R. Gatz, H. Breu, N. Levai, and W. Baumeister, *Ultramicroscopy* **25**, 111 (1988).

[39] M. Amrein, A. Stasiak, H. Gross, E. Stoll, and G. Travaglini, *Science* **240**, 514 (1988).

[40] M. Amrein, R. Dürr, A Stasiak, H. Gross, and G. Travaglini, *Science* **243**, 1708 (1989).

Sample Preparation

Buffers

Ionic Strength and Composition. Low salt or volatile salt buffers are employed when scanning in air and vacuum to minimize the changes in ionization that occur as the sample dries. Buffers such as 10 mM NaCl,[41,42] KCl,[43-45] or CH$_3$COONH$_4$,[44,46] pH 7–8, are used most frequently. Five times higher salt can be tolerated when scanning in solution,[14,47] but even then the distribution on and around the specimen cannot be controlled. "Dirty" patches are sometimes seen after electrochemically depositing the sample, especially when the buffer contains a heavy organic salt such as C$_4$H$_{11}$NO$_3$ (Tris).[47] Phosphate buffer, 50 mM Na$_2$HPO$_4$–NaH$_2$PO$_4$, pH 7, can be used without apparent problems.[14] All solutions should be made in double-distilled, deionized (18 MΩ cm) water and passed through 0.22 μm filters (Millipore, Bedford, MA) to remove small particulates. Heavy metal contaminants can be removed by the addition of NaEDTA to a final concentration of 10 mM, followed by passage through a gel filtration column of small pore size.

Spreading Agents. The spreadability of the sample may be enhanced by adding a dilute drop of detergent to the buffer,[45] or by adsorbing the molecules to a detergent film as described by Vollenweider *et al.*[48] The cationic detergent benzyldimethylalkylammonium chloride (BAC) is frequently used in preparing nucleic acids for electron microscopy and is small enough (MW ~350) to be used in STM without significantly increasing the granularity of the background. The dimensions of the specimen may be affected, however. Electron micrographs indicate an average diameter of the helix 2–3 times the expected value.[49] The procedure used by Cricenti *et al.*[50] is to dissolve BAC in CH$_3$NO (formamide) at a concentration of 2 mg/ml, dilute 1 : 50 into distilled water, then add 2 volumes to

[41] G. Lee, P. G. Arscott, V. A. Bloomfield, and D. F. Evans, *Science* **244**, 475 (1989).

[42] P. G. Arscott, G. Lee, V. A. Bloomfield, and D. F. Evans, *Nature (London)* **339**, 484 (1989).

[43] T. P. Beebe, Jr., T. E. Wilson, D. F. Ogletree, J. E. Katz, R. Balhorn, M. B. Salmeron, and W. J. Siekhaus, *Science* **243**, 370 (1989).

[44] D. Keller, C. Bustamante, and R. W. Keller, *Proc. Natl. Acad. Sci. U.S.A.* **86**, 5356 (1989).

[45] M. Salmeron, T. Beebe, J. Odriozola, T. Wilson, D. F. Ogletree, and W. Siekhaus. *J. Vac. Sci. Technol. A* **8**, 635 (1990).

[46] R. J. Driscoll, M. G. Youngquist, and J. D. Baldeschwieler, *Nature (London)* **346**, 294 (1990).

[47] S. M. Lindsay, *EMSA Bull.* **19**, 60 (1989).

[48] H. J. Vollenweider, J. M. Sogo, and Th. Koller, *Proc. Natl. Acad. Sci. U.S.A.* **72**, 83 (1975).

[49] C. Brack, *Crit. Rev. Biochem.* **10**, 113 (1981).

[50] A. Cricenti, S. Selci, G. Chiarotti, and F. Amaldi, *J. Vac. Sci. Technol. B*, **9**, 1285 (1991).

1 volume of the sample solution containing 100 μg/ml DNA and 0.1 mM NaCl. Seventy microliters is deposited on Parafilm, covered, and left for 1 hr at room temperature to let the DNA spread in the film of detergent that forms at the air–liquid interface. The STM substrate is then touched to the surface of the drop, blotted, and dried under vacuum for 5 min. Formamide solutions rapidly acidify after dilution with water and should be well buffered to avoid denaturing the DNA. A shorter incubation period is also advisable. The purity of the formamide is critical; the A_{270} must be less than 0.2 for reproducible results.[51]

Specimen Size and Concentration

Whether an image is obtained at tip-to-surface distances greater than the tunneling distance (~ 1 nm) depends on the conductivity and compressibility of the specimen. Multimolecular aggregates of DNA and RNA approximately 5–6 nm thick[41] and DNA–protein complexes about 10 nm thick[40] have been successfully imaged.

The optimal concentration depends on the size and stiffness of the molecules as well as on the method of deposition. At high concentrations, tangling and clumping are significant problems, which increase with molecular length and flexibility. There have been few good images of nucleic acids longer than several hundred base pairs (bp). Concentrations range from 10 to 50 μg/ml for those deposited on the substrate in 1 to 10-μl droplets. For those scanned under water, concentrations at the liquid–substrate interface are in the same range before voltage is applied. Somewhat higher concentrations may be used in the presence of cross-linking agents and multivalent cations that promote the orderly alignment of molecules,[41] or with intercalators and ligands that increase their stiffness.

Fixation

It is necessary to stabilize nucleic acid–protein complexes when scanning under conditions of uncontrolled salt deposition. A 10-fold increase in ionic strength can produce a 10^{10} decrease in protein binding strength[52,53] and cause complete dissociation of the complex. The standard procedure for cross-linking nucleic acids and proteins is to incubate the

[51] L. W. Coggins, *in* "Electron Microscopy in Molecular Biology: A Practical Approach" (J. Sommerville and U. Scheer, eds.), p. 1. IRL Press, Oxford, 1987.
[52] V. A. Bloomfield and P. G. Arscott, *in* "STM and SFM in Biology" (O. Marti and M. Amrein, eds.), in press. Academic Press, New York, 1992.
[53] M. T. Record, Jr., C. F. Anderson, P. Mills, M. Mossing, and J.-H. Roe, *Adv. Biophys.* **20**, 109 (1985).

sample for 10–15 min at 37° in an 0.1% buffered solution of formaldehyde or glutaraldehyde.[54]

Conductive Coating

Although it has been assumed that nucleic acids are insulators rather than conductors, the evidence thus far indicates that they can be successfully imaged in the absence of a conductive coating. There is apparently enough contrast in the work function from one part of the molecule to another to achieve atomic resolution of at least some structural features.[46]

Sample Deposition

Basic Method

The basic and most widely used method of preparing biological samples for scanning in air and vacuum is to pipette 1–100 μl of sample solution onto a substrate, cover, and let dry at room temperature. Variations include spraying the sample onto the substrate with an atomizer and floating the substrate on a 100-μl drop of the sample deposited on Parafilm.

Pulse Deposition

Allen and colleagues[55] dip the tip of the probe into the sample solution just before lowering it into position for tunneling. When it is within a few hundred nanometers of the surface, an electric field is generated in the tunneling gap by applying a potential of −4 V in air, using a pulse generator. A 10-μsec pulse deposits the sample in microdroplets directly under the tip. The sample is prepared by a variation of the Vollenweider procedure[48] (see Spreading Agents above): 2 volumes of BAC (0.001–0.1% in formamide) are added to 1 volume of DNA (2–200 μg/ml in 10 mM NaCl or CH_3COONH_4 buffer), gently mixed, and incubated at room temperature for 20 min. In the absence of BAC, no DNA is found on the substrate. The disadvantages of this method are several: the DNA is subjected to shear stress as it is flung onto the substrate, the detergent may add to the diameter of the DNA, and the substrate surface may be easily contaminated. If the electric field strength is of sufficient magnitude to ionize gases

[54] J. Sogo, A. Stasiak, W. de Bernardin, R. Losa, and Th. Koller, *in* "Electron Microscopy in Molecular Biology: A Practical Approach" (J. Sommerville and U. Scheer, eds.), p. 61. IRL Press, Oxford, 1987.

[55] M. J. Allen, R. J. Tench, J. A. Mazrimas, M. Balooch, W. J. Siekhaus, and R. Balhorn. *J. Vac. Sci. Technol. B*, 1272 (1991).

in the tunneling gap, as in glow discharge, the surface may become hydrophilic, but at the expense of significantly reduced conductivity.

Electrochemical Deposition

Samples scanned in water have to be more firmly anchored than those scanned in air or vacuum. The electroplating technique developed by Lindsay and colleagues[47,56] involves the use of a glass cell, 5 mm in diameter, which is set and sealed on top of a gold-coated substrate (cells of a slightly different design[14] are used with gold spheres). The surface of the substrate is first scanned in air to locate the flattest areas for plating. The probe is withdrawn, and 50 μl of sample solution is introduced into the cell through an inlet tube at the bottom. A Pt wire counterelectrode and Ag/AgCl reference electrode are lowered into solution, and voltage is applied to the Pt electrode for about 1 min. The tip is then put back into solution in preparation for scanning.

Negatively charged solutes are electrophoretically driven to the surface of the substrate as it becomes positively charged. DNA and RNA, at concentrations of $10-30$ μg/ml in 50 mM $Na_2HPO_4-NaH_2PO_4$, pH 7.5, form large aggregates which are stable for several hours. The packing density depends on the amount of voltage applied as well as on the size and stiffness of the molecules and their initial concentration. The mode of binding is also voltage dependent and may reflect the conformational state of the molecule. At -0.5 to -1 V, 150-bp fragments of calf thymus DNA bind with their long axes parallel to the surface, whereas at -2.3 V they bind with long segments perpendicular to the surface.[47]

Chemical Cross-Linking

Tris(1-aziridinyl)phosphine oxide (TAPO)[57] has been used to label polysaccharides in electron microscopy. Cricenti,[58] Selci,[59] and colleagues further exploit its affinity for sugars by using it both as a conductive stain and as a cross-linker to fix DNA molecules to a gold substrate. The three ethyleneimine groups react with sugars in the DNA backbone, and the phosphorus oxide with gold. Twenty microliters of activated TAPO is

[56] S. M. Lindsay, L. A. Nagahara, T. Thundat, U. Knipping, R. L. Rill, B. Drake, C. B. Prater, A. L. Weisenhorn, S. A. C. Gould, and P. K. Hansma, *J. Biomol. Struct. Dyn.* **7**, 279 (1989).
[57] Available from Polysciences, Inc., Warrenton, PA.
[58] A. Cricenti, S. Selci, A. C. Felici, R. Generosi, E. Gori, W. D. Djaczenko, and G. Chiarotti, *Science* **245**, 1226 (1989).
[59] S. Selci, A. Cricenti, A. C. Felici, F. R. Generosi, E. Gori, W. Djaczenko, and G. Chiarotti, *J. Vac. Sci. Technol. A* **8**, 642 (1990).

mixed with an equal volume of 1 mg/ml DNA in water and incubated for 1 hr at room temperature. The mixture is then deposited on a gold-plated aluminum stub and dried for several hours before scanning. The pitch and diameter of the helix do not appear to be changed by the treatment.

Scanning Parameters

Current and Voltage

The flow of electrons between the tip and the surface decreases by an order of magnitude for every 0.1 nm increase in gap distance. Nucleic acids are typically scanned at a set point $\leq$ 1 nA and $\pm$200 mV. No polarity differences have been noted, that is, the images appear to be the same whether the current flows from the tip to the surface or vice versa.[40,60,61] The lower the current and the higher the voltage, the less chance the tip will touch the specimen or substrate.

Rate

The scan rate selected depends on the scan size and on the topography and stability of the specimen. Weakly adsorbed molecules of DNA can move several nanometers per second under the force of the passing probe.[62] Moderate scan rates of 100–500 nm/sec[42,43,45] are generally used to avoid disturbing the specimen. Good images have been obtained at rates as slow as 10 nm/sec,[46] but at reduced rates there is a risk of inducing intramolecular vibrations that may blur the image.[61]

Mode

Constant Current and Constant Height Modes. In constant current mode, the flow of electrons between the tip and the surface is continuously monitored and the height of the tip rapidly adjusted to match a preset value. Because the current varies exponentially with gap distance, the gap can be kept nearly constant ($\pm$0.02 nm), thus allowing the contours of the conductive surface to be closely followed. In constant height mode, the gap distance is not modulated. Either the current or the voltage is monitored, and feedback controls respond to an average value. Data can be obtained at

[60] D. P. Allison, J. R. Thompson, K. B. Jacobson, R. J. Warmack, and T. L. Ferrell, *Scanning Microsc.* **4**, 517 (1990).
[61] P. G. Arscott, personal observations.
[62] R. D. Biggar, S. A. Major, A. S. McCormick, E. Spevak, and T. G. Vold, preprint (1991).

a 10 times faster rate than in constant current mode, but there is a considerable loss of information in the vertical dimension. Nucleic acids are therefore routinely imaged in constant current mode.

Hopping Mode. Jericho and co-workers[63] suggest using a hopping mode to keep from sweeping the specimen across the surface as the probe passes back and forth. The tip is periodically raised and lowered so that it is at a safe distance above the specimen for most of the scan. About 75% of the time it is in constant height mode and the other 25% in constant current mode. Most commercial instruments do not offer this option.

Spectroscopy Mode. STM can provide spectroscopic data as well as topographical maps of the electronic surface. In spectroscopy mode, the current is measured as a function of voltage, or the z or x position of the probe. In the first case, the current is measured as the tip is scanned at a constant height while the voltage is varied in discrete steps, and in the other two cases, the voltage is held constant while varying either the vertical or lateral displacement of the probe.[21]

Data Collection and Analysis

The scanning process is often tedious and time consuming. At least 4–5 hr is needed to equilibrate the piezo and adequately survey the sample. Each session begins with testing the probe on the substrate. Samples are often not prepared until there is a reasonable estimate of when the scan will begin. When scanning nucleic acids in air and vacuum, it is advisable to record the time of deposition on the substrate in order to assess the effects of dehydration.

Except for the first coarse positioning of the probe, which is done manually with the aid of a light microscope, all of the movements of the probe are controlled automatically or from the computer keyboard. The image displayed on the computer screen is made up of 200–400 scan lines representing the movement of the probe in the x and z directions. Each line is slightly shifted in the y direction to give a three-dimensional plot which can be gray- or color-scaled to simulate a photograph. Both homemade and commercial instruments are generally equipped with sophisticated software which allows a full range of filtering, signal averaging, and scaling to enhance the quality of the image. The standard routines also enable the image to be enlarged, viewed from different angles, and measured in a variety of ways. Data are labeled and stored on high density disks for later analysis.

Because the mechanism of image formation is still not entirely under-

---

[63] M. H. Jericho, B. L. Blackford, and D. C. Dahn, *J. Appl. Phys.* **65**, 5237 (1989).

stood, it is important to compare STM images and measurements with model structures based on X-ray diffraction or two-dimensional NMR coordinates whenever possible.[42,46] In experiments with nucleic acids, the pitch and diameter of the helix are the main consideration since they are indicative of the handedness, conformation, and hydration state of the molecules. The pitch, or half-pitch, is indicated by the periodicity of the rise and fall of the probe as it moves along the helix. This can be obtained directly by measurement of the peak-to-peak spacing in line plots of vertical versus translational movement or by fast Fourier transformation of the digitized data.[41,42] The latter method is faster, but it does not provide information about height or site-specific variations in the structure. Theoretically one could determine the diameter of the molecules either by width or height, but height measurements are unreliable. The values reported for double-stranded DNA are $0.9-1.2$ nm[41,44,46] instead of $2.0-2.5$ nm,[64] and for poly(dA) 0.84 nm[30] instead of 0.57 nm.[64] Anomalous heights have also been reported for various proteins.

Comments

The tunneling current, I_T, in the ideal case of a conductor or semiconductor[65] has the form

$$I_T(s) = I_0(V) \exp(-A\phi^{1/2} s)$$

where $I_0(V)$ is proportional to the availability of electrons and holes (the local density of states) in the two electrodes at bias voltage, V, A is 10.25 nm^{-1} eV$^{-1/2}$, ϕ is the height of the tunneling barrier (work function), and s is the width of the barrier (separation distance between electrodes). A topographical image is obtained only if the local density of states and work function do not vary from one part of the molecule to another. Nucleic acids, like most biological molecules, are made up of many different atoms in varied arrangement. One can therefore not expect the STM image to match the surface structure of a model based solely on the van der Waals radii of the atoms.[42]

Recent experiments with paired donor–receptor complexes bound to DNA suggest that electrons pass more freely through the σ-bonded sugar–phosphate chains than through the π orbital system of the bases, contrary to expectation.[66] In agreement with these findings, the most highly resolved images of DNA indicate that the probe "sees" the sugar–phosphate back-

[64] W. Saenger, "Principles of Nucleic Acid Structure." Springer-Verlag, New York, 1984.
[65] A. Baratoff, G. Binnig, H. Fuchs, F. Salvan, and E. Stoll, *Surf. Sci.* **168**, 734 (1986).
[66] M. D. Purugganan, C. V. Kumar, N. J. Turro, and J. K. Barton, *Science* **241**, 1645 (1988).

bone but not the bases in the grooves.[42] The ions and oriented water molecules around the phosphates may have a role in mediating electron transfer and must therefore be taken into account in determining both the local density of states and the work function.

The effects of pressure on the molecular structure are difficult to assess. Unless the scan is carried out in ultra high vacuum, the surface is likely to be covered with a layer of gaseous contaminants. The presence of this layer may help spread the force of the probe over a wide enough area to result in significant compression of both specimen and substrate.[8,26] As a result, the height as well as the width of the tunneling barrier may be affected. Both experimental and theoretical considerations indicate that the force is of sufficient magnitude to increase the overlap of electron orbitals and, under some conditions, bring the molecule into resonance with the Fermi level of the metals. Lindsay et al.[8] calculate that the energy gap may be shifted as much as 0.5 eV when the specimen is equidistant between the tip and the substrate.

The vertical dimensions of DNA imaged on graphite under ambient conditions are about one-half the value obtained by crystallography. It is unlikely that such a large difference can be due to either dehydration or pressure effects. Although the periodicity of the helix decreases with time of exposure to air,[41,42] the height of the molecule remains about the same.[61] One possibility is that the gaseous contaminants on the surface lower the conductivity of the specimen and make it appear closer to the substrate than it actually is. Even though the height of a double-stranded DNA or RNA molecule is greater than the tunneling distance, one need not invoke compression to explain STM imaging. Keller and colleagues[44] propose that the electrons are transferred in sequential steps, from the sugar–phosphate chain to the core region occupied by bases, to the sugar–phosphate chain on the opposite side of the molecule. Molecular orbital calculations indicate that such a scheme is feasible.[67]

Acknowledgments

This work was supported by grants from the National Institutes of Health and the National Science Foundation.

[67] P. Otto, E. Clementi, and J. Ladik, *J. Chem. Phys.* **78**, 4547 (1983).

[26] Cryoelectron Microscopy of DNA Molecules in Solution

By Jacques Dubochet, Marc Adrian, Isabelle Dustin, Patrick Furrer, and Andrzej Stasiak

Introduction

Since the initial studies of DNA, electron microscopy (EM) has provided crucial insights into the complexity of the tertiary structures of DNA molecules. For example, comprehension of DNA supercoiling required direct EM visualization[1]; sequence-dependent DNA bending was fully proved only after EM had provided unambiguous micrographs[2]; the handedness of DNA knots and catenanes was only possible to determine with the help of electron microscopy.[3] Nevertheless, all classic EM studies of the shape of DNA molecules suffered from a basic deficiency because the inherent three-dimensional structures of DNA molecules are lost when the molecules are mounted on the two-dimensional supporting film used in EM preparations. Such flattening of three-dimensional structures makes it difficult to conclude from micrographs what was the original structure of the analyzed molecules.[4] Interaction with the supporting film also changes the chemical and physical environment of DNA molecules and is likely to induce substantial changes in tertiary and secondary structure. Other routine steps of EM specimen preparation, such as staining and drying, can cause further changes of DNA structure.[5] There was therefore a great demand for an EM technique which would allow observation of DNA molecules in their unperturbed three-dimensional structure, freely suspended in the solution of interest and not affected by chemical staining artifacts.[6] This chapter shows that cryo-EM is such a method, and we describe here how cryo-EM can be used for the study of the tertiary structure of DNA molecules.

[1] J. Vinograd, J. Lebowitz, R. Radloff, R. Watson, and P. Laipis, *Proc. Natl. Acad. Sci. U.S.A.* **53**, 1104 (1965).

[2] J. D. Griffith, M. Bleyman, C. A. Rauch, P. A. Kitchin, and P. T. Englund, *Cell (Cambridge, Mass.)* **46**, 717 (1986).

[3] M. A. Krasnow, A. Stasiak, S. J. Spengler, F. Dean, T. Koller, and N. R. Cozzarelli, *Nature (London)* **304**, 559 (1983).

[4] T. C. Boles, J. H. White, and N. R. Cozzarelli, *J. Mol. Biol.* **213**, 931 (1990).

[5] H. J. Vollenweider, A. James, and W. Szybalski, *Proc. Natl. Acad. Sci. U.S.A.* **75**, 710 (1978).

[6] M. Adrian, B. ten Heggeler-Bordier, W. Wahli, A. Z. Stasiak, A. Stasiak, and J. Dubochet, *EMBO J.* **9**, 4551 (1990).

Physical Principle of Cryoelectron Microscopy

The underlying idea of cryoelectron microscopy of vitrified specimens is to observe the specimen with all its water immobilized by low temperature, the cooling being done in such a way that water has no time to crystallize. The native structure of macromolecules in solution is therefore essentially preserved. Freezing damage and sublimation of water in the vacuum of the EM are avoided.[7]

Water in Electron Microscopy

For 50 years electron microscopy was a science of the dry state because water was intentionally removed from specimens before they were introduced into the vacuum of the microscope. This has been an important limitation in the study of biological samples, which are composed of 70 to 80% water. To introduce hydrated specimens into the vacuum of the EM column, one has to prevent water from evaporating. Otherwise, rapid and uncontrolled water escape from the specimen would have adverse consequences for the observed material and for the EM.[8] Cryoelectron microscopy provides a possibility to overcome this difficulty. It makes use of the fact that the partial pressure of water decreases with temperature and, somewhere between -130 and $-110°$, water sublimation becomes negligible in the vacuum of the microscope.

The early work of Fernandez-Moran[9] showed that frozen specimens can, in principle, be observed in the EM, but the difficulty of the technique left it without practical use until Taylor and Glaeser[10,11] demonstrated that catalase crystals are much better preserved in the frozen, hydrated state, than in the dry form; they obtained images showing good resolution and contrast. Unfortunately, cryomethods were, for a long time, faced with the general drawback that severe freezing damage is produced when water is transformed into ice. Reducing or eliminating freezing damage has been the goal of cryobiology for several decades. The usual strategy, which has been remarkably successful, is to allow the formation of ice crystals but under conditions in which their damaging effects are minimized.[12]

It therefore came as a surprise when it was discovered, contrary to

---

[7] J. Dubochet, M. Adrian, J.-J. Chang, J.-C. Homo, J. Lepault, A. W. McDowwal, and P. Schultz, *Q. Rev. Biophy.* **21,** 129 (1988).

[8] D. F. Parsons, *Science* **186,** 407 (1974).

[9] H. Fernandez-Moran, *Ann. N.Y. Acad. Sci.* **85,** 689 (1960).

[10] K. A. Taylor and R. M. Glaeser, *Science* **106,** 1036 (1974).

[11] K. A. Taylor and R. M. Glaeser, *J. Ultrastruct. Res.* **55,** 448 (1976).

[12] P. Mazur, *Am. J. Physiol.* **247,** C125 (1984).

accepted ideas, that water can be vitrified.[13,14] The term vitrification must be correctly understood: vitreous is the state in which water is solidified without ice crystal formation. This does not mean that the crystals are too small to be observed, but that they do not exist at all.[7] This difference is important because crystallization is a strongly exothermic phase transition which, locally, is an all-or-none process. Water can be vitrified if it is cooled so that water molecules are immobilized before crystallization has started. For pure water, this can be done if the cooling speed is of the order of 10^6°/sec and if the temperature is then kept below -140°. Such cooling conditions can be achieved by rapid immersion in a good cryogen if the thickness of the specimen is only a fraction of a micrometer. It is the good fortune of electron microscopists that, if a specimen is thin enough to be observed in the electron microscope, it is also easy to vitrify. The cooling speed can be lower if water contains solutes which serve as a cryoprotectant. It is then often possible to vitrify biological samples down to a depth of approximately 10 μm by immersion in cryogen. The vitreous state can change to crystalline if the temperature rises above -140°. For pure water, devitrification takes place within a few minutes at -135°. The devitrification temperature of most biological samples is not well determined, but it is generally below -110°. The electron beam also helps in inducing crystallization. However, below -160°, a vitrified sample is stable enough for practical observations in the EM.

Once vitrification was found to be possible, most other problems which were previously thought to present considerable difficulties for the preparation and observation of frozen, hydrated specimens found simple and elegant solutions. Some of these solutions are discussed below, and they are presented in detail in review articles,[7,15] where the original references are cited.

Thin Film Formation

Forming a very thin layer of liquid for vitrification means that the surface-to-volume ratio must be very large. This is an energetically unfavorable situation. In practice this means that the liquid film tends to be too thick for electron microscopy, or it dries nearly instantaneously when it has the desired thickness. It was difficult to develop a supporting film with surface properties stabilizing the adsorbed liquid film. The solution came

[13] P. Brüggeller and E. Mayer, *Nature (London)* **288**, 569 (1980).
[14] J. Dubochet and A. W. McDowell, *J. Microsc.* **124**, RP3 (1981).
[15] N. Roos and A. J. Morgan, "Cryopreparation of Thin Biological Specimens for Electron Microscopy: Methods and Applications." Oxford University Press and Royal Microscopical Society, Oxford, 1990.

when it was realized that the best support is water itself, because a self-supporting water layer of submicron thickness has a remarkable stability which allows it to last for seconds spanning over small holes in a perforated supporting film.[7]

Cryoelectron Microscope

In the electron microscope, the vitrified specimen must be kept at a low temperature. This can be achieved in a conventional electron microscope equipped with a simple cryospecimen holder. The required temperature of $-160°$ is obtained by a convenient nitrogen cooling. A sophisticated electron microscope or the very low temperature that can be provided by helium is of little advantage. The only special feature that must be incorporated is an efficient anticontaminator.[16]

Contrast

Contrast of unstained objects is generally low, especially when the specimen is embedded in water rather than being dry. It was found, however, that an optimal use of phase contrast obtained by strong defocusing can produce sufficient contrast and that the very low value of the background noise of the vitrified unstained specimens results, in general, in an improved signal-to-noise ratio of the observed structure.[7]

Beam Damage

Beam damage is known to be the most severe limitation of conventional electron microscopy.[17] Vitreous specimens are also sensitive to beam damage; however, if methods minimizing irradiation are used, it is found that the fine structures of biological specimens are not more beam sensitive in water than in the dry state. On the contrary, the low temperature brings a reduction of beam damage by a factor of 2 to 10 at around $-160°$.[7]

Cryoelectron Microscopy: A Reliable and Increasingly Popular Method

During recent years simple and reliable methods have been developed for the preparation and observation of fully hydrated, vitrified biological suspensions. They are applied by a rapidly increasing number of laborato-

---

[16] J.-C. Homo, F. Booy, P. Labouesse, J. Lepault, and J. Dubochet, *J. Microsc.* **136,** 337 (1984).

[17] R. M. Glaeser, *J. Ultrastruct. Res.* **36,** 466 (1971).

ries all over the world and provide superior results with such specimens as viruses,[18] lipids,[19] two-dimensional protein crystals,[20] and protein filaments.[21,22] However, until now the majority of the studies have been made on specimens with various forms of symmetry. Only some studies were devoted to single molecules or asymmetrical structures.[23] DNA has been seen on various occasions,[24,25] but it was not clear at that time if the observed molecules were coated with proteins and adsorbed at the liquid interface as it is the case when they are prepared by the classic Kleinschmidt method.[26] Recent work,[6] however, has demonstrated that native DNA molecules can indeed be observed when they are suspended in their vitrified medium. The practical procedure for achieving these observations is described below.

Observation of DNA Molecules Suspended in Thin Layer of Vitrified Solution

The method for the preparation of vitrified thin layers of solution is described in detail elsewhere.[7,15] A short description is presented in the following paragraphs.

Specimen Preparation

Small double-stranded DNA molecules or DNA fragments at a concentration of approximately 200 μg/ml in low salt buffer (<20 mM salt) are well suited for observation. The preparation is made on a grid with a perforated carbon film[27] (average diameter of the holes 3 μm). The grid is placed in a pair of tweezers attached to an automatic plunger (such as illustrated in Fig. 1), and a 3-μl drop of suspension is put on the grid. A piece of blotting paper is pressed for about 2 sec on the grid. During this

[18] B. V. V. Prasad, J. W. Burns, E. Marietta, M. K. Estes, and W. Chiu, *Nature (London)* **343,** 476 (1990).

[19] Y. Talmon, J. L. Burns, M. H. Chestnut, and D. P. Siegel, *J. Electron Microsc. Tech.* **14,** 6 (1990).

[20] C. Toyoshima, P. N. T. Unwin, and E. Kubalek, *J. Cell Biol.* **107,** 1123 (1988).

[21] J.-F. Ménétret, W. Hofmann, and J. Lepault, *J. Mol. Biol.* **202,** 175 (1988).

[22] A. Milligan and P. F. Flicker, *J. Cell Biol.* **105,** 29 (1987).

[23] T. Wagenknecht, R. Grassucci, J. Frank, A. Saito, M. Inui, and S. Fleischer, *Nature (London)* **338,** 167 (1989).

[24] J. Dubochet, M. Adrian, J. Lepault, and A. W. McDowall, *Trends Biochem. Sci.* **10,** 143 (1985).

[25] J. Dubochet, M. Adrian, P. Schultz, and P. Oudet, *EMBO J.* **5,** 519 (1986).

[26] A. K. Kleinschmidt and R. K. Zahn, *Z. Naturforsch. B* **14,** 770 (1959).

[27] A. Fukami, K. Adachi, and M. Katoh, *J. Electron Microsc.* **21,** 99 (1972).

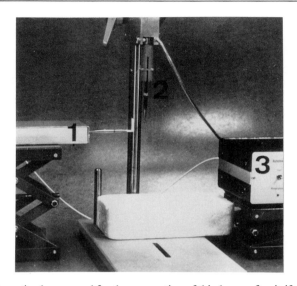

FIG. 1. Automatic plunger used for the preparation of thin layers of a vitrified suspension. The blotting paper is moved by an automatic device (1) toward the grid supporting a perforated carbon film with a drop of suspension (2). After approximately 1 sec, when most of the liquid is soaked up, the blotting paper is withdrawn while the plunger is released, thus falling in about 0.1 sec into ethane cooled near its freezing point. The thin liquid layer of suspension spanning the holes of the film is then vitrified in about 10^{-4} sec. Timing of the movements is controlled by an electronic box (3).

time, most of the liquid is withdrawn by the blotting paper from the surface of supporting film but not from the micron size holes in the film. The blotting paper is then removed and the specimen immediately plunged in a receptacle containing liquid ethane cooled near to its freezing temperature by liquid nitrogen. Vitrification of the submicron thick layer of suspension spanning the holes of the supporting film probably takes place within 10^{-4} sec. It is important that the time period between the end of the blotting process and vitrification be kept to a minimum, in order to avoid partial drying of the thin suspension layer. This could result in an increase of solute concentration by a factor of 3 or more if special precautions are not taken. An automatic plunger helps minimize this effect.[28]

[28] M. Cyrklaff, M. Adrian, and J. Dubochet, *J. Electron Microsc. Tech.* **16,** 351 (1990).

Transfer

After vitrification is achieved, the frozen specimen is transferred to the electron microscope or stored in liquid nitrogen for later observation. In general, a thin layer of ethane still covers the specimen after it is transferred to liquid nitrogen. Without rewarming, the specimen is mounted in the cryospecimen holder and rapidly transferred to the electron microscope. This transfer is made easier in modern commercial cryoholders by two cooled shutters protecting the specimen from direct contact with warm air and atmospheric moisture. Special attention must be given to water contaminating the specimen in the electron microscope. This special difficulty is due to the fact that microscopes, which are built to prevent contaminating material, like water, from reaching the specimen chamber from the rest of the column and in particular from the photographic chamber, are generally not designed to prevent contamination by water vapors originating from the specimen holder itself. This is difficult to avoid completely, since some of the regions of the specimen holder, cooled during transfer and coated with condensed water, rewarm slowly in the electron microscope and become an important source of water vapor in the immediate vicinity of the cooled specimen. The specimen itself then becomes an undesirable cold trap of water vapor. It is likely that many supposedly "vitrified" specimens were found to be of disappointing quality, simply because they were, in fact, air-dried specimens, covered in the electron microscope by a layer of vitrified condensed water vapors. Adequate protection against this effect is obtained by keeping the specimen shielded by the cooled protecting blades during the first few minutes after introduction and by using a special anticontaminator, placed in the immediate vicinity of the specimen.[16]

Observations

The observation conditions must be adapted to the low contrast of the specimen and to its sensitivity to beam damage. According to the theory of phase-contrast image formation, the best contrast is obtained when the first maximum of the transfer function corresponds to the dimension d of the inverse spatial frequency which is to be visualized. d is defined unambiguously for a periodic object, and this, in turn, defines the optimal defocus Δz. In the case where d is not very small (nanometers) the following relation holds: $\Delta z = d^2/2\lambda$, in which λ is the wavelength of the electron (3.2 pm for 100-kV electrons).[7] To observe side by side arrangements of filaments of 2 nm diameter, the optimal underfocus would then be 0.6 μm. The conditions are not so straightforward for individual DNA molecules which do not form periodic associates. Furthermore, inelastic scattering, which is prominent in aqueous specimens and which depends on specimen thickness, tends to increase the apparent underfocusing. Relying on experience, it is found that the best micrographs of DNA are generally obtained when Δz is between 0.6 and 1.5 μm.

Electron beam damage to DNA in vitreous water results in the spreading of the filament image and a reduction of its contrast. We estimate that a dose of approximately 3 kelectron/nm² reduces the contrast by a factor of 2. Nevertheless, this gives the possibility of recording two micrographs at 60,000× magnification under good conditions. (Speed of film 2 μm²/electron; OD 1.5; negligible pre- and intermediate irradiation).

Drift and/or specimen vibration are other limiting effects of beam damage. They are probably the consequence of electrostatic forces induced by charges building up in the irradiated specimen. They result in movements which blur the image and severely reduce the yield in good micrographs. The effect is especially noticeable in tilted specimens, suggesting that the movements in the direction perpendicular to the specimen plane cause most of the problems. The following measures were found to help reducing this effect: (1) A small objective aperture is probably acting as a source for an electron gas, neutralizing negative charges on the specimen. The absence of an objective aperture makes the specimen extremely sensitive to drift. (2) Irradiation should be limited to the recorded area. It is likely that spot scan irradiation[29] would be of considerable help. (3) The perforated supporting film should be thick, conducting, and mechanically resistant. (4) During image recording, irradiation should be limited to the central part of a hole, without touching the edge of the supporting film.

Bubbling, which is a typical effect of beam damage in more massive hydrated biological specimens, does not take place in a DNA solution unless it contains a large concentration of other organic material (sugar, glycerol, proteins).

DNA Images

A typical image of a vitrified sample of a naturally supercoiled plasmid preparation (pUC18 DNA, 2686 pb) suspended in low salt buffer is shown in Fig. 2. The right-hand side of the field presents a region where the molecules are quite clearly visible, whereas they are less visible at the upper left corner. The tendency of the molecules to concentrate toward the edge of the hole is demonstrated by the aggregate of molecules at lower right. As seen in Fig. 2, many molecules can be followed over their whole length.

29 K. H. Downing, *Science* **251,** 53 (1990).

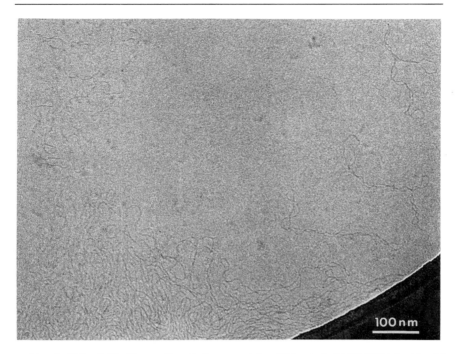

Fig. 2. Thin vitrified layer of a solution containing 200 μg/ml pUC18 DNA in 10 mM Tris-Cl (pH 8.0), 1 mM EDTA. The edge of a hole is visible at the lower right-hand corner of the micrograph. Several molecules are clearly visible on the right-hand side of the micrograph. Those at the upper left-hand corner are unsharp, probably because of drift. A large number of molecules, concentrated toward the edge of the hole, are visible at the bottom.

Similar molecules have been used to determine the three-dimensional shape of the supercoiled DNA molecules and to determine the twist and the writhing of individual DNA molecules.[6]

Figure 3a,b is a stereo pair of a natively supercoiled pUC9 molecule, taking the form of a tight superhelix induced by the presence of 10 mM MgCl$_2$ in the solution. This tight supercoiling causes two opposing double-stranded segments of supercoiled DNA molecules to approach each other longitudinally. The resulting molecules have an appearance of linear filaments with two end loops.[6] In the case of the photographed DNA molecule, *Escherichia coli* RNA polymerases are bound to the end loops. Tightly supercoiled regions show a better contrast than normal double-stranded regions (owing to the doubling of the mass per unit length of the filament) and are therefore more amenable to obtaining stereo impressions by viewing a stereo pair of micrographs. The stereo pair can also be used

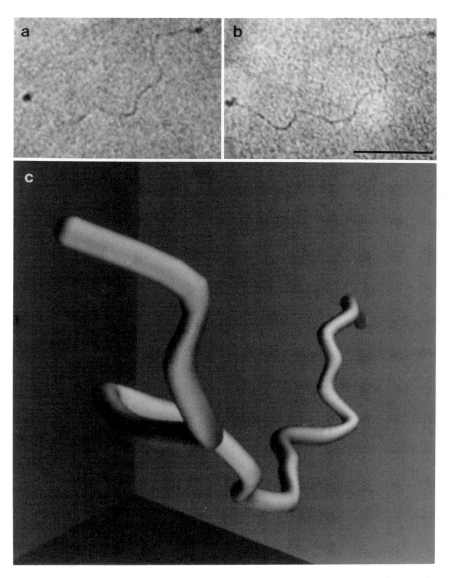

FIG. 3. Stereo pair of micrographs of a tightly supercoiled DNA molecule and the reconstructed model. (a,b) Stereo images of a supercoiled pUC9 DNA molecule with *E. coli* RNA polymerases bound to end loops. Bar: 0.1 μm. The tightly supercoiled region of the DNA molecule was used in a reconstruction procedure and for the generation of a three-dimensional computer model.[30] The reconstructed molecule in (c) is presented at higher magnification and from the other direction than the original stereo pair. The computer reconstruction provides a more objective and quantitative representation than the direct visual stereoscopic impression; it also permits direct measurements of such structural parameters as curvature, persistence length, and writhing of analyzed DNA molecules.

for the generation of a three-dimensional computer model[30] (Fig. 3c). It is obvious that the molecule is not flat but extends in the third dimension. This and many other similar pictures demonstrate that DNA molecules are freely suspended in the film of the vitrified suspension and disprove the hypothesis that observed DNA molecules can be adsorbed onto the surface of the film. Consequently, it appears that the method described above provides a faithful representation of the shape of individual DNA molecules in solution.

Potential Problems and Difficulties of Method

Thin Film

The fact that the specimens consist of a thin film creates a potential source of artifacts. Frequently, the DNA molecules tend to concentrate in the regions where the film is the thickest, generally toward the edge of the supporting film (Fig. 2). Consequently, the concentration of molecules in the thin film is not a good indicator of the concentration in the bulk. This is probably the reason why long DNA molecules [100 kilobases (kb)] are rarely incorporated in thin layers, either because they are not even incorporated in the film, or because they escape toward the regions of the specimen where the layer is too thick for good observations.

The film thickness ($\sim$100 nm) means that molecules have to be confined to this thin layer. For very small DNA molecules up to several hundred base pairs, this confinement should not cause any problem and molecules should be free to adopt any rotational position. Much longer DNA molecules might be restricted in adopting the preferred three-dimensional configuration as a result of confinement in the thin film.

Temperature

In order to be vitrified, the specimen must be cooled below $-140°$. Could it be that the shape of the DNA molecule changes during cooling before it is immobilized in vitreous ice? Our results suggest that this change is not considerable. It is known that the twist of naked DNA changes by $0.01°/°C$ per base pair (over a nondenaturating range of temperatures).[31] It could therefore be expected that for a molecule of 2686 base pairs, the change in temperature from $+20°$ to $-30°$ (approximate temperature

---

[30] I. Dustin, P. Furrer, A. Stasiak, J. Dubochet, J. Langowski, and E. H. Egelman, *J. Struct. Biol.* **107,** 15 (1991).

[31] R. H. Morse and C. R. Cantor, *Proc. Natl. Acad. Sci. U.S.A.* **82,** 4653 (1985).

where the growing viscosity of undercooled water would effectively block movements of DNA molecules) would change its twist by 4 turns, which would lead to the induction of 4 supercoils in a relaxed covalently closed DNA molecule. Instead, we observe that circular DNA molecules relaxed and covalently closed at 20° remain as relaxed, non-self-intersecting molecules in vitrified specimens.[6] Apparently rapid vitrification (0.1 msec) does not allow the molecules to change their shape significantly.

The question of how well the DNA molecules, observed in vitrified thin layers, represent the bulk sample investigated by other physical or biochemical methods cannot yet be answered fully. Cryoelectron microscopy has been shown to be a method providing extremely faithful representations of biological particles like viruses or protein crystals.[18] Preserving the preferred shape of very flexible and environment-sensitive DNA molecules might be more problematic, but the first results with DNA molecules are strongly encouraging.

Acknowledgments

We thank Beatrice ten Heggeler-Bordier for providing a sample of DNA with bound RNA polymerases, and Corinne Cottier for photographic service. This worth was supported in part by Swiss National Fund Grants 31-25694.88 and 31-27146.89 (to J. Dubochet and A. Stasiak).

[27] Solution Behavior of DNA Studied with Magnetically Induced Birefringence

By JIM TORBET

Introduction

In this chapter we describe how magnetically induced birefringence, widely referred to as the Cotton–Mouton effect, can be used to study the solution properties of nucleic acids. Results on DNA when linear, superhelical, crossed in four-way junctions, and compacted within spherical viruses will be discussed. The only known effect of the magnetic field on DNA is to induce molecular orientation. There is no evidence that the physical properties are modified in any other way, nor does the field cause chemical damage. Other orienting methods such as electric fields or shear are not so benign.

Biological samples have been found to manifest various forms of magnetism (ferromagnetism, paramagnetism, superparamagnetism), and the weakest and most common form, diamagnetism, is exhibited by nucleic acids. The potential energy due to the diamagnetic anisotropy of an axially symmetric molecule or complex in a magnetic field H is given by $-0.5\Delta\chi H^2 \cos^2 \theta$, where θ is the angle between the field direction and the axis of symmetry and $\Delta\chi$ is the diamagnetic anisotropy. The molecule experiences an orienting torque of $0.5\Delta\chi H^2 \sin 2\theta$. The orientation of molecules in solution is disrupted by the randomizing effect of Brownian motion kT, so that the average degree of orientation depends on the ratio $\Delta\chi H^2/kT$. Nearly complete orientation requires that $\Delta\chi H^2 > 20kT$ which is only attainable when many molecules act in an ordered cooperative fashion as in membrane sheets, liquid crystals, and macromolecular fibrous protein structures. When dealing with non-interacting molecules, in general $\Delta\chi H^2 < kT$ and orientation is poor even in the strongest available fields. Nevertheless this weak effect can be monitored by measuring the induced birefringence which depends on the molecular diamagnetic and optical anisotropies that are in turn a function of the molecular structure and flexibility. Hence insights into the latter molecular properties can be deduced from the magnetically induced birefringence.

Theory

General

Diamagnetism occurs because the movement of electrons about nuclei changes slightly when a magnetic field is applied, so inducing a very small magnetic field in opposition to the applied field. Because the electrons involved in chemical bonds have anisotropic displacements, they give rise to anisotropic diamagnetic properties. This anisotropic mobility of bonding electrons is also the source of the intrinsic component of the optical anisotropy (see below).

For a noninteracting monodisperse solution of rigid, axially symmetric molecules the following are defined: c, concentration; N_A, Avogadro's number; M_r, molecular weight; H, magnetic field strength; k, Boltzmann constant; T, absolute temperature; Δn, magnetic birefringence (i.e., the difference in the refractive indices of light linearly polarized parallel and perpendicular to the applied field direction); n, refractive index of the solution, which is 1.33 for dilute aqueous solutions; λ, the wavelength of light traveling at right angles to the direction of the magnetic field used to measure Δn. $\Delta\chi$ and $\Delta\alpha$ are, respectively, the molecular diamagnetic and optical anisotropies; Φ is the orientation factor ($1 \geq \Phi \geq -0.5$).

When $\Delta\chi$ is positive the molecules orient with their symmetry axis parallel to the field; at complete alignment $\Phi = 1$ and $\Delta n = \Delta n_{sat}$. Whereas when $\Delta\chi$ is negative, the symmetry axis rotates into the perpendicular plane, and Φ can reach a limiting value of -0.5, with $\Delta n = -0.5\Delta n_{sat}$. At intermediate orientations the birefringence is given by $\Delta n = \Phi\Delta n_{sat}$. In this chapter Δn_{sat} and $\Delta\alpha$ are related by $\Delta n_{sat} = 2\pi c N_A \Delta\alpha/nM_r$. In the regime of low orientation $\Phi = \Delta\chi H^2/15kT$ ($|\Phi| < 0.05$; $\Delta\chi H^2 < 0.6kT$).[1] The Cotton–Mouton constant, defined by $CM = \Delta n/\lambda H^2$, is given by

$$CM = \frac{2\pi c N_A \Delta\alpha\Delta\chi}{15nM_r\lambda kT} \tag{1}$$

Provided the concentration is accurately known, a measure of the Cotton–Mouton constant gives the value of the product $\Delta\alpha\Delta\chi$. The problem centers on relating this value to molecular properties.

There is no evidence to suggest that the applied magnetic field causes a significant change in the optical polarizability.[2] The major problem in dealing with birefringence measurements is that $\Delta\alpha$ has both an intrinsic and a form component.[3] The dependence of $\Delta\chi$ and the intrinsic component of $\Delta\alpha$ on DNA superstructure can be readily calculated to a good approximation by taking the geometric sum, assuming the anisotropic mobility of electrons in the DNA molecule originates principally from the base pairs.[4,5] In contrast the calculation of the form contribution to $\Delta\alpha$ remains a formidable theoretical problem.[6] Form birefringence arises when there is a difference in the optical frequency dielectric constant (or refractive index) between an anisotropically shaped molecule and its environment. Thus, when the molecule and its surroundings have similar refractive indices, form birefringence is small. This condition is approached in dehydrated fibers as the refractive index difference is minimized between individual molecules and their surroundings made up equally of like molecules and solvent. In dilute aqueous solution, however, the refractive indices of solvent and solute differ, and the form contribution might be significant. The interpretation of birefringence measurements in terms of structure is haunted by this potential dual origin of optical anisotropy.

The best experimental approach is to avoid having to deal with $\Delta\alpha$ directly. This can be done by measuring Δn_{sat} under conditions (solvent and concentration) similar to those used to make the CM measurements. Then by combining the values from both experiments $\Delta\chi$, which has a

[1] J. Torbet, *EMBO J.* **2**, 63 (1983).

[2] A. D. Buckingham and J. A. Pople, *Proc. Phys. Soc. London, Sect. B* **69**, 1133 (1956).

[3] M. Born and E. Wolf, "Principles of Optics." Pergamon, Oxford, 1980.

[4] M. F. Maestre and R. Kilkson, *Biopolymers* **5**, 275 (1965).

[5] R. L. Rill, *Biopolymers* **11**, 2929 (1972).

[6] R. Oldenbourg and T. Ruiz, *Biophys. J.* **56**, 195 (1989).

negligible form component, is obtained. However, this is rarely achieved and in some cases probably impossible, as with spherical viruses.

The Δn_{sat} for linear DNA in dilute aqueous solution has been obtained from electric field induced orientation.[7] The value obtained from DNA aligned in poorly hydrated (~40% solvent) fibers is approximately twice as negative.[6,8] This difference has been attributed to (although it is not yet been proved) positive form birefringence in the solution which is much smaller in fibers. One of the weaknesses in this analysis is that in fibers the aqueous space between aligned, juxtaposed, and highly charged molecules could also be optically anisotropic and so contribute to the effective $\Delta\alpha$. The difference in $\Delta\alpha$ determined in solution and in fibers suggests that the form birefringence is positive in solution and is equal to about one-half the modulus of the intrinsic birefringence.

DNA

In the analysis below the optical anisotropy of linear duplex DNA normalized to that of a base pair is used throughout. For the treatment to be wholly justified, the form component must either be relatively small or vary with conformation in a way similar to that of the intrinsic component. There is some limited support for this approach. First, the Cotton–Mouton constants of tRNA and a Holliday junction give values for the arm angles that are rather close to those found by other techniques. A second, more indirect argument is that as DNA form birefringence is always positive, if it were significant, conformations should exist which have a positive $\Delta\alpha$ while retaining a negative $\Delta\chi$, hence giving rise to a negative *CM*. However, a negative *CM* value from nucleic acids (when there is no potential protein contribution) has never been reported.

To consider DNA in detail the following parameters are introduced: N, the number of base pairs; m, M_r/N, average molecular weight of a base pair; $\Delta\chi_{bp}$ and $\Delta\alpha_{bp}$, the average diamagnetic and optical anisotropies per base pair, respectively; θ, the angle between the perpendicular to the base planes and the duplex long axis; β, the pitch angle of a regular superhelix. Using partly hydrated sheets,[8] $\Delta\chi_{bp}$ has been measured to be about $-17 \pm 3 \times 10^{-29}$ erg G^{-2}, which is 75% larger than the theoretical estimate[9] used previously.[1,10,11] The value of $\Delta\alpha_{bp}$ in solution[7,8] is about -12.5×10^{-24} cm^3. The uncertainty in the product $\Delta\chi_{bp}\Delta\alpha_{bp}$ is about $\pm 30\%$.[8] This leads

[7] N. C. Stellwagen, *Biopolymers* 20, 399 (1981).
[8] G. Maret and G. Weill, *Biopolymers* 22, 2727 (1983).
[9] A. Veillard, B. Pullman, and G. Berthier, C. R. *Hebd. Seances Acad. Sci.* 252, 2321 (1961).
[10] J. Torbet, P. A. Timmins, and Y. Lvov, *Virology* 155, 721 (1986).
[11] J. Torbet and G. Maret, *Biopolymers* 20, 2657 (1981).

to uncertainty in any derived quantities but has no consequence when relative values are compared. The Lorentz–Lorenz relation is not used to correct for the internal optical electric field, as it was not invoked in the determination of $\Delta\alpha_{bp}$.

Rigid DNA

For straight, rigid DNA the molecular anisotropies are given by $\Delta\alpha = N\Delta\alpha_{bp}$ and $\Delta\chi = N\Delta\chi_{bp}$ if the DNA is B form with θ approximately equal to $0°$; otherwise both anisotropies must be multiplied by a factor $(1.5\cos^2\theta - 0.5)$ to account for base tilt.[4] Substituting into Eq. (1) gives the expression for the Cotton–Mouton constant for a rigid DNA fragment:

$$CM_r = \frac{2\pi c N_A N\Delta\alpha_{bp}\Delta\chi_{bp}}{15nm\lambda kt} \tag{2}$$

Putting the values for the known quantities into this equation, $CM_r = 7.1 \times 10^{-11} N/T$. If the conformation differs from a simple straight rod the Cotton–Mouton constant is reduced, $CM = CM_r F$, where F describes the effect of the conformational change on the product $\Delta\alpha\Delta\chi$.

Virus Compacted DNA

As the coat protein of a icosahedral virus has spherical symmetry it cannot give rise to optical or magnetic anisotropy. Therefore, if magnetically induced birefringence is observed it can only be due to an anisotropic interior.[1] It is difficult to elaborate on the preceding conclusion when the core also contains protein, as it can contribute to both $\Delta\chi$ and $\Delta\alpha$ in an unpredictable way. In the absence of core protein it is possible, however, to go further: here, anisotropy can only be due to nucleic acid.[1,10] In this case $F = f^2$, where f is the average orientation factor for the bases, and the value of F gives an indication of the deviation of the packaging from spherical symmetry. Modeling the DNA as a helical spool with a fixed pitch angle β, $f = 1.5\sin^2\beta - 0.5$; $f = 0$ for isotropic packaging or if β is approximately $35°$, $f = 1$ if the base planes are all parallel, and $f = -0.5$ if the spool is a flat circle with the base planes perpendicular to the symmetry axis.[4,5] The bigger the value for f, the greater is the constraint put on the number of possible models.

Holliday Junction

For a bent, L-shaped rod such as tRNA,

$$F = 1 + 3(\gamma^2 - \gamma)\sin^2\phi \tag{3}$$

where ϕ is the angle between both arms and γ is the fractional length of the shortest arm.[12] When both arms are equal $\gamma = 0.5$ and $F = 1 - 0.75 \sin^2 \phi$, $1 \geq F \geq 0.25$. Equation (3) also applies to four-way crossed junctions provided each pair of arms is colinear and short enough to be nearly rigid. The reliability of the value of F is weakened by the uncertainty in $\Delta\alpha_{bp}\Delta\chi_{bp}$. The need to know the latter product can be circumvented by making comparative measurements, provided there is no contribution to the form birefringence from the overall shape. For example, using the Cotton–Mouton constant from a rod of the same length as one of the arms for normalization, then the relation $CM_H/CM_r = 2F$ can be used to determine ϕ accurately.

Flexible Linear

In solution large DNA molecules do not behave as rigid rods but as flexible chains. To deal with flexibility the following are introduced: p, persistence length; l, axial distance between base pairs in the duplex chain (~ 3.4 Å); L, contour length. For a flexible, wormlike chain where $L \gg p$, the Cotton–Mouton constant is independent of N:

$$CM_f = \frac{4\pi c N_A p \Delta\alpha_{bp}\Delta\chi_{bp}}{45 n l m \lambda k T} \tag{4}$$

Again, putting in the values for the known quantities, $CM_f/c = 1.39 \times 10^{-3}p/T$.

Although the derivation[8] of Eq. (4) was for linear, wormlike chain molecules, it also applies to large flexible rings such as relaxed plasmids ($F = 1$). However, it is probable that Eq. (4) will cease to be valid at larger values of L. Additionally, when L is small, the ring becomes a near rigid circle, the molecule rotates as a whole, and $F = 0.25$ (i.e., $\beta = 0$; $CM = 0.25CM_r$). However, for intermediate values of L, F is unknown ($1 < F < 0.25$), and the molecule is neither rigid nor wormlike. Consequently the use of Eq. (4) for intermediate values of L will probably result in underdetermination of p.

Superhelical DNA

Superhelical molecules are also flexible and are idealized as interwound consisting of two duplex helices winding about a common axis with a uniform pitch angle β. The superstructure, as in the spool model for virus compacted DNA above, changes the optical and magnetic anisotropies by

[12] P. Mellado and J. G. de la Torre, *Biopolymers* **21**, 1857 (1982).

a factor $1.5 \sin^2 \beta - 0.5$.[4,5] The axial base pair distance projected along the superhelical axis is $l \sin \beta$, and there are two superhelices. Now the Cotton–Mouton constant of a superhelical molecule, CM_{fs}, depends on both the pitch angle and the persistence length of the superhelix, $CM_{fs} = CM_f F$ where $F = 2(1.5 \sin^2 \beta - 10.5)^2/l \sin \beta$. Small changes in β can result in large variation in the function $(1.5 \sin^2 \beta - 0.5)^2/\sin \beta$ depending on the range; for example, the value doubles when β changes from $42°$ to only $45°$, whereas a similar change starting from $54°$ requires an increase to about $65°$. In contrast CM_{fs} is simply proportional to p, now the persistence length of the superhelix, which is likely to decrease in an undramatic way as the superhelical strain is reduced. Therefore, changes in CM_{fs} are probably primarily caused by variation in β.

Branching of the interwound superhelix probably has little consequence for the analysis because if they are short only a minor fraction of the DNA will be involved, whereas if they are large the condition $L \gg P$ will be approximately met. In any case, frozen hydrated samples, which capture best the solution conformation in electron micrographs, rarely show branching,[13] although they are common following negative staining.[14]

Experimental

Birefringence Measurements

The experimental setup shown in Fig. 1 is taken from Maret and Weill.[8] A horizontal beam of helium–neon laser light (λ 6328 Å) propagates perpendicular to the applied magnetic field. The polarizer and analyzer are crossed at $45°$ with respect to the field direction. A photoelastic modulator aligned with its optical axis parallel to the magnetic field direction introduces a sinusoidal variation in the phase difference, $\Delta\phi$, between the horizontally and vertically polarized light components at a frequency (ω) of 50 kHz, so $\Delta\phi = \Delta\varphi \sin \omega t$, where $\Delta\varphi \approx 0.2\pi$. A phase difference is also introduced by the passage of the beam through the walls of the sample cell, lenses, and Babinet–Soleil compensator, $\Delta\phi_e$, and by the field-induced orientation of the sample, $\Delta\phi_m$. The total phase difference is $\Delta\phi = \Delta\varphi \sin \omega t + \Delta\phi_o$ (where $\Delta\phi_o = \Delta\phi_m + \Delta\phi_e$). The light intensity on the photodiode is proportional to $\sin^2 \Delta\phi/2$ or to $\Delta\phi^2$ for small $\Delta\phi$ (i.e., to $\Delta\varphi^2 \sin^2 \omega t + \Delta\phi_o^2 + 2\Delta\phi_o\Delta\varphi \sin \omega t$).

[13] M. Adrian, B. T. Heggeler-Bordier, W. Wahli, A. Z. Stasiak, A. Stasiak, and J. Dubochet, *EMBO J.* **9**, 4551 (1990).

[14] T. C. Boles, J. H. White, and N. R. Cozzarelli, *J. Mol. Biol.* **213**, 931 (1990).

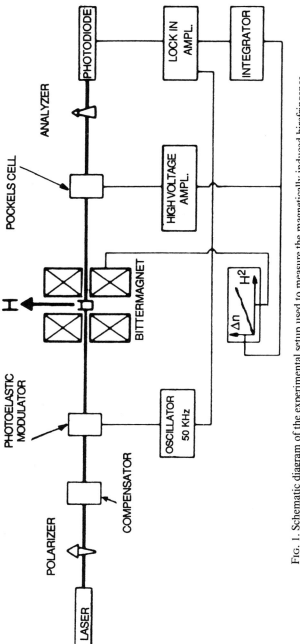

FIG. 1. Schematic diagram of the experimental setup used to measure the magnetically induced birefringence.

The part of the signal proportional to $\Delta\phi_o\Delta\varphi$ is detected by a phase-sensitive lock-in amplifier, and the dc output is used as an error signal in a feedback loop involving a polarity-converting integrator, a high-voltage amplifier, and a longitudinal Pockels cell. The birefringence of the Pockels cell varies linearly with the voltage applied across it and is used to continuously compensate for $\Delta\phi_o$. Thus if $\Delta\phi_e$ is relatively small or constant, the birefringence due to orientation during the field sweep can be measured in magnitude and sign, $\Delta n = \lambda\Delta\phi_m/2\pi l_c$ (l_c is the optical path length). Calibration is performed by measuring the voltage induced across the Pockels cell at a number of settings of the calibrated Babinet–Soleil compensator. The resolution of the apparatus in Δn is at best 10^{-10} or in terms of the *CM* about 10^{-16} for a 3 cm optical path length cell.

Magnet

The vertical magnetic field is generated by a water-cooled Bitter-type solenoid and has a maximum value of 130 kG (kilogauss). There is a vertical atmospheric bore of 5 cm diameter which contains the temperature stabilized ($\pm 0.1°$) sample chamber and a horizontal radial bore through which the light beam passes. The optical path length can be anything up to 3 cm. A more powerful magnet which will provide a maximum field of approximately 200 kG is currently under construction in the M.P.I. High Magnetic Field Laboratory (Grenoble, France). As the orientation depends on H^2, the maximum signal for weakly orienting samples will be increased by about 2.5 times, accuracy will be improved, and smaller sample volumes or concentrations will become measurable. The time to sweep the field up to the maximum value and down again is about 1 min. Each sample is measured several times, and the whole procedure takes about 15 min.

The maximum field that can be produced by permanent magnets is only 15 kG provided the gap between the poles is small. Conventional electromagnets go up as far as only 20 kG. Superconducting magnets can produce much stronger fields, but their sweep time is slow, which makes them unsuitable for measuring small Cotton–Mouton constants because of the difficulty in controlling the drift in birefringence from spurious sources $\Delta\phi_e$.

Results

Figure 2 shows two typical magnetically induced birefringence plots. The linear behavior indicates that the orientation is weak and the measurements are within the range of applicability of the Cotton–Mouton constant (see Table I).

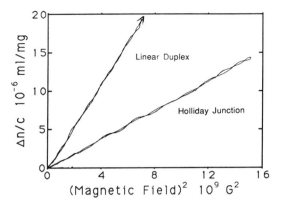

FIG. 2. Examples of continuously measured magnetically induced birefringence normalized with respect to total sample concentration obtained from solutions of linear duplex DNA (~6000 bp) and a Holliday junction (130 bp).

TABLE I
COTTON–MOUTON CONSTANTS[a]

Sample	M_r (×10⁶)	NA (%)	F (×10³)	±f	CM/c × 10¹² (G⁻² cm² g⁻¹)	CM/c_{NA} × 10¹² (G⁻² cm² g⁻¹)
TYMV	5.53	34	4.5	0.07	1.1	3.2
BMV	4.7	22	8.1	0.09	0.83	3.72
TBSV	8.7	17	7.9	0.09	0.77	4.47
TCV	8.8	17	8.9	0.095	0.85	5.0
CaMV	20	25	206.1	0.45	9.4	380
S_D	110	45	96.1	0.31	80	1780
Adenovirus type 2	175	14	—	—	12.5	87.5
Influenza Texas A	200	2.5	—	—	−0.47	−18.8
Yeast tRNA^Asp	0.024	—	104.5	0.48	1.9	1.9
Holliday junction	0.086	—	317	0.56	10	10
Linear duplex	—	—	—	—	29	29

[a] TYMV, Turnip yellow mosaic virus; BMV, bromegrass mosaic virus; TBSV, tomato bushy stunt virus; TCV, turnip crinkle virus; CaMV, cauliflower mosaic virus; S_D, an isometric bacteriophage. The first four samples listed contain single-stranded RNA, and the next two contain double-stranded DNA. A range of concentrations were measured in all cases, and only with the superhelical plasmid was the signal observed to be concentration dependent. M_r, Total molecular weight; NA, percent nucleic acid in virus; CM/c, Cotton–Mouton constant divided by total concentration; CM/c_{NA}, Cotton–Mouton constant divided by nucleic acid concentration. F and f are defined in the text.

Linear Duplex

One of the first applications of the Cotton–Mouton effect in biology was in the study of the flexibility of linear duplex DNAs of high molecular weight.[15] Later in a carefully controlled series of experiments the effect of ionic strength on erythrocyte DNA of an approximate average size of 6000 base pairs (bp) was measured as a function of ionic strength down to concentrations as low as 80 μg/ml.[8] Above about 10 mM NaCl added salt there is little ionic strength dependence of the value of CM/c and therefore of the persistence length. In this range $CM/c = 2.9 \times 10^{-11}$ G^{-2} g^{-1} cm^2, giving a persistence length of 670 Å. This is consistant with values reported using other techniques, although it is close to the upper limit found.[16] It has been suggested that this might be because the value of the product $\Delta\chi_{bp}\Delta\alpha_{bp}$ is too small.[8]

Below about 10 mM added Na$^+$, CM/c rises rapidly (the effect is smaller when Mg^{2+} counterion is used). Interaction between different molecules is shown to be small as there is little change in CM/c over a 7-fold range in concentration. In common with others, Maret and Weill[8] conclude that the main cause of the effect is the increase in persistence length due to intramolecular electrostatic repulsion, but they add that excluded volume effects might be more pronounced than hitherto thought. They compare their data with a number of theoretical predictions and find good fits when counterion condensation effects are included in the theory.

The liquid crystal properties of nucleic acids are of interest as a model for polyelectrolyte behavior and because the high concentrations required are to be found with DNA packaged *in vivo* into virus capsids and sperm heads. Magnetic birefringence provides a convenient way of studying the transition from the isotropic to the anisotropic liquid crystalline phase[11,17] because the signal is very sensitive to molecular cooperativity. These studies also demonstrated that the liquid crystalline phase can be highly magnetically oriented and so rendered more amenable to study with other physical techniques such as light diffraction, microscopy, and NMR.[17,18] Liquid crystalline nicked circular plasmids can be similarly oriented but apparently not their superhelical counterparts.[19]

[15] G. Maret, M. von Schickfus, A. Mayer, and K. Dransfeld, *Phys. Rev. Lett.* **35,** 397 (1975).
[16] P. J. Hagerman, *Annu. Rev. Biophys. Chem.* **17,** 265 (1988).
[17] E. Senechal, G. Maret, and K. Dransfeld, *Int. J. Biol. Macromol.* **2,** 256 (1980).
[18] R. Brandes and D. R. Kearns, *Biochemistry* **25,** 5890 (1986).
[19] J. Torbet and E. DiCapua, *EMBO J.* **8,** 4351 (1989).

Superhelical DNA

Most of the available evidence favors an interwound conformation for naked superhelical DNA in most circumstances down to superhelical densities at least as low as -0.019.[14] Cellular DNA has a range of superhelical densities, so it is desirable to know to what extent structural and physical properties are influenced by the degree of superhelical strain. We have recently used the variation in the Cotton–Mouton constant to study the solution behavior of superhelical and nicked circular plasmids.[20] A large nicked circular plasmid (5434 bp) was found to behave very like the linear molecules[8] discussed above when subjected to changes in ionic strength, namely, showing high CM values at low ionic strength which drop rapidly to reach a constant value as the ionic strength is increased. In dramatic contrast the CM of the same plasmid with natural superhelical density ($\sigma \sim -0.05$) actually grows nearly linearly with increasing ionic strength. As it would be surprising indeed if there were to be a rise in the persistence length as electrostatic shielding becomes more effective due to salt addition, we concluded that the pitch angle β is the dominant variable. Thus, in addition to making σ more negative,[21,22] salt has the effect of increasing the superhelical pitch angle. Furthermore it was also found that the variation in the Cotton–Mouton constant caused by added salt is more pronounced at high superhelical densities than at low.[20]

The sensitivity of the solution behavior of superhelical DNA to changes in superhelical density was explored by monitoring the CM during ethidium bromide titration. The curve (Fig. 3) resembles that observed with sedimentation, buoyant density, and viscometry.[23-25] Although the signal is in principle a function of both the pitch angle and the persistence length, the variation of the former is the main cause of the changes observed (see above and the section on theory). Therefore, the decrease in CM as the superhelical density drops from -0.05 to -0.032, followed by an abrupt rise as σ decreases further to -0.018, reflects a decrease followed by a rise in the superhelical pitch angle. Transitions in tertiary structure have been observed by Song *et al.*[26] at similar superhelical densities. From $\sigma = 0.018$ the CM decreases to a minimum at relaxation ($\sigma \approx 0$), and the structure

[20] U. Giesen, E. DiCapua, and J. Torbet, in preparation.

[21] J. C. Wang, *J. Mol. Biol.* **43**, 25 (1969).

[22] P. Anderson and W. Bauer, *Biochemistry* **17**, 594 (1987).

[23] M. Waring, *J. Mol. Biol.* **54**, 247 (1970).

[24] H. B. Gray, Jr., W. B. Upholt, and J. Vinograd, *J. Mol. Biol.* **62**, 1 (1971).

[25] B. M. J. Revet, M. S. Schmir, and J. Vinograd, *Nature (London) New Biol.* **229**, 10 (1971).

[26] L. Song, B. S. Fujimoto, P.-G. Wu, J. C. Thomas, J. H. Shibata, and J. M. Schurr, *J. Mol. Biol.* **214**, 307 (1990).

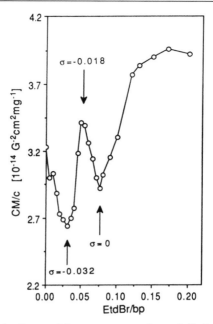

FIG. 3. Variation in the Cotton–Mouton constant of superhelical pUC8 (2717 bp) owing to titration with ethidium bromide (EtdBr). The DNA concentration at the outset was 3 mg/ml, and titration resulted in only minor dilution. Buffer: 100 mM NaCl, 0.5 mM EDTA, 10 mM Tris-HCl, pH 7.5.

makes the passage from interwound to relaxed flexible loop. As further ethidium bromide is added the superhelical density becomes positive and the *CM* rises to its maximum value. There is no symmetry about the relaxation condition, and the simple behavior on the positive superhelical density side contrasts with the successive transitions in tertiary structure on the negative density side. From recent electron micrographs of negatively stained samples it has been concluded that the pitch angle is rather insensitive to superhelical density.[14] Although this is at odds with our solution results, the change in β required to give rise to the large fluctuations in CM (Fig. 3) might be relatively small (see Theory).

Virus Compacted DNA

Filamentous viruses such as TMV[27] and bacteriophages Pf1 and fd[11,28] can be fully aligned when in the liquid crystalline phase even in the

[27] R. Oldenbourg, X. Wen, R. B. Meyer, and D. L. D. Caspar, *Phys. Rev. Lett.* **62**, 1851 (1988).
[28] J. Torbet and G. Maret, *J. Mol. Biol.* **134**, 843 (1979).

relatively weak fields produced by permanent magnets. This knowledge has been indispensible to the study of the high resolution structure of the coat protein of the two above bacteriophages by means of NMR[29] and X-ray fiber diffraction[28,30] (Work on bacterial flagella[31] provides a more recent example of how magnetic orientation has been used to advance X-ray structural studies.) Although a great deal is known about the structure of protein capsids of a number of spherical viruses from crystallography, it has proved difficult to delve into the inner organization. All spherical viruses studied so far give a measurable Cotton–Mouton constant.[1,10] (Table I), so it can be concluded that their cores are anisotropic. The signal from influenza is negative and is probably principally due to protein.[1] As adenoviruses and influenza viruses contain a significant amount of core protein, no further information can be extracted. However, as the Cotton–Mouton constant is very sensitive to structure, it might be valuable as a noninvasive method of monitoring internal structural changes in complex viruses.

The four small single-stranded RNA-containing icosahedral viruses (TYMV, BMV, TBSV, and TCV) are anisotropic but only to a small extent ($f \approx \pm 0.09$; Table I). In contrast, the two double-stranded DNA-containing viruses CaMV and S_D have a much larger degree of anisotropy which is consistent with the DNA being wound into a spoollike structure.[10] In cauliflower mosaic virus ($f = \pm 0.45$) the spool is rather flat ($\beta \approx 10°$). The isometric bacteriophage S_D is less anisotropic, and the pitch angle if the DNA is in the form of a regularly wound spool is larger ($\beta \approx 25°$).

The anisotropy is much greater in DNA-containing viruses than in the RNA viruses. Double-stranded DNA, although relatively stiff and therefore resistant to bending, does have a regular repeating structure which is amenable to ordered packing. Single-stranded RNA is inherently more flexible but has the potential to form a variety of double helical stems and loops of varying size which would be difficult to pack with systematic order inside the regularly repeating coat.

Holliday Junction

The general structure for the four-way junction or Holliday junction of genetic recombination has been proposed to be a stacked X structure. In the presence of metal ions, helices are believed to be stacked pairwise, generating two coaxial quasi-continuous helices crossing the junction core

[29] T. A. Cross and S. J. Opella, *J. Mol. Biol.* **182**, 367 (1987).
[30] D. A. Marvin, R. K. Bryan, and C. Nave, *J. Mol. Biol.* **217**, 293 (1991).
[31] I. Yamashita, F. Vonderviszt, T. Noguchi, and K. Namba, *J. Mol. Biol.* **193**, 315 (1987).

to form an X-shaped structure.[32] In the absence of appropriate metal ions the four arms of the junction remain fully extended in a square planar configuration. We (in collaboration with A. I. H. Murchie and D. M. J. Lilley) have performed exploratory experiments on one type of Holliday junction in the presence of 5 mM Mg^{2+} in order to assess if the Cotton–Mouton effect could contribute to the study of the structure. In this junction (type 2 as described by Murchie *et al.*[32]) each of the quasi-continuous helices has arms of unequal length (15 and 50 bp). The helices are much shorter than the normal duplex persistence length and can therefore be considered as rigid rods. From the CM we estimate $F = 0.32$ giving ϕ (acute) to be about 70° ($F = 1 - 0.75 \sin^2 \phi$), which is not very different from the value of ϕ found by fluorescent energy transfer, 60°.[32] This is an encouraging result which is backed up by a similar analysis of the data from tRNA,[1] taking into account the tilt ($\theta \approx 20°$) of the base planes, which gives a ϕ value of approximately 90°, which is again not far off the crystal value of about 100°. Clearly the best way to make use of the Cotton–Mouton effect would be to make comparative measurements on different junctions so to eliminate the error due to the uncertainty in $\Delta\chi_{bp}\Delta\alpha_{bp}$, provided of course the form component of $\Delta\alpha$ is not significantly influenced by the overall shape.

Summary

Magnetically induced birefringence can provide useful information on the physical properties of nucleic acids in a variety of forms. The technique is particularly sensitive to structural changes. Difficulties arise with drawing precise structural conclusions principally because of doubts about the relative contributions of the form and intrinsic components to the optical anisotropy. Nevertheless results so far suggest that this might not be a severe limitation. The induced birefringence is very sensitive to cooperative behavior; consequently the transition from an isotropic to a liquid crystalline phase can be followed in detail. Finally the high degree of magnetic orientation often achieved with liquid crystals assists their study with other physical techniques.

Acknowledgments

I thank the Max-Planck-Institute High Magnetic Field Laboratory, Grenoble, France, for making the magnetic birefringence measurements possible and the British Heart Foundation for support. I am indebted to G. Maret for his invaluable help and thank him and U. Giesen for commenting on the manuscript.

[32] A. I. H. Murchie, R. M. Clegg, E. von Kitzing, D. R. Duckett, S. Diekmann, and D. M. J. Lilley, *Nature (London)* **342**, 763 (1989).

[28] Calorimetry: A Tool for DNA and Ligand–DNA Studies

By KENNETH J. BRESLAUER, ERNESTO FREIRE, and MARTIN STRAUME

Introduction

Calorimetry occupies a venerable place in the history of science. In fact, during the 1700s, 1800s, and 1900s, the nature and measurement of heat has captured the attention of some of the giants of science, including Fahrenheit, Black, Watt, Lavoisier, Carnot, Cavendish, Joules, Kelvin, Clausius, Dalton, Hess, Bunsen, Thomsen, Berthelot, Gibbs, Helmholtz, Curie, and Nernst. Their interest in calorimetry derived, in part, from the belief that the heat change accompanying a reaction could serve as a measure of the affinity of the reacting species for one another, as well as a measure of the spontaneity of the reaction. These beliefs motivated the extensive thermodynamic studies of Thomsen and Berthelot. However, once the contributions of Guldberg, Waage, Gibbs, and Helmholtz made it clear that the spontaneity of a process depended on a decrease in *free energy* rather than *enthalpy,* the field of thermochemistry suffered neglect. A rebirth of interest in thermochemistry occurred at the beginning of the twentieth century when it was shown that enthalpy and heat capacity data could be used to calculate free energies.

As calorimeters are instruments that permit the direct and quantitative measurement of heat, it should not be surprising to learn that many of the giants of science noted above were the first to build calorimeters, thereby establishing the field of calorimetry. For example, in 1760, Black built the first "phase-transition calorimeter" and used it to demonstrate how the heat delivered to melting ice induces a phase transition (from ice to liquid water) rather than a temperature rise, thereby establishing the distinction between temperature and heat. In 1903, Pierre Curie developed a twin Dewar vessel "microcalorimeter" and used it to measure the heat output accompanying the decay of radium bromide.

For the most part, the early calorimeters were used to measure heat changes accompanying simple, well-defined processes (e.g., ice melting and radioactive decay). However, it is of interest to note that some of the earliest calorimetric studies were conducted on more complex "biological" systems. This observation is of particular interest in light of the frequently lamented rift between the two cultures of biology and chemistry.[1] However, a brief review of the history of calorimetry reveals that this current rift cannot be blamed on the past. For example, in 1780, Lavoisier and Laplace built an "ice calorimeter" in which they placed a guinea pig. They

METHODS IN ENZYMOLOGY, VOL. 211

measured the amount of ice melted and CO_2 produced by the animal over a 10-hr period and compared this with the heat produced in 10 hr on combustion of the equivalent amount of carbon. The experiment was inspired by their notion that animal respiration could be equated to simple combustion, so that the heat produced on exhalation of a specific amount of CO_2 should equal the heat produced by oxidation of the corresponding amount of carbon. In some sense, the measurements of Lavoisier and Laplace heralded the beginning of the field of biocalorimetry. Subsequent calorimetric studies on biological systems (including whole organisms) were pursued throughout the early nineteenth and twentieth centuries by scientists such as Dubrunfaut, Tangl, Meyerhof, Voit, Atwater, Bayne-Jones, and others.[2-5] In the aggregate, these studies led to the conclusion that the first law of thermodynamics was valid for living organisms as well as for the inorganic world, thereby placing the field of biocalorimetry on firm ground.

Despite this impressive history, during the first half of the twentieth century the field of biocalorimetry was not widely embraced by the scientific community. However, during the 1960s, 1970s, and 1980s, biocalorimetry has experienced an explosive rebirth. This renewed interest can be attributed to four important factors. First, major advances in electronics have vastly improved the sensitivity of modern calorimetrc instrumentation. Second, improvements in synthetic methodologies and biochemical separations/purifications now make available samples of sufficient purity and quantity for calorimetric measurements. Third, the development of theoretical formalisms coupled with advancements in computer technology now makes it possible to extract the maximum amount of information inherent in the raw calorimetric data. Finally, commercialization has made highly sensitive microcalorimeters more generally available to the scientific community so that the field of biocalorimetry no longer is restricted to those laboratories capable of building their own instruments.

The specific purpose of this chapter is to provide the reader with an overview of how calorimetric techniques can be used to study nucleic acid systems, particularly DNA. Consequently, no attempt will be made to present a comprehensive survey of the substantial literature that exists in this area. For such information the reader is referred to books on calorimetry,[6-9] previously published reviews,[10-18] as well as the original

[1] A. Kornberg, *Biochemistry* **26**, 6888 (1987).
[2] M. Debrunfaut, *C. R. Hebd. Seances Acad. Sci.* **42**, 945 (1856).
[3] O. Meyerhof, "Chemical Dynamics in Life Phenomena," Monograph, p. 110. 1924.
[4] S. Bayne-Jones and H. S. Rhees, *J. Bacteriol.* **17**, 123 (1929).
[5] S. Bayne-Jones, *J. Bacteriol.* **17**, 105 (1929).
[6] W. Hemminger and G. Höhne, "Calorimetry." Verlag Chemie, Weinheim, Germany, 1984.
[7] M. N. Jones, ed., "Biochemical Thermodynamics." Elsevier, Amsterdam, 1979.

research articles referenced in these books and review chapters. Instead, this chapter is designed to emphasize the unique information that can be obtained by application of selected calorimetric techniques to the study of DNA. To this end, we have organized the chapter into sections in which the presentation of each calorimetric technique is divided into three parts. The first two parts of each section provide, in nontechnical terms, a brief description of a particular technique followed by an outline of the information content of the method. The third and final part of each section presents one or more examples of how the technique has been used to study DNA. These examples have been selected for illustrative purposes only, and they generally reflect the tastes and prejudices of the authors rather than an assessment of the importance of the work.

Differential Scanning Calorimetry

Method

A differential scanning calorimeter (DSC) is an instrument that allows one to measure continuously the heat capacity of a system as a function of temperature. The technical features of DSC instruments previously have been described in detail[15,17] (see also references cited in both of these excellent reviews). In general terms, a typical DSC instrument contains two cells which are suspended within adiabatic shields and connected via a multijunction thermopile. Figure 1 shows schematics of such a DSC instrument. The cells are filled through capillary tubes that are accessed via external filling ports. In a typical experiment, the reaction and the reference cell are each filled completely (~ 1 ml), and the temperature is

[8] H. D. Brown, ed., "Biochemical Microcalorimetry." Academic Press, New York and London, 1969.

[9] A. E. Breezer, ed., "Biological Microcalorimetry." Academic Press, New York, 1980.

[10] H.-J. Hinz, in "Biochemical Thermodynamics" (M. N. Jones, ed.), p. 116. Elsevier, Amsterdam, 1979.

[11] W. Pfeil, in "Thermodynamic Data for Biochemistry and Biotechnology" (H.-J. Hinz, ed.), p. 349. Springer-Verlag, Berlin, 1986.

[12] K. J. Breslauer, in "Thermodynamic Data for Biochemistry and Biotechnology" (H.-J. Hinz, ed.), p. 402. Springer-Verlag, Berlin, 1986.

[13] G. Rialdi and R. Biltonen, in "MPT International Review of Science, Physical Chemistry." Medical and Technical Publ., and Butterworth, London, 1975.

[14] J. M. Sturtevant, Annu. Rev. Biophys. Bioeng. 3, 35 (1974).

[15] J. M. Sturtevant, Annu. Rev. 38, 463 (1987).

[16] I. Wadso, in "MPT International Review of Science, Physical Chemistry Series One" (H. A. Skinner, ed.), Vol. 10. Butterworth, London, 1972.

[17] P. L. Privalov and S. A. Potekhin, this series, Vol. 131, p. 4.

[18] P. D. Ross, in "Procedures in Nucleic Acid Research" (G. L. Cantoni and D. R. Davies, eds.), Vol. 2. Harper & Row, New York, 1971.

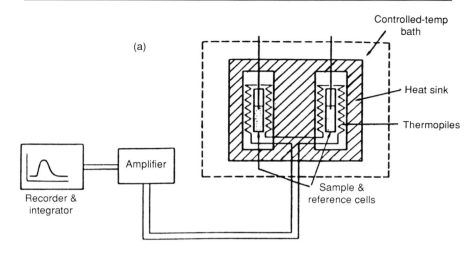

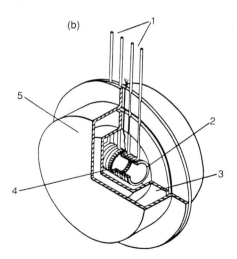

FIG. 1. (a) Schematic of the components of a typical differential scanning calorimeter. (b) Schematic representation of the cells and adiabatic shields of the DASM-4 differential scanning microcalorimeter. 1, Capillary inlets; 2, capillary cell; 3, inner shield; 4, thermopile between the cells; 5, outer shield. (By permission of Pergamon Press, Ltd.)

scanned from about 0 to 100° using electrical heaters which are in good thermal contact with the cells. When a thermally induced endothermic transition occurs in the sample cell, the temperature of this cell will lag behind that of the solvent reference cell, since some of the electrical energy is required to induce the transition rather than to increase the temperature of the solution. This temperature difference between the two cells is sensed by the thermopile and activates a circuit which adds additional electrical energy to the reaction cell so as to maintain it at the same temperature as the solvent reference cell. This compensating electrical energy, which is proportional to the energy associated with the thermally induced transition, is recorded. The instrument is calibrated by measuring the area produced by an electrical pulse of known energy.

These data, along with the known concentration of the solute, permit conversion of the observed electrical energy versus temperature curve to the corresponding C_p^{ex} versus temperature (T) curve, where C_p^{ex} is the excess heat capacity of the DNA solution relative to the reference buffer. A typical heat capacity curve measured for the thermally induced transition of an oligomeric DNA sample is shown in Fig. 2a. It is of interest to note

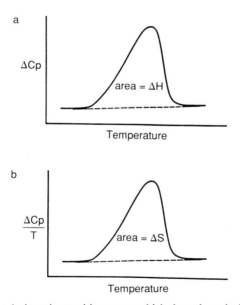

FIG. 2. (a) Typical calorimetric transition curve, which shows how the heat capacity, ΔC_p, changes with temperature. The area under the curve is equal to the transition enthalpy. (b) A $\Delta C_p/T$ versus temperature curve, which can be derived from the experimental calorimetric transition curve shown in (a). The area under the curve is equal to the transition entropy.

that, in most studies, the excess heat capacity at the peak of the transition is less than 1% of the total solution heat capacity. This feature emphasizes the need for DSC instruments of high sensitivity and precision.

Information Content

A single DSC profile provides a wealth of thermodynamic and extrathermodynamic information about a thermally induced DNA transition. Specifically, as described below, from a single heat capacity versus temperature curve, one can derive the following information: the transition free energy (ΔG^0), enthalpy (ΔH^0), entropy (ΔS^0), and heat capacity (ΔC_p); the "stateness" of the transition (two-state versus multistate); the average size of the molecule that melts as a single thermodynamic entity (e.g., the size of the cooperative unit); the cooperativity parameter, σ, as defined in the theoretical formalism of Zimm and Bragg[19] as well as Applequist[20]; and the partition function for the transition.[21-23] Knowledge of the partition function allows deconvolution or resolution of individual component transitions associated with complex DSC curves.[21-23] Such complex DSC curves can result from multiple two-state transitions, multiple cooperative transitions, sequential transitions, and transitions coupled to association–dissociation processes. By varying the DSC scan rate or a second independent variable (e.g., ligand concentration, pH, or ionic strength), one can obtain a family of heat capacity versus temperature curves. These data can be used to derive kinetic information (e.g., the mean relaxation times as a continuous function of temperature) as well as two-dimensional heat capacity surfaces.[23a-d] Such surfaces allow one to assess more strictly the validity of various DNA transition models.

It should be clear from the preceeding paragraphs that DSC measurements provide an extraordinary wealth of thermodynamic and extrathermodynamic information, much of which cannot be obtained from any other technique. In the paragraphs that follow, we describe how this information can be extracted from the experimental data.

From the area under the experimental C_p^{ex} versus T curve (see Fig. 2), one can calculate the transition enthalpy, ΔH^0, since $\int C_p dT = \Delta H^0$. Significantly, this calorimetrically determined transition enthalpy is model

[19] B. H. Zimm and J. K. Bragg, *J. Chem. Phys.* **31**, 526 (1959).
[20] J. Applequist, *J. Chem. Phys.* **38**, 934 (1963).
[21] E. Freire and R. L. Biltonen, *Biopolymers* **17**, 463 (1978).
[22] E. Freire and R. L. Biltonen, *Biopolymers* **17**, 481 (1978).
[23] E. Freire and R. L. Biltonen, *Biopolymers* **17**, 497 (1978).
[23a] E. Freire, *Comments Mol. Cell. Biophys.* **6**, 123 (1989).
[23b] G. Ramsay and E. Freire, *Biochemistry* **29**, 8677 (1990).
[23c] D. Xie, V. Bhakuni, and E. Freire, *Biochemistry* **30**, 10673 (1991).
[23d] M. Straume and E. Freire, *Anal. Biochem.,* in press (1992).

independent and therefore does not depend on the nature ("stateness") of the transition. This feature contrasts with model-dependent van't Hoff transition enthalpies which are derived indirectly from the temperature dependence of equilibrium properties.[15,17,24] Furthermore, from the difference between the initial and final baselines, the DSC measurement also provides a direct measure of the heat capacity change, ΔC_p, accompanying the transition. Consequently, one need not assume that ΔC_p is zero, as frequently is done when analyzing optical data. The experimental C_p^{ex} versus T curve also can be converted to a C_p^{ex}/T versus T curve.[24] Because $\Delta S = \int (C_p^{ex}/T) \, dT$, the area under such a curve provides a "direct measure" of the entropy change (see Fig. 2b). Thus, from a single DSC transition curve, one can obtain ΔC_p, ΔH^0, and ΔS^0. Using these data, the corresponding value of ΔG^0 can be calculated at any temperature (T) using the relationship $\Delta G^0 = \Delta H^0 - T\Delta S^0 = \Delta H_{T_m}[(T_m - T)/T_m] - \Delta C_p$ $(T_m - T) + \Delta C_p T \ln(T_m/T)$. This expression assumes that ΔC_p does not depend on temperature. Although this assumption usually is valid, it is not required to calculate ΔG^0 from the DSC data. It is used here only to simplify the expression.

The calorimetric approach just described for obtaining thermodynamic data on DNA transitions from DSC profiles has two significant advantages relative to van't Hoff methods.[15,17,24] First, one obtains a model-independent measure of the transition enthalpy and entropy rather than the model-dependent (usually "two-state") values obtained from van't Hoff analyses of equilibrium data. Second, the calorimetric experiment provides a direct measure of ΔC_p so that the temperature dependence of the stability can be assessed.

The DSC data also permit one to evaluate the cooperativity or stateness of a transition.[15,17,24] This represents a unique advantage of calorimetry. Specifically, comparison of the model-dependent $\Delta H_{v.H.}$ (which can be derived by analyzing the shape of the C_p^{ex} versus T curve) and the model-independent ΔH_{cal} allows one to conclude if the transition proceeds in an all-or-none fashion, thereby providing a test for the applicability of the two-state model to a given transition. If $\Delta H_{v.H.} = \Delta H_{cal}$, then the transition proceeds in a two-state manner. In such a case, meaningful thermodynamic data can be obtained by monitoring the temperature dependence of an equilibrium property and using the two-state van't Hoff relationship. However, if $\Delta H_{v.H.} < \Delta H_{cal}$, then the transition involves a significant population of intermediate states, thereby invalidating the use of the simple "all-or-none" analysis. On the other hand, if $\Delta H_{v.H.} > \Delta H_{cal}$, then intermolecular cooperation (e.g., aggregation) is implicated.

A quantitative comparison of the van't Hoff and calorimetric transition

[24] L. A. Marky and K. J. Breslauer, *Biopolymers* **26**, 1601 (1987).

enthalpies provides further insight into the nature of a transition. Specifically, the ratio $\Delta H_{\text{v.H.}}/\Delta H_{\text{cal}}$ provides a measure of the fraction of the structure that melts as a single thermodynamic entity, in other words, the size of the cooperative unit. This ability to define the size of the cooperative unit represents an important and unique advantage of the calorimetric measurement.

The DSC data also can be used to calculate σ, the cooperativity parameter relevant to the statistical treatments of Zimm and Bragg[19] as well as Applequist.[20] Specifically, the maximum value of $C_{\text{p,max}}^{\text{ex}}$ is proportional to $\sigma^{-1/2}$. The exact relationship[25] is given by

$$C_{\text{p,max}}^{\text{ex}} = (\Delta H)^2/4RT^2\sigma^{-1/2} \tag{1}$$

An alternative and somewhat simpler formulation is expressed in the relationship[14]

$$\sigma^{-1/2} = m_{\text{r}} \frac{\Delta h_{\text{cal}}}{\Delta H_{\text{v.H.}}} \tag{2}$$

where m_{r} is the residue weight of the repeating unit in the DNA polymer and Δh_{cal} is the calorimetric transition enthalpy per gram. Application of either Eq. (1) or Eq. (2) permits one to calculate a value for σ from the calorimetric data. This ability represents yet another unique feature of the information inherent in calorimetric data.

The development of sophisticated analytical procedures has further expanded the information that can be extracted from calorimetric data. In 1978, Freire and Biltonen,[21–23] showed that the partition function, Q, for a transition can be numerically calculated from the excess heat capacity function by a double integration procedure:

$$\langle \Delta H(T) \rangle = \int_{T_0}^{T} \langle \Delta C_{\text{p}} \rangle \, dT \tag{3}$$

$$\ln Q(T) = \int_{T_0}^{T} \frac{\langle \Delta H \rangle}{RT^2} \, dT \tag{4}$$

where $\langle \Delta H \rangle$ is the excess enthalpy function. It is well known from statistical thermodynamics that the partition function of a system contains all the information for a complete thermodynamic characterization of a system. This feature has been used to develop deconvolution procedures for evaluating transition mechanisms and for resolving individual components in complex DSC melting profiles. The resulting deconvolution procedure has been used successfully in the analysis of complex DSC profiles for multi-

[25] T. Ackermann, in "Biochemical Microcalorimetry" (H. D. Brown, ed.), p. 121 and 235. Academic Press, New York and London, 1969.

domain proteins, nucleic acids, and membrane proteins.[21-23,26-32a] The original procedure has been improved by the incorporation of multiple passes in the recursive algorithm, each one followed by nonlinear least-squares optimization (see Ref. 31 for a description of the procedure). The incorporation of the nonlinear least-squares routine has allowed the development of programs for the analysis and evaluation of different transition mechanisms such as multiple two-state transitions, multiple cooperative transitions, sequential transitions, and transitions coupled to association–dissociation processes.

Recently, Mayorga and Freire[33] have shown that, by performing DSC measurements at different scan rates, it is possible to estimate mean relaxation times as a continuous function of temperature. The resolution of this method is limited only by the available range of scanning rates and the instrument response time. With the present generation of DSC instruments, relaxation times as fast as 10 sec can be resolved.

Another important focus in the development of new procedures for analysis of DSC data is concerned with transition mechanisms of the form $A \rightleftharpoons B \rightarrow C$, in which the initial conformational equilibrium is followed by an *irreversible* step. Transition mechanisms of this type have been more frequently encountered in protein rather than DNA studies. Nevertheless, it should be noted that under favorable conditions one can extract meaningful thermodynamic information for the equilibrium unfolding step as well as kinetic information for the irreversible step.[34] For a critical discussion of methods for evaluating transition mechanisms of this type, see Ref. 34a.

Another important area of DSC data analysis focuses on the development of linked function analysis procedures. These procedures allow one to analyze two-dimensional heat capacity surfaces defined by multiple temperature scans obtained by systematically varying a second independent variable (e.g., ligand concentration, pH, or ionic strength). A global

[26] P. L. Privalov, P. L. Mateo, N. Khechinashvilli, V. M. Stepanov, and L. Revina, *J. Mol. Biol.* **152**, 445 (1981).
[27] P. L. Privalov, *Adv. Protein Chem.* **35**, 1 (1982).
[28] P. L. Privalov and L. V. Medved, *J. Mol. Biol.* **159**, 665 (1982).
[29] V. Tischenko, V. Zavyalov, G. Medgyesi, S. Potekhim, and P. L. Privalov, *Eur. J. Biochem.* **126**, 517 (1982).
[30] T. Tsalkova and P. L. Privalov, *J. Mol. Biol.* **181**, 533 (1985).
[31] C. Rigell, C. de Saussere, and E. Freire, *Biochemistry* **24**, 5638 (1985).
[32] C. Rigell and E. Freire, *Biochemistry* **26**, 4366 (1987).
[32a] E. Freire and R. L. Biltonen, *Biopolymers* **17**, 1257 (1978).
[33] O. L. Mayorga and E. Freire, *Biophys. Chem.* **87**, 87 (1987).
[34] V. Edge, N. M. Allewell, and J. M. Sturtevant, *Biochemistry* **24**, 5899 (1985).
[34a] E. Freire, W. W. van Osdol, O. L. Mayorga, and J. M. Sanchez-Ruiz, *Annu. Rev. Biophys. Biophys. Chem.* **19**, 159 (1990).

analysis of the data allows for discrimination between various possible transition models. A typical calorimetric scan with data points spaced at intervals of approximately $0.05°$ contains around 2000 data pairs. Adding a second dimension, for example, 20 different ligand concentrations, increases the total number of data points to 40,000, thereby permitting a far more demanding test of a given transition model.[23d]

A particularly interesting application in this area is in the study of ligand–DNA interactions. In the simple case where a ligand binds only to the initial duplex state, duplex melting is accompanied by ligand dissociation. Consequently, by comparing the DSC profiles of the ligand-free and ligand-bound duplexes one can determine the enthalpy change associated with ligand "debinding" as well as the influence of the ligand on the melting cooperativity of the duplex state.[35] Furthermore, under favorable conditions and in conjunction with the DSC measured transition enthalpy of the free duplex, it is possible to calculate ligand–DNA binding constants from the measured difference in transition temperatures (ΔT_m) of the ligand-free and ligand-bound duplexes.

The theoretical formalism required for such a calculation of binding constants from ΔT_m data has been described by Crothers.[36] For the simple case in which the ligand binds only to the initial duplex state the relevant equation is

$$1/T'_m - 1/T_m = R/n\Delta H \ln[1 + Ka_L] \qquad (5)$$

where T'_m and T_m are the Kelvin melting temperatures of the DNA duplex in the presence and absence of added ligand, respectively; n is the number of base pairs per ligand at saturation (e.g., $n = 1/r$, where r corresponds to the number of ligands per base pair); ΔH is the transition enthalpy per base pair for the melting of the DNA duplex in the absence of bound drug (a quantity that can be measured directly by DSC); K is the apparent ligand–duplex binding constant at $T = T_m$; a_L is the activity of the free ligand at $T = T_m$, which for strong binding constants ($K > 10^6$) can be approximated by one-half of the total concentration of added ligand (for smaller K values, a_L can be treated as a function of K); and R is the gas constant.

It should be emphasized that a number of implicit and explicit approximations are required for use of Eq. (5) as well as its more general form which describes the ΔT_m relationship for the case in which the ligand binds to both the initial duplex and the final single-stranded states (see Ref. 36). In particular, the ΔT_m equations have been derived assuming a model in which the ligand binds to separate, noninteracting sites on the DNA lattice. One must be cognizant that nonspecific binding as well as ligand aggrega-

[35] L. A. Marky, J. G. Snyder, D. P. Remeta, and K. J. Breslauer, *J. Biomol. Struct. Dyn.* **1**, 487 (1983).

[36] D. M. Crothers, *Biopolymers* **10**, 2147 (1971).

tion, both of which may depend on temperature and ionic strength, can introduce systematic errors in the ΔT_m analysis. Consequently, the appropriate use of ΔT_m equations requires a healthy respect for and understanding of the approximations employed in its theoretical derivation and its practical application. With an awareness of these considerations, Breslauer and co-workers have demonstrated the utility of the Crothers ΔT_m approach for determining ligand–DNA binding constants from DSC and optical melting data.[37-41]

Based on the descriptions provided above, it should be clear that DSC studies on DNA can provide an impressive amount of thermodynamic and extrathermodynamic information. To date, only a fraction of this information potential has been tapped in studies of DNA. However, with the increasing availability of high sensitivity DSC instruments coupled with improved methods for DNA synthesis, isolation, and purification, we expect this situation to change rapidly.

Applications

The laboratories of Sturtevant in the United States,[42-47] Ackermann in Germany,[48-53] and Privalov in the Soviet Union[54-56] were the first to recognize and to demonstrate the importance of DSC studies on nucleic

[37] L. A. Marky, J. Curry, and K. J. Breslauer, in "Molecular Basis of Cancer" (R. Rein, ed.), p. 155, Alan R. Liss, New York, 1985.
[38] J. Snyder, Ph.D. Thesis, Rutgers University, New Brunswick, New Jersey (1985).
[39] W. Y. Chou, L. A. Marky, D. Zaunczkowski, and K. J. Breslauer, J. Biomol. Struct. Dyn. 5, 345 (1987).
[40] D. P. Remeta, Ph.D. Thesis, Rutgers University, New Brunswick, New Jersey (1989).
[41] J. G. Snyder, O. Kennard, N. G. Hartman, B. L. D'Estantoit, D. P. Remeta, and K. J. Breslauer, Proc. Natl. Acad. Sci. U.S.A. 86, 3968 (1989).
[42] M. A. Rawitscher, P. D. Ross, and J. M. Sturtevant, J. Am. Chem. Soc. 85, 1915 (1963).
[43] I.E. Scheffler and J. M. Sturtevant, J. Mol. Biol. 42, 577 (1969).
[44] H. Krakauer and J. M. Sturtevant, Biopolymers 6, 491 (1968).
[45] J. M. Sturtevant and E. P. Geiduschek, J. Am. Chem. Soc. 80, 2911 (1969).
[46] L. G. Bunville, E. P. Geiduschek, M. H. Rawitscher, and J. M. Sturtevant, Biopolymers 3, 213 (1965).
[47] D. D. F. Shiao and J. M. Sturtevant, Biopolymers 12, 1829 (1973).
[48] T. Ackermann and H. Rueterjans, Ber. Bunsen-Ges. Phys. Chem. 68, 850 (1964).
[49] H. Rueterjans, Ph.D. Thesis, Westfälische Wilhelms Univresität, Muenster, Germany (1965).
[50] H. Klump and Th. Ackermann, Biopolymers 10, 513 (1971).
[51] H. Klump and W. Burkart, Biochim. Biophys. Acta 475, 601 (1977).
[52] E. Neumann and Th. Ackermann, J. Phys. Chem. 73, 2170 (1969).
[53] D. Bode, U. Schernau, and T. Ackermann, Biophys. Chem. 1, 214 (1973).
[54] P. L. Privalov, K. A. Kafiani, and D. P. Monaselidze, Biofizika 10, 393 (1965).
[55] P. L. Privalov, O. B. Ptitsyn, and T. M. Birshtein, Biopolymers 8, 559 (1967).
[56] P. L. Privalov, FEBS Lett. 40, 5140 (1974).

acid molecules (see Refs. 10–12, 25, and references cited therein). Their pioneering investigations laid the foundation for much of the subsequent work in this area. However, owing to the poor availability of substantial quantities of highly purified DNA polymers, as well as the near absence of techniques for synthesis of DNA oligomers, these early studies were limited to a small number of synthetic and natural polymers. Nevertheless, it is impressive to note that these original studies have withstood the test of time, despite the substantial technical improvements in DSC instrumentation.

More recently, Breslauer and co-workers have taken advantage of advances in synthetic methodologies to synthesize and to characterize thermodynamically the helix-to-coil transitions for a broad spectrum of specially designed oligomeric and polymeric DNA molecules.[37,39,41,57-61] (in particular, see Ref. 60) The data from these studies have provided a better understanding of the molecular forces that stabilize DNA structure and have established a thermodynamic library that is being used to predict from the primary sequence the most stable DNA structure under a given set of solution conditions, in other words, the establishment of a phase diagram for DNA. In addition, the resulting thermodynamic data have proved useful in a number of practical applications, such as (1) predicting the stability of probe–gene complexes; (2) selecting optimal conditions for hybridization experiments; (3) deciding on the minimum length of a probe required for hybridization; and (4) predicting the influence of a specific transversion or transition on the stability of an affected DNA region.

Differential scanning calorimetry also has been used to characterize thermodynamically the intramolecular B-to-Z conformational transition in DNA,[62-64] the thermally induced "unbending" of bent DNA,[65,65a] the melting of a DNA duplex containing an abasic site,[66] as well as the melting

[57] L. A. Marky, L. Canuel, R. A. Jones, and K. J. Breslauer, *Biophys. Chem.* **13**, 141 (1981).

[58] D. J. Patel, S. A. Kozlowski, L. A. Marky, C. Broka, J. A. Rice, K. Itakura, and K. J. Breslauer, *Biochemistry* **21**, 428 (1982).

[59] L. A. Marky and K. J. Breslauer, *Biopolymers* **21**, 2185 (1982).

[60] K. J. Breslauer, R. Frank, H. Blöcker, and L. A. Marky, *Proc. Natl. Acad. Sci. U.S.A.* **83**, 3746 (1986).

[61] M. M. Senior, R. A. Jones, and K. J. Breslauer, *Proc. Natl. Acad. Sci. U.S.A.* **85**, 6242 (1988).

[62] J. B. Chaires and J. M. Sturtevant, *Proc. Natl. Acad. Sci. U.S.A.* **83**, 5479 (1986).

[63] H. H. Klump and R. Löffler, *Hoppe-Seyler's Z. Physiol. Chem.* **366**, 345 (1985).

[64] H. H. Klump, *FEBS Lett.* **196**, 175 (1986).

[65] Y. W. Park and K. J. Breslauer, *Proc. Natl. Acad. Sci. U.S.A.* **88**, 1551 (1991).

[65a] K. J. Breslauer, *Curr. Biol.* **1**, 416 (1991).

[66] G. Vesnaver, C.-N. Chang, M. Eisenberg, A. P. Grollman, and K. J. Breslauer, *Proc. Natl. Acad. Sci. U.S.A.* **86**, 3614 (1989).

of single-stranded RNA and DNA sequences.[67-70] Such thermodynamic information should prove useful in developing a fundamental understanding of the forces that control DNA structures, as well as in evaluating the energetic role that DNA conformational switches may play in biochemical control mechanisms.

A recent DSC study on oligomeric DNA triplex structures illustrates how calorimetric data can be used to evaluate the influence of sequence, base modification, and solution conditions on triplex stability.[71] Such thermodynamic information is required for the rational design of third-strand oligonucleotides and for defining solution conditions that optimize the effective use of triplex structure formation as a tool for modulating biochemical events. Another recent calorimetric study of tetraplex formation reveals how calorimetric data can be used in conjunction with spectroscopic data to determine the molecularity of a multistrand DNA structure.[72] This application of DSC data may prove particularly useful in light of the recent excitement over and recognition of the potential importance that higher-order DNA structures may have in biological mechanisms.

As explained above, under favorable conditions, transition temperatures and enthalpies obtained from DSC measurements can be used in conjunction with binding density data to calculate ligand–DNA binding constants. Despite the considerable assumptions inherent in this approach,[36] Breslauer and co-workers have shown that the "ΔT_m method" can be used to calculate ligand–DNA binding constants that are in surprisingly good agreement with the values obtained from more traditional methods.[37-41] For example, the ΔT_m method yields a binding constant at 25° of $1.5 \times 10^7 \ M^{-1}$ for complexation of the anthracycline antibiotic daunomycin to the poly[d(AT)·poly[d(AT)] duplex.[40] This value is in very good agreement with the corresponding binding constant value of $1.7 \times 10^7 \ M^{-1}$ determined more conventionally by a Scatchard treatment of fluorescence quenching data.[40] Breslauer and co-workers have obtained similarly good agreement between K values derived from ΔT_m data and classic binding isotherm data for other DNA-binding ligands such as actinomycin D[38,41] and two steroid diamines.[38] More significantly, the ΔT_m method can be used to determine binding constants for strongly binding

[67] K. J. Breslauer and J. M. Sturtevant, *Biophys. Chem.* **7**, 205 (1977).

[68] J. Suurkuusk, J. Alvarez, E. Freire, and R. Biltonen, *Biopolymers* **16**, 2641 (1977).

[69] V. V. Filimonov and P. L. Privalov, *J. Mol. Biol.* **122**, 465 (1978).

[70] S. M. Freier, K. O. Hill, T. G. Dewey, L. A. Marky, K. J. Breslauer, and D. H. Turner, *Biochemistry* **20**, 1419 (1981).

[71] G. E. Plum, Y. W. Park, S. Singleton, P. B. Dervan, and K. J. Breslauer, *Pro. Natl. Acad. Sci. U.S.A.* **87**, 9436 (1990).

[72] R. Jin, K. J. Breslauer, R. A. Jones, and B. L. Gaffney, *Science* **250**, 543 (1990).

ligands for which traditional methods are precluded owing to the low concentration of the free ligand.[37]

Batch and Stopped-Flow Isothermal Mixing Calorimetry

Method

Until recently, isothermal calorimetric studies on nucleic acid systems primarily have employed batch type, heat conduction instruments. Such instruments basically consist of a bicompartment cell surrounded by thermoelectric elements embedded within a massive heat sink (see Ref. 73 and articles cited therein). The entire apparatus is kept in a temperature-controlled environment. A typical experiment involves filling each compartment of a bicompartment cell with aliquots of the reagents of interest. On mixing the two reagents, the liberated or absorbed heat is quantitatively conducted through the surrounding thermopiles to the heat sink. The output of the thermopiles, which measures the rate of heat transfer, is then amplified and recorded. Although quite useful, batch calorimeters suffer from three significant shortcomings: (1) long sample loading and equilibration times between runs ($\sim 1-2$ h), thereby limiting the throughput; (2) less than optimum effective sensitivities (~ 60 microjoules, μJ) owing to motion artifacts and mixing problems; and (3) substantial reagent volumes of at least 250 μl. Titration calorimetry overcomes some, but not all, of these limitations, while also allowing, under favorable circumstances, the determination of binding constants.

A recently developed stopped-flow microcalorimeter[74] surmounts all of the limitations of conventional batch calorimeters noted above, thereby making it particularly useful in biophysical studies which frequently require high sensitivity and small volumes. A typical stopped-flow microcalorimeter consists of a fluid delivery system, a mixing chamber, flow tubes, a temperature controller, and a massive heat sink. The reagent delivery system is comprised of four thermostatted Hamilton syringes; a computer-controlled stepping motor that delivers 80 μl of reagent from each syringe; four flow lines, each connected to a syringe via a three-way valve at its influent end, and welded to one of the two mixing chambers at its terminus; sample and reference mixing chambers embedded in the calorimetric block, each having a total volume of 160 μl; and two separate flow lines for transporting the reaction products from the mixing chambers

[73] C. P. Mudd, R. L. Berger, H. P. Hopkins, W. S. Friauf, and C. Gibson, *J. Biochem. Biophys. Methods* **6**, 179, (1982).

[74] C. P. Mudd and R. L. Berger, *J. Biochem. Biophys. Methods* **17**, 171 (1988).

to individual collection vials. The reaction is initiated by activating the syringe drive system which delivers 80 μl of each reagent within 0.6 sec into tantulum mixing chambers. Thermopiles surround each mixing chamber and detect the reaction heat that is either liberated or absorbed. As 80 μl of each reactant are used for a 1:1 mixing reaction, a syringe containing 2.5 ml is capable of yielding 30 reactions per filling. In addition, by varying the ratio of the delivered reagents, one can acquire data that correspond to discrete points on a continuous titration curve, as is obtained using isothermal titration calorimetry (see below). The entire instrument is thermostatted so that heat capacity changes (ΔC_p) can be determined by performing ΔH measurements at several temperatures. The technical details that have led these impressive improvements are described in Ref. 74.

Stopped-flow microcalorimeters have been specifically designed and constructed to surmount the limitations normally encountered in conventional batch microcalorimeters. To be specific, the stopped-flow instrument is nearly two orders of magnitude more sensitive than conventional batch microcalorimeters, exhibiting a characteristic sensitivity of 1.595 $J/V \cdot$ sec. The instrument has a rapid response time (i.e., 10 to 90% in 40 sec), negligible flow artifacts (i.e., less than 2 μJ), and a differential baseline stability of 100 nJ/sec over a 4-hr period. Relative to conventional batch calorimeters, the enhanced sensitivity of the stopped-flow instrument permits significant reductions in the total reaction volume (i.e., 160 versus 1000 μl), limiting reagent concentration (i.e., 20 versus 500 μM), and total run time (i.e., 200 versus 1800 sec). Furthermore, minimization of the reequilibration time between sample sets affords a throughput of 120 to 150 runs per day compared with the 3 to 4 runs typically completed in batch microcalorimetry.

In summary, batch, titration, and stopped-flow calorimeters are all forms of isothermal mixing calorimetry that differ formally only by the means by which one initiates the reaction under study. However, both the titration and stopped-flow modes exhibit significant advantages over the conventional batch instruments.

Information Content

Isothermal mixing calorimetry provides a direct and model-independent measure of the enthalpy change accompanying a chemical reaction. Figure 3 shows a typical "heat burst" curve that is produced on mixing of two reagents. The area under the curve reflects the output voltage of the thermopiles, which is proportional to the total reaction heat, Q_T. In the differential mode, Q_T is obtained by subtracting the measured heat response of the reference chamber from that of the sample chamber and

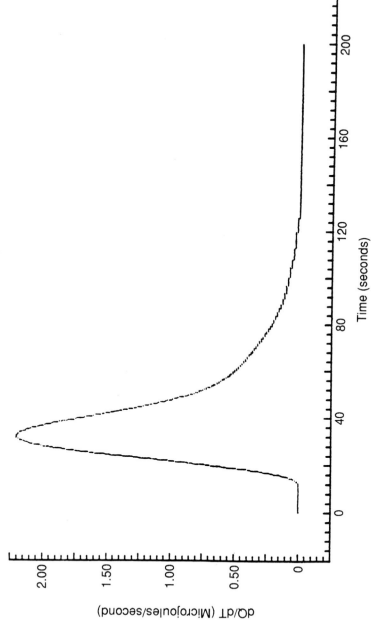

FIG. 3. Stopped-flow microcalorimetric thermogram delineating the exothermic heat evolved for the binding of daunomycin to poly[d(AC)]·poly[d(GT)]) at a ratio of 1 drug molecule per 20 phosphate residues (D:P = 1:20) at 25°.

integrating the area under the resultant voltage versus time curve according to the relationship

$$Q_T = k \times \text{Area} \qquad (6)$$

The proportionality constant, k, required to calculate Q_T is obtained by measuring the area produced by a reaction of known enthalpy. Typical calibration constants of 1.595 J/V·sec have been determined for this type of instrument. Knowledge of the reagent concentrations then permits conversion of the observed heat change, Q_T, to an enthalpy change, ΔH. Such directly measured calorimetric enthalpies are more reliable than indirectly derived, model-dependent van't Hoff values. By repeating the isothermal mixing measurement at several temperatures, one also can calculate ΔC_p, the heat capacity change, for the reaction. This much ne-glected thermodynamic parameter, which for all practical purposes only can be obtained calorimetrically, can provide important insights into the role of solvent in the reaction under study.

The enhanced sensitivity of the stopped-flow microcalorimeter makes it ideally suited not only for isothermal mixing experiments on systems composed of precious material, but also for systems in which one or more of the reagents exhibit concentration-dependent properties. As explained below in the Applications section, drug–DNA binding studies provide good examples of this behavior since many DNA-binding drugs (particu-larly intercalators) aggregate at the concentrations normally employed in conventional batch calorimetric studies. Under such circumstances, the observed binding heats have to be corrected for the binding-induced disag-gregation event. Frequently, this "correction" factor requires the use of van't Hoff enthalpies extracted from temperature-dependent optical stud-ies, thereby compromising the quality of the final "calorimetric" data. The enhanced sensitivity of the recently developed stopped-flow microcalo-rimeter precludes this problem.

In summary, independent of the mechanical mode by which one initi-ates a reaction (e.g., titration, batch, stopped-flow) isothermal mixing ca-lorimetry represents the only experimental method that can provide a *direct,* model-independent measure of the enthalpy change, ΔH, and the heat capacity change, ΔC_p, that accompanies an isothermal process.

Applications

Isothermal mixing calorimetry has been used in nucleic acid studies to characterize the influence of metal ion binding; to determine the enthalpy of duplex formation from the mixing of two complementary strands; to measure the energetics of pH-induced changes in nucleic acid structure;

and to determine the enthalpy of small ligand–DNA interactions and protein–DNA interactions.[35,37,39,40–42,75–93] The case of small ligand–DNA interactions provides a particularly good example for illustrating the advantages of the enhanced sensitivity of the stopped-flow versus the conventional batch calorimeter. Specifically, Breslauer and co-workers have used a conventional batch microcalorimeter to determine the enthalpy change associated with DNA complexation by daunomycin, an intercalating anthracycline antibiotic used in cancer chemotherapy regimens.[35,85] At the concentrations required for this study, daunomycin aggregates. Consequently, the observed heat changes had to be corrected for binding-induced disaggregation. This correction value was obtained from both temperature-dependent optical studies and temperature-dependent NMR studies.[35,40,85] Such indirect corrections (which require an assumed model for the aggregation event) compromise the quality of the final enthalpy values. By contrast, the enhanced sensitivity of the stopped-flow microcalorimeter has permitted Breslauer and co-workers to reinvestigate their daunomycin–DNA binding studies at concentrations where the dauno-

[75] G. Rialdi, J. Levy, and R. Biltonen, *Biochemistry* **11**, 2472 (1972).

[76] R. F. Steiner and C. Kitzinger, *Nature (London)* **194**, 1172 (1962).

[77] P. D. Ross and R. L. Scruggs, *Biopolymers* **3**, 491 (1965).

[78] P. D. Ross and R. L. Scruggs, *J. Mol. Biol.* **45**, 567 (1969).

[79] P. D. Ross, R. L. Scruggs, F. B. Howard, and H. T. Miles, *J. Mol. Biol.* **61**, 727 (1971).

[80] R. L. Scruggs and P. D. Ross, *J. Mol. Biol.* **47**, 29 (1970).

[80a] G. Vesnaver and K. J. Breslauer, *Proc. Natl. Acad. Sci. U.S.A.* **88**, 3569 (1991).

[81] K. J. Breslauer, G. Charko, D. Hrabar, and C. Oken, *Biophys. Chem.* **8**, 393 (1978).

[82] L. A. Marky, K. S. Blumenfeld, and K. J. Breslauer, *Nucleic Acids Res.* **11**, 2857 (1983).

[83] L. A. Marky, J. G. Snyder, and K. J. Breslauer, *Nucleic Acids Res.* **11**, 5701 (1983).

[84] L. A. Marky and K. J. Breslauer, *Proc. Natl. Acad. Sci. U.S.A.* **84**, 4359 (1987).

[85] K. J. Breslauer, W. Y. Chou, R. Ferrante, D. Zaunczkowski, J. Curry, D. P. Remeta, J. G. Snyder, and L. A. Marky, *Proc. Natl. Acad. Sci. U.S.A.* **84**, 8922 (1987).

[86] K. J. Breslauer, R. Ferrante, J. Curry, D. Zaunczkowski, R. S. Youngquist, P. B. Dervan, and L. A. Marky, *in* "Structure and Expression, Volume 2: DNA and Its Drug Complexes" (M. H. Sarma and R. H. Sarma, eds.), p. 273. Adenine Press, Albany, New York, 1988.

[87] C. P. Mudd, R. L. Berger, D. P. Remeta, and K. J. Breslauer, *Biophys. J.* **53**, 480a (1988).

[87a] D. P. Remeta, C. P. Mudd, R. L. Berger, and K. J. Breslauer, *Biochemistry* **30**, 9799 (1991).

[88] Y. Takeda, P. D. Ross, and C. P. Mudd, *Biophys. J.* **57**, 225a (1990).

[89] M. Lee, R. G. Shea, J. A. Hartley, K. Kissinger, R. T. Pon, G. Vesnaver, K. J. Breslauer, J. C. Dabrowiak, and J. W. Lown, *J. Am. Chem. Soc.* **111**, 345 (1989).

[90] R. Jin and K. J. Breslauer, *Biophys. J.* **55**, 374a (1989).

[91] D. P. Remeta and K. J. Breslauer, *Conv. Biomol. Stereodyn.* **6**, 275a (1989).

[92] M. Lee, R. G. Shea, J. A. Hartley, K. Kissinger, G. Vesnaver, K. J. Breslauer, R. T. Pon, J. C. Dabrowiak, and J. W. Lown, *J. Mol. Recognit.* **2**, 6 (1989).

[93] F. Quadrifoglio and V. Crescenzi, *Biophys. Chem.* **2**, 64 (1974).

mycin exists in the monomeric state, thereby precluding the need to correct the observed heat effects for drug dissociation.[40,85,87,87a]

The enhanced sensitivity of the stopped-flow instrument also has permitted Breslauer and colleagues to evaluate if the daunomycin binding enthalpies exhibit a dependence on the ligand-to-DNA ratio (the so-called r value).[87a] In cases of cooperative ligand binding, this capability provides an opportunity to evaluate the thermodynamic basis for this intriguing binding phenomenon. In fact, a recent stopped-flow microcalorimetric study on daunomycin binding to a variety of synthetic and natural DNA duplexes reveals the binding enthalpies to vary with the r values in a manner that depends on both base composition and base sequence.[40,87,87a] The measurements required to make such an assessment would have been difficult, if not impossible, to perform using conventional batch calorimeters.

A very recently published study[80a] illustrates the power of using both differential scanning and batch calorimetric techniques to characterize thermodynamically the forces that control DNA structures. Specifically, Vesnaver and Breslauer[80a] have used DSC to characterize the thermally induced *disruption* of a DNA duplex and its component single strands, while also using isothermal mixing calorimetry to characterize the *formation* of the same duplex from its component single strands. The data obtained from these measurements allowed construction of a thermodynamic cycle which revealed that single-stranded order can contribute significantly to the thermodynamics of duplex formation. This feature must be recognized and accounted for when designing single-strand probes for selective DNA hybridization experiments and when interpreting thermodynamic data associated with the formation of higher-order DNA structures (e.g., duplexes, triplexes, tetraplexes, etc.) from inter- and/or intramolecular associations of single strands.

Isothermal Titration Calorimetry

Method

In a differential isothermal titration calorimeter (ITC), two titration cells reside in the calorimeter assembly, one reference and one sample. Detection of heat effects in titration microcalorimeters is generally accomplished by use of semiconductor thermopiles interposed between the reaction cell(s) and a constant temperature heat sink. A typical calorimetric titration will involve having the reference cell filled with buffer only and the sample cell with buffer plus the reactant to which ligand will be titrated. Identical injections of ligand are introduced into both mechanically stirred titration cells by a dual-injection mechanism. Heat effects measured in the reference cell arise from injection and dilution of the ligand. Heat effects measured in the sample cell arise from injection and dilution of ligand as well as from the reaction of interest. The difference between the heat effects in both cells is then equal to the heat of reaction. This measured heat of reaction is proportional to the product of the amount of ligand bound and the enthalpy change for the binding process.

Heat flow between the titration cell(s) and the heat sink occurs only through the thermopile thermal detectors, the voltage output of which is directly proportional to the temperature difference across the faces of the thermopiles. This temperature difference, in turn, is proportional to the thermal power being exchanged between the titration cell(s) and heat sink [rate of heat transfer (cal sec^{-1})].[94]

The instrumental response time can be made approximately an order of magnitude faster by employing power compensation to drive the system actively toward the steady-state baseline condition. New ultrahigh sensitivity ITCs using this principle are characterized by instrument response times of the order of 10–15 sec compared to 100–300 sec for older instrumentation.[94] Because the total heat associated with reaction is the time integral of the observed thermal power released or absorbed (as shown in Fig. 4), reducing the time response provides more sensitive detection of total heat effects per injection. The thermal power signal-to-noise ratio is increased with power compensation because of the greater thermal power amplitude needed to generate the same total heat (area under the curve) in a shorter period of time.

In a power compensation ITC, the directly recorded quantity is the amount of thermal power that must be applied to actively compensate the heat of reaction induced by injection. This is accomplished by continuously regulating the amount of heat applied to the titration cell(s) so as to drive the measured thermal power amplitude toward the steady-state, baseline value. In practice, this is accomplished via an interfaced computer that monitors the output of a nanovoltmeter connected to record the voltage developed by the thermopiles. Real-time adjustments to the current applied to resistive cell feedback elements associated with the reaction cell(s) are then dynamically controlled by software to provide active power compensation of the detected change in thermopile output. The applied thermal power as a function of time required to return the ITC to its steady-state condition following injection then becomes the direct experimentally determined quantity that is proportional to the heat of reaction.

[94] E. Freire, O. L. Mayorga, and M. Straume, *Anal. Chem.* **52**, 950A (1990).

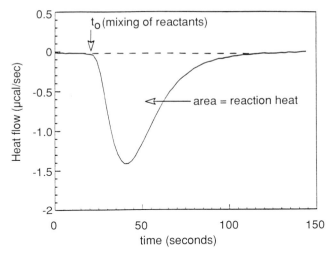

FIG. 4. Typical isothermal titration calorimetric trace. At time t_0, one of the reactants is injected into the calorimeter cell already containing the other reactant. The heat effects associated with the reaction are measured continuously as a function of time. The area under the curve is equal to the heat absorbed (positive) or released (negative) during the reaction. When normalized by the amount reacted, it yields directly the enthalpy for the reaction.

Information Content

The association of biological macromolecules to one another as well as their association to small ligand molecules play a central role in the structural assembly and functional regulation of many biological systems. Quantitative characterization of the thermodynamics of such molecular interactions is of fundamental importance for developing a thorough understanding of how biological function is regulated by ligand binding or molecular associations. A comprehensive understanding of the energetics associated with molecular interactions requires information about the thermodynamic parameters that define intrinsic binding events as well as any associated cooperative effects brought on by binding. Calorimetric titration experiments designed to characterize quantitatively the energetics of ligand-linked macromolecular cooperativity will provide considerable insight into the molecular basis of information transfer with macromolecules via cooperative effects.

Isothermal titration calorimetry measures the energetics of biochemical reactions or molecular interactions at constant temperature. The observed quantity is the amount of heat released or absorbed by the process. A broad range of molecular processes, such as ligand binding phenomena (e.g., protein–nucleic acid interactions, drug–nucleic acid interactions, hormone–receptor interactions, enzyme–substrate interactions, binding of substrates or substrate analogs, and cofactor binding) as well as interactions among components of multimolecular complexes (e.g., multisubunit enzyme complexes, multimolecular aggregates, and protein–membrane interactions), may be characterized by application of ITC.

Since the mid-1980s, ITC has experienced tremendous advances in sensitivity owing to the incorporation of microelectronics and new instrument designs. In the past, the use of ITC was somewhat limited owing to a lack of sufficient sensitivity for characterization of high affinity binding reactions. Such experiments require dilute reactant concentrations in order to determine equilibrium constants accurately. These limitations no longer exist given the recent development of instruments capable of measuring heat effects from reactions involving nanomoles or less of reactants.[94-101]

The new generation of titration calorimeters makes possible direct thermodynamic characterization of association processes exhibiting the very high affinity binding constants frequently encountered in biological systems, particularly those involved in regulation of function. Direct calorimetric titration experiments may be used to define fully the thermodynamics of both (1) intrinsic ligand binding as well as (2) any ligand-induced cooperative effects. The ability to measure small heats of reaction (of the order of $10^{-7}-10^{-6}$ cal/ml of solution) allows direct calorimetric examination of binding processes with K_a values as high as 10^8-10^9 M^{-1} (requiring dilute reactant concentrations of $\sim 10^{-6}$ M or less). Additionally, the recent development of an approach, total association at partial saturation (TAPS), for experimentally uncoupling intrinsic ligand binding energetics from those associated with ligand-induced macromolecular cooperative effects provides a direct calorimetric means for probing the molecular origins of information transduction within macromolecular structures.

Figure 4 presents a typical calorimetric titration response in which the thermal power versus time is measured. At time t_0 (the time of injection), titrant is introduced to the sample, producing an exothermic system response (i.e., a release of heat as indicated by the negative deflection of the thermal power). The area encompassed by the curve is the reaction heat

[95] R. B. Spokane and S. J. Gill, *Rev. Sci. Instrum.* **52**, 1728 (1981).
[96] J. Donner, M. H. Caruthers, and S. J. Gill, *J. Biol. Chem.* **257**, 14826 (1982).
[97] I. R. McKinnon, L. Fall, A. Parody, and S. J. Gill, *Anal. Biochem.* **139**, 134 (1984).
[98] G. Ramsay, R. Prabhu, and E. Freire, *Biochemistry* **25**, 2265 (1986).
[99] M. Myers, O. L. Mayorga, J. Emtage, and E. Freire, *Biochemistry* **26**, 4309 (1987).
[100] A. Schon and E. Freire, *Biochemistry* **28**, 5019 (1989).
[101] T. Wiseman, S. Williston, J. F. Brandts, and L.-N. Lin, *Anal. Biochem.* **179**, 131 (1989).

which, when normalized for the amount of material undergoing reaction, produces directly the enthalpy of reaction. In addition to defining the total heat associated with an injection event (as the area under the thermal power versus time curve), the time decay of the signal, after correction for instrument response, is related to the kinetics of the process.[33] The existence of multiple kinetic phenomena (e.g., slow structural rearrangement following binding) will result in a complex time dependence of the thermal power signal.

Quantitation of Binding Equilibria

Quantitative characterization of the energetics of molecular associations is commonly performed by analysis of binding isotherms. In this approach, ligand-binding macromolecules are sequentially titrated to completion under experimental conditions that permit extraction of all of the thermodynamic parameters that define the binding process. The characterization of the binding of a ligand molecule to a macromolecule requires the determination of the association constant, K_a, or the Gibbs free energy of association, $\Delta G^0 = -RT \ln K_a$, as well as the enthalpic and entropic contributions to the overall binding free energy ($\Delta G^0 = \Delta H^0 - T\Delta S^0$). Further determination of the temperature dependence of the enthalpy change permits estimation of the changes in heat capacity [$\Delta H = \Delta H^0 + \Delta C_p(T - T^0)$]. For those cases in which more than one binding site exists in the macromolecule, it is necessary to assess whether the sites are independent (or noninteracting) or whether the occupancy of one or more of them affects the ligand affinity of the remaining unligated sites. Interacting binding sites give rise to cooperative ligand binding behavior, a critically important general mechanism employed by a broad diversity of biological macromolecules to regulate functional responses. In these cases, it is important to determine the energetics of the cooperative interactions as defined by interaction free energy, enthalpy, and entropy changes. The major goal of the calorimetric approach to the analysis of ligand-binding phenomena is the determination of the intrinsic binding energetics as well as the cooperative energetics. Different strategies have been devised to achieve these goals, as described below.

The most common approach is the generation of a ligand-binding isotherm. In the calorimetric situation, the isotherm is defined in terms of the amount of heat that is released (exothermic) or absorbed (endothermic) as a function of the total concentration of ligand in the solution. Thermodynamic parameters (ΔG^0, ΔH^0, ΔS^0) are estimated by linear or nonlinear least-squares analysis of the binding isotherm, much the same as for other techniques. In the case of cooperative binding, the interaction parameters have to be included in the fitting as well. However, when analysis of ligand-binding isotherms is performed to extract cooperative interaction energy terms simultaneously with those characteristic of the intrinsic binding process, the mathematical correlation among model fitting parameters may become a formidable problem.[102] Experimental uncoupling of cooperative ligand binding energetics from those associated with intrinsic ligand binding, as is the case with TAPS (see below), permits a more robust parameter estimation free of the parameter correlation effects encountered when analyzing complete binding isotherms for intrinsic as well as cooperative effects.

Total association at partial saturation (TAPS), an approach unique to calorimetric titrations, alleviates the problems of parameter correlation mentioned above and directly quantifies ligand-induced cooperative effects independent of the influences of intrinsic binding.[103] Experiments in this case are performed under conditions of TAPS by titrating macromolecular ligand binding sites that are present at concentrations greater than K_d for the reaction. This ensures that all of the ligand introduced into the reaction cell by titration will bind to the available binding sites. At the point at which all acceptor sites are saturated, no further reaction occurs. These conditions make the dependence of the observed apparent binding enthalpy on extent of saturation with ligand a function of only the cooperative energetics of ligand binding. Because concentrations of macromolecular binding sites must be at least about 10^3 greater than K_d for the reaction, this technique provides direct experimental access to the cooperative binding energetics of systems with K_a exceeding approximately $10^9 - 10^{10}\ M^{-1}$ by uncoupling the energetics of cooperative interactions from those associated with intrinsic binding. For positive cooperativity, TAPS normally produces reliable parameter estimates, even if the initial concentration of binding sites is only 10^2 times larger than the K_d. For negative cooperativity, the situation is different and care must be taken because the drop in affinity may abolish the conditions for TAPS.

Binding Equilibrium Evaluated from Ligand-Binding Isotherms

A variety of mathematical procedures have been devised to estimate association constants, numbers of binding sites, and cooperative interactions from ligand-binding isotherms. Linearized transforms of ligand-binding data, such as the Scatchard plot (in which the ratio of bound to free ligand is plotted versus bound ligand), have been extremely widely used. Such transformations, however, produce biases in the analysis because the

[102] M. Straume and M. L. Johnson, *Biophys. J.* **56,** 15 (1989).
[103] G. Bains and E. Freire, *Anal. Biochem.* **192,** 203 (1991).

distinction between dependent and independent variables is obscured. Direct nonlinear least-squares analysis of the dependent versus independent variable(s) provides a statistically more accurate estimation of model parameter values free of the implicit biases that arise in analysis of (nonlinearly) transformed data.[102,104-106]

On binding of ligand to a macromolecule, heat is released or absorbed as a result of the binding event (owing to the enthalpy of ligand binding). The heat effects measured for each addition of ligand represent the experimentally observed response in ITC experiments and are described by

$$q = V\Delta H\Delta[L_B] \tag{7}$$

where q is the heat associated with the change in bound ligand concentration, $\Delta[L_B]$ is the change in bound ligand concentration, ΔH is the enthalpy of binding in (moles ligand)$^{-1}$, and V is the reaction volume.

Because the observed heat, q, is proportional to the increase in bound ligand concentration, the magnitude of q decreases as the system is titrated to completion. This is demonstrated in Fig. 5, where an example of a calorimetric titration of RNase A with 2'-CMP at 25° is presented. The heat effects associated with successive injections of 2'-CMP are seen to decrease in amplitude as the RNase A present in the sample cell is titrated to completion. The total cumulative heat released or absorbed is proportional to the total concentration of bound ligand:

$$Q = V\Delta H \sum \Delta[L_B] = V\Delta H[L_B] \tag{8}$$

where Q is the cumulative heat and $[L_B]$ is the concentration of bound ligand.

Analysis of binding isotherms in terms of the multiple sets of independent binding sites model[107,108] offers a general, flexible framework to extract information about a great variety of ligand-binding situations. Single or multiple binding sites, multiple sets of binding sites, as well as positive and negative cooperative effects may, in principle, all be accommodated by this model. The model assumes that the ligand-binding macromolecule possesses an arbitrary number of sets of noninteracting binding sites, and that all sites in the same set possess the same affinity for ligand. The concentration of ligand bound to each set of binding sites, as given by this

[104] M. L. Johnson, *Biophys. J.* **44**, 101 (1983).

[105] M. L. Johnson, *Anal. Biochem.* **148**, 471 (1985).

[106] M. Straume and M. L. Johnson, *Biochemistry* **27**, 1302 (1988).

[107] K. E. van Holde, *in* "Physical Biochemistry," 2nd Ed., Chap. 3. Prentice-Hall, Englewood Cliffs, New Jersey, 1985.

[108] C. R. Cantor and P. R. Schimmel, *in* "Biophysical Chemistry, Part III: The Behavior of Biological Macromolecules," Chap. 15. Freeman, New York, 1980.

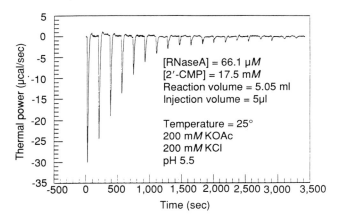

FIG. 5. Calorimetric titration of RNase A with 2'-CMP, showing the calorimetric response as successive injections of ligand are added to the reaction cell. As the binding sites become saturated with ligand, the magnitude of the heat effects decreases, becoming negligible as full saturation is approached.

model, is

$$[L_{B,i}] = [M] \frac{n_i K_i [L]}{1 + K_i [L]} \qquad (9)$$

where $[L_{B,i}]$ is the concentration of ligand bound to binding sites of set i, $[M]$ is the total concentration of macromolecule available for binding ligand, K_i is the intrinsic site association constant for binding sites of set i, n_i is the number of binding sites of set i on each macromolecule M, and $[L]$ is the concentration of free ligand. Because the heat associated with the binding event is proportional to the concentration of bound ligand, $[L_B]$, the cumulative amount of heat released or absorbed is then given by the sum of the heats corresponding to each set:

$$Q = V \sum_i \Delta H_i [L_{B,i}] = V[M] \sum_i \frac{n_i \Delta H_i K_i [L]}{1 + K_i [L]} \qquad (10)$$

where ΔH_i is the enthalpy of binding in (moles ligand)$^{-1}$ to binding sites of set i.

Equations (7) through (10) must be considered in terms of total ligand concentration by use of the expression $[L_T] = [L_B] + [L]$ (where $[L_T]$, $[L_B]$, and $[L]$ correspond to the concentrations of total, bound, and free ligand, respectively). The variable model parameters n_i, K_i, and ΔH_i are then estimated by nonlinear least-squares analysis of the calorimetric titration data. Either the cumulative heat, Q, or the individual heat, q, may be considered in the parameter estimation process. However, analysis of the individual heats, q, is preferable because the uncertainties associated with each injection, σ_q, are not propagated as is the case with cumulative heat data. Figure 6 presents an analysis of the calorimetric titration experiment presented in Fig. 5 for the binding of 2′-CMP to RNase A by application of the above model. In this case, the model parameters that must be estimated are the binding enthalpy (ΔH) and the binding affinity (as determined by the ligand association constant, K_a, or the free energy of ligand binding, $\Delta G^0 = -RT \ln K_a$).

Cooperative Energetics as Determined by Total Association at Partial Saturation

Calorimetric titrations conducted under conditions of total association at partial saturation allow direct calorimetric visualization of cooperative

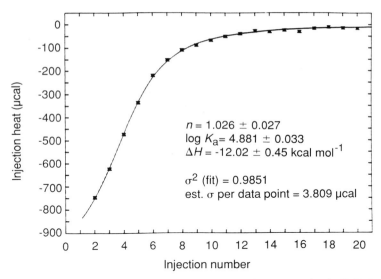

$n = 1.026 \pm 0.027$
$\log K_a = 4.881 \pm 0.033$
$\Delta H = -12.02 \pm 0.45$ kcal mol^{-1}

σ^2 (fit) $= 0.9851$
est. σ per data point $= 3.809$ μcal

FIG. 6. Results of analysis of the calorimetric titration of RNase A with 2′-CMP shown in Fig. 5 presented as individual injection heats as a function of injection number. The curve represents the nonlinear least-squares best fit of the data. The results indicate the existence of one 2′-CMP binding site (1.026 ± 0.027) with a binding association constant K_a of 7.60 (± 0.58) $\times 10^4$ M^{-1} at 25° and a reaction enthalpy of -12.02 ± 0.45 kcal mol^{-1}. The derived variance of fit is 0.9851, indicating a satisfactory fit (i.e., the variance of fit is less than 1). The estimated standard deviation per data point was 3.809 μcal as determined from electrical calibration measurements performed in conjunction with the calorimetric titration. Because the reaction volume is 5.05 ml, this translates to a specific instrumental sensitivity of 0.754 μcal ml^{-1}.

interactions without interferences from the binding event per se.[103] Because relatively high concentrations of the ligand-binding macromolecule are needed to satisfy this condition ($\sim K_d \times 10^3$ or more), cooperative interaction energies may be determined directly even for those cases in which the ligand affinity is extremely high and cannot be determined directly (i.e., for those cases with K_a values greater than $\sim 10^9\ M^{-1}$).

The example of a macromolecule possessing two ligand binding sites will suffice to demonstrate the principle and application of TAPS. A model for ligand binding to such a macromolecule must account for the intrinsic binding properties of the binding sites ($-RT \ln K_a = \Delta G^0 = \Delta H^0 - T\Delta S^0$) as well as allow for ligand-linked cooperative interactions ($\Delta g = \Delta h - T\Delta s$) induced by binding of the first ligand. As shown in Fig. 7, the relative free energies of all accessible states of this ligand-binding system are given by (1) a singly degenerate unliganded reference state with $\Delta G_0 = 0$, (2) a doubly degenerate singly liganded state with $\Delta G_1 = \Delta G^0$, and (3) a singly degenerate fully liganded state with $\Delta G_2 = 2\Delta G^0 + \Delta g$. The corresponding relative enthalpies of each of these system states are given by (1) $\Delta H_0 = 0$ for the singly degenerate unliganded reference state, (2) $\Delta H_1 =$

STATE	LIGANCY	FREE ENERGY	ω	S.W.
0	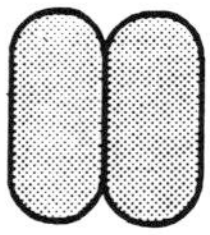	Ref.	1	1
1	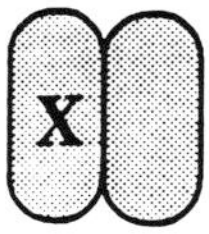 X	ΔG	2	$2\ K\ [X]$
2	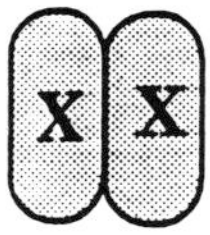 X X	$2\Delta G + \Delta g$	1	$k\ K^2 [X]^2$

FIG. 7. Complete representation of the thermodynamic parameters necessary to define fully both the intrinsic and cooperative ligand-binding energetics of a ligand-binding macromolecule possessing two, interacting binding sites. The free energy, degeneracy (ω), and statistical weight for each accessible state of the system are presented. The objective of calorimetric titrations under conditions of TAPS is to quantify directly, experimentally the cooperative energetics associated with ligand binding (i.e., Δg, Δh, and Δs; see text for additional details). Reproduced from Fig. 1 of Ref. 103.

ΔH^0 for the doubly degenerate singly liganded state, and (3) $\Delta H_2 = 2\Delta H^0 + \Delta h$ for the singly degenerate fully liganded state. An interaction free energy (Δg) of zero produces two independent sites, the affinities of which are unaffected by the state of ligand occupancy of the other site. However, a negative Δg produces positive cooperativity by increasing the affinity for the second ligand, whereas a positive Δg causes reduced affinity for the second ligand, resulting in negatively cooperative binding.

Both the interaction free energy (Δg) and the interaction enthalpy (Δh) can be determined from the dependence of the observed apparent binding enthalpy ($\langle \Delta H \rangle$) on the degree of binding saturation under conditions of total association at partial saturation (TAPS). As demonstrated in Fig. 8, the observed enthalpy of binding approaches the intrinsic binding enthalpy (ΔH^0 assumed to be -30 kcal mol^{-1}) at low degrees of saturation, whereas at full saturation the observed enthalpy of binding is the average over both sites [$(2\Delta H^0 + \Delta h)/2$, with a Δh of -10 kcal mol^{-1}]. The dependence of $\langle \Delta H \rangle$ on the degree of saturation is determined solely by the magnitude

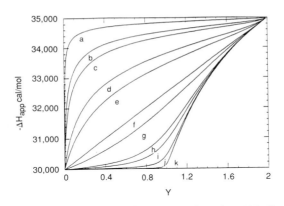

FIG. 8. Dependence of the apparent binding enthalpy for a ligand-binding macromolecule possessing two, interacting binding sites presented as a function of the degree of ligand saturation for a variety of cooperative interaction free energies. The intrinsic enthalpy of binding (ΔH) is assumed for this simulation to be -30 kcal mol^{-1}, whereas a cooperative interaction enthalpy (Δh) of -10 kcal mol^{-1} is assumed. The statistical thermodynamic partition function for this system is given by $Z = 1 + 2K[X] + kK^2[X]^2$ where $K = \exp(-\Delta G/RT)$ and $k = \exp(-\Delta g/RT)$. The ligand saturation function is then $Y = (1/Z)(2K[X] + 2kK^2[X]^2)$, and the apparent binding enthalpy ($\langle \Delta H \rangle$) is given by $\langle \Delta H \rangle = (1/Z)\{2K[X]\Delta H + kK^2[X]^2(2\Delta H + \Delta h)\}$. A range of k values from 1000 (positive cooperativity, negative Δg) to 0.001 (negative cooperativity, positive Δg) are considered with $k = 1$ representing the case of independent sites (i.e., no cooperativity associated with ligand binding). The values of k are as follows: (a) 1000, (b) 100, (c) 50, (d) 10, (e) 5, (f) 1, (g) 0.5, (h) 0.1, (i) 0.05, (j) 0.01, and (k) 0.001. Reproduced from Fig. 3 of Ref. 103.

and sign of the interaction free energy (Δg), whereas the magnitude of $\langle \Delta H \rangle$ is related to both ΔH^0 and Δh. Thus, at any given degree of saturation, the distribution of accessible system states will be determined by the energetics of the cooperative interactions. This distribution may be obtained experimentally by nonlinear least-squares analysis of the observed dependence of $\langle \Delta H \rangle$ on degree of macromolecular ligand saturation. Knowledge of the parameter values for ΔH, Δh, and Δs then fully defines the experimental observations under conditions of TAPS in terms of the ligand-linked cooperative binding energetics characteristic of the ligand-binding macromolecule.

Cooperative effects associated with ligand binding are thought to arise from ligand-induced protein conformational changes. Isothermal titration calorimetry, by employing TAPS, allows direct measurement of the energetics of the cooperative interactions and putative conformational changes independent of the intrinsic binding energetics. Complementary analysis of complete ligand-binding isotherms provides quantitation of the thermodynamics of the intrinsic binding process(es), as well. Comprehensive studies of this type will permit direct characterization of the nature and magnitude of the forces associated with many different biomolecular interactions. Calorimetric titrations of the sort described may prove to be particularly useful, for example, in screening of mutant forms of ligand-binding macromolecules created by genetic means as well as in characterization of the effects of amino acid substitutions on the thermodynamics of protein stability and interaction with ligands or other proteins. Ultimately, this type of comprehensive thermodynamic information will contribute to identification of the sequence of molecular events involved in transduction and regulation of biological information in a wide range of systems.

Applications

Many nucleic acid-binding proteins and macromolecular assemblies exhibit very high binding affinities as well as cooperative effects associated with the binding process. These may take the form of cooperative macromolecular structural changes induced by binding of small regulatory ligands which in turn modify the macromolecular binding affinity for a nucleic acid regulatory site, as in the case of regulation by *lac* repressor, for example.[109] Other systems exhibit long-range cooperative effects as a result of nearest-neighbor interactions among adjacent protein molecules bound to single-stranded nucleic acids (e.g., bacteriophage T4 gp32[110]). The *Esch-*

[109] A. Revzin, *in* "The Biology of Nonspecific DNA–Protein Interactions" (A. Revzin, ed.), Chap. 4. CRC Press, Boca Raton, Florida, 1990.
[110] R. L. Karpel, *in* "The Biology of Nonspecific DNA–Protein Interactions" (A. Revzin, ed.), Chap. 5. CRC Press, Boca Raton, Florida, 1990.

erichia coli single-stranded binding protein (*Eco*SSB) has been quite extensively characterized[111-113] and has been shown, by spectroscopic methods, to exhibit multiple types of cooperative single-stranded nucleic acid binding behavior that is strongly coupled to environmental ionic strength.

A higher order level of molecular association phenomena must be considered in cases of regulation in multimolecular aggregates. The affinity of cAMP receptor protein (CRP) for nucleic acids is influenced by binding of cAMP (a regulatory ligand that modifies the affinity of CRP for nucleic acids).[114] CRP, in turn, is implicated as potentially interacting in a regulatory manner with other nucleic acid-binding enzymes, thus modifying their activity, possibly through communication via protein–protein contacts. A great many examples of protein–nucleic acid interacting systems exist for which knowledge of the thermodynamic mechanisms underlying ligand binding or macromolecular association will contribute to elucidating the molecular basis for functional regulation of genetic expression.[115-117] Thermodynamic characterization of both the intrinsic binding or association process(es) and the cooperative energetics associated with any cooperative structural changes will shed light on the molecular origins of regulatory communication pathways.

Multifrequency Calorimetry

Method

The calorimetric techniques discussed above are all designed to measure the equilibrium thermodynamics associated with structural transformations or macromolecular interactions with ligands or other macromolecules. Recently, a new type of calorimetry aimed at measuring the dynamics or kinetics of the energetics of macromolecular transitions has been introduced.[118-120] This technique, multifrequency calorimetry, per-

[111] T. M. Lohman and W. Bujalowski, *in* "The Biology of Nonspecific DNA–Protein Interactions" (A. Revzin, ed.), Chap. 6. CRC Press, Boca Raton, Florida, 1990.

[112] J. Greipel, C. Urbanke, and G. Maass, *in* "Protein–Nucleic Acid Interactions" (W. Saenger and U. Heinemann, eds.), Chap. 4. CRC Press, Boca Raton, Florida, 1989.

[113] J. W. Chase and K. R. Williams, *Annu. Rev. Biochem.* **55**, 103 (1986).

[114] S. Garges and S. Adhya, *Cell (Cambridge, Mass.)* **41**, 745 (1985).

[115] P. H. von Hippel and O. G. Berg, *in* "Protein–Nucleic Acid Interactions" (W. Saenger and U. Heinemann, eds.), Chap. 1. CRC Press, Boca Raton, Florida, 1989.

[116] C. R. Vinson, P. B. Sigler, and S. L. McKnight, *Science* **246**, 911 (1989).

[117] R. Schleif, *Science* **241**, 1182 (1988).

[118] O. L. Mayorga, W. van Osdol, J. L. Lacomba, and E. Freire, *Proc. Natl. Acad. Sci. U.S.A.* **85**, 9514 (1988).

[119] E. Freire, *Comments Mol. Cell. Biophys.* **6**, 123 (1990).

[120] O. L. Mayorga, W. van Osdol, and E. Freire, *Biophys. J.* **59**, 48 (1991).

mits a determination of the time regime in which the energy transformations take place, allowing the assignment of time constants to the enthalpic events associated with a reaction.

The idea behind multifrequency calorimetry is simple even though the mathematical formulation is somewhat cumbersome and will not be discussed here (the reader is referred to Ref. 34a for a complete review). Basically, a multifrequency calorimeter is composed of a very thin, disk-shaped, sample compartment (0.1 – 1 mm thick) in which the macromolecular solution is placed. On one face of the disk-shaped cell (the excitation side) a battery of miniature Peltier elements generate a periodic temperature oscillation. This oscillation can be a simple sine wave, a superposition of harmonics, or a more complex waveform. The amplitude of the temperature oscillation is usually of the order of 0.05°. On the opposite side of the sample cell (the measurement side) the response temperature oscillation after the wave has traveled through the solution is measured. The response wave is attenuated in amplitude and phase shifted with regard to the excitation wave. The magnitude of these changes are a function of the excitation frequency and depend on instrument parameters (e.g., cell thickness), intrinsic solution parameters (e.g., heat capacity of the solution), and the thermodynamic and kinetic parameters associated with the transition under consideration. Instrumental designs and data analysis procedures are directed to extract the parameters associated with the transition.

Information Content

Multifrequency calorimetry is particularly important for the analysis of complex transitions involving intermediate states. In protein folding studies, for example, conventional calorimetry has become the technique of choice for thermodynamic analysis owing to its ability to provide extremely precise and model-independent estimates of the overall energetics of the process. Contrary to the thermodynamic analysis, the kinetic analysis of protein folding has relied primarily on a variety of observables that report time-dependent changes of some system property as the system evolves from the folded to the unfolded state or vice versa. Because physical observables are often sensitive only to local properties of the molecule, it is a common occurrence for different physical observables to report distinct numbers of relaxation processes and amplitudes. In general, those relaxation processes cannot be correlated directly with the dynamics of the enthalpic events associated with the folding/unfolding transition. The goal of multifrequency calorimetry is precisely to partition the overall enthalpy change of a reaction into the time frame in which it takes place and to assign relaxation times to the enthalpic components of the reaction.

The nature of the response function measured by a multifrequency calorimeter can be intuitively understood by considering a macromolecular system at the equilibrium midpoint between two states A and B (A $\rightleftharpoons$ B). Under those conditions one imposes on the system a temperature oscillation that shifts the equilibrium alternatively to the left and right of the reaction. Simultaneously, the amount of heat that is absorbed and released as the reaction shifts to the right and left is measured. If the frequency of the excitation temperature oscillation is sufficiently slow when compared to the relaxation time of the reaction, then the macromolecular system will essentially "follow" the wave. On the other hand, at faster excitation frequencies the macromolecular system will be progressively unable to follow the temperature oscillation, and the amount of heat absorbed or released will diminish.

In a multifrequency calorimetry experiment, the amplitude (or phase shift) of the temperature oscillations on the response side of the sample compartment are compared to those that would have been observed if there were not a transition as a function both of temperature and frequency. The fundamental quantity of interest is the ratio of the response amplitude in the absence of a transition ($A°$) to that in the presence of a transition (A). For the case of multiple independent transitions this quantity is given by the following equation:

$$A°/A = \exp\left[x(vB/C_p°)^{1/2}\sum \frac{C_{p,j}^{ex}(1 + 2\pi v\tau_j)}{1 + (2\pi v\tau_j)^2}\right] \quad (11)$$

where x is the cell thickness, $C_p°$ is the bulk heat capacity of the solution, $C_{p,j}^{ex}$ is the excess heat capacity associated with the jth transition, and τ_j the relaxation time associated with that transition. Equation (11) can be analyzed by a nonlinear least-squares fitting algorithm to give the relaxation times and their corresponding contributions to the thermodynamic excess heat capacity function. In Eq. (11), B includes all the constant terms including the instrument calibration constant.[34a]

The quantity $A°/A$ has very important features, as demonstrated for the case of a single relaxation process. In this case, the temperature–frequency surface defined by $A°/A$ has a maximum located at a temperature equal to T_m and at a frequency equal to $\frac{1}{2}\pi\tau$. This is due to the excess heat capacity being maximal at T_m and the system being excited at its natural frequency. In general, the relaxation modes of the frequency-dependent heat capacity are characterized by the amplitudes, $C_{p,j}^{ex}$, and the relaxation times, τ_j, characteristic of the folding–unfolding transition. Figure 9 shows a com-

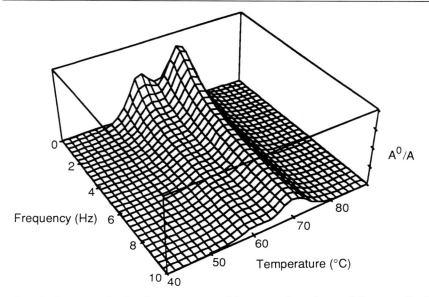

FIG. 9. Computer-simulated temperature and frequency dependence of the normalized multifrequency calorimetry response amplitude for a hypothetical transition composed of two peaks characterized by relaxation times of 200 and 500 msec.

puter-simulated response amplitude frequency–temperature surface for the case of two independent transitions characterized by two different relaxation times.

Applications

Multifrequency calorimetry has been applied to the analysis of membrane phase transitions[118] and to the folding–unfolding transition of the protein cytochrome *c*.[120] It is expected that in the years to come it will find an important niche in the field of nucleic acids research. It is well known that nucleic acids produce complex calorimetric unfolding profiles characterized by the presence of multiple components corresponding to different structural regions of the molecule. Multifrequency calorimetry offers an extra dimension to the analysis, and as such it should improve the resolution of complex nucleic acid transitions.

Concluding Remarks

We began this chapter by noting that calorimetry occupies a venerable place in the history of science. We conclude with the hope that the contents of this chapter have demonstrated calorimetry to be a powerful tool in biophysical research, capable of uniquely providing thermodynamic and extrathermodynamic characterizations of DNA structure, conformational transitions, and ligand interactions. It should be appreciated that the calorimetric techniques described here are general and therefore also can be applied to the study of other biopolymer systems and their interactions. Only recently, however, has the scientific community begun to appreciate the impressive information content of calorimetric techniques. We hope this chapter serves to advertise the full potential of calorimetry as a powerful experimental tool in the arsenal of techniques available to biophysical chemists.

Acknowledgments

This work was supported by grants from the National Institutes of Health: GM23509 (K.J.B.), GM34469 (K.J.B.), CA47795 (K.J.B.), RR04328 (E.F.), NS24520 (E.F.), and GM37911 (E.F.).

Author Index

Numbers in parentheses are footnote reference numbers and indicate that an author's work is referred to although the name is not cited in the text.

E

F

R

Rabinovich, D., 105, 106, 107, 108, 221, 222, 223(10, 11, 12, 15, 21), 409, 411, 419, 422, 423, 427, 428, 461
Rackwitz, H. R., 44
Radloff, R., 507
Ragg, E., 269
Raghuraman, M. K., 192, 194(16), 197(16), 405
Rahmouni, A. R., 153, 157(72)
Raizance, J. H., 390, 398(13)
Rajagopal, P., 239, 242(20), 243(20), 248(20), 249, 332
Ramalho-Ortigo, G., 3
Ramesh, N., 157
Ramm, P. J., 50
Ramsay, G., 538, 554
Ramsing, N. B., 199, 200, 201(4, 9), 202(4, 5, 9, 12), 203(12), 205(12), 206(9, 12), 207(2, 9, 12), 208(9, 15), 209(9, 15), 210(4, 12), 214(2, 4, 5, 7, 12, 15), 215(2, 15), 217(2, 4, 5, 7, 12), 219(2, 4, 7), 220(12), 401
Ramstein, J., 77, 313
Rance, M., 245
Rao, G. S., 457
Rao, S. T., 70
Rasovska, E., 153, 156(71), 172
Ratliff, R. L., 390, 396, 398, 401, 403, 405
Ratter, J. B., 142
Rauch, C. A., 507
Rava, R. P., 337, 346(13)
Rawitscher, M. A., 543, 550(42)
Ray, W. J., Jr., 268
Record, M. T., Jr., 209, 500
Reddy, M. P., 64, 274
Redfield, A. G., 236, 242(5)
Reese, C. B., 4, 9(11), 37
Regbert, T., 8
Reichelt, K., 498
Reid, B. R., 165, 236, 237, 238, 245(9, 12), 249
Reid, B., 254
Reinhardt, C. G., 276
Reitz, M., 274
Remeta, D. P., 542, 543, 545(40), 550, 551(40)
Rentzeperis, D., 200, 207(13), 208(13)
Revankar, G. R., 40, 42(20)

Revet, B. M. J., 529
Revet, B., 137, 447
Revina, L., 541
Revzin, A., 562
Rhees, H. S., 534
Rhodes, D., 451
Rialdi, G., 535, 550
Rice, J. A., 544
Rich, A., 44, 76, 105, 106, 107, 108, 128, 129, 131, 132, 133(23), 134, 135(27), 136, 138, 139, 140(25, 26), 141, 143(9), 144, 149(7, 9, 17), 151, 153(7, 17), 155(7), 157, 158, 221, 223(5, 6, 8), 247, 248, 249(53), 337, 342(21), 390, 399, 409, 412, 413, 419, 423, 424, 425, 427, 428, 451
Richards, F. M., 455
Richards, R. E., 241
Richlin, D. L., 456
Riddle, R. R., 5
Ridoux, J. P., 313, 321, 323, 331, 338, 339(29), 341(29)
Rigell, C., 541
Rikhirev, M. E., 192, 405
Riley, M., 215
Rill, R. L., 316, 502, 520, 522(5), 524(5)
Rinkel, L. J., 292, 295, 303, 306(18)
Rinnooy Kan, A. H. G., 456
Ripoll, D. R., 456
Rippe, K., 199, 200, 201(9), 202(8, 9, 12), 203(8, 12), 205(12), 206(9, 2), 207(2, 13), 208, 209(9), 210(12), 214(2, 6, 12, 15), 215(2, 15), 217(2, 12), 218(8), 219, 220(12, 23, 33), 352
Ris, H., 486
Rivalle, C., 323, 327(27)
Robbins, R. K., 40, 42(20)
Robert-Nicoud, M., 76, 137, 336, 342(6), 348(6)
Robertus, J. D., 451
Robins, M. J., 40
Robinson, H., 107, 108
Robinson, P., 222, 223(18), 427
Roe, J.-H., 500
Rohrer, H., 491, 497(2)
Roongta, V. A., 255, 257, 269(22, 28), 272(22, 28), 279(22), 281(22, 28), 284(22), 285(22)
Roongta, V., 261, 262(44), 272(44), 273(44), 278, 279(86)

Subject Index

of nucleic acids
  general features of, 308–311
  information obtained by, 307
  of poly(dA)·poly(dT)–distamycin in
    D$_2$O, 334–335
  of poly(dA)·poly(dT) in D$_2$O, 334
  of poly(dA)·poly(dT) in H$_2$O, 315, 317
  of poly(dA)·poly(dT)–netropsin com-
    plexes, 331–333
  of poly(dA)·poly(rU) in H$_2$O, 315, 317
  of poly[d(A-T)]–netropsin complexes,
    331–333
  of poly(rA)·poly(dT) in H$_2$O, 315, 317
  of poly(rA)·poly(rU) in H$_2$O, 315, 317
  of 22R–22Y double helix in D$_2$O, 329
  of 22R–22Y–22T triple helix in D$_2$O, 329
  of triple-strand structures, 325–327
Fourier transform NMR spectrometers,
  multinuclear, 254
FRET. *See* Fluorescence resonance energy
  transfer
Furanose–phosphate backbone, Raman
  bands of, 346–347
Furanose sugar conformations
  couplings, sums of, measurement of,
    295–296
  determination of, from NMR coupling
    constants, 286–306
  general background, 286–291
  Φ_m parameter, pseudorotational wheel
    representation of, 287–288
  P parameter, pseudorotational wheel rep-
    resentation of, 287–288
  pseudorotational state of, proton–proton
    torsion angles and, geometrical rela-
    tion between, 291
  validity of, 296
Furanose sugar rings
  N-type, 287–289
  puckering modes of, 287–288
  S-type, 287–289
β-D-Furanose sugar rings, *A, B,* and phase
  parameters for, 291–292

G

G·A mispair, structure of, 228–229
G(anti)·A(anti) mispair, stick representa-
  tion of, 230
G(anti)·A(syn) mispair

stick representation of, 230
structure of, 229
G(syn)·A(anti) mispair, stick representation
  of, 230
G,C-containing polymers, Raman spectroscopic changes for, 341–342
G·C mispair
  stick representation of, 222, 230
  structure of, 222, 246
G′2-DNA, 192–194
  formation of, 196–197
G4-DNA, 192–194
  complex motifs, formation of, 195–196
  influence of pH and melting of, 199
  simple motifs, formation of, 195
  stability of, 199
G4′-DNA, 192–194
  formation of, 197
Gel electrophoresis
  detection of cruciform structures by,
    167–169
  two-dimensional, detection of intramolec-
    ular protonated DNA structures by,
    185–187
G·G mispair, crystallization of, 228
Gold, 497–498
Gold films, epitaxially grown on mica, 498
Gold foil, 497–498
Gold spheres, faceted, 498
G-quartet structures. *See* Guanine quartet
  structures
G-rich sequences, quartet structures with,
  194
G·T mispair, 227–228, 231
  hydration network of, 228
  omit electron density map of, 227
  stick representation of, 230
  structure of, 229
Guanidine hydrochloride, inhibition of B–
  A transition of DNA by, 124–125
Guanine, protecting groups for, 6–7
Guanine bands, 347
Guanine quartet structures, 191–199
  formation of, methods for, 194–198
  G-G base-paired, formation of, 197–199
  G-rich sequences with, 194
  identification of, 198
  major, 193
    classification of, 193
  molecularity of, determination of, 198–
    199